THE ULTIMATE GUIDE TO POOL MAINTENANCE

About the Author

Terry Tamminen is a leading expert on pools and spas and consults on water technology around the world. He is the founder of Waterkeeper programs, activist organizations dedicated to preserving and protecting coastal resources, throughout California. Mr. Tamminen has served as the Secretary of the California Environmental Protection Agency and is also the author of *The Ultimate Guide to Above-Ground Pools* and *The Ultimate Guide to Spas and Hot Tubs*, both published by McGraw-Hill.

THE ULTIMATE GUIDE TO POOL MAINTENANCE

Terry Tamminen

THIRD EDITION

McGraw-Hill

New York Chicago San Francisco Lisbon London Madrid Mexico City Milan New Delhi San Juan Seoul Singapore Sydney Toronto

The McGraw-Hill Companies

1 2 3 4 5 6 7 8 9 0 DOC/DOC 0 1 3 2 1 0 9 8 7

ISBN-13: 978-0-07-147017-9
ISBN-10: 0-07-147017-4

Sponsoring Editor: Larry S. Hager
Production Supervisor: Richard C. Ruzycka
Editing Supervisor: Stephen M. Smith
Project Manager: Patricia Wallenburg
Copy Editor: Marcia Baker
Proofreader: Paul Tyler
Indexer: Karin Arrigoni
Art Director, Cover: Handel Low
Composition: TypeWriting

Printed and bound by RR Donnelley.

Previously published as *The Ultimate Pool Maintenance Manual,* copyright © 2001.
Previously published as *The Pool Maintenance Manual,* copyright © 1996.
Originally published as *The Professional Pool Maintenance Manual,* copyright © 1995.

To the pool and spa service pros everywhere who toil under the hot sun each day to keep our water clean

CONTENTS

Acknowledgments **xvii**

Introduction **xix**

Chapter 1 **The Pool and Spa** **1**

How It Works 1

How Much Water Does It Hold? 3

Square or Rectangular 3

Circular 5

Kidney or Irregular Shapes 7

Parts per Million (ppm) 8

Types of Pools and Spas 9

Concrete and Plaster 11

Vinyl-Lined 12

Fiberglass 15

Above-Ground Pools 15

Wood 19

Pool and Spa Design and Construction 20

Plans and Permits 21

Excavation 21

Plumbing 27

Steel 27

Electrical 28

Gunite 28

Tile 28

Rock, Brick, or Stone 30

Coping, Decks, and Expansion Joints 31

Equipment Set 32

Cleanup 32

Plaster 33
Start-up 34

Chapter 2 Basic Plumbing Systems **37**

Skimmers 37
Main Drains 41
General Plumbing Guidelines 43
PVC Plumbing 47
Plumbing Methods 50
Copper Plumbing 53
Plumbing Methods 53
Miscellaneous Plumbing 57
Sizing of Plumbing 58

Chapter 3 Advanced Plumbing Systems **61**

Manual Three-Port Valves 61
Operation 61
Construction 63
Maintenance and Repair 64
Motorized/Automated Three-Port Valve Systems 66
Operation 67
Construction 67
Maintenance and Repair 67
Reverse Flow and Heater Plumbing 69
Unions 70
Gate and Ball Valves 71
Check Valves 72
Solar Heating Systems 76
Types of Solar Heating Systems 76
Plumbing 78
To Solar or Not to Solar? 80
Installation 82
Maintenance and Repair 85
Water Level Controls 85

Chapter 4 Pumps and Motors **93**

Overview 93
Strainer Pot and Basket 96
Volute 97
Impeller 98
Seal Plate and Adapter Bracket 100
Shaft and Shaft Extender 101
Seal 102
Motor 103
Types 104
Voltage 105
Housing Design 105
Ratings 106
Nameplate 106
Horsepower and Hydraulics Equals Sizing 108
Hydraulics 108
Sizing 118
Maintenance and Repairs 119
Strainer Pots 119
Gaskets and O-Rings 120
Changing a Seal 122
Pump and/or Motor Removal and Reinstallation 133
New Installation 135
Replacing a Pump or Motor 137
Troubleshooting Motors 139
Priming the Pump 142
T-Handles 145
Motor Covers 146
Submersible Pumps and Motors 146
High-Volume Pump-Out Units 146
Low-Volume Pumps and Motors 147
Cost of Operation 148
Booster Pumps and Motors for Spas 148
Basic Electricity 149
Electrical Terms 149

Electrical Theory 150
Electrical Panel 152
Circuit Breakers 153
Wiring 156
Gauge and Type 156
Ground Fault Interrupter (GFI) 157
Switches 159
Safety 159
Testing 161
Something Better 163

Chapter 5 Filters 165

Types 165
Diatomaceous Earth (DE) Filters 165
Sand Filters 169
Cartridge Filters 173
Makes and Models 174
Sizing and Selection 174
Backwash Valves 179
Backwash Hoses 182
Pressure Gauges and Air Relief Valves 184
Sight Glasses 185
Repair and Maintenance 186
Installation 186
Filter Cleaning and Media Replacement 188
Leaks 199

Chapter 6 Heaters 207

Gas-Fueled Heaters 207
The Millivolt or Standing Pilot Heater 211
The Control Circuit 212
Natural versus Propane Gas 221
Electric-Fueled Heaters 222
Solar-Fueled Heaters 223
Heat Pumps 225

Oil-Fueled Heaters 227
Makes and Models 228
Selection 228
Sizing 228
Cost of Operation 232
Installation, Repairs, and Maintenance 233
Installation 235
Repairs 246
Preventive Maintenance 267

Chapter 7 Additional Equipment 269

Time Clocks 269
Electromechanical Timers 269
Twist Timers 273
Electronic Timers 274
Repairs 275
Remote Controls 277
Air Switches 278
Troubleshooting 280
Wireless Remote Control 281
Hardwired Remote Control 283
Flow Meters 289
Diving Boards, Slides, Ladders, and Rails 291
Diving Boards 291
Slides 295
Ladders and Rails 297
Safety Barriers 298
Automatic Pool Cleaners 299
Electric Robot 299
Booster Pump Systems 300
Suction-Side Systems 309
Lighting 309
Standard 120/240-Volt Lighting 310
Low-Voltage Lights 318
Fiberoptics 318
Covers 320

Bubble Solar Covers 320
Foam 323
Sheet Vinyl 323
Electric Covers 324

Chapter 8 Water Chemistry **329**

Demand and Balance 330
Components of Water Chemistry 331
Sanitizers 331
pH 347
Total Alkalinity 349
Hardness 350
Total Dissolved Solids (TDS) 350
Cyanuric Acid 351
Weather 352
Algae 353
Forms of Algae 354
Algae Elimination Techniques 355
Water Testing 362
Test Methods 362
Chlorine 365
pH 367
Total Alkalinity 368
Hardness 369
Total Dissolved Solids (TDS) 369
Heavy Metals 369
Cyanuric Acid 370
Test Procedures 370
Langlier Index 371
Water Treatment 373
Liquids 373
Granulars 374
Tabs and Floaters 375
Mechanical Delivery Devices 375
Salt Chlorine Generators 377

Chapter 9 Cleaning and Servicing **383**

Tools 383

Telepoles 383

Leaf Rake 387

Wall Brush 388

Vacuum Head and Hose 388

Leaf Vacuum and Garden Hose 390

Tile Brush and Tile Soap 392

Test Kit and Thermometer 393

Spa Vacuum 393

Pumice Stones 394

Pool Cleaning Procedures 394

Deck and Cover Cleaning 394

Water Level 396

Surface Skimming 396

Tiles 398

Equipment Check 399

Vacuuming 400

Chemical Testing and Application 408

Brushing 408

Winterizing 409

Temperate Climates 410

Colder Climates 410

Chapter 10 Special Procedures **419**

Draining a Pool 419

Breaking-in New Plaster 423

Break-in Step by Step 423

Leak Repair 428

Leak Detection Made Easy: Four Tests 428

Patching and Repairing 432

Remodeling Techniques 440

Plastering and Replastering 441

Fiberglass Coatings 444

Inexpensive Pool Face-Lifts: Paint 445

Acid Washing 450
Remodeling the Deck 456

Chapter 11 Water Features 459

Fountains 459
Koi Ponds 460
Installing a Koi Pond 464
Rockscapes 468
Waterfalls and the Vanishing Edge 471

Chapter 12 Commercial Pools 475

Types of Commercial Pools 476
Volume Calculations 477
Slope Calculations 477
Bather Loads 479
Bather Displacement 479
Commercial Equipment 480
Surge Chamber 480
Slurry Feeder 481
Filter 481
Gas Chlorinator 483
Chlorine Generators 484
High-Capacity Automatic Chlorine Feeders 485
The Commercial Equipment Room 487
Safety Equipment 487
Toss Rings 492
Life Hooks 492
Thermometers 492
Dehumidification of Indoor Commercial Pools 493
Health Issues 494

Chapter 13 50 Things Your Pool or Spa Can Do for Our Environment 499

Chemicals 499
Energy Conservation 500

Water Conservation 502
Recycle 503
Miscellaneous 504

Labor Reference Guide 507

Glossary 511

Reference Sources and Websites 539

Index 545

ACKNOWLEDGMENTS

There are many opinions about the "right" way to do things in pool and spa maintenance and certainly many competitors in the equipment and supply realm, but the one thing everyone agrees on is that both the water technician and the homeowner need good information to properly and safely maintain a pool or spa. Many manufacturers, builders, and pool/spa owners helped me with this book in that spirit. To them I extend sincere thanks. Many are mentioned in the text, but two deserve extra credit:

- My partner, Ritchie Creevy of Southern California Water Technologies, for endless advice and training.
- Owen W. Smith, Professor, California State University, Northridge, for illustrations in the First Edition of this book. Many of those are repeated in this volume.

INTRODUCTION

Dustin Hoffman, Madonna, Stacy Keach, Dick Clark, Barbra Streisand, Charles Bronson, David Letterman, Rich Little, Carroll O'Connor, Lou Gossett, Dyan Cannon, Kareem Abdul-Jabbar, Walt Disney, Martin Sheen, Sting, Goldie Hawn, Olivia Newton-John, Roy Orbison, Diana Ross, George C. Scott, Dick Van Dyke, Bruce Willis.

That is a partial list of the celebrities for whom I have done pool or spa work in the past 30 years (actually, the list of agents, producers, artists, and authors is even more impressive, but you might not recognize the names). These celebrities are, after all, just people who want a clean pool, spa, or water feature like anybody else.

Welcome to the world of water maintenance. Yes, water maintenance, not pool or spa maintenance. A pool or spa is merely the chosen vessel, but the water is the real product. It might be flowing in a pool, spa, fountain, or commercial application, but the point is that you are dealing with water and its effects on the vessel, plumbing, and related equipment. By accepting this basic premise, you will approach every aspect of this book in a way that will give you understanding of the *why* and *how* and not just procedures. In short, you will achieve better results in all aspects of pool, spa, and fountain *water* maintenance.

This book is designed for the pool and spa professional, but because levels of understanding in this industry vary so greatly, it assumes little knowledge of each topic and presents material from the most basic conceptual discussion, graduating to the most complex. I hope this book will aid those just starting in the field and the do-it-yourself homeowner. The glossary at the end of the book will assist you in gaining a thorough understanding of water maintenance.

This new edition goes well beyond the earlier ones, reflecting the rapid advances in the pool and spa industry over the past decade, and includes:

- The latest in chlorine alternatives, including the growing trend of using salt water for sanitizing your pool water.
- "Quick Start" guides that allow you to assess if this is a job for you or one better left to a pro. You can also use these guides as a handy checklist when performing the task.
- Difficulty ratings for each procedure—"Easy," "Advanced," or "Pro"—tells you right away if this is a task for your skill level.
- Lots of great web references and handy Internet tools that will help you keep your knowledge of pool maintenance up to date.
- Up-to-date information on robotic pool cleaners—you may be surprised to learn how effective and inexpensive these labor-saving devices can be!
- Frequently Asked Questions (FAQs) after each chapter, providing a handy reference for some of the most basic—but important—information for keeping your pool in tip-top shape.
- A guide to purchasing the right pool for your needs, including the latest in traditional in-ground pools, fiberglass pools, inflatables, above-ground pools, and more.
- How to heat your pool for free with inexpensive solar-heating devices.
- More photos and "Tricks of the Trade" that make anyone a pool maintenance pro.

Now all you have to do is open the book and get started. Don't let the length of this book deter you. It's full of details on every subject, but you can easily find just the information you need by checking the Index and turning right to that section. Once there, you'll find all the diagrams and step-by-step descriptions needed to help you get the job done right—the first time!

Good luck.

Terry Tamminen

CHAPTER

1

The Pool and Spa

This is a book about water before it is a book about pools, spas, fountains, or other water containers (Fig. 1-1). If you wanted to be a banker, it would be nice to understand something about the bank, vault, and cash drawers, but the real business is the money and how it is used. Similarly, this book will contain appropriate information about the "containers," but the fact is, most of us will never build a pool, spa, fountain, hot tub, or other such container. Our focus is on the water and the related products and equipment that move or change it.

How It Works

Let's begin with a basic overview of the typical container and water system. Figure 1-2 shows a typical pool and spa and its related equipment. A hot tub or spa alone is plumbed and serviced in a similar manner. To understand a pool or spa, we must follow the path of the water. That is also how this chapter (and the entire book) is outlined—in the logical pattern that the water travels from pool through plumbing to the pump/motor, filter, heater, and back to the pool.

Follow the arrows in Fig. 1-2 to follow the path of the water. The water enters the plumbing through a main drain and/or a surface skimmer (components of either a pool or a spa). It does this thanks to suction created by a pump and motor. After passing through the

FIGURE 1-1 Typical pool and spa.

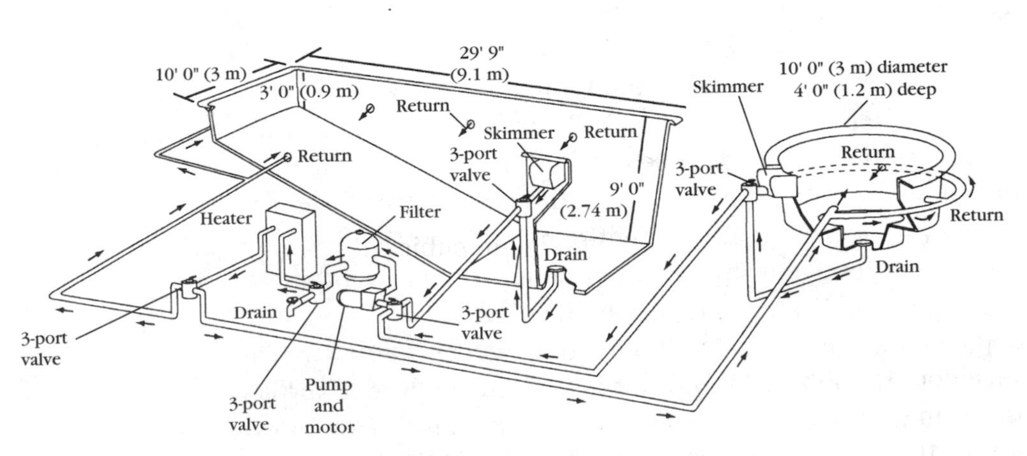

FIGURE 1-2 Typical pool and spa with equipment.

pump, the water is cleansed by a filter, warmed by a heater, and returned to the pool or spa through return outlets.

How Much Water Does It Hold?

Because many of the calculations in this book depend on knowing the quantity of water involved, here is how to calculate the volume of your pool or spa.

Square or Rectangular

The formula is simple:

$$\text{Length} \times \text{width} \times \text{average depth} \times 7.5 = \text{volume (in gallons)}$$

Let's first examine the parts of the formula. Length times width gives the surface area of the pool. Multiplying that by the average depth gives the volume in cubic feet. Since there are 7.5 gallons in each cubic foot, you multiply the cubic feet of the pool by 7.5 to arrive at the volume of the pool (expressed in gallons).

The formula is simple and so is the procedure. Measure the length, width, and average depth of the pool, rounding each measurement off to the nearest foot or percentage of one foot. If math was not your strong suit in school, remember one inch equals 0.0833 feet. Therefore, multiply the number of inches in your measurements by 0.0833 to get the appropriate percentage of one foot. Example:

$$\begin{aligned} 29\text{ ft, }9\text{ in.} &= 29\text{ ft} + (9\text{ in.} \times 0.0833) \\ &= 29 + 0.75 \\ &= 29.75\text{ ft} \end{aligned}$$

The same formula works in metric:

$$\text{Length} \times \text{width} \times \text{average depth} = \text{volume in cubic meters}$$

That's as far as you will probably need to take the equation since things like pool chemical dosages in metric measurements will be based on a certain amount per cubic meter. If it's a smaller volume of water like a spa, dosages may be expressed in certain amounts per liter of water. Since there are 1000 liters in 1 cubic meter, the formula becomes

Length × width × average depth × 1000 = volume in liters

Since metric units are already based on a decimal system in units of 10, there is no need to convert anything. As with standard units, round off to the nearest decimal for ease of calculation.

Average depth will be only an estimate, but obviously if the shallow end is 3 feet and the deep end is 9 feet, and assuming the slope of the pool bottom is gradual and even, then the average depth is 6 feet. If most of the pool is only 3 or 4 feet and then a small area drops off suddenly to 10 feet, you will have a different average depth. In such a case, you might want to treat the pool as two parts. Measure the length, width, and average depth of the shallow section, then take the same measurements for the deeper section. Calculate the volume of the shallow section and add that to the volume you calculate for the deeper section.

In either case, be sure to use the actual water depth in your calculations, not the depth of the container. For example, the hot tub depicted in Fig. 1-3 is 4 feet deep, but the water is only filled to about 3 feet. Using 4 feet in this calculation will result in a volume 33 percent greater than the actual amount of water. This could mean serious errors when adding chemicals, for example, which are administered based on the volume of water in question. There might be a time when you want to know the potential volume, if filled to the brim. Then, of

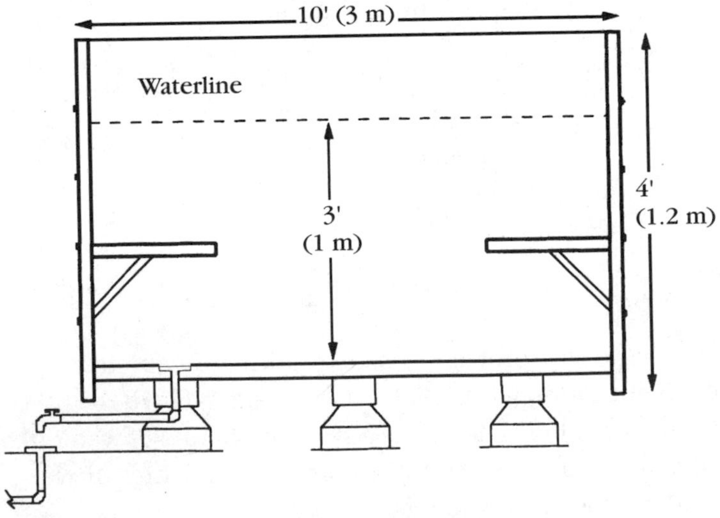

FIGURE 1-3 Cross-section of a typical hot tub.

course, you would use the actual depth (or average depth) measurement. In the example, that was 4 feet.

Try to calculate the volume of the pool in Fig. 1-2:

$$\text{Length} \times \text{width} \times \text{average depth} \times 7.5 = \text{volume (in gallons)}$$

$$29.75 \text{ ft} \times 10 \text{ ft} \times 6 \text{ ft} \times 7.5 = 13{,}387.5 \text{ gal}$$

$$9.1 \text{ m} \times 3 \text{ m} \times 1.8 \text{ m} \times 1000 = 49{,}140 \text{ L (or 49 kL)}$$

Circular

The formula:

$$3.14 \times \text{radius squared} \times \text{average depth} \times 7.5 = \text{volume (in gallons)}$$

The calculations in metric units will be the same, except remember to multiply by 1000 instead of 7.5 to determine volume in liters.

The first part, 3.14, refers to pi, which is a mathematical constant. It doesn't matter why it is 3.14 (actually the exact value of pi cannot be calculated but who cares?). For our purposes, we need only accept this as fact.

The radius is one-half the diameter, so measure the distance across the broadest part of the circle and divide it in half to arrive at the radius. Squared means multiplied by itself, so multiply the radius by itself. For example, if you measure the radius as 5 feet, multiply 5 feet by 5 feet to arrive at 25 feet. The rest of the equation was explained in the square or rectangular calculation.

Use the hot tub in Fig. 1-3 to calculate the volume of a round container. Let's do the tricky part first. The diameter of the tub is 10 feet. Half of that is 5 feet. Squared (multiplied by itself) means 5 feet times 5 feet equals 25 square feet. Knowing this, you can return to the formula:

$$3.14 \times \text{radius squared} \times \text{average depth} \times 7.5 = \text{volume (in gallons)}$$

$$3.14 \times 25 \text{ ft} \times 3 \text{ ft} \times 7.5 = 1766.25 \text{ gal}$$

In metric, the radius of the same spa measures 1.52 meters. Multiplied by itself, this equals 2.3 meters. The average depth is 0.9 meter, so the equation looks like this:

$$3.14 \times 2.3 \text{ m} \times 0.9 \text{ m} \times 1000 = 6500 \text{ L}$$

Note that in measuring the capacity of a circular spa, you might need to calculate two or three areas within the spa and add them together to arrive at a total volume. An empty circular spa looks like an upside-down wedding cake, because of the seats, as in Figs. 1-4A and B. Therefore, you might want to treat it as two separate volumes—the volume above the seat line and the volume below. In the wooden hot

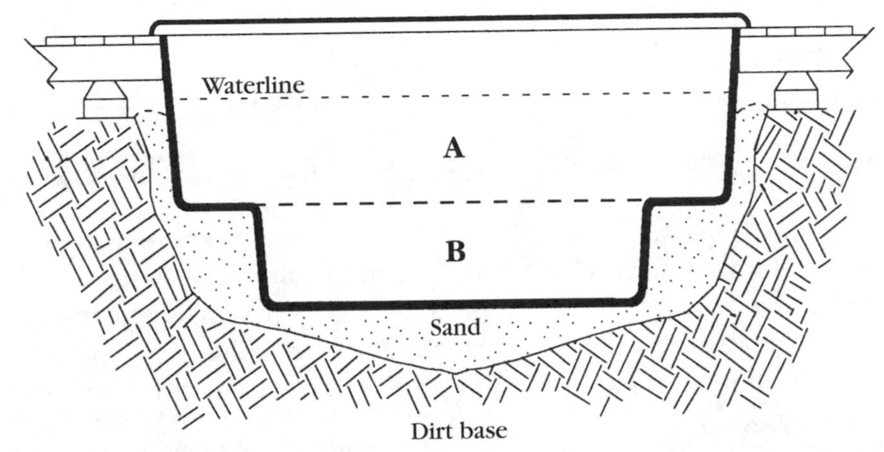

FIGURE 1-4A Cross-section of a typical spa.

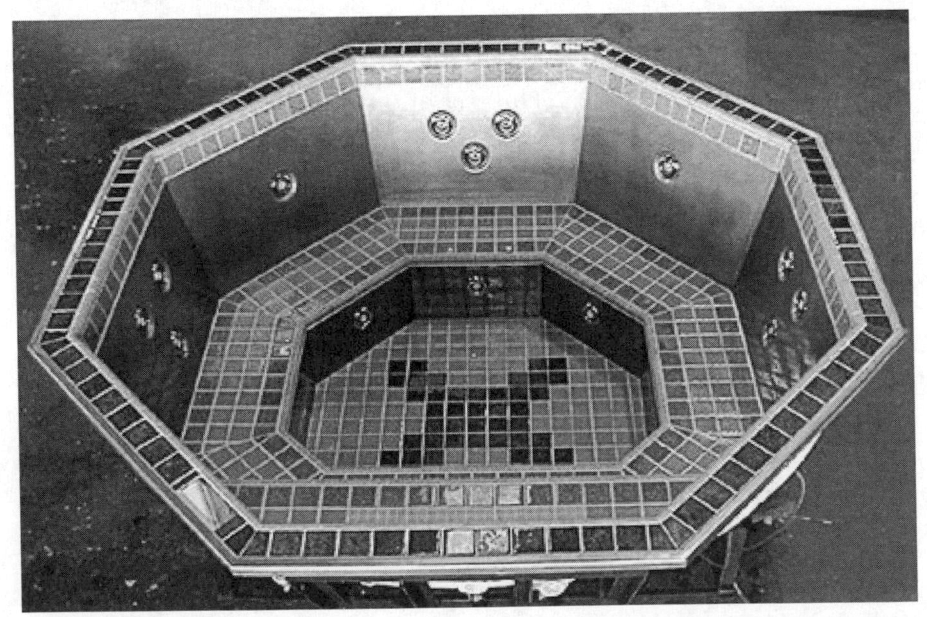

FIGURE 1-4B Typical spa shell. *Bradford Spas*

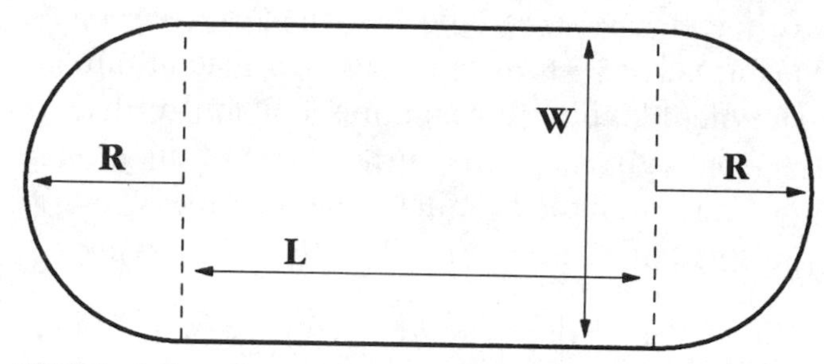

FIGURE 1-5 Volume of irregular shapes.

tub depicted in Fig. 1-3, where there is actually water above and below the seats, the tub can be measured as if there are no seats because this difference is negligible.

Kidney or Irregular Shapes

There are two methods used to calculate the capacity of irregular shapes. First, in Fig. 1-5, you can imagine the pool or spa as a combination of smaller, regular shapes. Measure these various areas and use the calculations described previously for each square or rectangular area and for each circular area. Add these volumes together to determine the total capacity. Figure 1-5 contains one rectangle and one circle (shown in two halves).

The second method is as diagrammed in Fig. 1-6:

$$0.45 \times (A+B) \times \text{length} \times \text{average depth} \times 7.5 = \text{volume (in gallons)}$$

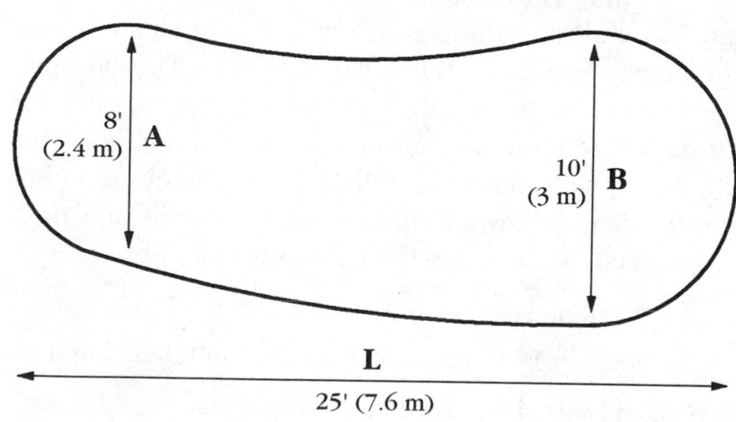

FIGURE 1-6 Volume of kidney shapes.

Again, the calculations in metric units will be the same, except remember to multiply by 1000 instead of 7.5 to arrive at liters instead of gallons.

The total of measurement A plus measurement B multiplied by 0.45 multiplied by the length gives you the surface area of the kidney shape (A + B = 18 feet). The rest of the calculations you are now familiar with. Try this volume calculation:

$$0.45 \times (A+B) \times \text{length} \times \text{average depth} \times 7.5 = \text{volume (in gallons)}$$

$$0.45 \times 18\text{ ft} \times 25\text{ ft} \times 5\text{ ft} \times 7.5 = 7593.75\text{ gal}$$

$$0.45 \times 5.4\text{ m} \times 7.6\text{ m} \times 1.5\text{ m} \times 1000 = 27{,}702\text{ liters}$$

Now, for you math majors who look over these calculations and discover that 7593.75 gallons equals 28,742 liters, not 27,702, don't worry. I rounded to the nearest tenth in each case, so my results are slightly different. But this does illustrate an important point. The difference is almost 1000 liters (or 264 gallons). This is less than a 3 percent error, but if you are applying chemical treatments to your pool or spa, it could make a difference, so the moral of the story is be as accurate as you can with the original measurements. Then a bit of rounding in the calculations won't be so critical.

Parts per Million (ppm)

One other important calculation you will use is parts per million (ppm). The amount of solids and liquids in the water is measured in parts per million, as in three parts of chlorine in every one million parts of water (or 3 ppm). However, one gallon of chlorine, for example, poured into one million gallons of water does not equal 1 ppm. That is because the two liquids are not of equal density. This becomes obvious when you discover that a gallon of water weighs 8.3 pounds (3.8 kilograms) but a gallon of chlorine weighs 10 pounds (4.5 kilograms) in a 15 percent solution, as described later. The chlorine is a more dense liquid—there's more of it than an equal volume of water.

To calculate parts per million, use the following example:

1 gal of chlorine in 25,000 gal of water

= 10 lb of chlorine in 207,500 lb of water

Now dividing each by 10 gives you:

1 lb of chlorine in 20,750 lb of water

So you see that 1 part of chlorine is in each 20,750 parts of water. But how does that translate to parts of chlorine per one million parts of water? To learn that, you must find out how many 20,750s there are in a million.

$$1{,}000{,}000 \div 20{,}750 = 48.19$$

$$48.19 \times 1 \text{ part of chlorine} = 48.19$$

There are 48.19 parts of chlorine in each million parts of water, expressed as 48.19 ppm.

Using the same formula without first translating the two liquids into pounds would give an answer of 40 ppm. Obviously this great discrepancy can result in substantial errors in treating water chemistry problems. But we're not through just yet.

If chlorine were 100 percent strength as it comes out of the bottle, that would be all there is to this calculation. As you will see in later chapters, that is not the case. In fact, liquid chlorine is produced in 10 to 15 percent solution, meaning 10 to 15 percent of what comes out of the bottle is chlorine and the rest is filler. Therefore, to really know how many parts of chlorine are in each million parts of water, you must adjust your result for the real amount of chlorine. Usually liquid chlorine is 15 percent strength (common laundry bleach is the same product, but around 3 percent strength), so:

$$48.19 \times 0.15 = 7.23 \text{ ppm}$$

Therefore, 7.23 ppm is our true chlorine strength in the example of 25,000 gallons (94.6 cubic meters or 94,625 liters) of water.

Types of Pools and Spas

Ever since the invention of the creekside swimming hole, complete with swinging rope or tire, pool builders have invented new and creative ways to capture water in our backyards. Today, because of modern materials, engineering, and building techniques, there are countless types of pools, spas, fountains, and ponds. Here are some of the most common. You will find that others are variations of these basic types.

QUICK START GUIDE: PURCHASING THE RIGHT POOL—AS EASY AS 1 - 2 - 3

Here is a simple checklist of three key features that will direct you to the right pool choice, listed in order of priority for most consumers:

1) SIZE

- *Family use:* A good rule of thumb is to allow 15 square feet (1.4 square meters) of water surface for each bather. A 10' x 20' (3 x 6 meters) pool has 200 square feet of surface (18 square meters), or enough room for up to 13 bathers.
- *Diving:* If you need a pool large and deep enough to dive or jump in, buy an in-ground pool. There is no above-ground pool designed for these activities and, in fact, every manufacturer strongly warns against using their products for these purposes. It just isn't safe, even with the optional deeper swimming end that is designed into some above-ground models.
- *Swimming laps:* If length is more important than bather load, you may want to choose a rectangular pool of sufficient length for a good workout. Lap pools are typically shallow.
- *Yard space:* Regardless of the intended use, you are limited by the size of your yard, especially the area that is mostly flat. You also need to allow at least 3' (about 1 meter) on all sides of the pool. Measure twice before buying the pool!
- *Size matters:* There's nothing worse than an undersized pool. You may have a small family, but watch how fast it grows when you have a pool. Allow for the largest party or gathering you are likely to have when selecting a pool.

2) PRICE

- A good quality 20' x 35' (6 x 10 meters) metal-sided, above-ground pool will cost around $4,000, including standard filtration equipment, ladder, and sand or other materials to prepare the ground. Professional installation costs up to an additional $1000, with some installers charging $25 per inch (2.5 cm) to level the ground beyond the first 3 inches (7.6 cm). Smaller versions, easier to manage for the do-it-yourself owner, can cost as little as $300 for a 15' (4.6 meters) diameter (round) shallow model, including a simple filter/pump circulation unit.
- A good quality 20' x 35' x 4' average depth (6 x 10 x 1.2 meters) in-ground gunite pool will cost around $20,000. This varies greatly, depending on the access to your yard, type of equipment you may want included, and whether the pool includes a spa. You can lower the total price by at least $5000 by choosing a fiberglass shell instead of gunite construction.
- A good quality soft-sided/framed pool of 10' x 20' (3 x 6 meters) will cost around $3000, requires little or no ground preparation or materials, and is easily installed by the consumer. A basic equipment package and ladder is typically included.

- **A frameless, round o-ground pool of 16' in diameter (5 meters) will cost under $1500, including basic equipment and ladder. A 10'-diameter (3 meters) shallow model with a small circulation pump can cost less than $200 at major mass-market retailers. Of course, these are the easiest to set up for the average consumer and require no other preparation.**

3) EXTRAS

- **The larger the pool, the larger the pump, plumbing, filter, and heater. These can add a few hundred dollars to any choice of pool.**
- **Decks, fencing, and landscaping can add thousands of dollars to the project, so be sure to estimate those costs before proceeding.**
- **Annual closure and mobility: If you need to tear down the pool each winter or plan to move, you may prefer the superior portability of the soft-sided on-ground pools.**
- **Don't forget that larger volumes of water require more maintenance costs every year, so be sure the upkeep and the initial installation fit your budget.**

Concrete and Plaster

Concrete and plaster pools are the most typical "hole-in-the-ground" pool, using steel-reinforced concrete to form a shell (Fig. 1-7). Because concrete is porous, the shell is coated with plaster to hold water and

FIGURE 1-7 Reinforcing steel bars (rebar) laid out for pool construction. *Questar Pools and Spas, Escondido, Calif.*

FIGURE 1-8 In-ground concrete pool.

for cosmetic purposes. Concrete can be sprayed over this latticework of reinforcing steel bars (rebar) or forms, or it can be poured from a mixing truck. The sprayed types are the most common because the material is easier to work to create free-form shapes (Fig. 1-8).

Sprayed concrete types include gunite (an almost dry mix of sand and cement) or shotcrete (a wetter version of gunite). Poured concrete requires forms into which the wet concrete is poured.

A cutaway of these types of construction is shown in Fig. 1-9.

Vinyl-Lined

Vinyl-lined pools or spas are built as metal or plastic frames (less frequently masonry blocks or pressure-treated wood is used) above the ground or set into a hole in the ground. Prefabricated panels of plastic, aluminum, steel, or (rarely) wood are joined to the frame making a form that is then lined with heavy vinyl to create the actual pool shell (Fig. 1-10). These pools require somewhat different treatment and maintenance methods than a concrete pool. Prefabrication and easier assembly make these pools less costly than concrete styles. Some small units are made as do-it-yourself backyard kits consisting of self-supporting aluminum or steel frames with a vinyl liner, sometimes with stairs, decks, and equipment all packaged together.

The choice of framing material is guided by your needs, location, and budget. A full description of above-ground pools, including their

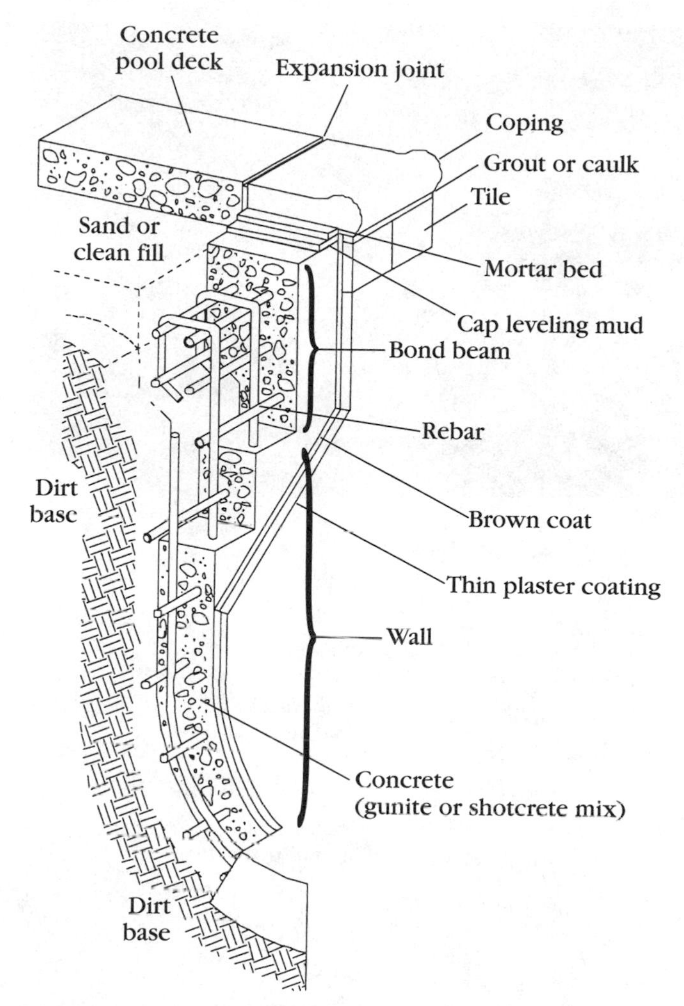

FIGURE 1-9 Cross-section of a typical gunite pool wall.

installation and maintenance, can be found in *The Ultimate Guide to Above-Ground Pools* (McGraw-Hill, 2004), but here are the most common styles and materials used:

- *Steel.* Sheets of steel, reinforced with vertical and horizontal reinforcing bars, are cut to form panels which are connected together to create the form of the finished pool. Panels and braces are galvanized or coated to prevent rust. Steel is favored for its ability to withstand expansion and contraction in freezing climates and for its ability to be shaped at the factory to create free-form pools. Steel is also the least expensive choice.

A

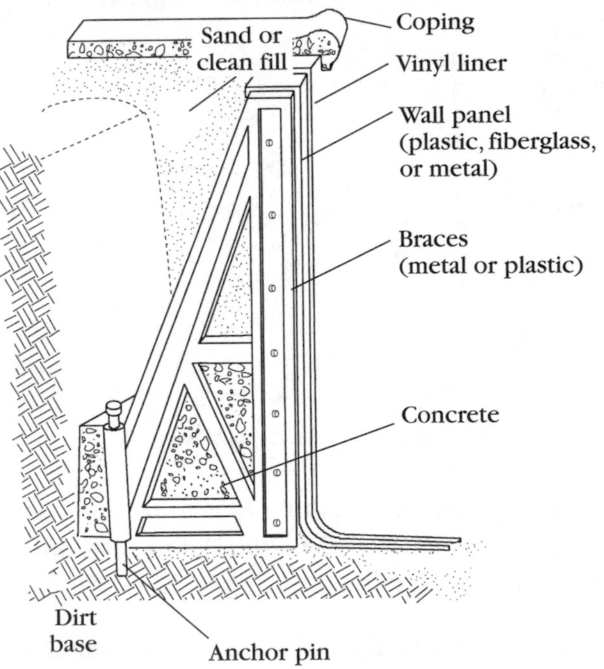

B

FIGURE 1-10 Construction (A) and cross-section (B) of a typical vinyl-lined pool.
Sentry Pool, Inc., Moline, Ill.

- *Aluminum.* Aluminum panels are lighter (a consideration when installing) and generally resist corrosion better than steel. The drawbacks of aluminum are the higher cost and the fact that panels are not as easily bent into creative shapes for free-form pools. The latter problem is the result of the way aluminum panels are made—extrusion, with bracing built into the design of each panel.
- *Polymers.* The most costly, polymer wall panels are the lightest and, of course, don't corrode. Like aluminum, however, polymer panels are constructed (molded in this case) with the bracing built in, so bending to create free-form pools is not possible without creating new molds that create the desired shape.

Fiberglass

Fiberglass pools or spas consist of a fiberglass or acrylic shell resting in a hole in the ground. Above-ground models are framed with metal or plastic like vinyl-lined pools. Some concrete pools are lined with fiberglass instead of plaster and, in fact, this process is becoming more popular because fiberglass requires less maintenance than plaster.

Some fiberglass pools or spas are assembled on-site from panels, some are an entire molded shell, and some are fiberglass walls mounted on a concrete pool bottom.

Spas, if not built with a pool or as an architectural feature in the home, are almost all some form of fiberglass shell (Fig. 1-11). These can range from a shell set directly into the ground to small portable units (Figs. 1-12 and 1-13) that are self-contained wooden frames with wooden skirting containing the fiberglass spa and all the equipment in one package.

Special maintenance procedures for spas are discussed in Chap. 9 and a full explanation of all types of spas and hot tubs, including construction and installation, can be found in *The Ultimate Guide to Spas and Hot Tubs* (McGraw-Hill, 2005).

Above-Ground Pools

Perhaps no other area of pool construction is as innovative as the above-ground pool. As previously mentioned, vinyl-lined pools can be constructed above or below the ground, but there are many other materials and designs of pools that are meant to be used above ground (Fig. 1-14).

FIGURE 1-11 Typical acrylic-over-fiberglass preplumbed spa.

FIGURE 1-12 Typical portable spa.

FIGURE 1-13 Self-contained spa. *Caldera Spas and Baths, El Cajon, Calif.*

FIGURE 1-14A Above-ground and portable pools. *Delair Group, LLC.*

FIGURE 1-14B Above-ground and portable pools. *Splash SuperPools.*

The two major advantages of above-ground pools are their low cost and portability. Above-ground pools are generally designed to be packed away in winter or taken along with the family furniture when moving from one home to another.

Zodiac, makers of the famous inflatable boat, and several other manufacturers are applying PVC technology to pools and spas. Inno-

FIGURE 1-14C Above-ground and portable pools. *Sofpool, LLC.*

FIGURE 1-14D Inflatable above-ground pools. *Zodiac Pools.*

vative bracing allows manufacturers to create free-standing pools of vinyl or other plastics without massive wall panels. These pools are also valued for their ease of installation. Following the manufacturers' directions, above-ground pools can be assembled and enjoyed on the same day. Plumbing and equipment are equally easy to assemble, usually provided by the pool manufacturer as a preassembled package with threaded or snap-together fittings and flexible PVC plumbing.

Some above-ground pool packages include decks. Indeed, some manufacturers cleverly outfit the decking with solar heating panels for no-cost water heating.

Wood

Hot tubs and some pools (Fig. 1-15) are made from a variety of woods, most commonly redwood (but cedar, teak, mahogany, and more exotic woods are also used). In fact, they can be made from any wood, but those mentioned are most resistant to rot.

FIGURE 1-15 Wooden pool. *Technobois.*

Pool and Spa Design and Construction

This book is not designed to make you a pool or spa builder or design expert. In the reference sources at the end of the book you will find a list of excellent resources to help you do those things, taking into consideration your ground plan, slope and geology, wind and sun conditions, landscaping, home value, use and need, and a host of other planning and building aspects. Ask four pool builders their opinion of this process and you will get at least five answers and discover that no two are alike in approach or priority. An entire book could be written on that subject alone, but unless you are planning to study for a pool contractor's license or are a homeowner with a self-destructive mentality, you will not be building or designing pools. Generally you will be dealing with pools or spas that are already in use.

Still it is worth understanding how that pool or spa came into being. As a pool technician or someone buying a home with a pool, you will want to recognize good (or bad) construction. This knowledge will help you estimate the maintenance costs, future potential repair expenses, or the problems involved in upgrading the pool or spa in question.

The construction of a pool or spa is basically approached in the following way (usually in this order).

Plans and Permits

The architect or pool builder supplies plans (Fig. 1-16) and hires an engineer to provide the steel (rebar) schedule based on calculations of soil stability, geology, etc. The schedule includes the specifications of thickness, tensile strength, and how close each bar will be to the next based on how strong the final product needs to be. This will differ from one part of the pool to another. For example, the bottom needs to support more weight than the sides.

A local building permit is issued when plans and steel schedules are approved. Careful planning should include consideration of backyard access. Be sure heavy equipment will not be running through fences or over fragile planting areas, septic tanks, pipes not deeply buried, etc. Also make sure the area to be excavated (including pipe trenches) does not traverse underground gas pipes, water pipes, phone lines, or electrical conduit. If so, make plans to reroute these and excavate them carefully.

Excavation

The excavator lays out the pool diagram, based on the plans, using bender board or wooden planks (Fig. 1-17A). He digs according to the plan plus one foot to allow for the thickness of material and plumbing. He digs straight down approximately 3 feet (1 meter), then slopes the remainder to the bottom. The actual slope and contour of the pool is determined by the rebar sculpting. The excavator also cuts out a 2-cubic-foot (5.7-decaliter) area for each skimmer and for each light to be installed. He also trenches from the poolside to the equipment area for laying pipe. It should also be the responsibility of the excavator to remove the dirt from the job site, as well as tamp down, compact, and add a layer of gravel to the finished hole. The gravel aids drainage if groundwater is present from below or if leaks occur in the structure.

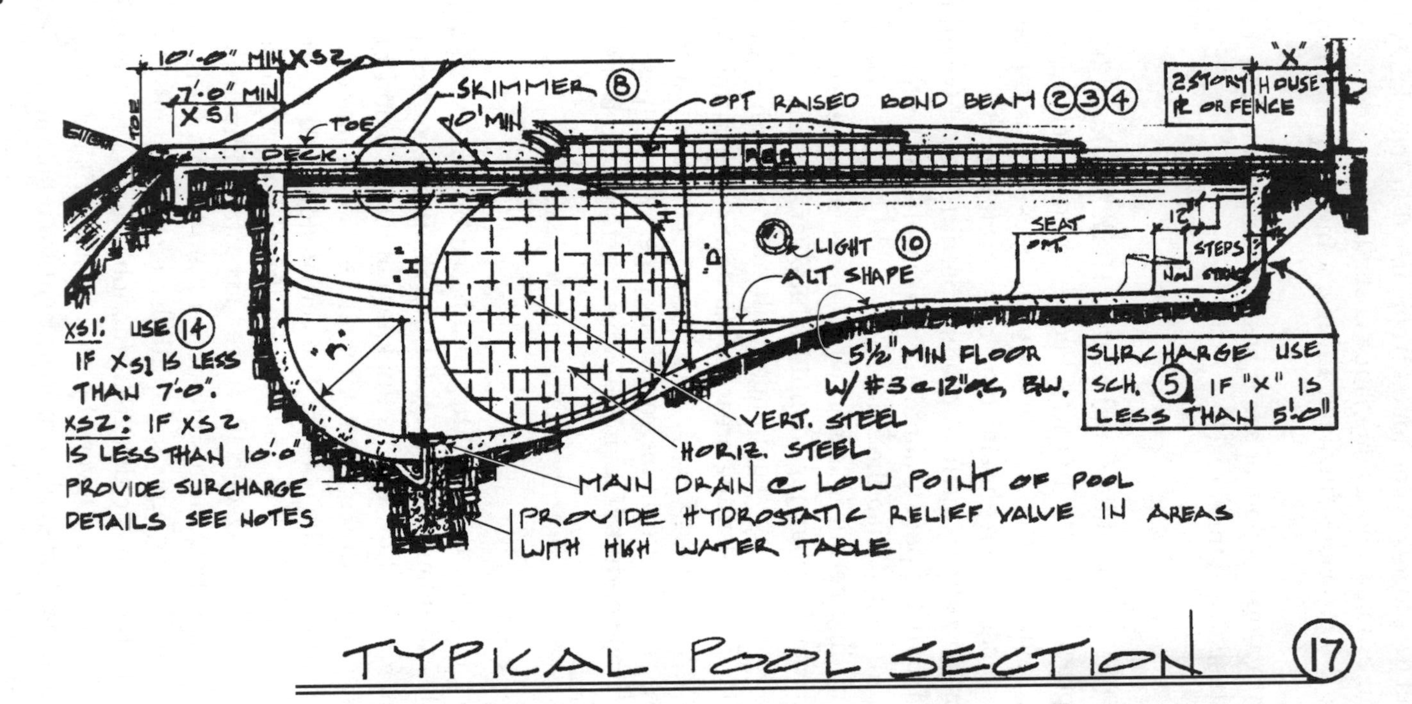

FIGURE 1-16 **Typical pool and spa plans.** *Pool Plans, Inc.*

A

B

FIGURE 1-17 Typical phases of pool construction. *Anthony & Sylvan Pools, Doylestown, Pa.*

C

D

FIGURE 1-17 (*Continued*)

E

F

FIGURE 1-17 *(Continued)*

G

H

FIGURE 1-17 *(Continued)*

Plumbing

Sometimes the steel rebar is laid first (Fig. 1-17B), with the steel man laying in the main drain and leaving room for the plumber to work around the steel, but often the plumber lays in his plumbing first. Skimmers are laid in where the prevailing wind in that area will push debris toward them (a change that should be made on the job if it was not figured into the original plan; Fig. 1-17C). Plumbing must be at least 18 inches (46 centimeters) below ground and gas lines 12 inches (30 centimeters) below (unless PVC gas line is used, which must be 18 inches below as well).

The plumbing includes the main drain and skimmers, pipes to the equipment area and back to the pool, return outlets, automatic cleaner piping, waterfall lines, a water filler line, or other design requirements. All plumbing is then sealed off and tested under pressure to test for leaks. Some large pools or commercial installations might have water sent back to the bottom of the pool for even distribution of filtered water, chemicals, and heat.

Steel

The rebar is laid in a crisscross pattern and formed or sculpted to the final contour and design of the pool (Fig. 1-17C). Generally, rebar is centered 12 inches (30 centimeters) apart on pool walls and 6 inches (15 centimeters) apart on the bottom and stress points (it does no good to locate steel any closer than this because there will not be enough room for the concrete mix, which needs space to build up strength). Wherever the rebar crosses, it is tied together with heavy wire to create a large, single, mesh bowl. The steel man must be sure to closely cut the ends after making these ties. Long, loose tails that stick up will rust and later show through the gunite and plaster layers.

Heavier steel is often used for the top 12 inches (30 centimeters) and over the edge, called the *bond beam*. The bond beam supports the coping and sometimes the edge of the deck so it must be extra strong (often engineered to support up to 1000 pounds per linear foot or about 1400 kilograms per meter). Sometimes the bond beam extends up several feet above the waterline for waterfalls or tile areas. If the job is a vinyl-lined pool, this steel process will instead be the layout of the support structure.

Electrical

The electrician grounds or bonds the steel [and any other metal within 5 feet (1.5 meters) of the water's edge]. He also adds light fixtures, which must usually be at least 24 inches (60 centimeters) below the water surface. Leaks in pools often occur at the light fixture cutout area or niche. The electrician is concerned that his wiring is completed to local code and connected properly, but often cares little about waterproofing. That is not his job! Therefore, builders often end up filling conduit and the joints where the light niche and pool wall meet with silicone sealant.

Gunite

The exact gunite or shotcrete mix is specified by the engineer based on strength and weight-bearing needs (as you have seen, water weighs a lot!). Often the mix will be five parts sand or gravel to one part cement.

The mixture is shot under pressure with a hose and nozzle (Fig. 1-17D) from the mixing truck in and around the rebar to create the pool or spa shell, usually 4 to 5 inches (10 to 13 centimeters) thick for walls, 6 inches on the deep end floor, and 11 inches (28 centimeters) for the bond beam. Like spraying water from a garden hose, some waste, splashing, and overspray occurs.

Some of the gunite does not adhere to the rest of the mix and falls away from the surface being sprayed. This waste is called *rebound* and should be cleaned up and thrown away. Some builders use it to fill in step, love-seat, or other contour areas. This is a poor practice because rebound hardens quickly and when it is used as filler it creates air pockets, which later settle and cause leaks. This is why a large percentage of cracks and leaks are found on or near pool steps (Fig. 1-17E). In a vinyl-lined pool, this gunite process will instead be the installation of the panels that form the shell.

Tile

Tile is added, usually at the waterline (Fig. 1-17F) to create an aesthetically pleasing finish and to provide a material at the surface (where oil and scum accumulate) that is not porous and is therefore easier to clean. Sometimes decorative tile patterns or racing lanes are installed to match the tile surface line (Fig. 1-18A).

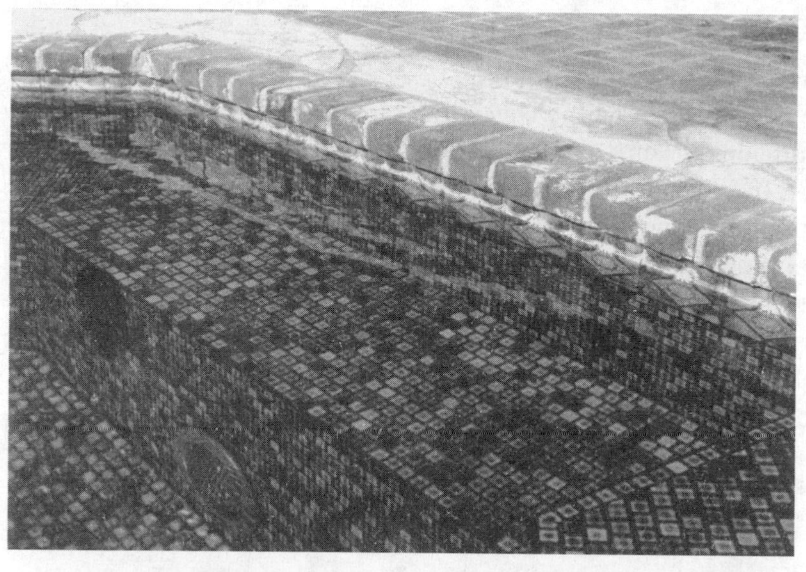

A

B

FIGURE 1-18 Tile at the waterline (A) in an all-tile pool (B).

Of course, the entire pool can be tiled (Fig. 1-18B). I have serviced several such pools, and although the original cost is high, I can enthusiastically recommend this design. An all-tile pool is strikingly beautiful, holds a pH better than plaster pools, never needs refinishing, and does not stain or etch.

Rock, Brick, or Stone

Rock, brick, or stone brought to the edge of the pool, or in some cases over the edge and below the waterline, create unique designs and natural pond looks. Until a few years ago, rocks were trucked in and cemented in place around the pool. Unfortunately such shipping and installation was very expensive and the bond beam needed extra, costly reinforcement to support the weight.

Since the late 1980s, however, *rocks* are formed with light rebar or chicken wire sculpting, then covered with a special plaster, concrete, and sand mix colored to look like natural rock (Fig. 1-19). Made onsite, such artificial rock is created to conform exactly to design specifications and is far cheaper and lighter than real rock. Rock, slate, and waterfalls are added using Thoroseal (a waterproof concrete sealer) to set them in place. More information about "rockscapes" is provided in Chap. 11.

The only drawback to these and natural rock is that they are porous and you will soon see unsightly white scale forming at the waterline.

FIGURE 1-19 Artificial rock pools, spas, and waterfalls. *California Pools & Spas, West Covina, Calif.*

FIGURE 1-20 Scale deposits on tile.

Even if you maintain perfect chemistry in the pool, natural evaporation leaves behind any mineral present in the water as *scale* (Fig. 1-20), which appears as white scum, mostly calcium, around the waterline. Although this can be scrubbed off of nonporous tile, it must be sandblasted off rocks and will reappear in short order. One solution is to keep a constant water level, replacing water as it evaporates so the scale line is hidden under the waterline. Of course, maintaining proper water chemistry balance also reduces scale. These aspects are discussed in later chapters.

Coping, Decks, and Expansion Joints

Coping (Fig. 1-21) is the finish work done to the top of the pool wall, usually attached directly to the bond beam. Coping stones are often precast and made of porous material to provide better traction for the wet feet and hands of swimmers entering or leaving the pool.

After the coping is laid on, the deck is poured (or deck brick or stone work is done as in Fig. 1-17G) up to the edge of the coping, leaving a ¼- to ½-inch (6- to 13-millimeter) gap. The gap, which allows expansion or contraction of the deck and coping materials in hot and

FIGURE 1-21 Typical pool coping, deck, and expansion joints. *Questar Pools and Spas, Escondido, Calif.*

cold temperatures, is filled with silicone caulking to keep out water. All of this work should be done before plastering or finish work, because it is usually the messiest procedure. I have seen builders forced to drain and replaster newly finished pools because they completed the pool before the deck work was done, only to find sloppy deck workers scatter the fresh, soft pool plaster with cement, gravel, and stone chips.

Equipment Set

The pump(s), filter, heater, and ancillary equipment is set on its concrete pad and connected to the plumbing (Fig. 1-22). The electrician and plumber finish these hookups.

Cleanup

A smart builder cleans the area of loose debris, rebound, concrete dust, other debris—in short, anything that might end up in the pool after the water goes in or which might stain the new plaster. Even dirt taken from the excavation will stain fresh plaster.

FIGURE 1-22 **Typical pool equipment set.** *Questar Pools and Spas, Escondido, Calif.*

Plaster

The fine white or colored plaster, also called marcite in some regions, is now troweled over the gunite (Fig. 1-17H), about ½ inch (13 millimeters) thick. Plaster can now be mixed to just about any color you can imagine. Rails, ladders, drain covers, rope hooks, and so on are added and plastered in place. Water is added immediately because the plaster hardens and cures underwater. If the plaster is allowed to dry out before water is added, the weight of the water can create stress and cracks.

Plaster is discussed at greater length in the chapters on chemistry and special procedures; however, it is worth noting that pool builders and plasterers have differing opinions regarding just about every aspect of plaster application, care, and maintenance. There are general standards created by industry organizations (see Reference Sources and Websites), but experience, local variations, and tricks of the trade often create differing methods.

If this installation is a vinyl-lined pool, the laying out of the liner takes the place of plastering.

Startup

Plaster requires a break-in to prevent staining. The water also must be treated before swimming. This process is treated in detail in a later chapter.

Depending on how you intend to use your new pool, you might add a diving board, water slide, fountain, special landscaping, sundeck, or other accessories. These options are described in later chapters. In any case, it's now time to enjoy your new pool (Figure 1-23)!

FIGURE 1-23A The finished pool. *Zodiac Pools.*

FIGURE 1-23B The finished pool and spa.

FAQs: POOLS AND SPAS

Will a Pool or Spa Add Value to My Home?

- An in-ground pool or spa is more likely to add value than an above-ground pool or portable spa. If the yard is small and the pool or spa takes up most of the space, that could hurt your property's value to someone who values a yard, so make sure the installation is appropriately sized for the available area.

Should I Hire a Contractor or Do It Myself?

- Building an in-ground pool or spa is best left to the professionals, although if you are handy, you could probably save money by installing some of the equipment or doing the deck and landscaping yourself. Above-ground pools are fairly easy to install on your own, but you may want to watch someone else do it first to be sure you are up to the task.

A Pool Is a Big Investment—How Many Months of the Year Can I Use It?

- Even in cold climates, you should be able to enjoy your pool at least six months of the year by using a cover in spring and fall to retain the warmth. Of course, if you can afford to heat or enclose the pool, you can enjoy it year-round. If you do plan to heat your pool, you might invest in the smallest one that meets your needs to avoid heating extra water that you really won't use.

CHAPTER 2

Basic Plumbing Systems

To understand a pool or spa, you must follow the path of the water. This and all following chapters have been arranged in this manner and you will find it is easy to troubleshoot a pool or spa problem by following this path.

As can be seen in Fig. 2-1, water from the pool or spa (not both at the same time) enters the equipment system through a main drain on the floor, through a surface skimmer, or through a combination of both main drain and skimmer. It travels to a three-port valve (if there is no spa, there will be no such valve) and into the pump, which is driven by the attached motor. From the pump, the water travels through a filter, up to solar panels (if part of the installation), then to the heater, and back through three-port valves to the pool or spa return lines.

Skimmers

Some pools have more than one skimmer. The purpose of the skimmer, as the name implies, is to pull water into the system from the surface with a skimming action, pulling in leaves, oil, and dirt before they can sink to the bottom of thc pool. It also provides a conveniently located suction line for vacuuming the pool. Most skimmers today are molded, one-piece plastic units. Older pools have built-in-place concrete skimmers.

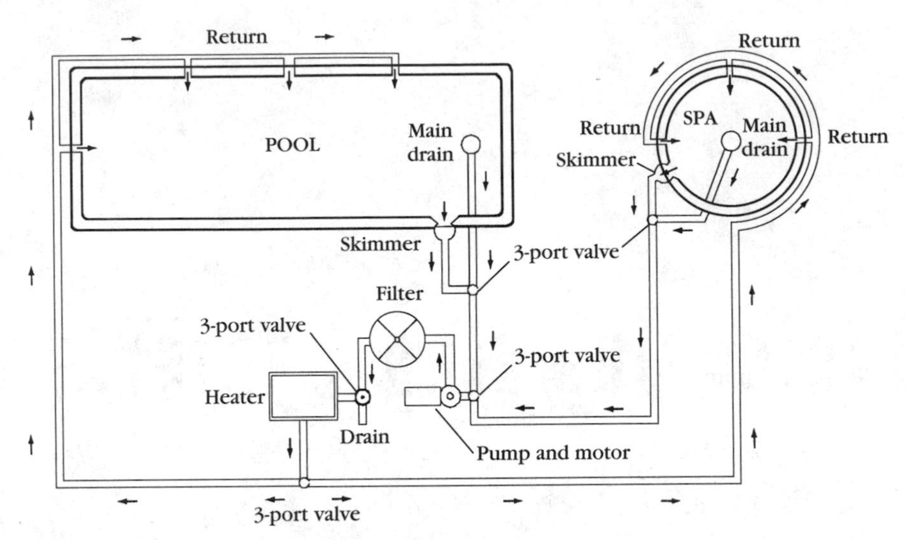

FIGURE 2-1 Typical pool and spa plumbing layout.

In either case, the skimmer is accessed through a cover on top that sits on the deck at the edge of the pool (the cover will be plastic or concrete) or by reaching into the skimmer through the opening that faces the pool itself. As shown in Fig. 2-2, most skimmers are built into the deck, connecting to the pool out of sight. Some, as with portable or above-ground pools, are separate units that hang on the edge of the pool (in the water or outside of it). Redwood hot tubs use a flat, vertical skimmer that has no basket but skims the surface and pulls any floating debris to a plastic screen. Some portable spa skimmers have a cartridge filter built in. Some pool skimmers include automatic water level controls and automatic chlorinators.

The water pours over a floating weir (Fig. 2-2) that allows debris to enter, but when the pump is shut off and the suction stops, the weir floats into a vertical position, preventing debris from floating back into the pool. Some skimmers have no such weir (although spring-loaded weirs are available that can be fitted into any skimmer mouth) and use a floating barrel as part of the skimmer basket. The purpose of the basket is to collect leaves and large debris so they can then be easily removed.

The disadvantage of both types of weirs is that leaves can cause them to jam in a fixed position, thus preventing water from flowing

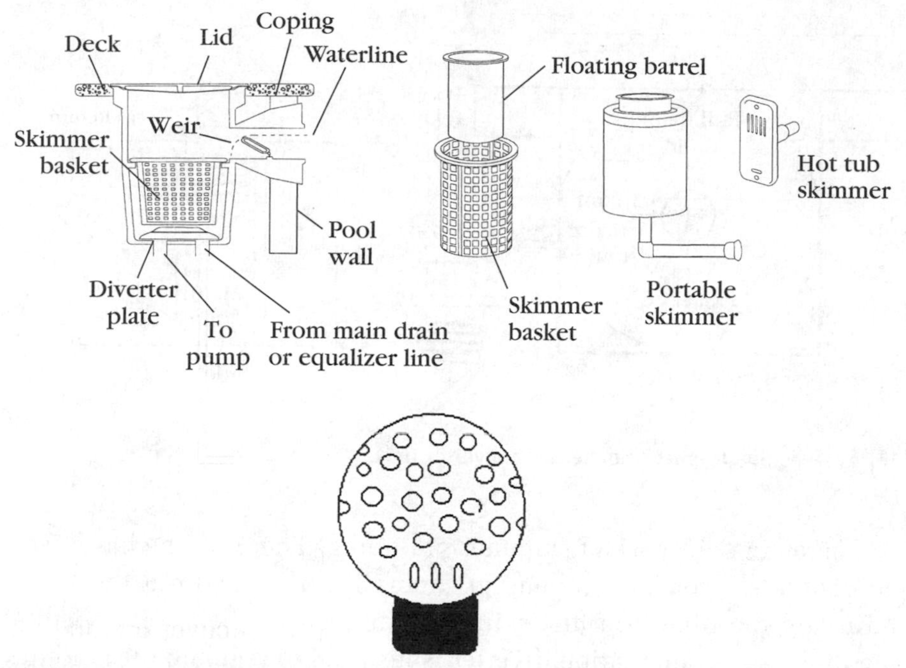

FIGURE 2-2 **Top: typical skimmers. Bottom: whiffle ball skimmer.**

into the skimmer. When this happens, the pump will lose prime (water flow) and run dry, causing damage to its components. Therefore, during windy periods it might be better to remove the weir from the skimmer to prevent such problems.

Another style of debris collector is the plastic ball with holes in it, like a "whiffle ball." A nipple on the ball inserts into the suction port of the skimmer so that debris is collected as a result of suction on the ball as water flows through. This style of collector is useful when the skimmer can't accommodate an actual basket.

By the way, you should exercise care when working around the skimmer when the pump is on. I have nearly had fingers broken when placing my hand over a skimmer suction opening and have lost various pieces of equipment, T-shirts, bolts, and plastic parts, which invariably end up clogged in the pipe at some turning point where leaves, hair, and debris later catch and close off the pipe completely. Keep small objects away from the skimmer opening when the basket is removed and especially keep your hands from covering that suction hole.

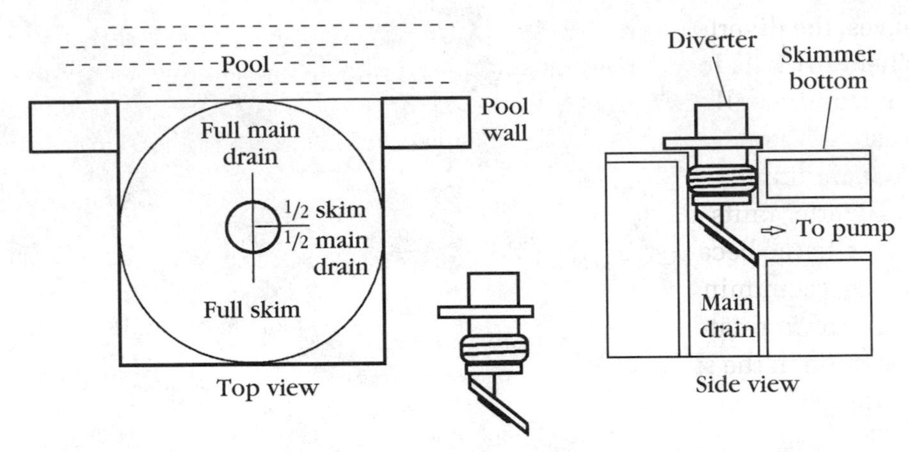

FIGURE 2-3 Single-port skimmer with diverter unit.

There are two types of skimmer plumbing. The first one has a single visible suction port (opening). Actually the pipe from the main drain and the pipe from the skimmer connect just below the visible opening and a combination diverter is inserted to regulate the amount of suction from one or the other (Fig. 2-3). A neck on the diverter extends up from the skimmer bottom for attaching your vacuum hose when cleaning the pool. The value of this system is that when vacuuming the pool, you can divert all of the pump suction to the skimmer bottom, in effect, shutting off the main drain.

The diverter also has a nipple aligned horizontally to the opening. Usually when the nipple faces away from the pool, the flare on the bottom of the diverter closes the main drain pipe and all of the suction from the pump is now at the skimmer. When the nipple faces toward the pool, the body of the diverter closes the pipe from the skimmer and all of the suction is now at the main drain. Obviously various degrees between these two settings will divide suction between the skimmer and main drain.

Each pool needs its own setting to compensate for various factors such as wind conditions, equipment efficiency, and type of cleaning conditions. For example, if the pool gets more leaves than dirt, the diverter should be set to make the suction in the skimmer stronger than in the main drain. That will help the skimmer pull the leaves into the skimmer basket. If the pool tends to get more dirt or sand than leaves, the diverter should be set to strengthen the main drain suction. When dirt falls to the bottom of the pool, the strong suction from the main drain will pull the dirt toward it. This is also helpful when brushing the pool bottom, because suspended dirt will be pulled into the main drain.

Diverter units are made of plastic or bronze. I always carry a bronze unit with me because the plastic ones tend to come loose and float out during vacuuming if the suction from the pump is not strong. They also tend to rotate in the skimmer as you work, changing the amount of suction in the skimmer from what you have set. The bronze diverter, obviously heavier, solves these problems.

The other type of skimmer plumbing (see Fig. 2-2) has two separate ports—one is a pipe that goes directly to the main drain or to an equalizer line, while the other goes directly to the pump. In this type of skimmer, a diverter plate regulates the suction between the main drain and the skimmer. Usually the port farthest from the pool edge is the pipe that goes to the pump, and the port closest to the pool goes to the main drain or equalizer line. An *equalizer line* is simply a pipe that extends from the skimmer bottom down 18 to 24 inches (46 to 61 millimeters) and through the pool wall just below the skimmer.

In both styles of skimmer, the idea is that if the pool runs low on water, the pump can pull water from the bottom of the pool via the main drain instead of the empty skimmer (or from the side of the pool below the skimmer in the case of the equalizer line) so the pump will not run dry.

Some older (often concrete) skimmers have odd-sized ports that can't accommodate your vacuum hose. In these cases, a special cover plate (Fig. 2-4) can create a generic adapter.

Main Drains

The main drain has one or more plumbing ports. One port feeds a pipe to the pump. In a spa, there might be several ports for several pipes leading to different pumps (for jet action).

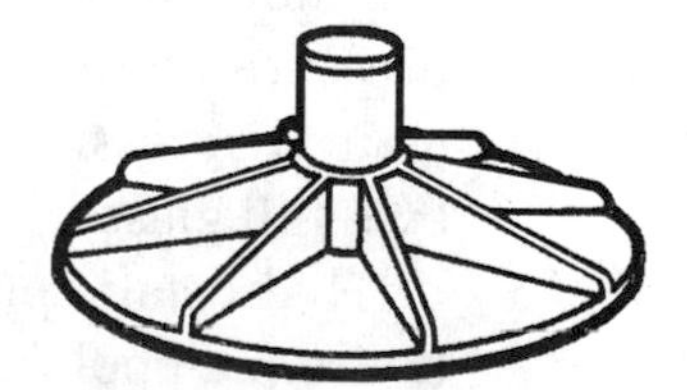

FIGURE 2-4 Skimmer vacuum adapter plate.

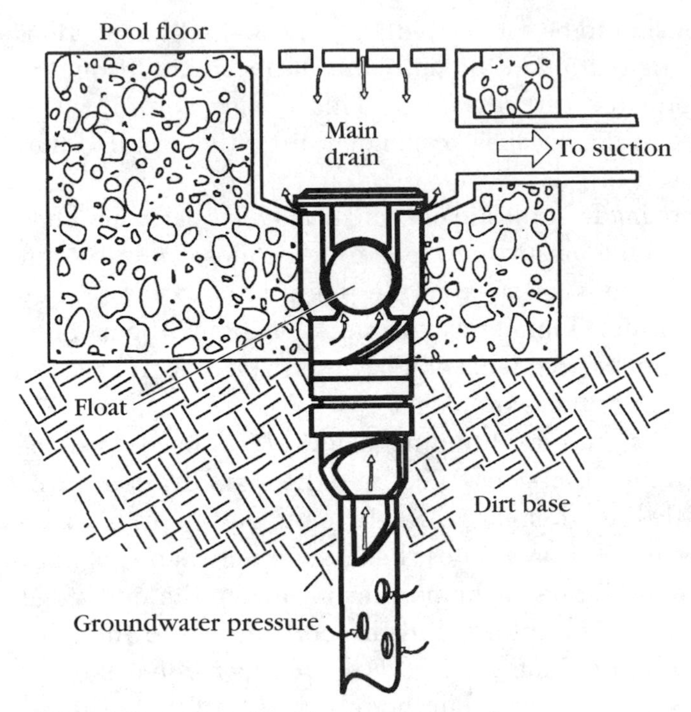

FIGURE 2-5 **Main drain with a hydrostatic valve.**

Another port is a one-way valve (check valve) that allows water that might collect under the pool to enter the pool, but no water can flow out. Figure 2-5 depicts a main drain with a *hydrostatic valve*.

Water collecting under the pool creates extreme upward pressure that can crack the pool. This pressure, called *hydrostatic pressure*, is relieved by this valve. To illustrate this yourself, take an empty bucket and, holding it upright, try to press the bottom of the bucket into a tub of water. The hydrostatic pressure makes it nearly impossible. Now try the same experiment with a full bucket. It sinks easily into the larger container of water. The water inside compensates for the hydrostatic pressure on the outside. You can try a smaller version of this experiment by pressing an empty glass down into a bowl of water, followed by a full glass.

Hydrostatic pressure is an important consideration when planning to drain a pool for any reason. Obviously it is not wise to drain a pool completely during the rainy season or if there is any other suspicion of groundwater.

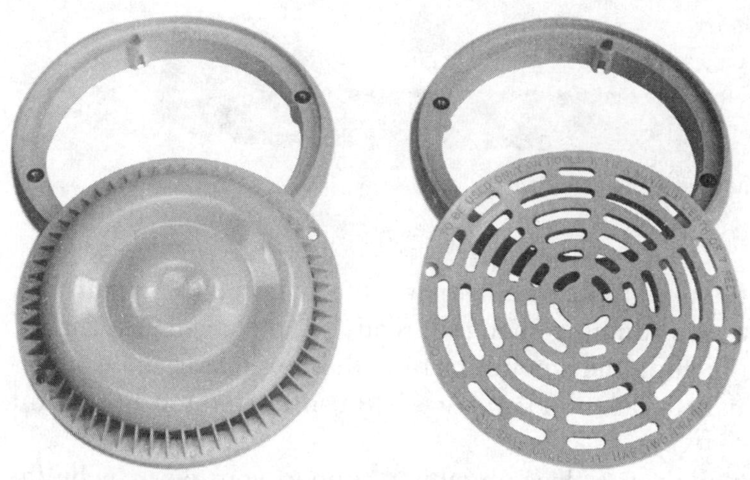

FIGURE 2-6 Main drain antivortex cover and standard cover.

In some spas, there might be more than one main drain so that if one becomes covered with a foot or hand, water is pulled from the other, avoiding injury to the bather. These drains are usually located at least 12 inches (30 centimeters) apart. Obviously in a pool where the main drain is very deep, this is not a concern, so safety suction lines are not added. Also, the suction in a pool is usually divided between the main drain and the skimmer, so one is not dangerously stronger than the other.

Because spas are relatively shallow, strong suction can create a whirlpool effect. To prevent this, many spa main drains are fitted with antivortex drain covers which are slightly dome-shaped with the openings located around the sides of the dome (Fig. 2-6). Pool main drain covers are flat with the openings on top.

In any case, the drain area is covered by a grate, usually 8 to 12 inches (20 to 30 centimeters) in diameter, that screws or twist-locks into a ring that has been plastered into the pool bottom.

General Plumbing Guidelines

RATING: EASY

Before proceeding to specific instructions on working with PVC plastic, galvanized, or copper plumbing, here are a few general guidelines that I think are important regardless of the material you are using.

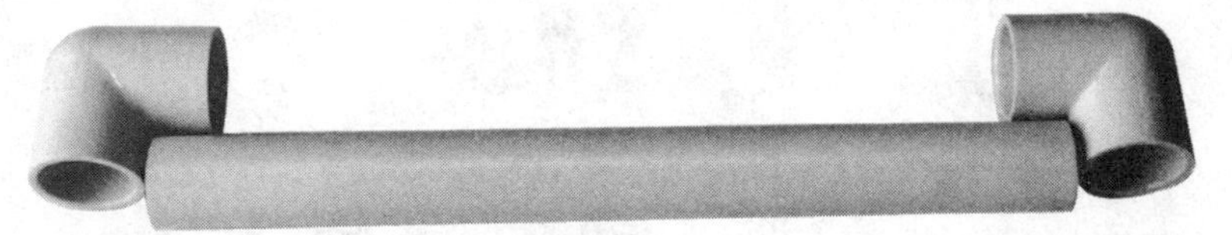

FIGURE 2-7 Pipe measuring and fitting.

Measure the pipe run carefully, particularly if you are repairing a section between plumbing that is already in place. In measuring, remember to include the amount of pipe that fits inside the connection fitting, usually about 1½ inches (3.8 centimeters) at each joint (Fig. 2-7).

When working on in-place plumbing, support your work by building up wood or bricks under the pipe on each side of your work area. This prevents vibration as you cut, which can damage pipes or joints further down the line. Also, unsupported pipe sags and binds when you cut it. That is, as you cut, it pinches the saw blade, making cutting difficult, and straight, clean cuts impossible!

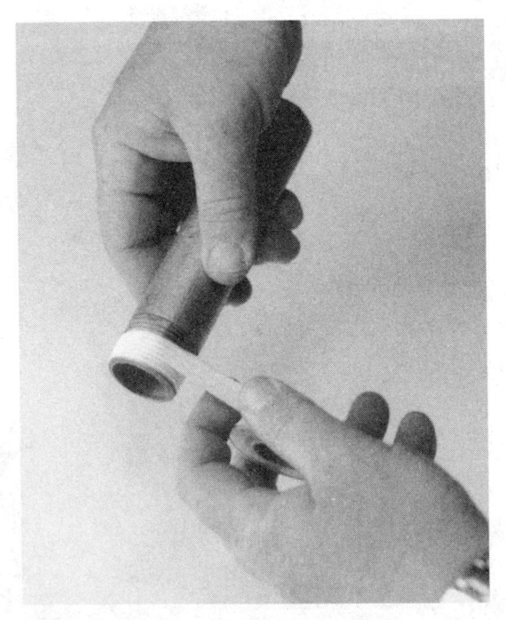

FIGURE 2-8 Correct application of Teflon tape.

Threaded fittings are obvious and simple; however, leaks occur most often in these connections. The key is to carefully cover the male threads with Teflon tape and to tighten the fitting as far as possible without cracking.

Teflon tape fills the gaps between the threads to prevent leaking. Apply the tape over each thread twice, pulling it tight as you go so you can see the threads. Apply the tape clockwise (Fig. 2-8) as you face the open end of the male threaded fitting. If you apply the tape backwards, when you screw on the female fitting, the tape will skid off the joint. Try it both ways to see what I mean and you will only make that mistake once.

Another method of sealing threads is to apply joint stick or pipe dope. These are odd names for useful products that are

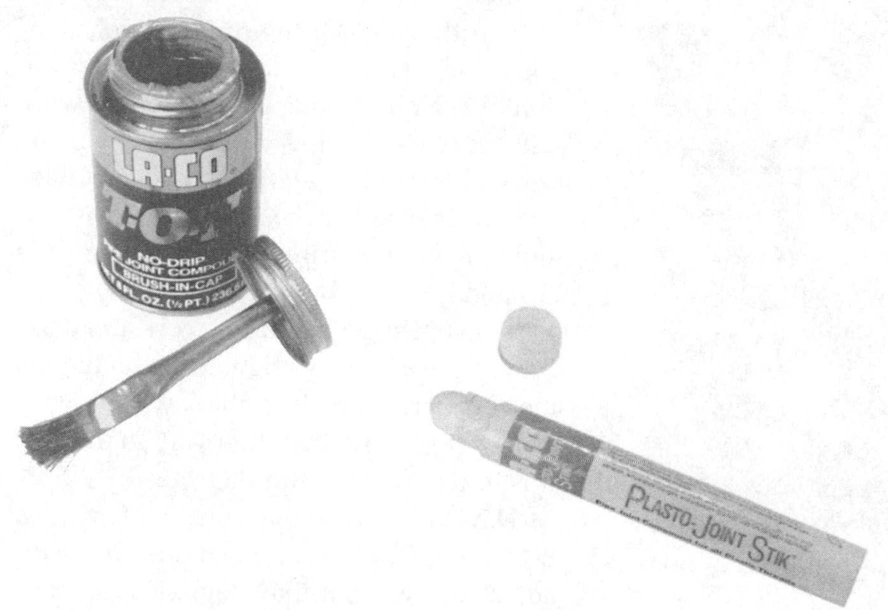

FIGURE 2-9 Pipe dope and joint stick.

applied in similar ways (Fig. 2-9). Joint stick is a crayon-type stick of a gum-like substance that works like Teflon tape. Rub the joint stick over the threads so that the gum fills the threads. Apply pipe dope the same way. The only difference is that dope comes in a can with a brush and is slightly more fluid than joint stick. The key to success with joint stick or pipe dope is to apply it liberally and around all sides of the male threaded fitting, so that you have even coverage when you finally screw the fittings together. Some product will ooze out as you tighten the fittings together, but that proves that you have applied enough.

I use Teflon tape because I know upon application that it is an even and complete coverage of the threads. Pipe dope or joint stick might not apply evenly or can bunch up when threading the joint together. If you use dope or stick, be sure it is a nonpetroleum-based material (such as silicone). Petroleum-based products will dissolve plastic over time, creating leaks.

When working with PVC pipe and fittings, tighten threaded fittings with channel lock pliers of adequate size to grip the pipe. Using pipe wrenches usually results in application of too much force and cracking

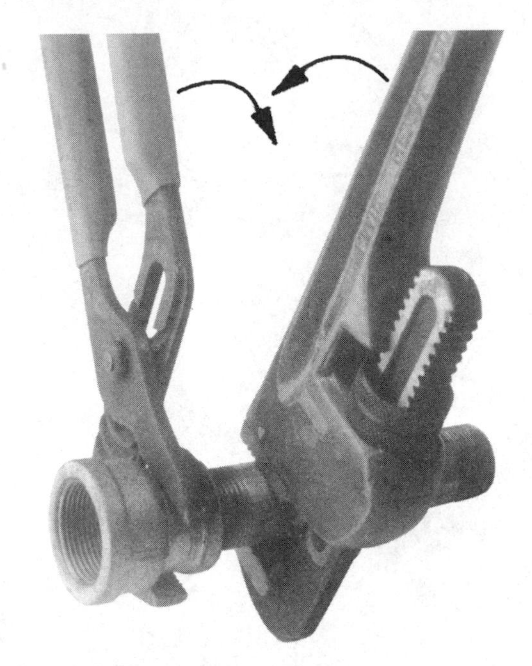

FIGURE 2-10 Correct wrench use.

of the fittings. Save the pipe wrenches for copper or galvanized plumbing. If you don't have pliers large enough for the work and must use a pipe wrench, tighten the work slowly and gently—not bad advice with copper either, because copper is soft and will bend or crimp if too much force is applied.

To avoid slipping off the work and damaging fittings, always tighten with channel locks or wrenches into the jaw, not away from it (Fig. 2-10). Another way to think of it is into their base, not their head.

Now a word about pipe cutters (Fig. 2-11). Made for PVC or metal pipe, these are adjustable wrench-like devices that have cutting wheels. You lock the device around the pipe and rotate it, constantly tightening it as you go, until the pipe is cut. They provide the straightest, cleanest cut of all. However, in pool work you will be dealing mostly with 1½- to 2-inch-diameter (40- to 50-millimeter) pipe in close quarters. You rarely have the luxury of enough space to get around the entire pipe, and these cutters take far longer than a good, fresh hacksaw blade. Use them if you like, but I think you will soon abandon them in favor of simple old Mr. Hacksaw.

The most important advice I can give you on odd-job or tight-quarters plumbing, indeed on any of this plumbing, is to ask questions. There are so many different, unique fittings and fixtures for cutting, joining, and repairing plumbing that they can fill a book of their own, and it would still be out of date because of constant revision and new products. So if you run across a tough connection of odd pipe or different materials, ask questions at your local plumbing supply house. Most of the counter help is knowledgeable and willing to advise you because their advice sells their products. A good idea is to take Polaroid snapshots of the job (or bring the materials in with you if you must cut them out anyway) so they will thoroughly understand your specific needs.

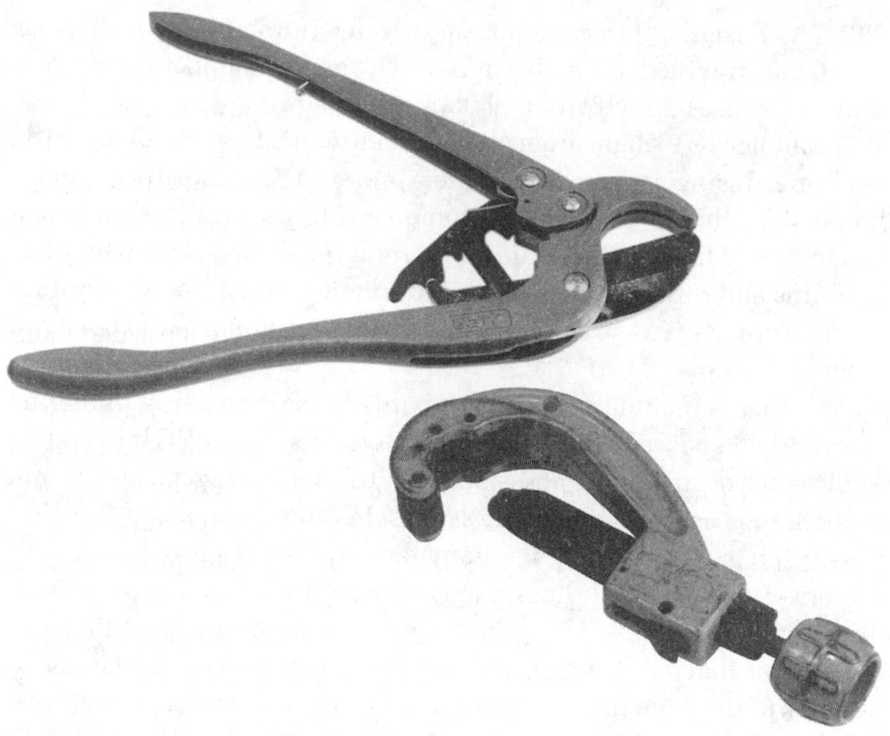

FIGURE 2-11 Pipe cutters.

PVC Plumbing

If you played with Tinker Toys, Lego blocks, or Lincoln Logs as a kid, you will find working with PVC plumbing literally, well, a snap.

Pool plumbing is prepared with plastic or metal lengths of pipe and connection fittings that join those lengths together. The pipe acts as the *male* which fits and is glued into the *female* openings of these connection fittings. Alternatively, connection is made by each side having threads, joined by screwing them together. The plastic pipe used is PVC (polyvinyl chloride) and it is manufactured in a variety of different strengths depending on the intended use.

To help identify the relative strength of PVC, it is labeled by a *schedule* number; the higher the number, the heavier and stronger the pipe. Pool plumbing is done with PVC schedule 40. Some gas lines are plumbed with PVC schedule 80.

PVC is designed to carry unheated water (under 100°F). CPVC is formulated to withstand higher temperatures for connection close to (or in some cases directly to) a pool or spa heater.

Ultraviolet (UV) light from the sun causes PVC to become brittle over time, losing its strength under pressure and creating cracks. Chemical inhibitors are added to some PVC to prevent this, the most common and cheapest being simple carbon black (which is why plastic pumps and other pool and spa equipment is often black). Another common preventive measure is to simply paint any pipe exposed to sunlight.

PVC pipe is manufactured in a flexible version, making plumbing easier in tight spaces and for spas and jetted tubs. Flex PVC is available in colors for cosmetic purposes and has the same characteristics and specifications as rigid PVC of the same schedule and size.

All pipe is measured by its diameter, expressed in inches or millimeters. Typically pool plumbing is done with 1½- or 2-inch (40- or 50-millimeter) pipe, referring to the interior diameter (the diameter of the pipe that is in actual contact with the water). The exterior diameter of the pipe differs depending on the material. For example, the exterior diameter of 2-inch PVC pipe is greater than that of 2-inch copper pipe because the PVC pipe walls are thicker than those of copper.

All pipe is connected with fittings (Fig. 2-12). Fittings allow connection of pipe along a straight run (called couplings), right angles (called 90-degree couplings or elbows), 45-degree angles, T fittings, and a variety of other formats. In the case of PVC, such fittings are most often smooth-fitted and glued together (called *slip* fittings).

Some fittings are threaded (called threaded fittings) with a standard plumbing thread size so they can be screwed into comparable connecting fittings in pumps or other plumbing parts. National Pipe Thread (NPT) standards are used in the United States so different products of various materials by different manufacturers will all work together. The NPT standard includes a slight tapering between the male and female connections. The importance of this is that because of this taper, it is easy to overtighten plastic threaded fittings and crack them. Great Britain, Europe, and Asia not only operate on metric measurements, but also have their own unique thread standards. Fittings with male (external) threads are called *mip* and fittings with female

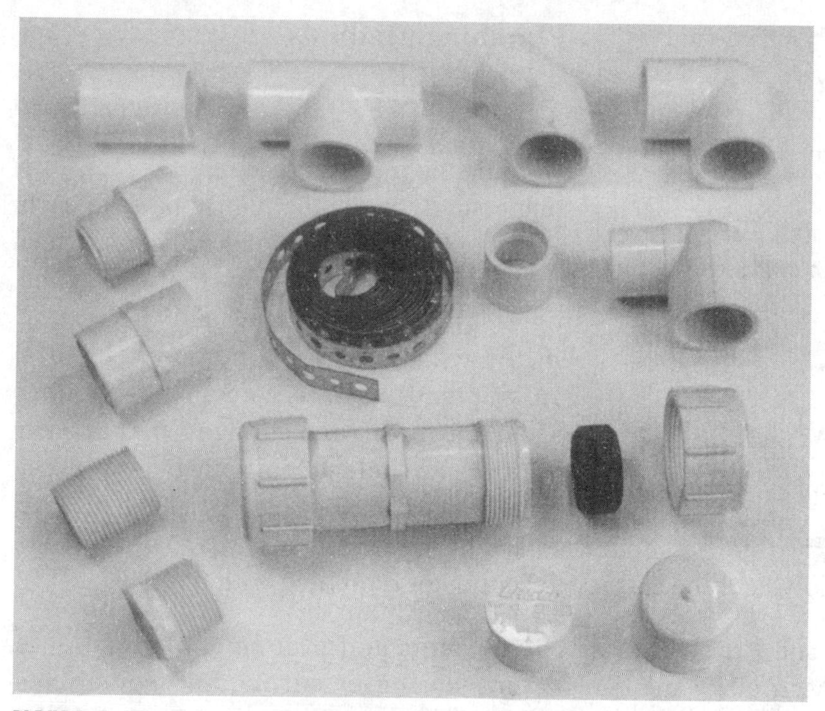

FIGURE 2-12 Pipe connection fittings. Top row (left to right): straight slip coupling, T coupling, 45-degree slip coupling, 90-degree slip coupling. Second row: MIP, FIP, plumbers strap tape, reducer bushing, 90-degree slip street coupling. Third row: close nipple, compression coupling. Bottom row: male threaded plug, male slip plug, female slip cap.

(internal) threads are called *fip*. If one side of the fitting is mip and one side is slip, you order it as *mip by slip*, and so on.

In most cases with pool and spa plumbing, the long runs of pipe will be underground. Sometimes, however, horizontal runs will be under a house or deck or over a slope where support is needed. In this case, pipe should be supported every 6 to 8 feet (2 to 2.5 meters), hung with plumber's tape to joists or supported with wooden bracing. PVC does not require support on vertical runs because of its stiffness, but common sense and local building codes might require strapping it to walls or vertical beams to keep it from shifting or falling over. Remember, the pipe becomes considerably heavier when it is filled with water and might vibrate along with pump vibration.

Plumbing Methods

RATING: EASY

The concept of joining PVC pipe involves welding the material together by using glue that actually melts the plastic parts to each other. In truth, each joint will have an area that is slightly tighter than the rest. In the tightest parts, this welding actually occurs. In the remainder, the glue bonds to each surface and itself becomes the bonding agent. Obviously the strongest part of each joint is the welded portion; but in either case, the key is to use enough glue to ensure total coverage of the surfaces to be joined.

TOOLS OF THE TRADE: PVC PLUMBING

The supplies and tools you need for PVC plumbing are

- **Hacksaw with spare blades (coarse: 12 to 18 teeth per inch or 2.5 cm)**
- **PVC glue and primer**
- **Cleanup rags**
- **Fine sandpaper**
- **Teflon tape or joint stick**
- **Waterproof marker**

Following is the correct procedure for plumbing with PVC (Fig. 2-13):

1. **Cut and Fit** Cut and dry fit all joints and plumbing planned. It is easy to make mistakes in measuring or cutting and sometimes fittings are not uniform so they don't fit well. Dry fitting ensures the job is right before gluing. If you need the fitting and pipe to line up exactly for alignment with other parts, make a line on the fitting and pipe (Fig. 2-13A) with a marker when dry fitting so you have a reference when you glue them together.
2. **Sand** Lightly sand the pipe (Fig. 2-13B) and inside the fittings so they are free of burrs. The slightly rough surface will also help the glue adhere better.
3. **Prime** You might need to apply a preparation material, called *primer*, to the areas to be joined before gluing (Fig. 2-13C). Some PVC glues are solvent/glue combinations and no primer is required. In some states, however, use of primer might be required by building code, so check that before selecting an all-in-one product. If you are using primer, apply it with the swab provided to both the pipe and the inside of the fitting. Read and follow the directions on the can.
4. **Glue** Before gluing, be ready to fit the components together quickly because PVC glue sets up in 5 to 10 seconds. Apply glue to the pipe and inside of the fitting (Fig. 2-13D).

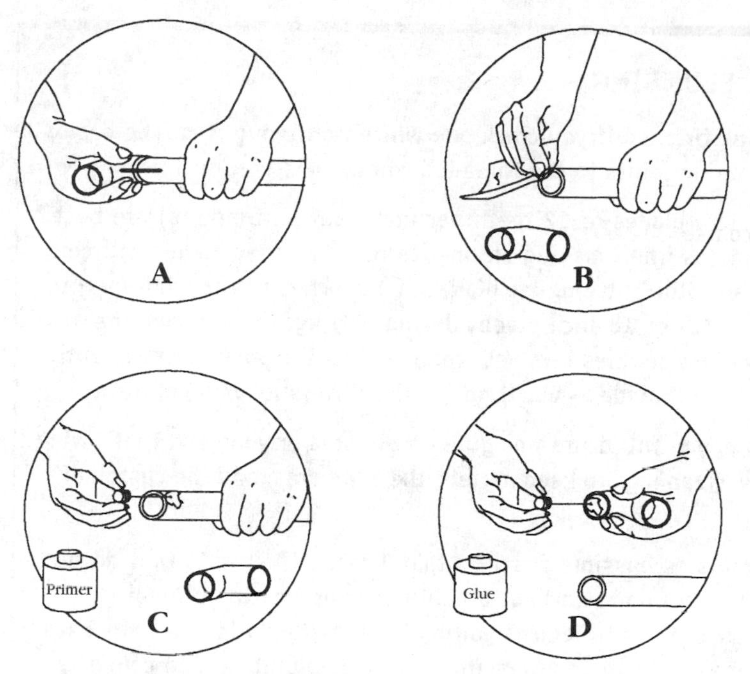

FIGURE 2-13 Step-by-step PVC plumbing.

5. **Join** Fit the pipe and fitting together, duplicating your dry fit, and twist about a half turn to help distribute the glue evenly, realigning the lines drawn on the pipe and on the fitting. If using flexible PVC, because it is made by coiling a thin piece of material and bonding it together, do not twist it clockwise. This can make the material swell and push the pipe out of the fitting. Get in the habit of twisting all pipe counterclockwise (even though it makes no difference with rigid PVC) and you will never make that mistake.
6. **Seal** With rigid PVC, hold the joint together about a minute to ensure a tight fit; about two minutes with flex PVC. Although the joint will hold the required working pressure in a few minutes (and long before the glue is totally dry), allow overnight drying before running water through the pipe to be sure. I have seen demonstrations with some products (notably Pool-Tite solvent/glue) where the gluing was done underwater, put immediately under pressure, and it held just fine. I don't, however, recommend this procedure as I have gone back on too many plumbing jobs to fix leaks a few weeks later because I hastily fired up the system after allowing only a few minutes drying time.

TRICKS OF THE TRADE: PVC PLUMBING

1. Make all threaded connections first, so if you crack one while tightening it can be easily removed. Then glue the remaining joints to the threaded work.
2. When cutting PVC pipe, hacksaw blades of 12 teeth per inch (2.5 centimeters) are best, particularly if the pipe is wet (as when making an on-site repair). Finer blades will clog with soggy, plastic particles and stop cutting. Use blades of 10 inches (25 centimeters) in length. They wobble less than 12- or 18-inch blades during cutting. In all cases, the key is a fresh, sharp blade. For the few pennies involved, change blades in your saw frequently rather than hacking away with dull blades—you'll notice the difference immediately.
3. No matter how careful you are, you will drip some glue on the area or yourself. That's why I always carry a supply of dry, clean rags to keep myself, the work area, and the customer's equipment clean of glue.
4. Try to make as many *free* joints as possible first. By that I mean the joints that do not require an exact angle or which are not attached to equipment or existing plumbing. The free joints are those that you can easily redo if you make a mistake. Do the hard ones last—those that commit your work to the equipment or existing plumbing and cannot be undone without cutting out the entire thing and starting over.
5. Use as much glue as you need to be sure there is enough in the joint. It's easier to wipe off excess glue than to discover that a small portion of the joint has no glue and leaks.
6. Practice. PVC pipe and fittings are relatively cheap, so make several practice joints and test them for leaks in the shop before working on someone's equipment in tight quarters in the field.
7. Flexible PVC is the same as rigid, but when you insert the pipe into a fitting, hold it in place for a minute or longer because flex PVC has a habit of backing out somewhat, causing leaks.
8. In cold weather, more time is required to obtain a pressure-tight joint, so be patient and hold each joint together longer before going on to the next.
9. Bring extra fittings and pipe to each job site. Bring extras of the types you expect to use, as well as types you don't expect to use, because you just might need them. Nothing is worse than completing a difficult plumbing job and being short just one fitting, or needing to cut out some of your work and not having the fittings or a few feet of pipe to replace them. It is often several miles back to the office or the nearest hardware store to grab that extra fitting that should have been in your truck in the first place. Bring extra glue, sandpaper, and rags, too.

Copper Plumbing

RATING: PRO

Copper plumbing is quickly disappearing from the pool and spa scene for a number of reasons. Unlike PVC, metal plumbing such as copper will corrode, especially in the presence of caustic pool and spa chemicals moving at high speed and under pressure through the pipes. Copper plumbing is also more difficult to install and repair, and it has become extremely expensive in recent years.

Copper was more recently still in use where dissipation of heat was important, such as in plumbing directly connected to the heater. Stainless steel heat risers, CPVC, and threaded galvanized plumbing have replaced that function, making copper plumbing for pools and spas obsolete.

In fact, copper pipes are something of a hazard, because even the most careful pool technician can make chemistry mistakes. When water becomes acidic, copper oxidizes and deposits greenish black metals on porous plaster surfaces. Not only is this unsightly and difficult to remove, but the copper pipe becomes increasingly thin every time such oxidization occurs until, finally, leaks begin.

Having said all this, copper is still widely present in various older installations. Understanding copper and its repair techniques is therefore an important part of a complete education in this field. Copper pipe and fittings look similar to those made of PVC and come in similar sizes and fitting types.

Copper pipe is made in three thicknesses, designated by the letters K (thick wall), L (medium), and M (thin). I know, M should have been medium, but I didn't invent this system. Most pool plumbing is done with L (medium thickness) material.

Plumbing Methods

RATING: PRO

Be sure to read the general plumbing guidelines section and the one on PVC plumbing because some of the commonsense methods apply to copper work as well.

TOOLS OF THE TRADE: COPPER PLUMBING

The supplies and tools you need for copper plumbing are

- **Hacksaw with spare blades (fine: 32 teeth per inch or 2.5 centimeters)**
- **50/50 solid wire solder**
- **Flux with application brush**
- **Emery cloth or fine steel wool**
- **Butane torch (self-igniting, or lighter with propane torch)**

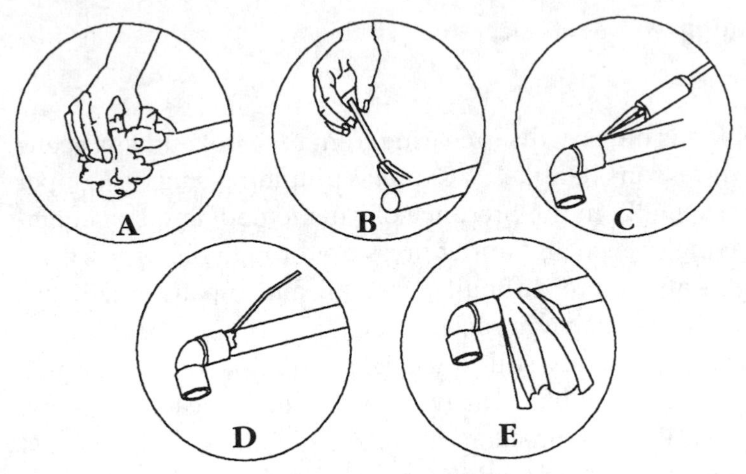

FIGURE 2-14 Step-by-step copper plumbing.

Following is the correct procedure for plumbing with copper (Fig. 2-14):

1. **Cut, Sand, and Flux** The secret to successful copper soldering (called *sweating*) is clean pipe and fittings. Start by cutting and dry fitting all intended connections, then clean the ends of the pipe (Fig. 2-14A) and inside the fittings with the emery cloth (or fine steel wool), sanding until the copper is shiny and bright.

   Apply a thin coating of flux to the pipe (Fig. 2-14B) and inside the fitting with a small, stiff brush, coating the entire area where you want solder to make a connection.

2. **Sweat** Fit the parts together and twist a half turn to evenly and thoroughly distribute the flux. Heat the joint with your torch (Fig. 2-14C), moving the flame back and forth and around the joint area to distribute the heat evenly until the flux starts to bubble. Be sure the entire joint is hot. A good way to know is to keep touching the solder to the joint until it melts. When it does, you have the right temperature.

   Here is where three plumbers will give you four opinions. Some say when that temperature is reached, remove the torch and solder the connection quickly. Others say move the heat source an inch (2.5 centimeters) behind the joint to keep some heat but do not overheat. The problem is that if the joint cools, the solder will not melt, but if the joint gets too hot, the flux will burn off (and the solder only goes where the flux is present). The correct way? Practice it yourself on scrap material and develop your own approach. Whatever you are comfortable with is the correct way.

3. **Solder** Touch the solder to the joint (Fig. 2-14D). Solder with lead melts and flows more easily. Recently many states have restricted the use of lead in solder, at least when soldering plumbing that is used for drinking water, so it might not apply to pool installations in your state. I find 50/50 (50 percent lead, 50 percent amalgams) works best, but if no-lead solder is used, even more heat must be applied and maintained during the soldering operation to ensure melting and even distribution.

   My method is to apply solder to the fitting while continuing to heat the joint, about an inch behind the joint. Solder is drawn toward the heat. The solder melts and enters the joint wherever there is flux, drawn in by capillary action, forming a seal. Work around the joint, making sure solder is drawn into the entire joint.

4. **Clean** Clean excess solder away with a damp cloth (Fig. 2-14E) before it cools; but be careful, the heat turns the moisture on the rag to steam which burns you worse than grabbing the hot pipe directly. Allow the work to cool.

Two last items relating to copper plumbing. First, threaded fittings are handled like PVC threaded fittings (see previous section) using pipe dope or Teflon tape to ensure leak-free connections.

Second, another type of copper connection fitting is the compression fitting (Fig. 2-15). This fitting is used in small diameter

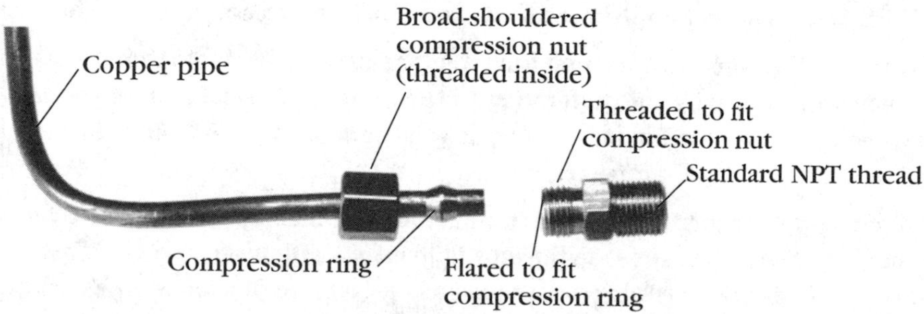

FIGURE 2-15 Copper compression fittings.

TRICKS OF THE TRADE: COPPER PLUMBING

1. Copper pipe is cut with a hacksaw like PVC. I prefer a fine blade for copper [at least 24 teeth per inch (2.5 centimeters) or preferably 32] as it cuts faster with each stroke.
2. When making several connections with various fittings, I like to sweat as many joints as possible while they are not connected to the equipment (just like the free joints I mentioned in the PVC section). This way, I can test each joint as it is completed. Hopefully only one or two joints will then have to be soldered in-place and trusted to the luck and skill of the work.

   To check if the joint is leak free, hold the work with one hand over one end of the pipe or fitting and blow hard through the other open end. No air should escape through the joint—if air doesn't leak, neither will water. When sweating a joint attached to equipment or other in-place plumbing, the only leak testing will be when the job is complete and the system is started up with water.
3. Solder will not seal if there is any moisture. If you must solder where some water is weeping from a pipe connected to equipment, stuff the line with bread to absorb the water. When the system is turned on again, the force of the water will break down the bread and allow it to be filtered or removed.
4. Most leaks in sweating are caused by moisture, overheating the joint, and most of all, unclean pipes or fittings.
5. Unlike PVC, sweating can be undone if you need to repair a bad job or take old work apart to perform new installations or add-ons. Apply heat to the joint, just as in sweating, until the solder melts and pull the joint apart. If you intend to reuse the fittings or that part of the pipe, carefully clean and sand the copper to a good shine before reuse.
6. Practice makes perfect. Although copper is expensive, it is more costly to make errors in the field. Take some pipe and a fitting and try this process several times at your shop. After sweating and testing your work for leaks, heat the joint until the solder melts and take it apart, clean it, and do it again. Because of tight quarters and odd angles, it only gets tougher in the field, so if you can't do it in the shop, you won't be successful on the job.
7. Read the directions on the torch, the can of flux, and the spool of solder. Sometimes reading another person's directions for performing the same task will make more sense than mine. Don't worry, I won't be offended, as long as your work doesn't leak. Also, labels can provide helpful hints that make for better, quicker jobs.
8. When supporting copper pipe to joists for long horizontal runs, use plumber's tape every 6 to 8 feet (2 to 2.5 meters) as with PVC, but wrap the pipe in that spot with insulating tape first. If you fail to do this, the different metals (copper pipe and galvanized plumber's tape) will cause electrolysis, corrosion, and leaks.

($\frac{1}{4}$- to $\frac{1}{2}$-inch or 6- to 13-millimeter) pipe, often inside the heater (see the pressure switch section in the heater chapter). The compression nut is placed on the pipe followed by the compression ring. The pipe is placed inside the opposing fitting and when the nut is screwed onto that fitting, the compression ring tightens around the pipe and seals it.

Miscellaneous Plumbing

Sometimes in tight quarters or for temporary connections you can use rubber connection fittings called mission clamps, balloon fittings, or no-hub connectors (Fig. 2-16).

These fittings are handy for connecting pipes of different sizes or types, clamping directly onto the pipe or fitting without gluing, threading, or sweating. The hazard is that these can leak, wear out, or fail under extreme pressure as when there is a restriction in the system from debris or a dirty filter.

I don't recommend these fittings, but I carry several different sizes anyway in case I need to make a quick repair (usually at 4:30 p.m. on a Friday afternoon) that will be improved later.

FIGURE 2-16 Mission clamps, balloon fittings, and no-hub connectors.

Sizing of Plumbing

Most building codes restrict the speed of flow through pipes to prevent stripping, breakdown, and erosion of the pipe material. Typically water may not move faster than 8 feet (2.5 meters) per second through copper pipe, and 10 feet (3 meters) per second through PVC. Suction pipes of any type are typically restricted to 8 feet (2.5 meters) per second.

Obviously the larger the pipe, the better. There is less restriction and therefore less strain on all equipment and plumbing. Use the largest diameter pipe and fittings you can for the job.

Typically pool and spa equipment is already built and plumbed for 1½- or 2-inch (40- or 50-millimeter) plumbing, and while you can adapt 2-inch (50-millimeter) pipe to a pump that is designed for 1½-inch (40-millimeter) fittings, you don't want to use the reverse. In that case, the pump will be trying to push the proverbial 10 pounds of potatoes into a 5-pound bag.

I discuss sizing in more detail in the chapter on pumps, which largely dictates the plumbing sizing; however, the considerations in all cases are

- Desired flow rate of water (measured in gallons per minute),
- Length of plumbing runs,
- Number and angles of connection fittings,
- Pump efficiency and capacity, and
- Equipment and restrictions after the pump.

FAQs: BASIC POOL PLUMBING

Can I Attach PVC Plumbing to Copper or Pipes of Other Metals?

- Generally speaking, yes, you can transition from one pipe material to another with threaded fittings. If the plumbing will carry heated water, it is not recommended to use different plumbing materials, because the heating and cooling will cause them to expand and contract differently, resulting in loose fittings and leaks.

Are Plumbing Materials, Like PVC Glue and Copper Solder, Toxic or Harmful in Any Way?

- There are fumes when you work with glues or solder that should be avoided. Always work in well-ventilated areas or use a small fan to circulate the air. You should avoid prolonged inhalation or skin contact with these materials and clean up promptly after each plumbing job.

Can I Get Everything I Need to Repair My Pool Plumbing at Any Hardware Store?

- Most hardware stores that carry PVC pipe will also carry the fittings, materials, and tools described in this chapter. The most important tool is a hacksaw—available at any hardware store—so be sure to stock up on fresh blades!

CHAPTER

3

Advanced Plumbing Systems

Perhaps the fastest growing segment of the pool and spa industry is in advanced plumbing systems, including automated valves, reverse-flow heating, solar heating, and the use of space-age materials. In this chapter, I discuss some of the more common applications of advanced plumbing and the maintenance of these systems.

Manual Three-Port Valves

The design of three-port valves takes water flow from one direction and divides it into a choice of two other directions. Picture a Y, for example, with the water coming up the stem, then a diverter allows a choice between one of two directions (or a combination thereof). Conversely, the water flow might be coming from the top of the Y, from two different sources, and the diverter decides which source will continue down the stem or mixes some from each together.

Figure 3-1 shows a typical Y or three-port valve, which is very common and easy to use and maintain. Several manufacturers make similar units based on the same concept.

Operation

Whether the three-port valve is Noryl plastic (a type of PVC) or brass, the concept is the same with them all. A housing (Fig. 3-1, item 1), built

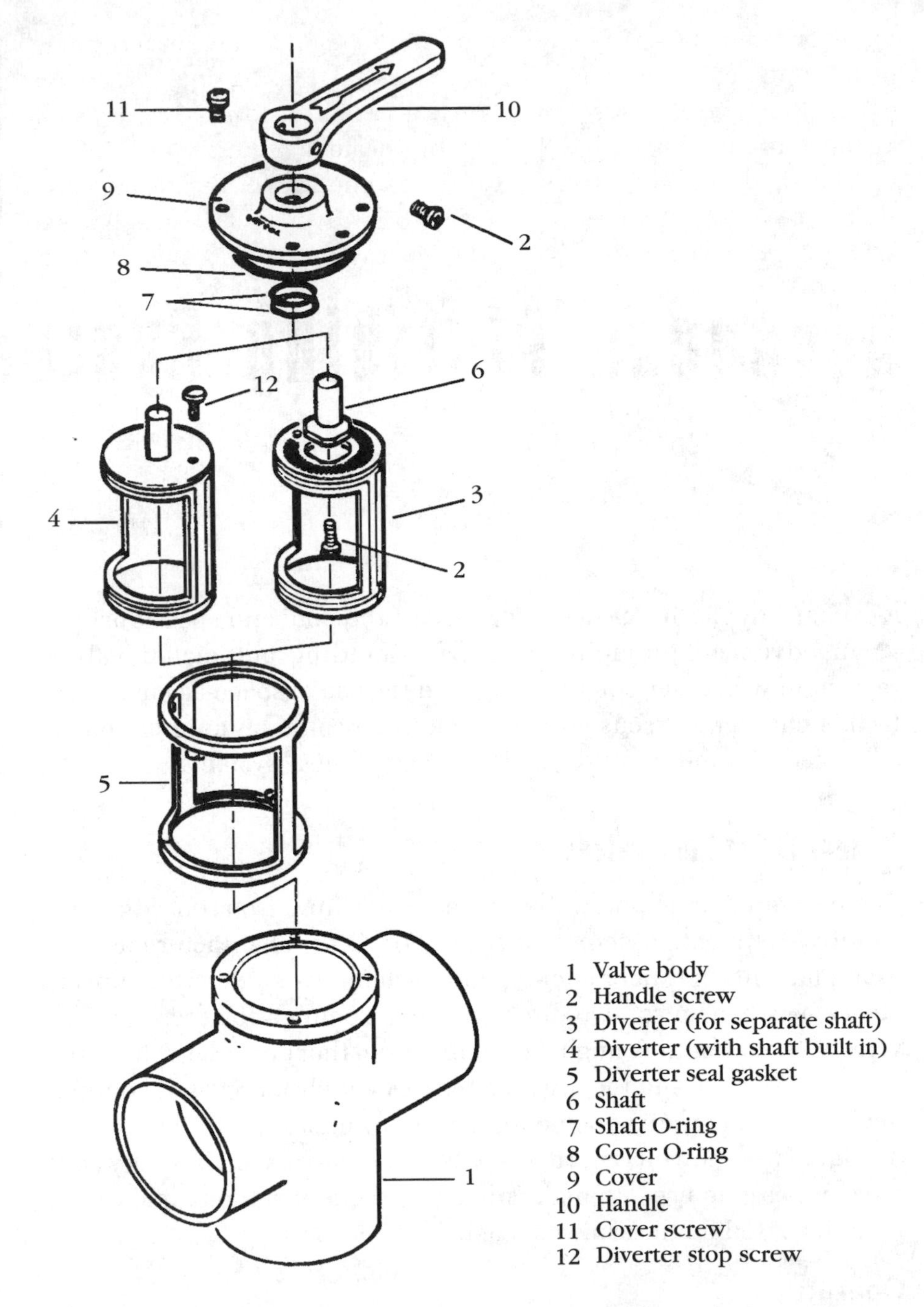

FIGURE 3-1 **Construction of a typical three-port valve.** *Pentar Pool Products, Inc.*

to 1½- or 2-inch (40- or 50-millimeter) plumbing size, houses a diverter (Fig. 3-1, item 3 or 4) that is moved by a handle (Fig. 3-1, item 10) on top of the unit. Typically these valves are used when a pool and spa are both operated from the same pump, filter, and heater equipment.

The suction line from the pool enters one arm of the valve body; suction from the spa enters the other. The diverter between the two arms determines which line is connected with the stem, from which the water continues to the pump.

Conversely, when the water leaves the equipment, it passes through another three-port valve. The water this time passes up the stem and the diverter determines if the water flows to the arm plumbed into the pool return or the one plumbed into the spa return. By setting the diverter equally between the two, water from each side is mixed. Sometimes this creates a pool or spa draining problem that must be corrected. These repairs are dealt with in the following sections.

Construction

The diverter is surrounded by a custom-made gasket (Fig. 3-1, item 5) so that no water can bypass the intended direction. A valve with this type of diverter and gasket is called a *positive seal* valve.

Some valves, for use where such water bypass is not considered a problem, have no such gaskets and, in fact, the diverter is designed more like a shovel head than a barrel. These divert most of the water in the desired direction with a lesser amount going in the other direction. These are called *nonpositive* valves.

The diverter is held in the valve housing by a cover (Fig. 3-1, item 9) that attaches to the housing with sheet metal screws (Fig. 3-1, item 11), and is sealed with an O-ring (Fig. 3-1, item 8) to make it watertight.

Notice that besides the four screw holes in the cover and housing, there is a fifth hole in the cover that corresponds to a post on the housing. This is to ensure that the cover lines up properly with the housing, because on the underside of the cover are specially molded stops. A small screw (Fig. 3-1, item 12) on top of the diverter hits these stops molded into the underside of the cover. This allows the diverter to be turned only 180 degrees (one-half turn), in either direction, ensuring that the diverter stops turning when facing precisely one side or the other.

This screw is removed when the valve is motorized because the motors only rotate in one direction and are already precise in stopping every half-turn. Small machine screws (Fig. 3-1, item 2) hold the handle on the shaft. A hole in the center of the cover allows a shaft from the diverter to attach to the handle for manual operation of the valve. To make the shaft hole in the cover watertight, two small O-rings (Fig. 3-1, item 7) slide on the shaft in a groove under the cover.

Maintenance and Repair

RATING: EASY

As the simple parts suggest, there's not too much that can go wrong with manual three-port valves.

INSTALLATION

Noryl three-port valves are glued directly to PVC pipe using regular PVC cement like any other PVC fitting. Care should be exercised not to use too much glue, as excess glue can spill onto the diverter and cement it to the housing. Excess glue can also dry hard and sharp, cutting into the gasket each time the diverter is turned, creating leaks from one side to the other.

Brass valves are sweated onto copper pipe like any other fitting. Be sure to remove the diverter when sweating so the heat doesn't melt the gasket.

LUBRICATION

For smooth operation, the gasket must be lubricated with pure silicone lubricant. Vaseline-like in consistency, this lubricant cannot be substituted. Most other lubricants are petroleum-based which will dissolve the gasket material and cause leaks. Lubrication should be done every six months or when operation feels stiff. This preventive maintenance is particularly important with motorized valves because the motor will continue to fight against the sticky valve until either the diverter and shaft break apart or, more often, an expensive valve motor burns out.

Lubrication is the most important maintenance item with any three-port valve. When the valve becomes stiff to turn it places stress on the shaft. On older models, the shaft is a separate piece that bolts onto the diverter (Fig 3-1, items 2, 3, and 6). Particularly on these models, but actually on any model, the stress of forcing a sticky valve will separate the stem from the diverter. If this happens, turning the handle and stem does not affect the diverter. To repair this, remove the handle and cover, pull out the diverter, and replace it with a one-piece unit (Fig. 3-1, item 4). If the gasket looks worn, replace it and lube it generously before reassembly.

REPAIRS

Few things go wrong with these valves, but the breakdowns that do occur are annoying and recurrent. Before disassembling any valve, check its location in relation to the pool or spa water level. If it is below the water level, opening the valve will flood the area. You must first shut off both the suction and return lines. When installations are made below water level, they are usually equipped with shutoff valves to isolate the equipment for just such repair or maintenance work. If the valves are above the water level, you will need to reprime the system after making repairs (see the section on priming).

Leaks are the most common failure in these valves. The valve will sometimes leak from under the cover. Either the cover gasket is too compressed and needs replacement or the cover is loose. The cover is attached to the Noryl valve housing with sheet metal screws. If tightened too much, the screw strips out the hole and you will be unable to tighten it. The only remedy is to use a slightly larger or longer screw to get a new grip on the plastic material of the housing. Be sure to use stainless steel screws or the screw will rust and break down, causing a new leak. If new screws have already been used and there is not enough material left in the housing for the screw to grip, you must replace the housing. I have managed to fill the enlarged hole with super-type glue or PVC glue and, after it dries, replace the screw. These repairs usually leak and are only temporary measures. You can also fill the hole with fiberglass resin, which usually lasts longer.

Leaks also occur where the shaft comes through the cover. Remove the handle and cover and replace the two small O-rings. Apply some silicone lubricant to the shaft before reassembly. This lubricates the operation of the valve, decreasing friction that can wear out the O-rings. The lubricant also acts as a sealant.

Leaks can occur where the pipes join to the housing ports. In this case, the only solution is replacing the housing. I have tried to reglue the leaking area by removing the diverter and gluing from both inside and outside of the joint. This has never worked! Try if you will, but I think you'll be wasting your time.

Finally, leaks occur inside the valve with no visible external evidence. By this I mean that water is not completely diverted in the intended direction, but slips past the diverter seal to the closed side of the valve. The symptom will be a spa that drains or overflows for no apparent reason. The cause might be a diverter that is not aligned precisely toward the intended port. Remove the diverter and make sure the shaft has not separated or become loose from the diverter.

With a motorized unit, be sure the motor is clean, free of rust, and able to turn its precise one-half turn each time. The other and most usual cause, however, is that the diverter gasket has worn out or become too compressed to stop all water from getting past. You might visually inspect the gasket and find that it looks good. Replace it anyway. It takes very little deterioration or compression to cause these bypass-type leaks. Again, lubricate the gasket well before reassembly for smooth operation and because the lube acts as a sealant.

If a new gasket doesn't stop the water bypass, you might find the diverter itself has shrunk or warped slightly. This doesn't occur often, but particularly with hot spa water or if the system has been allowed to run dry and heat up, you might be looking at a diverter that is not large enough to contain the water flow to one side only. It is difficult to see because such shrinkage is minimal, but it only takes a little to allow the bypass problem. Replace the diverter and see if this solves the leakage problem.

In even rarer circumstances, I have seen valve housings that have expanded or warped from overheating, usually when the system has been allowed to run dry and extreme heat builds up in the plumbing. If the new diverter and gasket seem loose, this might be the problem.

Such overheating and warping is a more common problem in systems with solar heating. Water in solar panels often exceeds 200°F (93°C), so if the system backflows or any of this superheated water is allowed to sit in the valves, they will warp in a short time. Even a slight warp is enough to allow leaks to occur.

Motorized/Automated Three-Port Valve Systems

The three-port valves just described are manually operated. These same valves can have small motors mounted in place of the manual handle for automatic or remote operation (Fig. 3-2).

Operation

The value of motorization is that the pool or spa equipment is usually located away from the pool and spa, making manual operation inconvenient. Some builders place the manual valves near the spa rather than in the equipment area; however, a small switch that operates the valve motors is often preferred. A variation of that concept is to locate the motor switch with the equipment and operate it with a remote control unit. The remote might also operate switches for lights, spa booster motors and blowers, or other optional accessories.

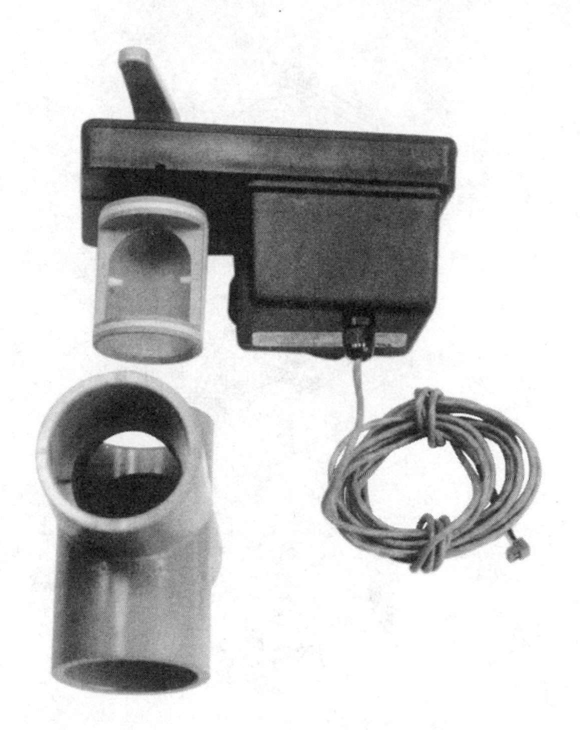

FIGURE 3-2A Typical motorized three-port valve.

Construction

To motorize a manual three-port valve, the cover and diverter are removed and replaced with a motorized unit (also called a *valve actuator*) that includes those parts (Fig. 3-2). Simple instructions provided with each make of valve motor show how to secure the unit to the valve body.

Wiring diagrams are provided with each type of valve motor and they are designed to operate on standard 110 volt, 220 volt, or from an automated system that has been transformed to 12 or 24 volts. Remote and automated systems are dealt with in more detail in a later chapter.

Maintenance and Repair

RATING: ADVANCED

Few problems occur with motorized valves (beyond those discussed in the section on manual valves). As mentioned, if the valves are not properly lubricated or become jammed with debris, the motor will continue to try to rotate the valve, finally burning itself out.

To determine if the motor has burned out, using your electrical multimeter, verify that current is getting to the motor. Obviously if there is no current, the problem is in the switch or power supply and

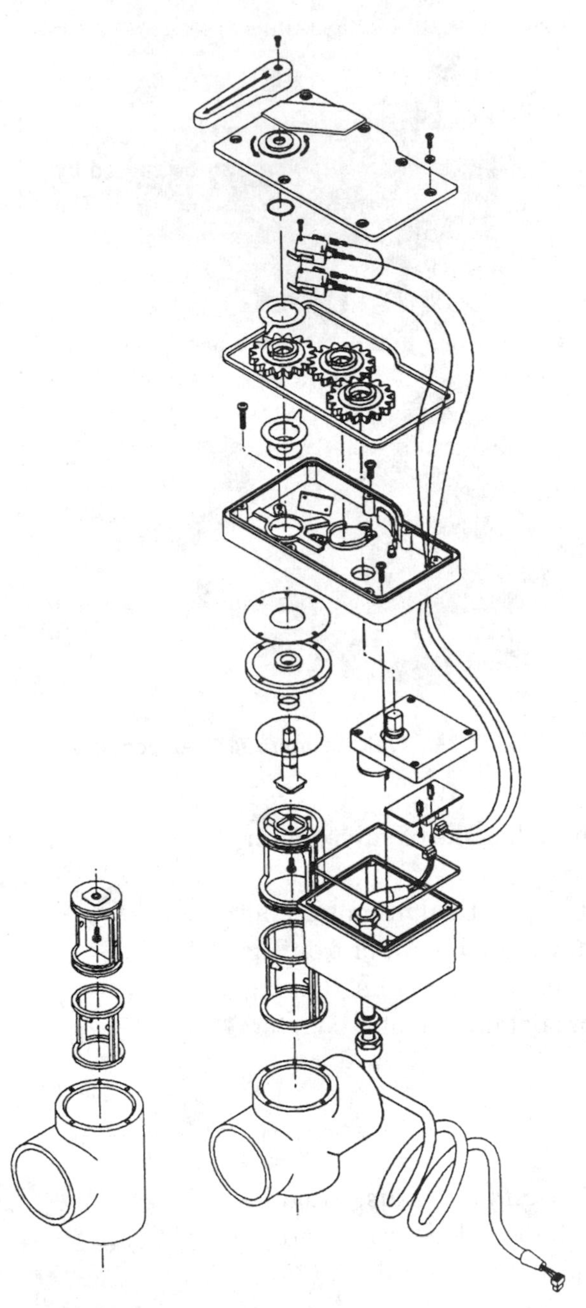

FIGURE 3-2B Exploded view of motorized three-port valve. *Pentar Pool Products, Inc.*

probably not the motor. If you are not familiar with basic electricity, call an electrician to help you or study the basic electricity chapter later in this book.

If current is present, remove the motor unit from the valve and try to operate the system. If the motor rotates normally, then the problem is a stuck valve and not a burnt motor. Tear down and repair the valve as described in the previous section.

If the motor is burnt out, it can easily be replaced without replacing the entire unit or valve. Although slightly different with each manufacturer, the process is usually no more than four screws and three wires and will be obvious when opening the motor housing.

Another problem that can occur with motorized valves is that if the mounting bracket or screws holding the unit together become loose, the unit will not align correctly with the valve. The motor will then rotate its 180 degrees, but it will not fully rotate the valve diverter to match. The solution is to tighten all hardware and replace any rusted screws.

I have also seen salt in the moist air near the ocean eat away the motor shaft. The only solution in this case is to replace the motorized unit and be sure all leaks are sealed. Near the ocean or in damp climates, enclosing motorized units is a good idea. If it is not practical to enclose the unit, cover it with plastic and sealing tape. Obviously don't tape anything to the shaft itself that will bind up the unit.

Reverse Flow and Heater Plumbing

Motorized valves are often installed in plumbing systems designed with reverse flow. During normal circulation of the pool, water is taken from the skimmer and main drain and returned to outlets located about 18 inches (46 centimeters) below the water surface. The concept of reverse flow is that when the heater is turned on to heat the water, motorized valves reverse the flow. Water is taken through these shallow outlets and returned through the main drain or specially installed return outlets in the floor of the pool (Fig. 3-3). The thought is that because hot water (like air) rises through cooler water, the heated water will rise through the pool and heat the pool more uniformly. When the warm water is returned in a normal system to the shallow outlets, only the top 2 or 3 feet (60 or 100 centimeters) of the pool water is heated.

I'm not a fan of these systems, because they require many more motorized valves and other moving parts than a traditional system, more initial expense, more maintenance expense, and more that can go wrong over time. The argument in favor is that the reverse flow system takes warmer water from the surface of the pool (already warmed by the sun) and returns it through the bottom of the pool where the colder water is displaced toward the surface. The claim is that this process uniformly warms the entire pool at a faster pace. Proponents further note that in traditional circulation systems, the coldest water is taken from the bottom of the pool to be warmed and returned to the surface of the pool. Since warm water rises through colder water, it will take much longer to uniformly warm the pool in this manner.

ELECTROLYSIS WHERE YOU LEAST EXPECT IT

A less frequent problem can be caused by electrolysis or simply a leaking valve. The motor shaft is usually made of galvanized metal or aluminum. If the valve is leaking or the motor housing is not watertight and a combination of moisture and electricity is present as a result, electrolysis will disintegrate the soft metal of the motor shaft. The motor might continue to operate, but as the shaft dissolves, it will not turn (or not completely turn) the valve diverter.

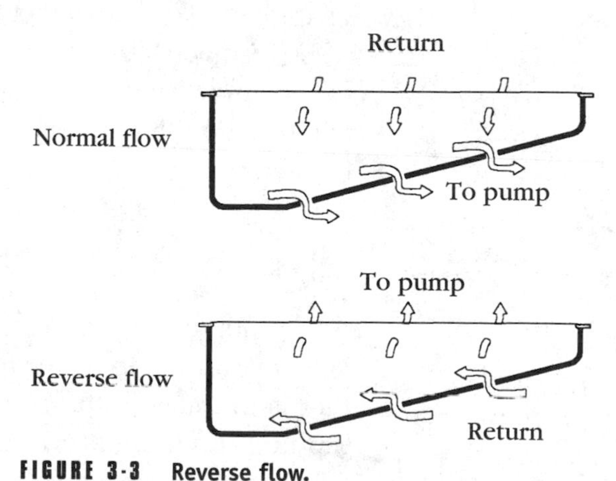

FIGURE 3-3 Reverse flow.

It is important here to understand how pool heaters work (a subject discussed in more detail in a later chapter). Water passes through the typical gas-fueled pool heater, increasing in temperature around 10°F (6°C) on its way back to the pool. So if you take 60°F (15°C) water from the bottom of the pool, it will be around 70°F (21°C) when it returns at the surface in a traditional system (for this illustration, we need not factor in heat loss as the water travels through the pipes underground). But if you take water from the surface of the pool that is 70°F (21°C), raise the temperature to 80°F (27°C), and then return it to the bottom of the pool, that warmer water will rise, displacing the cooler water above it toward the surface, thus actually cooling the water near the surface.

In short, either system requires many passes of the water through the heater, each time raising the temperature until the overall desired temperature is reached, before you will be swimming in significantly warmer water. I have seen no credible studies to show that one method is more energy- or time-efficient than the other. The one obvious difference, then, is the fact that reverse heating systems require expensive, complicated additions to the pool's plumbing system.

Moreover, even manufacturers of the reverse flow systems agree that for general cleaning and filtration, the traditional circulation pattern is best, thus the need for automated valves to reverse the system when the heater comes on as opposed to using the reverse system all of the time. Therefore, some pool owners may wish to disconnect the reverse flow system rather than pay to maintain it.

Don't try to leave the units intact and simply disconnect the wiring—you might also disconnect the heater on/off or pump switches that are associated with the remote control system (if part of the installation). It is easier (and you'll create spare parts) to manually remove the motors while leaving the valves in a normal circulation position, clip off and cap the wiring to those motors, and leave the rest alone.

Unions

An improvement for installing equipment is the plumbing union. When you need to repair or replace a pump, filter, or other equipment that is plumbed into the system, you must cut out the plumbing and replumb upon reinstallation. The concept of the union is that when you remove a particular piece of equipment, you need only unscrew the plumbing and reinstall it later the same, simple way.

Although unions add a few dollars to your initial installation, they allow you to easily remove and replace equipment without doing any new plumbing. Unions, like other plumbing, are made of plastic or metal in standard diameters and are adapted to plumbing like any other component (gluing, threading, or sweating).

Figure 3-4 shows a typical plumbing union. A nut is placed over the end of one pipe, then male and female fittings (called *shoulders*) are plumbed onto each end of the pipes to be joined. As can be seen, the joint is made by screwing the nut down on the male fitting. Teflon tape or other sealants are not needed as the design of the union prevents leaking (either by the lip design as shown or by use of an O-ring seated between the shoulders).

FIGURE 3-4 Plumbing union.

Unions are made for direct adaptation to pool and spa equipment, where the pipe with the nut and female shoulder is male threaded at its other end for direct attachment to the pump, filter, or any other female threaded equipment. Then, only the male shoulder need be added to the next pipe and the piece of equipment can be screwed into place.

Gate and Ball Valves

Gate and ball valves are designed to shut off the flow of water in a pipe and are used to isolate equipment or regulate water flow. You might see these in systems where the equipment is installed below the water level of the pool. Without them, when you open or remove a piece of equipment or plumbing you will flood out the neighborhood.

Some fountains where water pressure and flow must be precisely regulated use shutoff valves to adjust the water flow in the plumbing. Finally, on older plumbing systems, before the development of three-port valves, shutoff valves were used on each pipe to manually determine water flow from and to pools and spas.

There are basically two types of shutoff valves. Figure 3-5B shows the *gate valve*. As the name implies, it has a disc-shaped gate inside a housing that screws into place across the diameter of the pipe, shutting off water flow. A variation of this is the *slide valve* (Fig. 3-5A), where a simple guillotine-like plate is pushed into place across the diameter of the pipe.

The other design is the *ball valve* (Fig. 3-5A), where the valve housing contains a ball with a hole in it of similar diameter as the pipe. A handle on the valve turns the ball so the hole aligns with the pipe, allowing water flow, or aligns across the pipe, blocking flow. In each of these designs, flow can be controlled by degree as well as total on or total off.

The gate valve is operated by a handle that drives a worm screw-style shaft inside a threaded gate. If the gate sticks from obstruction or rust and too much force is applied to the handle, the screw threads will strip out, making the valve useless. The valve cap (also called the bonnet) can be removed (unscrewed) and the drive gear and gate can be removed and repaired or replaced without removing the entire valve housing. Most plumbing supply houses sell these replacement guts, but the parts from one manufacturer are not interchangeable with another.

Also notice the packing gland (Fig. 3-5C) that prevents leaks where the shaft enters the valve body. If leaks occur here, the packing material can be replaced by unscrewing the cap nut, removing the old twine (specially treated graphite-impregnated twine), and rewinding new twine. Sometimes just tightening the cap nut will stop the leak, but it also tightens the packing material on the shaft, making it more difficult to turn. Plastic gate and ball valves use O-rings to prevent leaks in this location.

Check Valves

The purpose of the check valve is to check the water flow—to allow it to go only in one direction. The uses are many and will be noted in each equipment chapter where they are employed; however, four common uses are

- In heater plumbing (to keep hot water from flowing back into the filter)
- In spa air blower plumbing (to make sure air is blown into the pipe but water cannot flow back up the pipe and into the blower machinery)

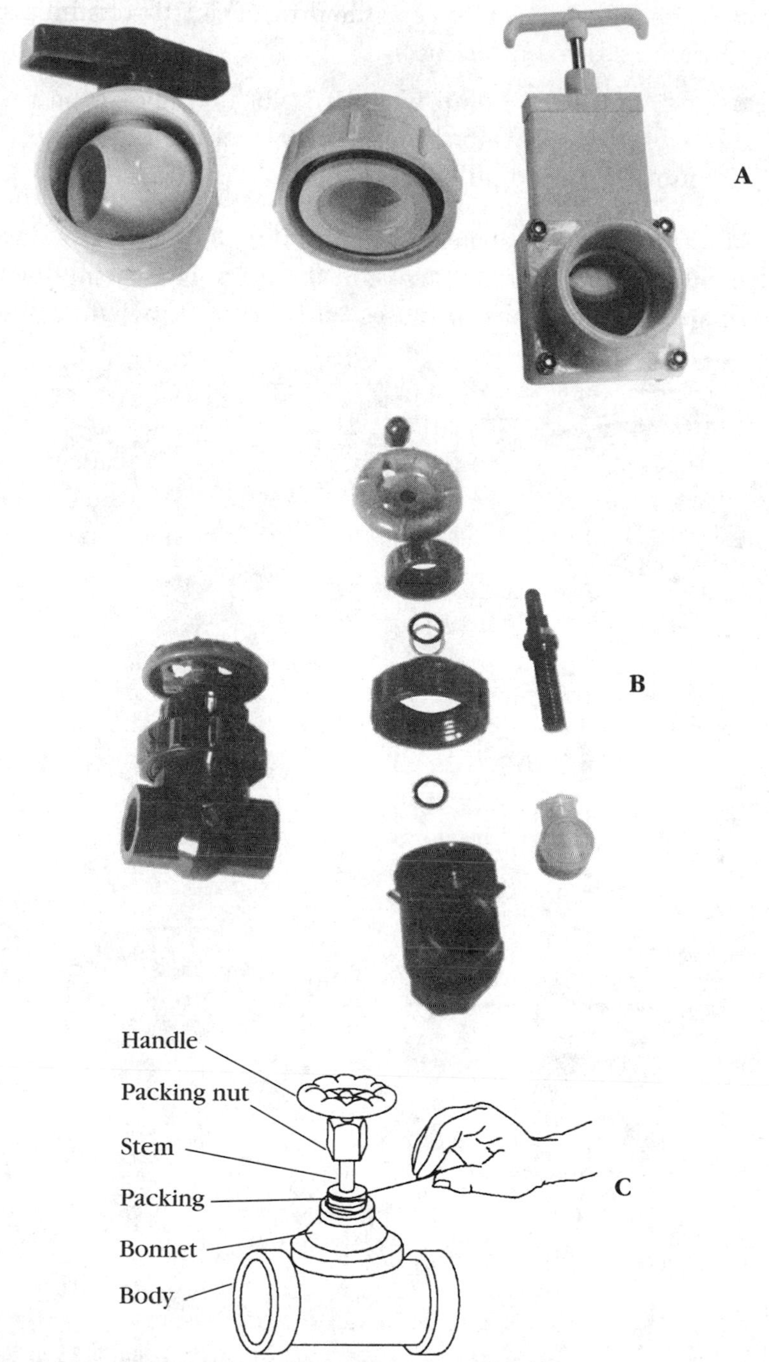

FIGURE 3-5 Ball (A left), slide (A right), and gate (B) valves and valve packing (C).

- With chlorinators (to keep the flow of caustic chemicals moving in the desired direction)
- In front of the pump when it is located above the pool water level (to keep water from flowing back into the pool when the pump is turned off)

There are two types of check valves (Fig. 3-6). One is a flapper gate (also called a *swing gate valve*) and the other is a spring-loaded gate. The flapper style opens or closes with water flow, while the spring-

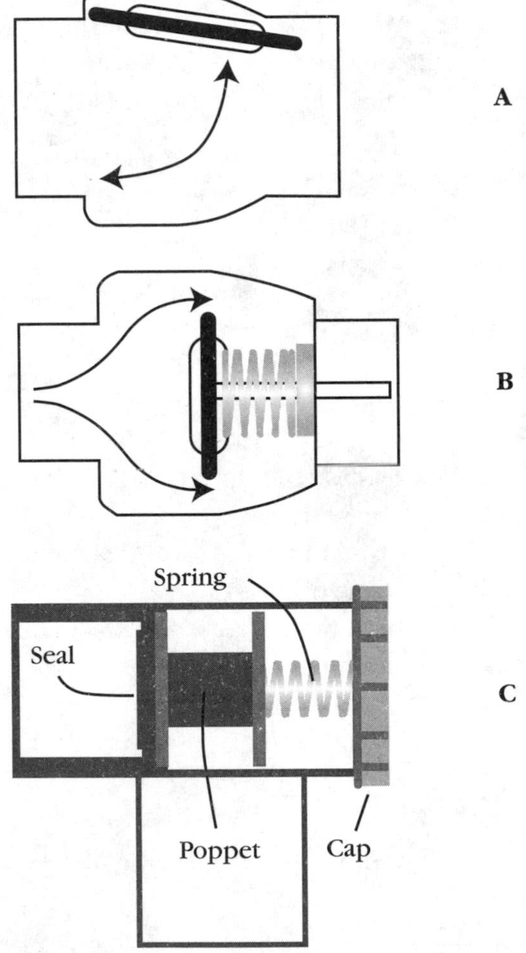

FIGURE 3-6 Inline swing gate (A), inline spring-loaded gate (B), and 90-degree spring-loaded (C) check valves.

loaded style can be designed to respond to certain water pressure. Depending on the strength of the spring, it might require one, two, or more pounds of pressure before the spring-loaded gate opens. As with other valves, check valves are made of plastic or metal in standard plumbing sizes and are plumbed in place with standard glue, thread, or sweat methods.

A problem I have encountered with swing gate–type check valves, especially metal ones, is that the gate comes off the hinge from rust or obstruction damage. In this case, the valve must be replaced. The spring-loaded valve rarely breaks.

NOISY VALVES?

The flapper-style valve is simple and little can go wrong with it. It must be installed with the hinge of the flapper on top. If it is installed on the bottom, gravity will pull the flapper open all the time. Sometimes with metal flapper check valves you will hear them chatter as the flapper opens and closes, particularly if there is air in the system. This is not a malfunction or a problem of the valve (see the section on priming).

The only real weakness of all check valves is that they clog easily with debris, remaining permanently open or permanently closed. Because of the extra parts inside a spring-loaded check valve, they are more prone to failure from any debris allowed in the line. If the valve is threaded or installed with unions, it is easy to remove it, clear the obstruction, and reinstall it.

Another solution is to use the 90-degree check valve. This valve allows you to unscrew the cap, remove the spring and gate, remove any obstruction, and reassemble. Be careful not to overtighten the cap—they crack easily on older models; newer models are made with beefier caps to prevent this problem. These units have an O ring in the cap to prevent leaks. It is wise to clean these out every few months (or more frequently, depending on how dirty the pool or spa normally gets) and lubricate the gate (using silicone lube only).

These valves are great because of their ease of cleaning and repair, but I don't recommend using them unless you need to make the 90-degree turn in your plumbing anyway. Otherwise you are adding more angles to your plumbing, which restricts water flow unnecessarily. Maybe someday they'll invent a nonangled unit that can be cleaned out as easily. Why don't you come up with such a design and retire on the profits?

Some check valves are made of clear PVC, which allows you to see if they are operating normally.

Solar Heating Systems

Solar heating systems are discussed here in the advanced plumbing chapter because it is the plumbing of these systems that most concerns the pool and spa technician.

Certainly an entire book could be written about solar heating and installations. This is particularly true in today's market where new technologies, alloys, and plastics are being used to manufacture solar panels. For example, modern solar panels with sensors track the sun and actually rotate around one or more axes to receive maximum sun exposure. Such systems are fairly costly and complex. Therefore, you might be advised to leave solar heating system installation and repair to the experts—call in a subcontractor and earn a referral fee.

Having said that, many repairs and some installations are simple and profitable, so if you need the work, don't be shy. As mentioned, the plumbing is the same as regular pool and spa plumbing, so leak repair is easy, although some of it might be on your customer's roof. In any case, here are some guidelines for approaching solar heating systems.

Types of Solar Heating Systems

The function of the solar panel is to absorb heat from the sun which is transferred to the liquid as it passes through. Designs and materials are hotly debated (pun intended) among various manufacturers, but efficiency of a solar heating system is less a factor of panel design than a factor of system setup. Exposure to direct sunlight, hours of sunlight, and amount of wind, clouds, or fog are all important factors that will impact on efficiency when setting up a system.

Solar panels are made from plastic or metal and are then glazed (covered in glass) or left unglazed. Obviously the glazed panels are heavier and more expensive; however, they absorb and retain more solar heat and are therefore more efficient (fewer panels are required to accomplish the same amount of solar heating).

OPEN LOOP SYSTEMS

The first system, shown in Fig. 3-7A, is called an *open loop* system, meaning it is open to the pool water. This is the most common type you will encounter. A variation on this is a system that is not connected to the pool or spa equipment, but rather which has its own plumbing from the pool and back, along with its own circulating pump.

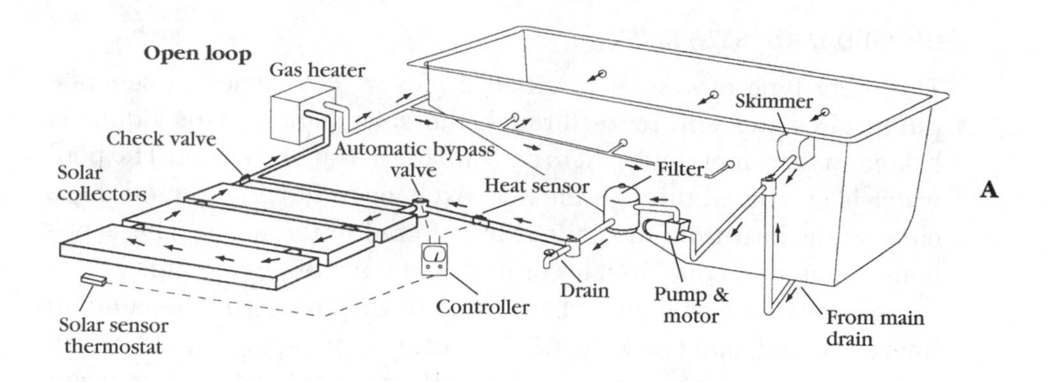

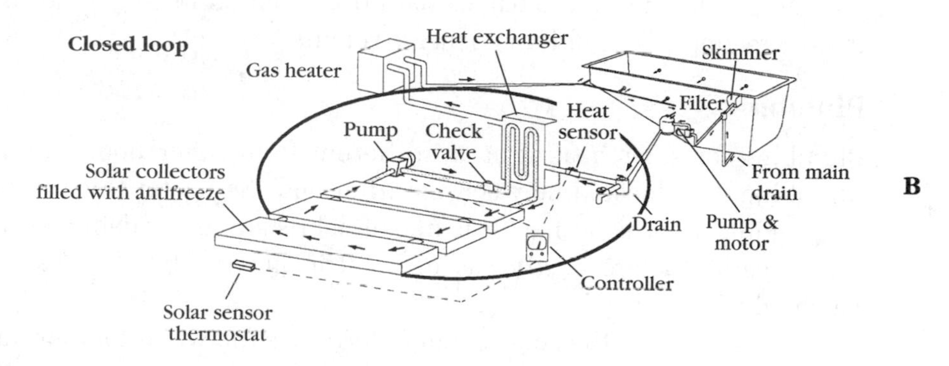

FIGURE 3-7A and B Typical solar plumbing and heating systems.

FIGURE 3-7C Solar heating on rooftop.

CLOSED LOOP SYSTEMS

The other type of system is called a *closed loop*, where a separate pump circulates antifreeze through the solar panels. This liquid is heated in the panels and sent to coils inside a heat exchanger. The pool water is circulated through the heat exchanger, flowing around these coils so the heat from the coils is transferred to the water. These systems are used in cold climates or in desert climates, where it is hot by day but very cold by night, where water in an open loop system might freeze, expand, and crack the panels. Another advantage of the closed loop is that harsh chemicals in the pool water or hard, scaling water are not circulated into the panels with the potential to clog or corrode them, creating the need for expensive repairs.

Plumbing

Plumbing for solar heating is no different from other pool and spa plumbing. It is located between the filter and the heater (Fig. 3-7) so water going to the solar panels is free of debris and is available for free solar heating before costly gas or electric heating by the system's mechanical heater.

A thermostat on the solar panel determines the water temperature and if it is warmer than the water coming out of the filter, a three-port motorized valve (called an *automatic bypass valve* in solar installations—it's the same motorized valve described earlier in this chapter) sends the water to the solar panels for heating and returns it to the plumbing that enters the heater. The heater thermostat senses the temperature of this solar heated water and if it is still not as hot as desired, the heater will come on to heat it further before returning it to the pool.

Therefore a main component of solar heat plumbing is the three-port valve that either sends the water from the filter directly to the heater or sends it first to the solar panels and then the heater. A check valve is installed on the pipe that returns water from the solar panel to the heater to prevent water from entering this return line when the solar panels are not in use. This might instead be another three-port valve that performs the same function as the check valve but also ensures that when not in use, the solar panels will not drain out. This might not be important where solar panels are installed at or below the water level of the equipment and pool, but most installations of panels are on rooftops, high above the water level. Ball and check valves should be used so that the solar heating system can be completely and easily isolated from the circulation system (Fig. 3-8), allowing normal pool operation when repairing the solar panels.

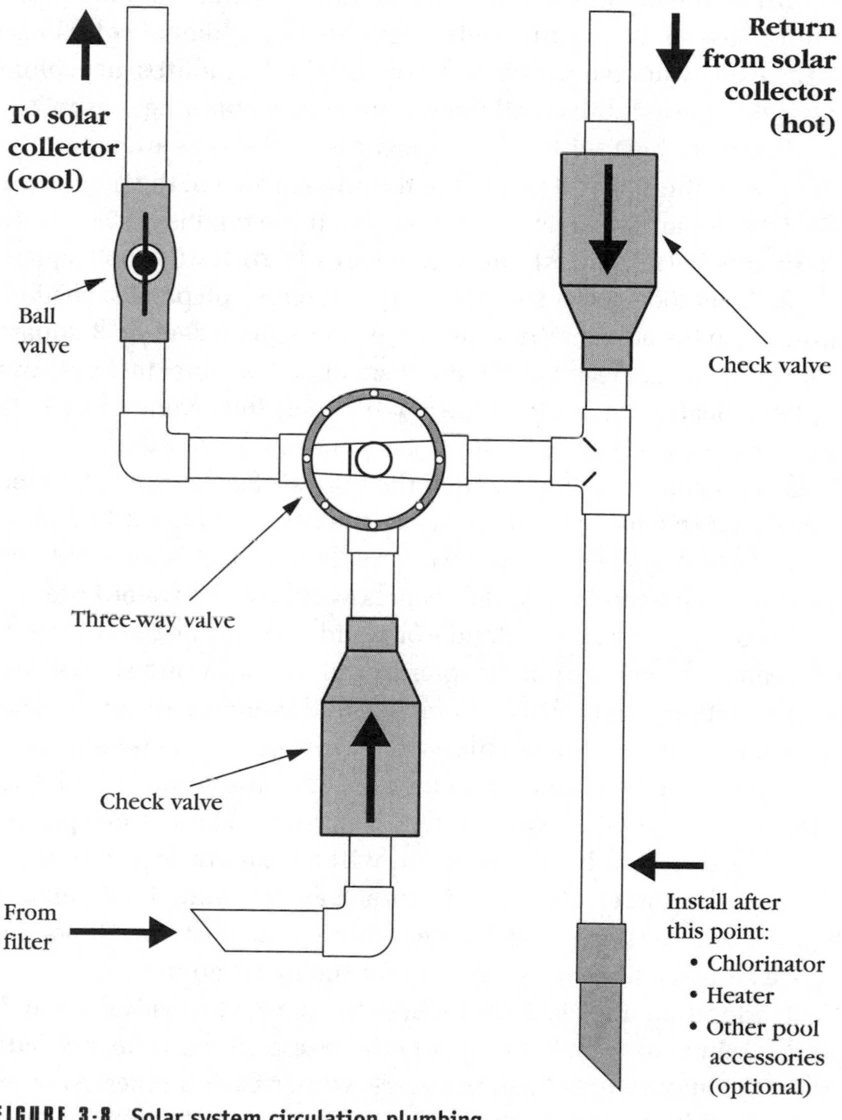

FIGURE 3-8 Solar system circulation plumbing.

To Solar or Not to Solar?

The decision to invest in a solar heating system will be based on the desired length of the "swimming season" and the desired swimming temperature. If nothing is done to prevent heat loss (or to add heat) to the pool, the water temperature will closely resemble the average air temperature. Therefore, if you want a swimming temperature above 70°F (21°C) and the average air temperature in your area meets that criterion only in the months of June through September, then that is your swimming season. It may take only a few hours per day of solar heating to raise the water temperature above 70°F in the months just before and after that period, thus easily doubling your swimming season.

To effectively heat with solar, regardless of the type of panel, the general rule of thumb is 75 percent of the surface area of the pool is the surface area of panels needed. For example, if the pool is 20 feet by 40 feet (6 meters by 12 meters), the surface area is 20 × 40 = 800 square feet × 0.75 = 600 square feet (56 square meters) of panels needed. Because panels are generally 4 by 8 feet (32 square feet or 3 square meters) or 4 by 10 feet (40 square feet or 3.7 square meters), our example pool would need 19 of the small panels (600 square feet ÷ 32 square feet per panel) or 15 of the larger panels.

Some say as little as 60 percent of the pool surface area can be used for these calculations, but I have found it is better to have a few more square feet of panels because you can't really have too many—but you can certainly have too few. Added panels will heat the water faster or, at least, more effectively on cloudy or windy days. For the few extra dollars, the customer will be happier in the long run. Figure 3-9 will assist with estimating probable efficiency and therefore overall sizing. Orientation due west, for example, will require solar panels equal to at least 85 percent of the pool's surface area. But the same installation oriented due south will require only 70 percent. These concepts are based on the northern hemisphere and will be exactly opposite in the southern hemisphere. The manufacturer can tell you the weight of each panel with water, so you can determine if the customer's roof has the space and weight-bearing capacity for the installation.

Next, a location must be found where the panels can face the sun. It might differ in your area, but generally the best position is facing south to obtain the most hours of sun per year—winter and summer. Another factor is prevailing wind. High winds can tear panels from the roof or

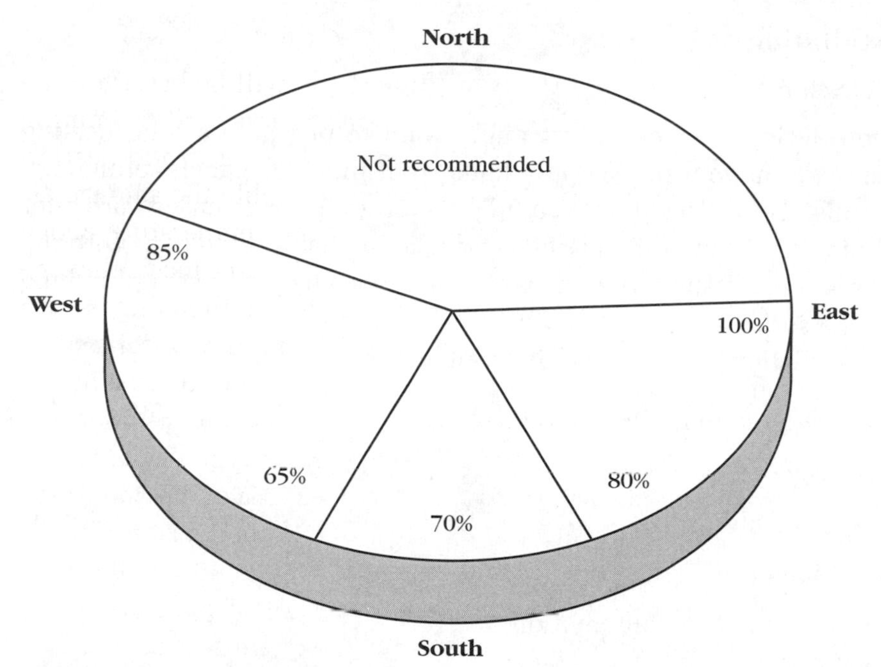

FIGURE 3-9 **Solar panel sizing guide. (1) Determine the area of the water surface of your pool. (2) Locate on the chart the direction your solar panels will face. (3) Multiply the percentage taken from the chart by the surface area of the pool to determine the total area of panels needed to effectively heat the pool. Example: The pool's water surface is 500 square feet (46 square meters) and the panels will face due west. 85 percent of 500 = 425 square feet (40 square meters) of panel surface required.**

create so much cooling that the panels will not be effective. Such concerns will also dictate how many panels you need to install.

Of course, a solar heating system designed to heat a pool will very quickly raise the temperature of a spa. If the pool also has a spa, this may be another good reason to invest in solar. Finally, to estimate the cost of the initial investment, as a general rule of thumb it will cost about $8 to $12 per square foot or 930 square centimeters (installed) for a solar heating system.

Knowing only these facts, you can help your customer determine if solar is likely to be practical. The cost of the system will be paid off with energy savings and tax benefits depending on how much gas or electricity would otherwise be used, but your customer might be more influenced by ecological concerns, added value to the home, or other personal concerns.

Installation

RATING: PRO

If you decide to proceed, you might want to purchase a solar heating package from your pool supply house that includes panels, plumbing, controls, and instructions. You might want to hire a licensed carpenter (let him get the building permits and take the liability) to help with the installation and support of the panels while you complete the plumbing into the system.

Installation of a solar heating system will need to consider:

- Orientation, pitch, and location
- Size
- Hydraulics
- Mounting
- Controls and automation
- Monitoring and isolation

We have already reviewed orientation. The angle of the panels ("pitch") as they sit on the roof or ground is also important, because the more the sun's rays strike the panels at a 90-degree angle, the more heat will be absorbed into the water. Therefore, a 20- to 30-degree pitch helps the efficiency of the system in winter when the sun tends to be on the horizon rather than directly overhead.

The existing pump will probably be adequate for circulation when adding solar equipment, because the gravity and siphon effect balances the additional pressure the pump experiences trying to push the water up to the panels. You do, however, have to calculate the effect of the length of pipe and fittings as with any plumbing installation (see the section on hydraulics).

Remember too that in order to achieve the balance of pressure and siphon, the pump must be able to get the water up to and through the panels when the circulation system is first turned on each day, so make careful calculations before determining that the existing pump is adequate. Generally, for every 10 square feet (0.9 square meter) of solar panel, the system will require 1 gallon (3.8 liters) per minute of flow.

Plumbing is always arranged so that water flows from the bottom of the solar panels toward the top and no more than 400 square feet

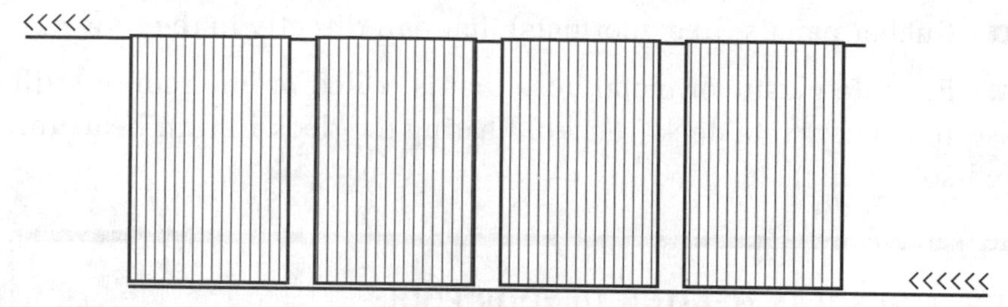

FIGURE 3-10A Water flows through solar panels from bottom toward top.

FIGURE 3-10B Solar panel on above-ground pool. *SmartPool, Inc.*

(37 square meters) in any one array. Both of these measures assure even flow of water through the panels. If more than 400 square feet of panels are needed, they can be plumbed as shown in Fig. 3-10.

As mentioned, there are numerous manufacturers and styles of panels and controls, far more than can be outlined here. Do some homework on what is available in your area. Just so you know what to look for, here are a few types:

- Plastic panels (glazed or unglazed)
- Metal panels (glazed or unglazed)
- Thin, lightweight aluminum panels

- Rubber panels (like doormats) that nail directly to the roof
- Flexible plastic or metal hose that is coiled on the roof or built into a concrete deck (the sun heating the deck in turn heats the solar coils)

QUICK START GUIDE: ADD SOLAR HEATING TO YOUR POOL

Rating: Pro

1. PREP

- **Unpack solar heating kit: panels, plumbing, connectors. Read owner's manual.**
- **Shut off pump and tape over switch or breaker, so no one can turn it on before you finish.**
- **Isolate equipment plumbing by closing valves at suction line (at skimmer and/or main drain connection before pump) and return line (at pool discharge outlet).**

2. SETUP

- **Mount panels per owner's manual instructions on ground, deck, prefabricated rack, or roof.**

3. PLUMB

- **Cut pool equipment plumbing AFTER the filter, but BEFORE the heater (if your system has one).**
- **Plumb solar panel "Intake" line to discharge pipe from filter; plumb solar panel "Outlet" line to pool return line (or "Intake" line of heater if you have one). Use shutoff valves at each location.**

4. STARTUP

- **Reopen pool plumbing valves (and valves on solar panel plumbing).**
- **Start pump, purge air, and check for leaks. The solar panels are now in operation.**

Whatever the style, remember when planning an installation that the pipes to and from the panels should be insulated so heat is not lost along the way. A good idea is to attend the next pool and spa industry convention in your area and check out the wide variety of makes and models. You will receive literature and even small panel samples that can help you and your customers make decisions. Much like portable spa manufacturers, however, solar panel makers have swiftly come and gone out of business.

Although the panels themselves are fairly breakproof, choose a simple style that doesn't require replacement parts from manufacturers who might not be in business next year when you need to make repairs. Better yet, choose a manufacturer who has been around awhile.

Maintenance and Repair

RATING: EASY

Once installed, most homeowners and pool technicians tend to forget about the solar heating system. Inspection every two or three months should be made to check for leaks. Leaks can easily occur because of the extremes of hot and cold temperatures that cause the panel materials to expand and contract. Leak repair depends on the type of material in the panel or plumbing, and each manufacturer makes leak repair kits with instructions. The plumbing to and from the panels can be repaired as needed using the techniques outlined in the chapter on basic plumbing.

The second common problem is dirty panels. Dirt prevents the panels from absorbing heat and can cut efficiency by as much as 50 percent. A pool technician can make good profits by charging customers for regular solar panel cleaning, requiring no more than soap and water.

Finally, panels can become clogged with scale from hard pool water and chemicals. Poor circulation is the tip-off, and the solution is to disassemble the panels from each end, exposing the pipes of the panel that actually carry the water, and reaming these out with special brushes attached to your power drill. Again, how you make this repair depends on the maker of the panel and its style. The maker should provide instructions and special tools for this procedure. As with leak repair or cleaning of solar panels, reaming is simple to perform using techniques and skills learned elsewhere in this book.

Water Level Controls

The most failureproof (read idiotproof) method of replacing evaporated water in the pool or spa is to turn on the hose. Unfortunately the pool technician doesn't have the hour or two to stand around waiting for the level to come back up, and your customer will most likely forget to turn the hose off even though you tell them to do it in an hour or two.

Alternatively there are two kinds of automated water fill systems and variations on those themes. If the pool or spa was built with a water fill line plumbed in place, the on/off antisiphon valve

TRICKS OF THE TRADE: SOLAR HEATING TROUBLESHOOTING

Rating: Easy

Solar Heating Problem:	Check and Correct:
Pool/spa not as warm as it should be	• Panels too small or incorrectly oriented • Circulation through panels not long enough each day • Circulation at wrong time of day (if water goes through the panels at night, the water may be cooling instead of heating) • Auto controls not working properly • Water flowing through panels too fast • Panels dirty
Air bubbles at pool return lines only when solar is operating	• Check for clean filter • Vacuum relief valve not operating properly or clogged
Some panels warm to the touch, others cool	• Check for circulation problems • Check that each array is no more than 400 square feet • Check flow rate • Check for any valves between panels
Leaks in panels or plumbing	• Check water chemistry • Check that panels and plumbing are secure

(Fig. 3-11C) can be replaced with a mechanically timed valve. In this way, you can set the water to run 1 to 60 minutes.

Instructions for removing the manual valve and installing this timer valve are in the package with the valve and are simple to follow. It does mean you must turn off the household main water supply, usually at the meter at the street, to make the swap. This is probably the only pool or spa repair requiring shutoff of the household water supply.

To perform this task, locate the supply meter, usually in front of or alongside the house, in the ground, mounted in a concrete box. Inside the box you will see the meter with a gate valve (Fig. 3-12) on the outflow side of the meter. Turn this off. If it is stuck or rusted, there is a shutoff valve on the inflow side of the meter, but not with a standard handle. This valve can be turned off with a channel lock pliers or a

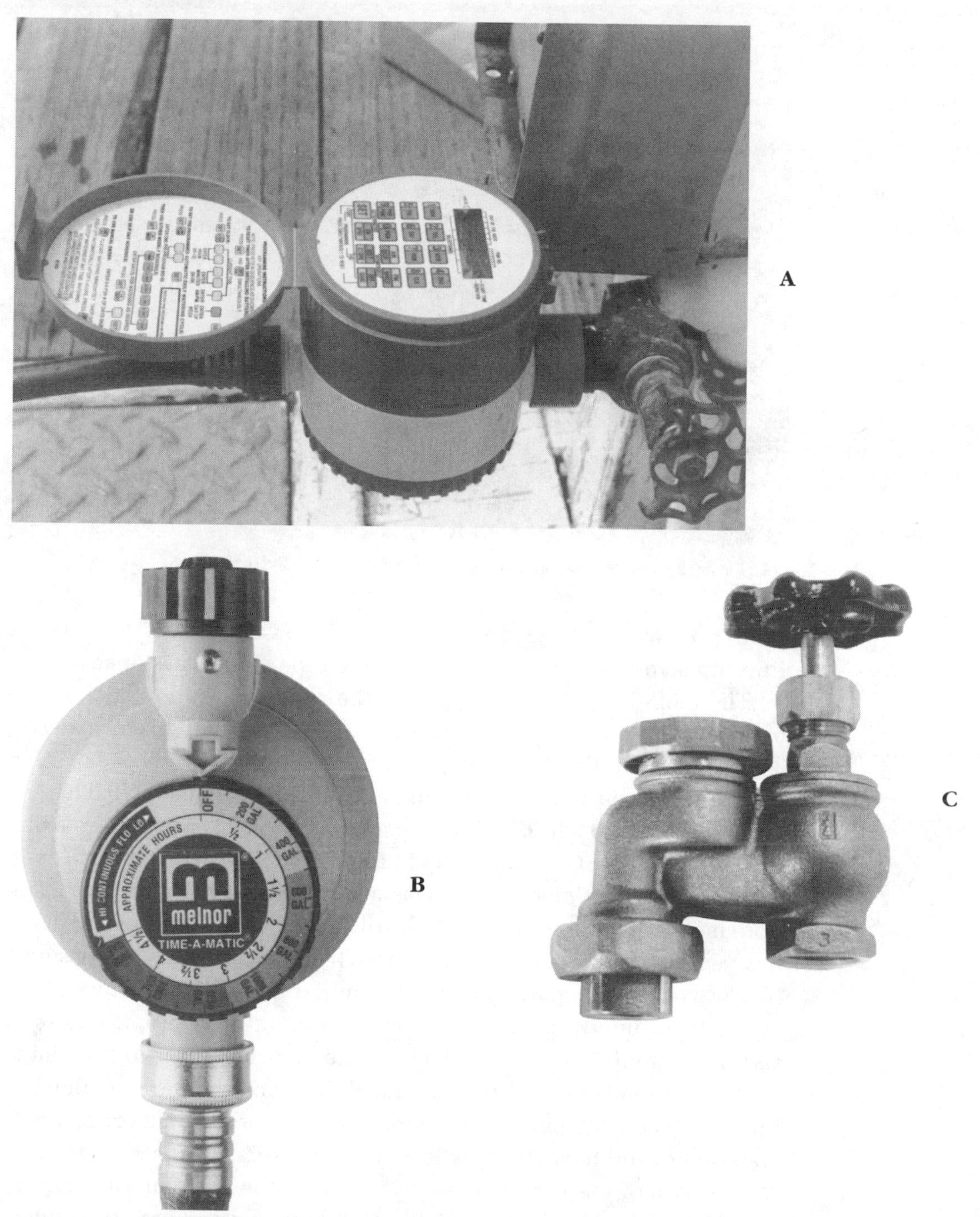

FIGURE 3-11 Water level control devices: by volume or by time. *B: Melnor, Inc., Moonachie, N.J.*

FIGURE 3-12 Typical household water main meter and shutoff.

small pipe wrench by gripping the post of the valve and turning it so it lines up across the pipe (in line with the pipe means it is open).

These fill units are reliable, but as they age they tend to stick in the open position and not shut off. Actor George C. Scott and his lovely wife, actress Trish Van Devere, were greatly upset one night when one of these valves stuck open and flooded their backyard. They thought I had left the water running and left some hot messages on my answering machine that would have made General Patton blush. In the morning when I arrived and replaced the timer, George was gracious in transferring the blame from me to the faulty device, but I learned to replace these devices every two years for all my customers that have them as a preventive measure.

A variation on the timer valve is a similar unit calibrated by gallons rather than time (Fig. 3-11B). It screws onto a hose bib, the hose is screwed onto the timer unit, the hose bib is turned on, and the dial is set for the number of gallons you need. You determine the gallons or liters needed by calculating how many inches or centimeters of water are needed, and how many gallons are in those inches (see Chap. 1).

The idea of these units is terrific because the addition of water by gallons or liters is more precise than unmeasured gallons by minutes.

Unfortunately these units sometimes fail to shut off, and I have stopped using them because even the newer models break down too often. Perhaps in future models, the problems will be eliminated.

There are, however, some models available in gardening supply shops that include solid-state components for regularly scheduled timed water flow or preset volume flow (Fig. 3-11A). These require batteries and careful setting and work well if you carefully check and recheck your settings and change the batteries frequently so the system doesn't fail. Because of these limitations, they are more suited to a homeowner who is there every day to keep an eye on the system, rather than a pool technician who depends on it to do the job while being checked only once a week.

The other type of water level control works much like the float valve in your toilet. A float (Fig. 3-13), which opens and closes a valve attached to a water supply line, is located at the water level desired. When the level drops, the float drops, opening the valve. As the level rises, the opposite happens. The float and plumbing can be located in the skimmer, but are more often located in a small separate tank (called a reservoir), perhaps not even near the pool. The tank must be set at the same level as the pool, so the water in it imitates the water level of the pool. A pipe connects this tank with the pool so the actual water and its level are the same in each. Also, as the level drops, the water fills the pool through this common pipe.

These units are reliable and can be adjusted for water level by bending the arm on the float to the desired level or by setting the elbow in the float arm accordingly. A setscrew loosens the elbow to allow adjustment. The small valve is threaded, so if it rusts, clogs, or fails, it can be unscrewed and easily replaced.

Water level controls serve another valuable purpose. As water evaporates, it leaves minerals (mostly calcium) behind as scale that builds up on tiles and artificial rocks. In later chapters I review prevention and removal techniques, but the simplest method is to keep a constant water level in the pool. As water evaporates and leaves scale, fresh water is introduced to refill the pool to cover the scale line. Simple and effective!

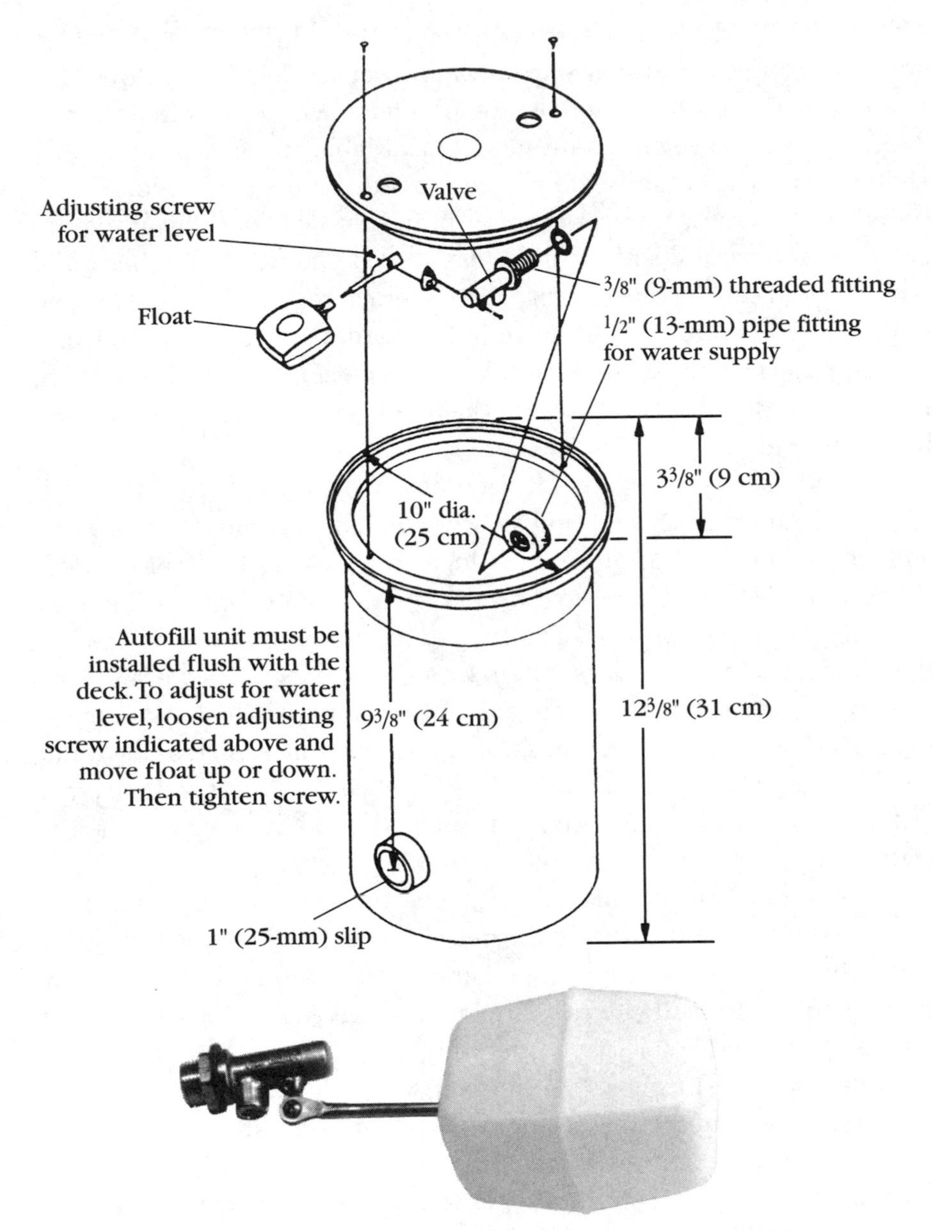

FIGURE 3-13 Float water level control. *MP Industries, Garden Grove, Calif.*

FAQs: ADVANCED PLUMBING SYSTEMS

Should I Add Valves to My Plumbing System?

- Yes. Shutoff valves at the suction and discharge points of your plumbing (as close to the pool as possible) will allow you to isolate plumbing and equipment for maintenance and repair. They also help if you discover a leak and need to determine if it is in the pool itself, or in the plumbing and equipment.

Will a Solar Heating System Pay for Itself?

- If you are heating your pool with gas or electricity, a solar heating system will pay for itself in two to five years, depending on your use of the pool and the heating fuel. If it is your sole source of pool heat, it pays for itself by significantly extending the swimming season.

Must Solar Panels Be Mounted on a Rooftop to Be Effective?

- No. Most solar panels for above-ground pools (and some for in-ground pools) are set up on the ground near the pool. The key to efficiency is angle and exposure to the sun, regardless of where they are mounted. Wind is another important factor—panels on a roof might be cooled sooner than panels set on the ground in a more wind-protected area.

CHAPTER

4

Pumps and Motors

Let's begin by eliminating a common error in terminology. The pump and motor are two different elements of the water circulation system, not the same thing, not interchangeable. The *motor* is the device that converts electrical energy into mechanical energy. It powers the *pump*, which is the device that actually causes the water to move. One is not much use without the other.

While built of various metals or plastics, all pump and motor combinations are composed of essentially the same components. If you understand the basic concept and components, you can find your way around almost any pump or motor. Before discussing the components of a pump and motor, let's understand the concept of what they do and how they work.

Overview

Pool and spa pumps are classified as *centrifugal* pumps. That is, they accomplish their task of moving water thanks to the principle of centrifugal force. To imagine this concept, hold a bucket with some water in it at the end of your arm and spin it around in a big sweeping circle (Fig. 4-1).

Centrifugal force is what keeps the water in the bucket as you spin it. If you poke a hole in the bucket and spin it again, that same force,

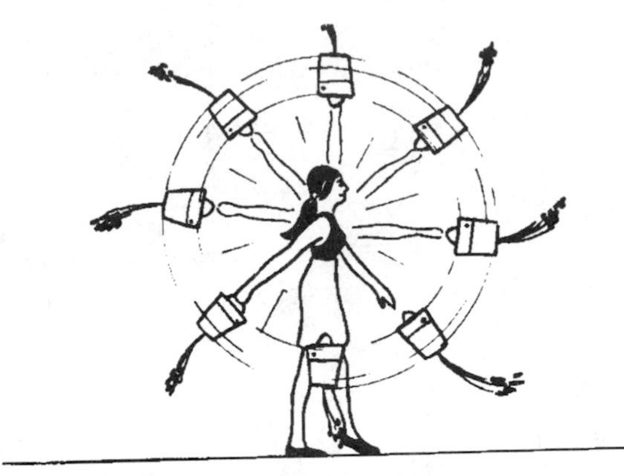

FIGURE 4-1 Centrifugal force. *Sta-Rite Industries, Delevan, Wis.*

pushing the water to the bottom of the bucket, sends it shooting out the hole. If you spin the bucket faster, the water shoots out with more force. Obviously, if you make a larger hole, more water will shoot out as you spin it around.

The pump operates the same way (Fig. 4-2). The impeller in the pump spins, shooting water out of it. As the water escapes, a vacuum is created that demands more water to equalize this force. Water is pulled from the pool or spa and sent on its way through the circulation plumbing. Just as various designs in your swinging bucket and its hole determine the amount of water and how fast it escapes, so too the various designs of impellers, diffusers, and volutes determine the same features in a pool pump. This is discussed in more detail in later sections.

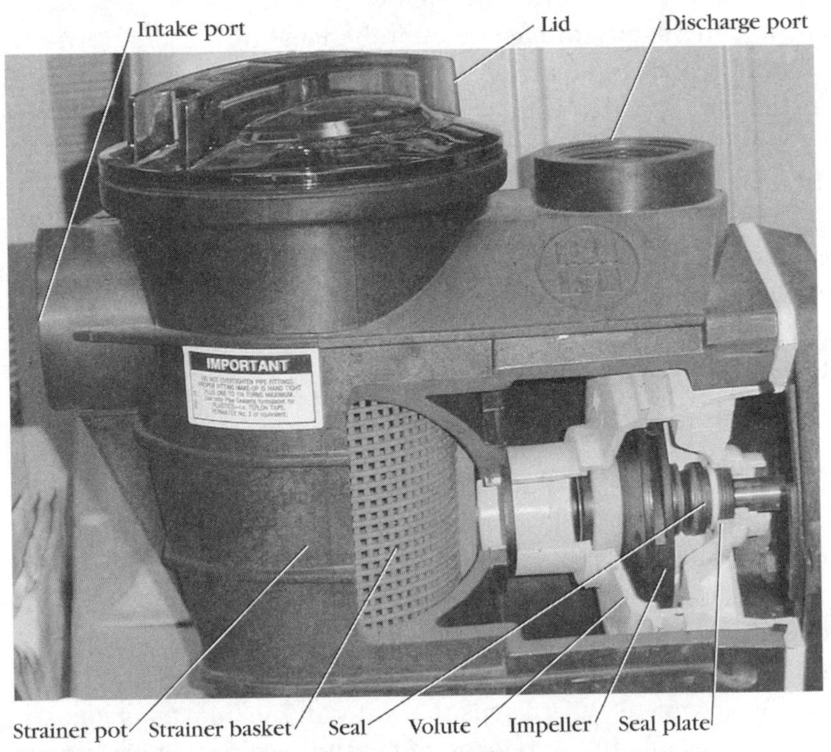

FIGURE 4-2 Typical pool pump (cutaway view).

Pumps used for pools are *self-priming*; that is, they expel the air inside upon startup, creating a vacuum that starts suction. Once water is flowing through the pump, if you close a valve on the outflow side of the pump, restricting all flow, maximum possible pressure is created. However, unlike a piston or gear pump, there is no destructive force created—the impeller simply spins in the liquid indefinitely.

Let's examine the components of the pump and motor partnership (Figs. 4-3 and 4-4).

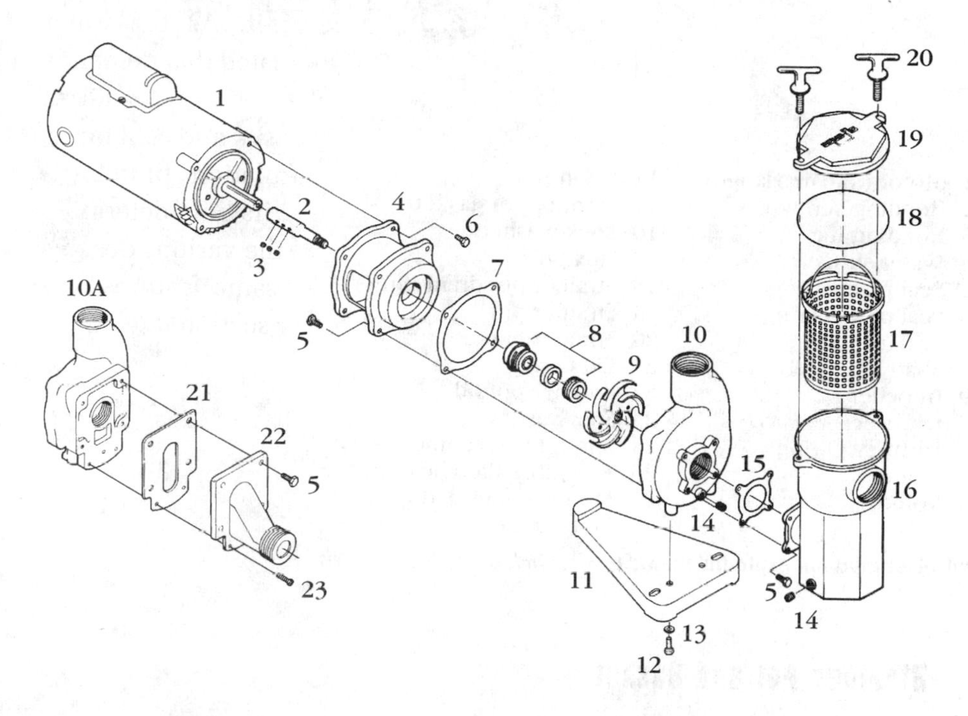

1 Motor (C frame)
2 Shaft extender
3 Allen setscrew
4 Bracket
5/6 Hex bolt
7 Volute gasket
8 Shaft seal
9 Impeller
10 Volute
10A Alternate volute (for use without strainer pot)
11 Base
12 Hex bolt
13 Lock washer
14 Drain plug
15 Trap gasket
16 Strainer pot
17 Strainer basket
18 Lid O-ring
19 Strainer pot lid
20 T-handle
21 Gasket
22 Suction flange (for use without strainer pot)
23 Assembly screw

FIGURE 4-3 Typical bronze pump and motor, exploded view. *Aqua-Flo, Inc.*

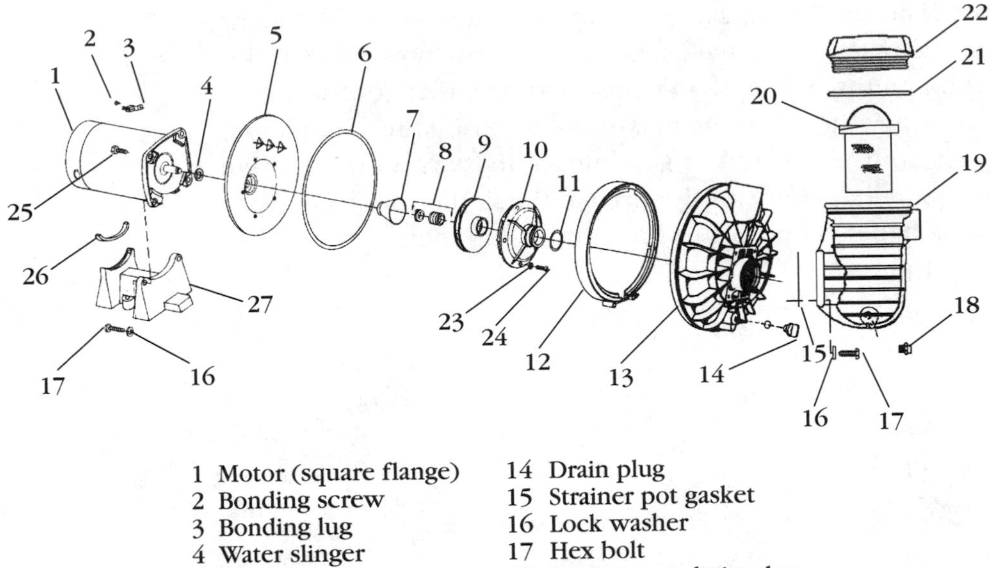

1 Motor (square flange)
2 Bonding screw
3 Bonding lug
4 Water slinger
5 Seal plate
6 Seal plate O-ring
7 Seal insert
8 Shaft seal
9 Impeller
10 Diffuser
11 Diffuser O-ring
12 Clamp
13 Volute
14 Drain plug
15 Strainer pot gasket
16 Lock washer
17 Hex bolt
18 Strainer pot drain plug
19 Strainer pot
20 Strainer basket
21 Lid O-ring
22 Strainer pot lid
23 Star washer
24 Assembly machine screw
25 Assembly allen-head bolt
26 Motor mount pad
27 Motor mount

FIGURE 4-4 Typical plastic pump, exploded view. *Sta-Rite Industries, Delevan, Wis.*

Strainer Pot and Basket

The plumbing from the pool or spa main drain and skimmer runs to the inlet port of the pump, which is usually female threaded for easy plumbing, although some designs are male and female threaded.

Water flows into a chamber, called the *strainer pot* or *hair and lint trap*, which holds a basket (generally 4 to 6 inches or 10 to 15 centimeters in diameter and 5 to 9 inches or 13 to 23 centimeters deep) of plastic mesh that permits passage of water but traps small debris. Some baskets simply rest in the pot, others twist-lock in place.

Most have handles to make them easier to remove, although I have yet to see a design where the handle is not so flimsy that it doesn't break off the second or third time you use it. The strainer basket is similar to the skimmer basket which traps larger debris.

The strainer pot is a separate component in some pumps that bolts to the volute with a gasket or O-ring in between to prevent leaks. Sometimes the pot includes a male threaded port that screws into a female threaded port in the volute. In some pumps, it is a component molded together with the volute as one piece (Fig. 4-2). In bathtub spas or booster pumps, where debris is not a problem, there is no strainer pot and basket at all.

To clean out the strainer basket, an access is provided. The strainer cover is often made of clear plastic so you can see if the basket needs emptying. It is held in place by two bolts that have a T top (Fig. 4-3) or plastic handle (Fig. 4-4) for easy gripping and turning. Some pumps have a metal clamp that fits around the edge of the cover and strainer pot. These are tightened with a bolt and nut combination. On others the cover is male threaded (Fig. 4-4) and screws into the female threads of the pot.

In all styles of pot, the strainer cover has an O-ring that seats between it and the lip of the strainer pot, preventing suction leaks. If this O-ring fails, the pump sucks air through this leaking area instead of pulling water from the pool or spa.

Notice in Figs. 4-3 and 4-4 that the pot has a small threaded plug that screws into the bottom. This plug is designed to allow complete drainage of the pot when winterizing the pump. It is made of a weaker material than the pot (on metal pots, for example, the plug is made of plastic, soft lead, or brass). If the water in the pot freezes, this sacrificial plug pops out as the freezing water expands, relieving the pressure in the pot. Otherwise, of course, the pot itself will crack.

Volute

The *volute* is the chamber in which the impeller spins. Combined with the impeller, the volute forces water out of the pump and into the plumbing that takes the water to the filter (or directly back to the pool, spa, or fountain if the system is not filtered or heated). The outlet port is usually female threaded for easy plumbing. When the impeller (Fig. 4-5) moves water, it sucks it from the strainer pot.

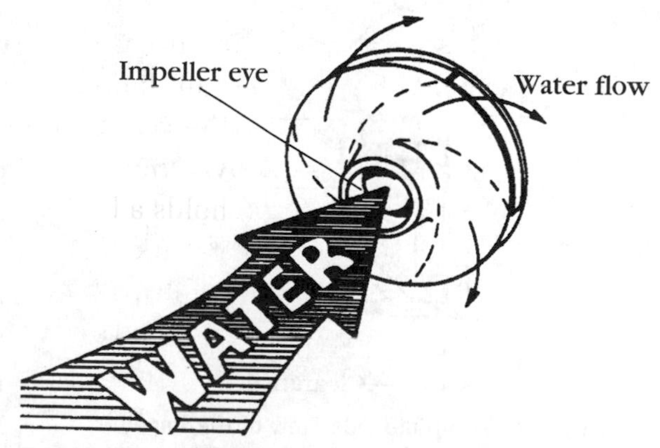

FIGURE 4-5 Water's-eye view of the impeller. *Sta-Rite Industries, Delevan, Wis.*

The resulting vacuum in the pot is compensated for by water filling the void. The rushing water is contained by the volute which directs it out of the pump. Therefore the pot can be considered a vacuum chamber and the volute a pressure chamber.

The impeller by itself will move the water, but it cannot create a strong vacuum by itself to make the water flow begin. The area immediately around the impeller must be limited to eliminate air and help start the water flow. A diffuser (Fig. 4-4) and/or closed-face impeller help this process, but in many pump designs, the volute serves this purpose.

Figure 4-2 shows how the volute closely encircles the impeller. Figure 4-4 shows a design with a separate diffuser that houses the impeller. In some designs, the inside of the volute is ribbed to improve the flow efficiency.

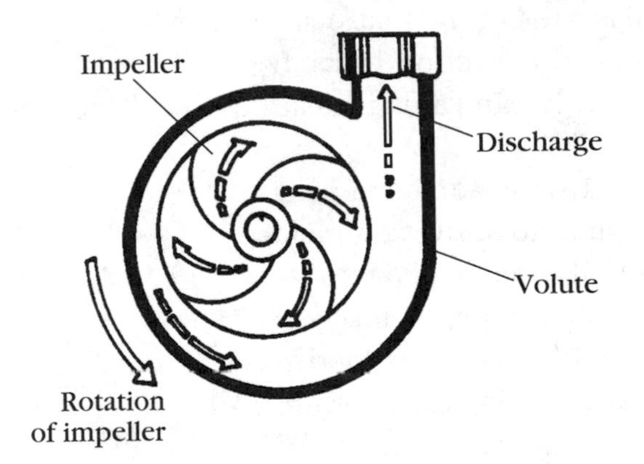

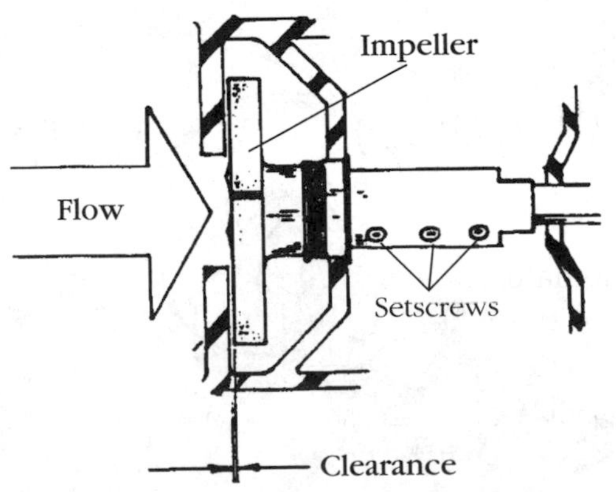

FIGURE 4-6 Front and side view of the impeller inside the volute. *Sta-Rite Industries, Delevan, Wis.*

Impeller

The impeller is the ribbed disk (the curved ribs are called *vanes* and the disk is called a *shroud*) that spins inside the volute. As water enters the center or eye of the impeller (Fig. 4-5), it is forced by the vanes to the outside edge of the disk, just like our spinning bucket example. As the water is moved to the edge, there is a resulting drop in pressure at the eye, creating a vacuum that is the suction of the pump. The amount of suction is determined by the design of the impeller and pump components and the strength of the motor that spins the impeller.

There are essentially two types of impellers: closed-face (Fig. 4-5) and semiopen-face (Fig. 4-6). Some publications call the semiopen-face impeller an open face. This is not accurate, because for the pump to be self-priming, which most pool and spa pumps are, it needs a disk (shroud) on the front face as well. Therefore, although a particular impeller itself has no front shroud and might be called open when it stands alone, it does in fact make use of some sort of front shroud, either a diffuser located closely around the impeller or the interior surface of the volute.

In Fig. 4-6, you get another water's-eye view of a volute and impeller. The side view shows how close the "open-face" impeller is to the interior side of the volute, effectively forming a front shroud. The clearance between the volute interior and impeller face is critical. Too far away and there will be insufficient pressure created in the volute, causing weak or no suction. Too close and the impeller might rub against the volute or jam if small debris lodges between the two. As discussed later in the section on the shaft, the semiopen impeller pump can be adjusted for optimum efficiency of the impeller.

In the closed-face impeller, as the name implies, the vanes of the impeller are covered in both front and back. Water flows into the hole in the center and is forced out at the end of each vane along the edge of the impeller. This type of impeller, especially in connection with a diverter, is extremely efficient at moving water.

If the closed-face impeller is so much more efficient and requires no shaft extender or adjustment, why have the semiopen-face impeller designs survived? Because the downside of the closed face is that small stones, pine needles, and other fine debris can get past both the skimmer and strainer baskets and clog the closed vanes, slowing or completely shutting off water movement. The semiopen-face design allows this small debris to pass (actually it is usually pulverized) to the filter, although it is not impossible for a heavy volume of small debris to clog the open vanes as well.

One of the chief culprits in clogging of both semiopen and closed impeller designs is DE (diatomaceous earth). As is described in the chapter on filters, this white, powdery material is used to precoat some designs of filter grids. If introduced too quickly, DE will clog any restricted area—plumbing elbows, strainer baskets, and impeller vanes.

Most impellers on pool and spa pumps have a female threaded hole on the center back side (the side facing away from the water source) that screws onto the male threaded end of the motor shaft (or shaft extender). The rotation of the shaft is just like a bolt being threaded into a nut. As the shaft turns, it tightens the impeller on itself. Others, like the old Purex AH-8 models, are fitted with setscrews that clamp the impeller onto the end of the motor shaft.

Impellers are rated by horsepower to match the motor horsepower that is used. This, in turn, determines the horsepower rating of the pump or pump and motor you have. What if you used a 2-hp motor and a 1-hp impeller? The pump would still only pump the volume of the 1-hp impeller, but the motor would not have to work as hard as it was designed to work. No problem.

But what about the reverse? A 2-hp impeller will move its rated volume and speed of water even with a 1-hp motor, but the motor works harder than it is designed to and soon overheats and burns out. Big problem! Moral of the story—if you must assemble miscellaneous unmatched parts, always make the impeller rating equal to or less than the rating of the motor.

Let's say you have a box of spare impellers and don't know what their rated horsepower is or what pump they came from. Start by asking at your pump rebuilding shop or supply house. Another way is to examine the impeller for codes or markings. On many bronze semi-open-face impellers, a 0.5, 1, 2, etc., is engraved on the inside of one of the vanes, telling you the horsepower. On Sta-Rite plastic closed-face impellers, a code is used, such as 137-PD. By checking your supply house catalog for Sta-Rite pumps, you will see the model code 137-PD refers to a 2-hp pump.

Failing any identifying marks such as these, don't try to guess at the rating. It is cheaper to buy a new impeller for a particular pump than to install one of greater horsepower than the motor and, ultimately, damage the motor.

Seal Plate and Adapter Bracket

The volute is the pressure chamber in which the impeller spins to create suction. If this were all one piece, there would be no way to remove the impeller or to access the seal. Therefore this chamber is divided into two sections. The actual curved housing is called the *volute*, while its other half is called the *seal plate* or *adapter bracket*.

The seal plate (Fig. 4-4, item 5) is joined to the volute with a clamp. An O-ring between them makes this joint watertight. The motor is bolted directly onto this type of seal plate. In other designs, the seal plate is molded together with an adapter bracket that supports the motor (Fig. 4-3, item 4) and bolts to the volute, with a paper or rubber gasket between them to create a watertight joint. In yet another style, the pump sections are joined with a threaded union type of clamp, like the lid of a jar. This allows disassembly by hand.

In both cases, the shaft of the motor passes through a hole in the center of the seal plate and the impeller is attached, threaded onto the shaft. The bracket allows access to the shaft extender (see following section) for adjusting the clearance between the impeller and volute (Fig. 4-6). The pump design shown in Fig. 4-4 is a closed face and needs no such adjustment, so the shaft need not be exposed.

TOUGH NUT TO CRACK

Note in Fig. 4-4, the clamp (item 12) that joins the volute and seal plate is made tight by a bolt and nut. Never assemble the pump with this bolt underneath the pump. When the pump is bolted to the deck, it makes it a knuckle-busting, cussword of a job to remove the clamp. Moreover, if the pump leaks at all, the bolt gets and stays wet, causing it to rust, making unscrewing it even tougher, or the bolt breaks altogether. If you come across one, move the clamp bolt to the top of the pump before installation.

Shaft and Shaft Extender

The shaft of the motor is the part that turns the impeller, creating water flow. Figure 4-3 shows a motor with shaft. In this design of pump, the impeller needs to be adjusted in relation to the volute, so a shaft extender has been created. The extender, made of brass or bronze, slides over the motor shaft and is secured by three allen-head setscrews (Fig. 4-6). The male threaded end of the shaft extender then fits through the seal plate and the impeller is screwed into place.

Note that the extender is round, but a flat area has been created on two sides. In this way, a ¾-inch (19-millimeter) box wrench can be used to prevent the extender from spinning when performing maintenance. Note also that some designs require an O-ring near the threads of the extender to ensure a watertight seal. Other designs rely only on the pump seal. Once assembled, the clearance between the impeller and volute can be adjusted as shown in Fig. 4-6 (side view).

The shaft of the motor in this example (Fig. 4-3) is called a *keyed shaft*. This means the cylindrical shaft has a groove running the length of the shaft to accept the setscrews. In this way, when the setscrews are in place, they prevent the motor shaft from slipping or skidding inside the extender, thus failing to turn the impeller. Figure 4-4 shows a pump style that requires no shaft extension. The motor shaft, already engineered to the exact length required, has a threaded end to accept the impeller.

The shaft should never be in contact with electric current, but water is a great conductor and wet conditions around pool equipment can circumvent the best of designs. Because of this, most motor shafts today are designed with a special internal sleeve to insulate the electricity in the motor from the water in the pump.

Seal

Obviously if the shaft passed through the large hole of the seal plate without some kind of sealing, the pump would leak water profusely. If the hole was made small and tight, perhaps of tight-fitting rubber, the high-speed spinning of the shaft would create friction and burn up the components or the shaft would bind up and not turn at all. Some clever engineer devised a solution to this problem called a *seal*.

The seal allows the shaft to turn freely while keeping the water from leaking out of the pump. In Fig. 4-3, the seal (item 8) is in two parts. The right half of the seal is composed of a rubber gasket or O-ring around a ceramic ring. This assembly fits into a groove in the back of the impeller. The ceramic ring can withstand the heat created by friction. The left half fits into a groove in the seal plate and is composed of a metal bushing containing a spring. A heat-resistant graphite facing material is added to the end of the spring that faces the ceramic ring in the other half.

The tight fit of the seal halves prevents water from leaking out of the pump. The spring puts pressure on the two halves to prevent them from leaking. As the shaft turns, these two halves spin against each other but do not burn up because their materials are heat-resistant and the entire seal is cooled by the water around it. Therefore, if the pump is allowed to run dry, the seal is the first component to overheat and fail.

The pump design in Fig. 4-4 includes an additional seal housing or insert. If the pump runs dry, the heat buildup not only melts the seal, but also the plastic seal plate in which it is mounted. The inset helps isolate the seal plate from this heat and the cone-shaped unit diffuses heat.

Pump makers are always improving the heat dissipation (heat sink) capabilities of their pumps so that dry operation will result in little or no damage to the seal or pump components. Still, pumps are not designed to run without water for more than a few minutes while priming.

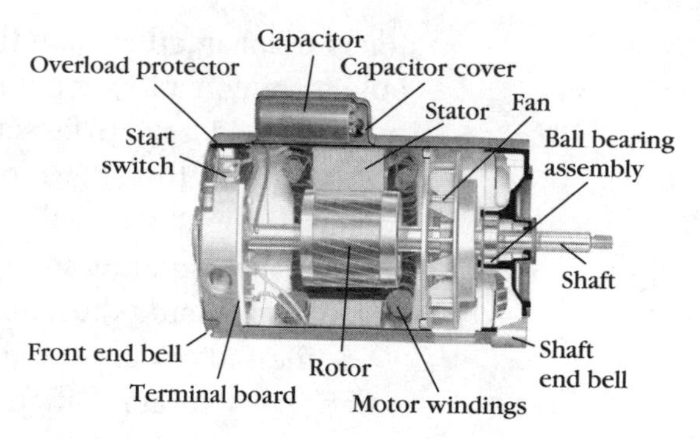

FIGURE 4-7 **Typical pool and spa motor.** *Franklin Electric.*

Motor

Before reading this section, a basic knowledge of electricity is helpful, so you might want to review the section on basic electricity at the end of this chapter. After all, the motor is the device that converts electricity into mechanical power.

Motors, like the pumps they drive, are rated by horsepower, usually in pool and spa work as ½, ¾, 1.0, 1.5, and 2.0 horsepower. Commercial installations might use higher rated systems, but these are the most common.

As shown in Fig. 4-7, electricity flows through the motor windings, which are thin strands of coiled copper or aluminum wire. The windings magnetize the iron stator. If you paid attention in your first grade science class, you recall that opposite poles of a magnet attract each other, but like poles repel. Using this concept, the rotor spins, turning the shaft. Some designs employ one set of windings for the startup phase where greater turning power (torque) is needed and another set for normal running.

The shaft rides on ball bearings at each end. A built-in fan cools the windings because some of the electrical energy is lost as heat. The caps on each end of the motor housing are called *end bells*. A starting switch is mounted on one end with a small removable panel for electrical connection and maintenance access.

This is where you will also find the thermal overload protector. This heat-sensitive switch is like a small circuit breaker. If the internal temperature gets too hot, it shuts off the flow of electricity to the motor to prevent greater damage. As this protector cools, it automatically restarts the motor, but if the unit overheats again, it will continue to cycle on and off until the problem is solved or the protector burns out.

It takes a great deal of electricity to start a motor but far less to keep it going (in fact, about five to six times as much). The capacitor, as the name implies, has a capacity to store an electrical charge. The capacitor is discharged to give the motor enough of a jolt to start, then it is able to run on a lower amount of electricity. Without the capacitor, the motor would need to be served by very heavy wiring and high-amp circuit breakers to carry the starting amps. The startup amperage of a motor is about twice that of its running amperage. The capacitor is located in a separate housing mounted atop the motor housing (as in Fig. 4-7) or inside the front end bell.

Some motors are designed to operate at two speeds. For example, some spas operate at high speed for jet action, but lower speed for circulation and heating. In pool and spa work, the normal rotation speed is 3450 revolutions per minute (rpm) and the low speed is 1750 rpm.

Types

Now that you understand these basic concepts, I will discuss the three main types of motors you will find in water work.

SPLIT PHASE

These motors are ¾-hp or less, and are found in small fountain applications where startup power requirements are very low—there is no capacitor.

CAPACITOR START, INDUCTION RUN (CSI)

The most typical motor used in the pool business. This motor uses a capacitor and starting windings to start up, then these are shut down and a running winding takes over. As noted previously, the capacitor and startup windings allow faster, stronger torque to overcome the initial resistance of the impeller against standing water, then when the water is moving and less power is needed to keep it moving, the system shuts off and the lighter running winding takes over.

CAPACITOR START, CAPACITOR RUN (CSR)

This is the concept of the energy efficient motor. The difference between a CSR and CSI motor is that the CSR motor employs a capacitor on the running windings as well. This smoothes out the variations in the alternating current (ac) power that helps reduce heat loss in the winding (remember, heat loss is electricity wasted). In short, CSR motors are more efficient but cost more because of the added parts. These motors are also called switchless, because on some designs the run capacitor makes a start switch unnecessary. This is a good thing because start switches get dirty and fail and tend to be fragile parts that readily break when brought into sharp contact with a screwdriver.

ENERGY EFFICIENT

Energy efficient motors are CSR motors that have heavier wire in the windings to lower the electricity wasted from heat loss.

A good way to compare energy efficiency between two motors is to compare the gallons pumped to kilowatts used. Let's say one pump produces a flow rate of 50 gallons (189 liters) per minute, which is 3000 gallons (11,355 liters) per hour. Divide that by the kilowatts used per hour. The higher the resulting number, the more efficient is the pump and motor.

By the way, kilowattage is determined by multiplying amps by voltage. Lets say the unit runs at 9 amps at 220 volts—9 × 220 = 1980 watts. Kilowatts (meaning 1000 watts) would then be 1980 divided by 1000. So this unit uses 1.98 kilowatts each hour. The 3000 gallons divided by 1.98 equals a rating of 1515 gallons (5734 liters) pumped for every kilowatt used. You can now make similar calculations for other pump and motor units for comparison.

Voltage

I have been discussing typical 110/220-volt motors. In fact, most motors are designed to be connected to either power source. By changing a wire or two internally, you determine which voltage is used. The instructions for this conversion are printed on the motor housing or inside the access cover.

If your motor is wired for 220 volts and you feed it 110 volts, it will run slowly or not start. If your motor is wired for 110 volts and you feed it 220 volts, the thermal overload protector should overheat and cut the circuit. Higher horsepower motors might run on three-phase current. You don't want to fool with that. Call an electrician.

Housing Design

The housing of the motor is designed to adapt to various pump designs. Figure 4-3 shows a motor called a *C frame*, because the face of the motor resembles a C. All this means is that it will fit certain kinds of pumps. Figure 4-4 shows a motor called a *square flange*, for equally obvious reasons. There are other types, such as the 48, uniseal flange, and those designed for automatic pool cleaner booster pumps. When replacing a motor, you need to buy the proper housing type.

Ratings

The *service factor* of a motor is a multiplier, a number. When you multiply the service factor number by the rated horsepower number, you get the real horsepower at which the motor is designed to operate on a continuous basis. As an example, a motor rated at 1 hp with a service factor of 1.5 can actually safely run a 1.5-hp pump ($1.0 \times 1.5 = 1.5$ hp).

The motor on the fish pond at comedian Rich Little's house burned out one day. It was a 2-hp motor. It was too late in the day to get to the supply house and buy a new one, and if the pump didn't run, the very large and very expensive koi would end up in goldfish heaven by morning. All I had in my shop was a 1.5-hp motor, but it was rated with a service factor of 1.5, meaning 1.5×1.5 equals 2.25 hp. I could safely use this motor on the existing pump and expect it to do the job. It worked!

Nameplate

All of what you need to know about a motor is printed on the nameplate (Fig. 4-8), a sticker applied to the housing. Here's what you can learn from the nameplate.

ELECTRICAL SPECIFICATIONS

A diagram shows how to wire the starting switch plate for 110- or 220-volt supply. If this diagram is not on the outside of the motor, remove the small access door in the end bell and it should be printed on a sticker in there. These stickers frequently come off as the motor gets older, so if no diagram is available, refer to the manufacturer's website or guidebook available at your supply house.

The nameplate also tells you the amperage. It might say "10.5/5.2." This means the startup draw is 10.5 amps, and the normal running draw is 5.2 amps. Make sure you are reading this information from the area that says "Maximum Load" or "Maximum Amps." Some makers publish the amps required to power the nominal horsepower as well. These are lower than the maximum and should not be used when sizing wire or circuit breakers.

The nameplate also lists the electrical phasing (single phase or 1) and cycle frequency (called *hertz*). Alternating current in the United States runs on 60 hertz. In Europe it is 50 hertz, and that's why you can't use some appliances from one country in another, even with a voltage converter, because things like VCRs and TVs rely on the cycles as well as the voltage. "Stupid" appliances like toasters or shavers only care about voltage. Why? Go read the toaster and shaver books. It only hertz once. (Sorry, I couldn't resist that one.)

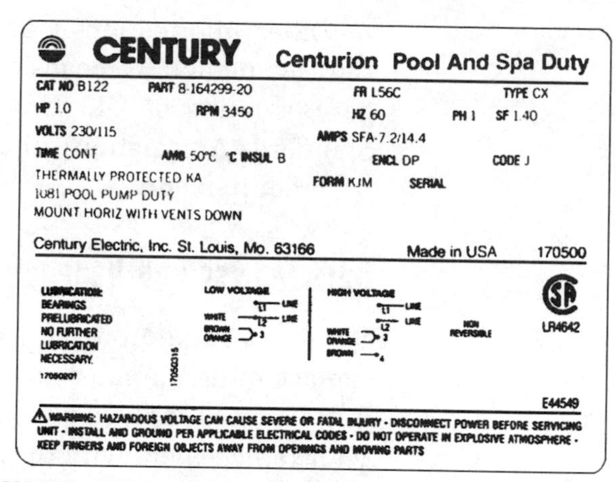

FIGURE 4-8 Typical motor nameplate.

MANUFACTURER AND DATE

The manufacturer's name appears on the rating plate, along with the model and serial numbers, and includes the day, month, and year the motor was built.

HORSEPOWER

The relative strength of the motor is expressed in horsepower and corresponds to the specifications of the pump that is to be driven by the motor.

INSTALLATION

As noted previously, a diagram of the wiring connections is printed on the nameplate. There will also be a few words about mounting, such as "mount horizontally" or "mount with vents down."

DUTY RATING

Pool and spa motors are designed for continuous duty, meaning they can run 24 hours a day for their entire service life without stopping. The nameplate shows this by the rating "Continuous Duty." The horsepower, service factor, rpm, and frame style of the housing are listed. If the motor has a thermal overload protector, the nameplate will indicate it.

Other information is also shown on the nameplate, such as the starting method (C means capacitor, for example), the insulation category, a rating of UL (Underwriter's Laboratory) or CSA (Canadian Standards Association) approval, and ambient (surrounding) temperature requirements.

Horsepower and Hydraulics Equals Sizing

So now you know what a pump and motor are, and if you need to replace either, in most cases you will assume the original designer or builder used the correct size pump and motor for the job and make your replacement with the same size. Or will you?

What if the original equipment was too small or too large? What if the plumbing has been repaired (which might have added or deleted pipe and fittings) or equipment has been added or deleted, thus changing the system and requiring the pump to work more (or less)? What if the identifying rating plates have been removed or are so weatherworn that you can't tell what size the existing equipment is? Finally, what if it is a brand new installation? How do you decide what is the right pump and motor for the job?

Well I'm glad you asked all of those intelligent questions. The answer is that you need to know a little about the pool or spa system's needs and hydraulics.

Hydraulics

RATING: ADVANCED

Hydraulics, the study of water flow and the factors affecting that flow, is important to understand because its principles affect plumbing and equipment sizing choices. Understanding hydraulics as it applies here is actually quite simple and the math involved is very basic. It only looks tough because there are so many factors to consider. Before starting on this complex series of calculations, think about this. Unless you are building a pool or spa with clean components able to operate as the manufacturer recommends and you have blueprints to know what plumbing components have been included underground, then all of this section is theoretical. In fact, most of the time you won't know what plumbing exists out of sight or how much resistance to water flow is being created by old filters, heaters, solar panels, and so on.

So why learn hydraulics? First, because understanding it helps you estimate the right pump for a replacement or when water circulation is poor in a poll or spa. It helps you avoid adding plumbing or other components that might aggravate an already bad situation. Second, in some cases, you might be the designer or installer of a pond, spa, or pool, so this information is essential.

TERMS

First, a few terms used in this section must be explained.

Head and Flow Rate: *Head* is the resistance of water flow through plumbing and equipment expressed in feet. (The lower the better.) *Flow rate* is the volume of water moved in a given period of time. Here's an example.

Let's say you have a pump and motor with a 1-foot pipe attached to the outflow port sticking straight up in the air. For the moment ignore the source of the water on the suction side. You turn on the pump and water flows out at a rate of 10 gallons per minute (oh yes, you had a 10-gallon bucket and stopwatch nearby).

The 1-foot vertical distance is the head and the 10 gallons per minute (gpm) is the flow rate. The pump is rated at 10 gpm at 1 foot of head (or 1 foot of resistance). But what if that resistance is increased? What will happen to the flow rate?

Suppose you are moving your furniture around the living room for fun and profit. You can raise the 100-pound couch over your head by 2 feet. You ask your friend to get on the couch, making it now weigh 200 pounds. Now you are able to lift it only 1 foot over your head.

This rather silly example tells you what to expect when adding resistance to the pump. Let's go back to our pump with the pipe sticking straight up and make the outflow pipe 8 feet, then measure the flow rate. It is down to 5 gpm. The additional resistance (head) of that added pipe means that the pump cannot push the water as fast. This loss of flow, as head increases, is called *head loss*, a somewhat deceptive term because it is actually flow that is lost. The term actually means flow loss caused by head increase, but just to confuse us, engineers call this process head loss. Got it?

If you continue to increase the head (resistance) by adding more vertical pipe, the flow rate will continue to decrease until, at last, no water comes out at all. In the example, let's say that happens when you add 10 feet of pipe. This pump can now be charted on a graph (Fig. 4-9), allowing you to study its performance characteristics. With such a graph, you can choose any flow rate for your hypothetical pump and learn what the maximum amount of resistance can be if you are to maintain that flow rate. Conversely, you can choose any amount of head you think a certain plumbing system might create, then learn the flow rate expected out of that system.

Figure 4-9A shows the graph for this hypothetical (and rather small) pump. On the left side of the graph is the possible feet of head, from 0 to 10 feet. On the bottom is the possible flow rate, from 0 to 10 gpm. By finding where the head and flow rates intersect, you can create a pump curve for your pump. A metric version of this graph compares liters per minute with head expressed in meters (Fig. 4-9B).

Because you generally have no way to measure these factors, the manufacturer provides a pump curve for each pump. If you know the amount of head (resistance) in your pool or spa system and you know the desired flow rate, then you can determine which pump will satisfy those needs by referring to the manufacturer's pump curves.

Pumps are designated low, medium, high, or ultra-high head. The higher the head designation, the less strain is placed on the pump and motor components:

- Low head: suck well, push poorly.
- Medium head: suck well, push well.
- High head: suck poorly, push well (most common in pools and spas).
- Ultra-high head: suck poorly, push well (pool sweeps).

One factor affecting which type a particular pump will be is its impeller. Thin vents on the face (closed or semiopen) result in greater push but poor suck; in other words, poor self-priming capabilities but good circulating flows.

Suction Head: So far I have only discussed the head created by adding resistance to the outflow side of the pump. By restricting the

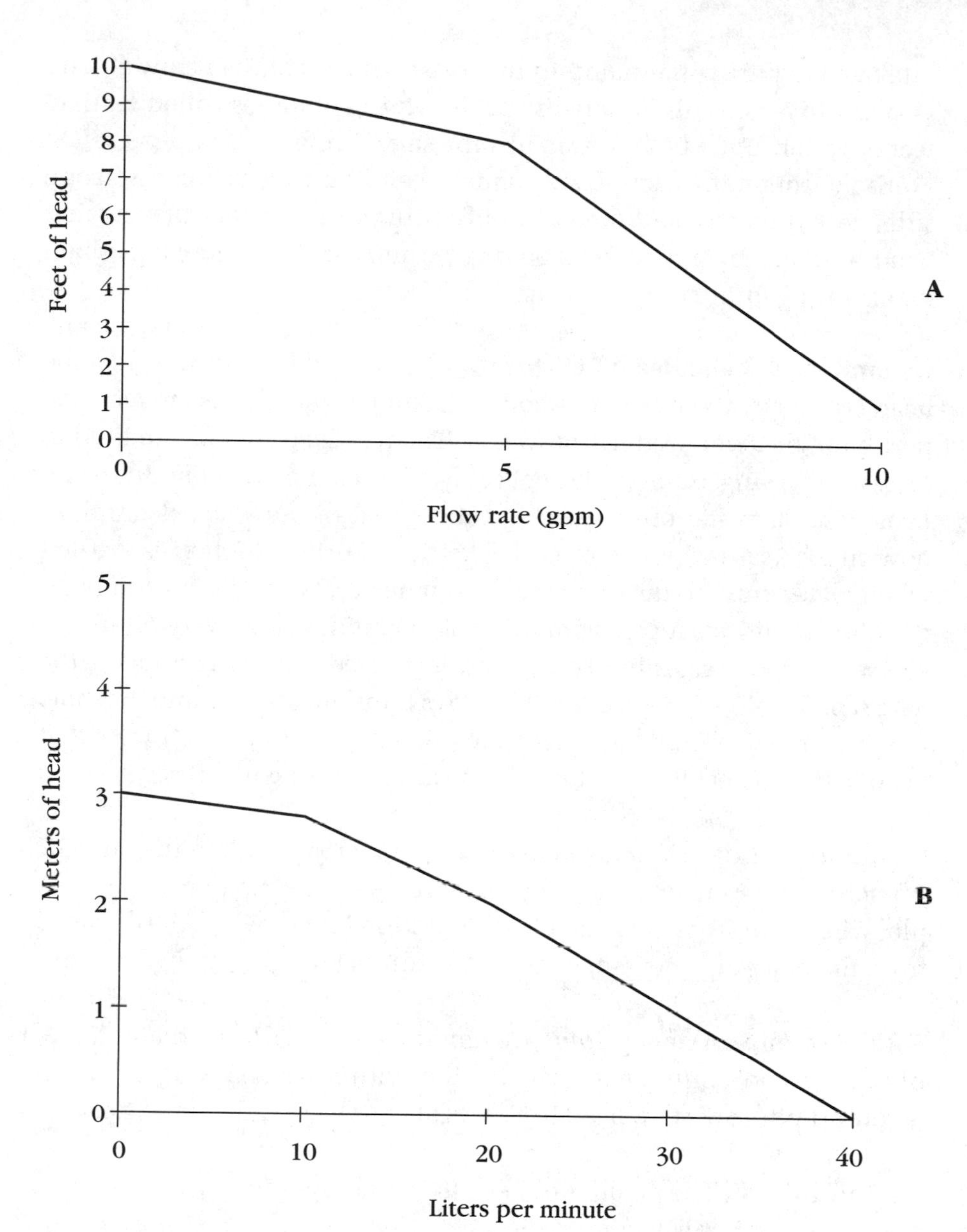

FIGURE 4-9 Sample pump performance curve: (A) U.S. system, (B) metric system.

intake or requiring the pump to lift water from a source below it, you also create head. This is called *suction head*, sometimes called vertical feet of water. Don't be confused, it's the same thing.

Each foot on the suction side equals a similar foot on the discharge side, called *discharge head*. The only thing to remember here is that head (resistance) is created on both sides and must be calculated when determining pump size.

Dynamic and Static Head: Up to now I have described *static head*—the head created by the weight of standing (static) water. This is only a small portion of the total head in the system. The rest is created by the friction of water flowing through the entire system, called *dynamic* (moving) *head*. The diameter of the pipe and the speed of the water determines how much resistance is created by friction. Further friction is created when water must go through or around other obstacles, such as through the filter, heater, solar panels, and plumbing fittings (yes, every plumbing elbow or bend creates head too). And just to be clear at the outset, the length of the pipe (in feet) does not always translate directly into the same number of feet in head loss. There are reference tables that tell you what head loss to expect for each foot of pipe or each fitting (read on).

Cavitation: *Cavitation* refers to the vacuum created when the outflow capacity of a pump exceeds the suction intake. This happens, for example, when a pump is oversized for the suction line or when the distance from the body of water is too far. The result is bubbling and vibration.

Total Dynamic Head: *Total dynamic head* (TDH) is the total of plumbing and equipment head for the entire system. Vacuum head (suction) plus pressure head (discharge) equals total dynamic head.

Shutoff Head: The amount of head at which the pump can no longer circulate water. It is 0 gpm.

CALCULATIONS

Here are a few general numbers to use in your calculations.

Pipe Fittings: To make it easier to calculate head in your plumbing system, it is measured for every 100 feet of pipe or the equivalent (standard friction charts are available at the supply store for each diameter of pipe when you purchase the pipe). Plumbing connections, fittings, and valves have different amounts of resistance than straight pipe, so these must first be converted to the equivalent length of straight pipe. Unions and straight connectors act like additional lengths of straight pipe, so no special calculations are needed. Going around corners is what creates head. Here are the values for the most common PVC fittings you will use:

- 1½-inch (40-millimeter) × 90-degree elbow = 7.5 feet of straight 1½-inch pipe (2.3 meters of 40-millimeter pipe)
- 2-inch (50-millimeter) × 90-degree elbow = 8.6 feet of straight 2-inch pipe (2.6 meters of 50-millimeter pipe)
- 1½-inch (40-millimeter) × 45-degree elbow = 2.2 feet of straight 1½-inch pipe (67 centimeters of 40-millimeter pipe)
- 2-inch (50-millimeter) × 45-degree elbow = 2.8 feet of straight 2-inch pipe (85 centimeters of 50-millimeter pipe)

It is interesting to see that three times as much resistance is created when a 90-degree fitting is used instead of a 45. That is particularly significant, because there are times when you have a choice between using two 45-degree fittings in a job rather than one 90. The combination of two 45-degree fittings creates less resistance than one 90.

Also note that in a T fitting, the turn around its 90-degree bend is sharper than the more gradual sweep through 90 degrees created by a typical 90 fitting. Thus, more resistance is created in a T fitting, even though the water is being bent 90 degrees in either case.

Filters: Filters create 5 to 7 feet (1.5 to 2 meters) of head. The manufacturer will tell you in the literature that accompanies the product how many feet of head the unit creates. You can also measure the amount by placing a pressure gauge on the pipe leading into the filter and one on the pipe going out. The difference, measured in pounds per square inch (psi) and multiplied by 2.31, tells you the feet of head.

As dirt builds up in a filter, however, head increases. A clean filter will have no more than 3 psi (207 millibars) difference between the input pressure and the output pressure. Since 1 psi (69 millibars) equals 2.31 feet (70 centimeters) of head, a new, clean filter should add no more than 6.9 feet (2 meters) of head.

As you will see in the chapter on filters, manufacturers recommend cleaning a filter when the operating pressure (as read on the pressure gauge, expressed in psi) builds up to more than 10 psi (689 millibars) over clean operating pressure. Since 10 psi equals 23.1 feet of head, you can see that the resistance caused by dirt added to the normal amount of head for the filter itself can total over 30 feet (almost 10 meters) of head for this component alone.

This is another factor that tends to make these calculations more art than science, and why providing a slightly larger horsepower pump than required on a system is always a good idea.

Heaters: Heaters create 8 to 15 feet (2.5 to 4.5 meters) of head. Like filters, the manufacturer will tell you in the literature that comes with the unit what the head loss is for the unit at a given flow rate. Also like filters, as scale (lime and other minerals) builds up in the heat exchanger (explained in more detail in the chapter on heaters), more friction is created and therefore more head. Unlike filters, all water flowing through the heater does not pass through the same components of the unit. Heaters have bypass plumbing (also discussed later), so not all water flows through the heat exchanger. Thus restrictions in the exchanger may not lead to as much resistance as you might expect. This is another good reason to beef up your pump when doing the sizing calculations for an installation.

Poolside Hardware: Main drain covers, skimmers, and return outlets all add head. To know exactly how much, you must refer to each manufacturer's specifications. A general rule of thumb is to add 5 feet (1.5 meters) of head to allow for the total of such components in your system.

Pumps: Pumps also create head, but the manufacturer's charts allow for this, so your calculations need not consider it. When you look at the TDH for the system on the pump curve, the pump head loss is already figured in the performance ability.

TURNOVER RATE

The *turnover rate* of a body of water is how long it takes to run all the water through the system. It is desirable for the water to completely circulate through the filter one to two times per day, but local codes generally require a specific time period. In Los Angeles, for example, it is

- Pools must turn over in 6 hours,
- Spas must turn over in ½ hour, and
- Wading pools must turn over in 1 hour.

I have mentioned that various components offer more or less resistance at different speeds (expressed in gallons per minute). To calculate the TDH of a system, you must know that speed. To decide what speed is needed (and therefore what size pump is needed to deliver that speed in your system) you must establish a turnover rate.

Let's say you've calculated the volume of water in the pool (see Chap. 1) as 18,000 gallons (68,000 liters).

18,000 gal ÷ 6 hours = 3000 gal (11,333 L) per hour

3000 gph ÷ 60 min = 50 gpm (188 lpm)

Therefore, you need a pump capable of delivering a flow rate of 50 gpm (188 lpm) under the TDH of the system. The manufacturer's pump curves described previously will tell us which pump can do this (Fig. 4-10).

METHODS OF CALCULATING TDH

As I mentioned earlier, unless you are the pool or spa builder, you don't know exactly what plumbing is included in the system, so TDH is an educated guess at best. Here, however, are the three methods for calculating TDH.

Method 1: Exact Values: If you have the exact specifications of the pool as built or as proposed, measure all the pipe from the pool, through the equipment, and back to the pool. Add the equivalent feet of pipe for all the fittings. Add the feet of head at the desired flow rate for the filter, heater, and any other components to arrive at the TDH for the system.

Method 2: Estimated Values

1. Suction-side head. Assume 2 feet (60 centimeters) of head for each 10 feet (3 meters) the equipment is away from the pool.

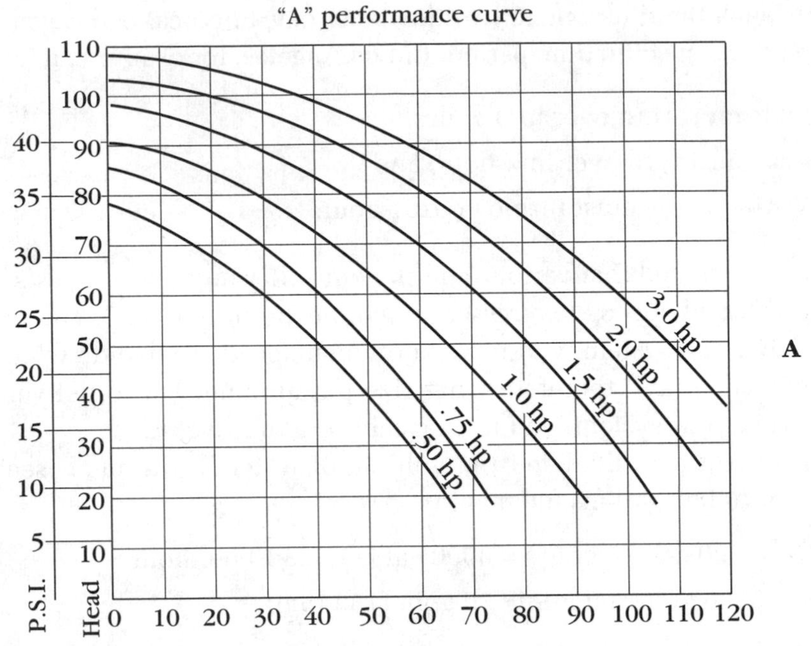

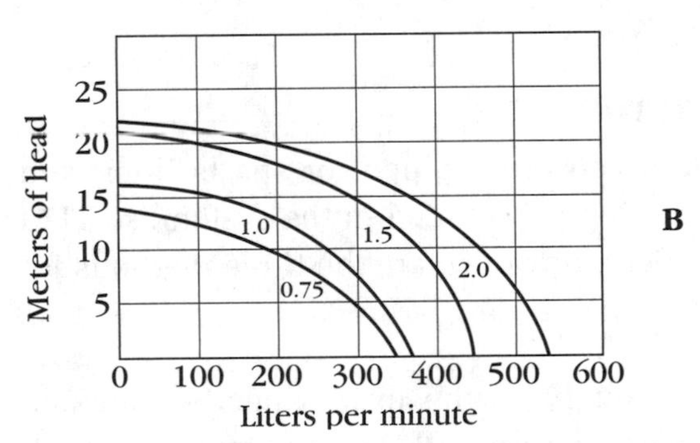

FIGURE 4-10 Actual pump performance curve: (A) U.S. system, (B) metric system. *A: Aqua-Flo, Inc. B: Hurlcon Pty., Ltd.*

2. Discharge-side head. Estimate how many feet or meters of pipe are in the system back to the pool. Double that estimate to allow for fittings.

3. Using the table of friction loss that you picked up at the supply store, calculate the feet or meters of head, calculate the feet or meters of head for the total amount of pipe on the discharge side.

4. Equipment head. Consult manufacturer's tables and charts for the desired flow rate (in the example, 50 gpm or 188 lpm).
5. Add these three parts together to get the TDH.

Let's try a simple example. Let's say our equipment is 30 feet (9 meters) from the pool [at 2 feet of head per 10 feet of distance (60 centimeters per 3 meters), that makes 6 feet (1.8 meters) of head]. There is about 60 feet of 1½-inch plumbing (18 meters of 40-millimeter pipe) between the equipment and the run back to the pool (doubled is 120 feet or 36 meters). The table says 13.5 feet of head per 100 feet of pipe equals 1.2 × 13.5 or 16.2 feet (4.9 meters) of head. The filter manufacturer says our sample filter has 7 feet (2.1 meters) of head; the heater manufacturer says 15 feet (4.6 meters) of head. Therefore

- Suction-side estimate: 6.0 feet (1.8 meters)
- Discharge-side estimate: 16.2 feet (4.9 meters)
- Main drain and skimmer estimate 5.0 feet (1.5 meters)
- Filter: 7.0 feet (2.1 meters)
- Heater: 15.0 feet (4.6 meters)

for a total dynamic head of 49.2 feet or 14.9 meters.

Now you can consult various manufacturers' pump charts to decide which pump will deliver the desired 50 gpm at 149.2 feet of TDH or 188 lpm at 14.9 meters of head.

Method 3: Measured Values: An easier and more accurate way to estimate all of this, if the existing pump is operating, is to measure the vacuum on the suction side of the pump and the pressure on the discharge side. Plumb a vacuum gauge on the pipe entering the pump. It measures inches of mercury or millibars. Every 1 inch of mercury equals 1.13 feet of head (every 1 millibar = 1 centimeter of head). Plumb a pressure gauge on the pipe coming out of the pump. It measures pounds per square inch (psi) or millibars per square centimeter. Every 1 psi of pressure equals 2.31 feet of head (every 1 millibar = 1 centimeter of head).

Multiply the gauges out accordingly and the sum of the two gives you the TDH of the system. This might sound like work, plumbing in two separate gauges, but it really isn't, and it gives you the most accurate TDH calculation because it takes into account the dirty filter, the limed-up heater, all the unseen plumbing . . . everything. It also allows you to keep an eye on the TDH in the system at any time and more easily troubleshoot poor performance in the equipment.

Sizing

So now you've chosen a pump. Is bigger better? Well, yes, because as I mentioned, TDH estimating is more art than science and it changes every minute as the system gets dirty or clogged. So you do want a pump that offers at least a little more capacity than absolutely required.

The only caveat here is that running water not only encounters friction created by pipes and equipment, but the water itself is creating friction. This friction will strip copper from pipes and heater components causing all kinds of havoc (see the chemistry chapter), damages filter grids, and makes diatomaceous earth or sand inefficient (see the filter chapter).

Because of this, most building codes set maximum flow rates of 8 feet (2.5 meters) per second through copper pipe and 10 feet (3 meters) per second through PVC. Since heaters all use copper heat exchangers, use 8 feet per second even if the plumbing is PVC. Los Angeles County, for example, allows a maximum flow rate of 8 feet per second on suction pipes of any type. What is feet per second in terms of gallons per minute? Refer to the standard friction loss chart at the supply store to learn that (rounded to the nearest tenth):

- 50 gpm in 1½-inch pipe (189 lpm in 40-millimeter pipe) = 7.9 feet per second (2.4 meters per second)
- 50 gpm in 2-inch pipe (189 lpm in 50-millimeter pipe) = 4.8 feet per second (1.5 meters per second)
- 60 gpm in 1½-inch pipe (227 lpm in 40-millimeter pipe) = 9.5 feet per second (2.9 meters per second)
- 60 gpm in 2-inch pipe (227 lpm in 50-millimeter pipe) = 5.7 feet per second (1.7 meters per second)

By the way, there are a few exceptions to the rules. Los Angeles County requires pumps to deliver the desired gallons per minute at 60 feet (18 meters) of head. When sizing pumps, you must assume at least 60 feet of head regardless of the actual calculations. In filters, on the other hand, you must use the actual head as measured—go figure.

Altitude also affects these calculations. Over 3300 feet (1000 meters) above sea level a motor runs hotter, so you will want to upgrade to the next horsepower.

Maintenance and Repairs

Since the pump and motor are the heart of the system, if they fail or don't perform efficiently, the abilities of the other components won't much matter. Keeping the motor in good working order is a matter of keeping it dry and cool. The best detection tool for motor problems is your ears. Laboring motors or those with bad bearings will let you know.

Keeping the pump in good order is also a matter of sight. Seeing leaks tips you that the pump needs attention. If the motor needs to stay dry, but problems with pumps most often result in leaks, the potential for pump and motor breakdown is high. Therefore, keeping an eye and ear on your pump and motor will pay dividends in a pool or spa that keeps running.

The basic repairs and maintenance of the pump and motor unit, starting from the front, the first place the water encounters, are discussed in the following paragraphs.

Strainer Pots

RATING: EASY

Clean out the strainer basket often. Even small amounts of hair or debris can clog the fine mesh of the basket and substantially reduce flow. To be honest, this job is a pain in the valve seat. You have to shut down the system, struggle with tight cover bolts or clamps, clean out hair and filth from the basket, put the basket back, find a water source to fill the pot so the pump will reprime easily, check the

TRICKS OF THE TRADE: PUMP AND MOTOR HEALTH CHECKLIST

Look

- **Motor dry**
- **Vents free of leaves or other debris**
- **No pump leaks**
- **Strainer pot clean**

Listen

- **Steady, normal hum**
- **No laboring, cavitating, or grinding noises**

Feel

- **Motor warm, but not hot**
- **No major vibration**

TOOLS OF THE TRADE: PUMPS AND MOTORS

- Flat-blade screwdriver
- Phillips screwdriver
- Allen-head wrench set
- Open end/box wrench set
- Hacksaw
- 5⁄16-inch (8-millimeter) nut driver
- Impeller wrenches
- Teflon tape
- Silicone lube
- Needle-nose pliers
- Hammer
- Seal driver
- Emery cloth or fine sandpaper
- Impeller gauge
- Tap and die set

O-ring, replace the cover, tighten the bolts or clamps, restart the system, and most often, reprime . . . whew!

But this, along with keeping a clean skimmer basket, are the two most simple and important elements to keep a pool clean and the other components working. If the water can't flow adequately, it can't filter or heat adequately either. It will turn cloudy, allow algae growth, and make vacuuming difficult.

The only other problems you might encounter at the strainer pot are broken baskets or a crack in the pot itself. If the basket is cracked it will soon break, so replace it. If allowed to operate with a hole in it, the basket will permit large debris and hair to clog the impeller or the plumbing between the equipment components.

Cracks might develop in the pot itself, especially if you live where it gets cold enough to freeze the water in the pot. Again, the only remedy is replacement. Follow the directions described in the following paragraphs for changing a gasket, because it requires the same disassembly and assembly techniques.

Gaskets and O-Rings

Most problems occur in strainer pots when the pump is operated dry. The air heats in the case as the impeller turns without water to cool it. The strainer basket will melt; the pot cover, if plastic, will warp; and the O-ring will melt or deform. Usually, replacement of the overheated parts solves the problem.

GASKETS

RATING: ADVANCED

When gaskets leak, or in extreme cases, if the strainer pot itself must be replaced (Fig. 4-3, items 15 and 16, and Fig. 4-4, items 15 and 19), the replacement process is the same. Remove the strainer pot [take out the four bolts, usually using a ½- or 9⁄16-inch (13- or 14-millimeter) box wrench]. Clean out the old gasket thoroughly. Failure to do this will leave gaps in the new gasket that will eventually leak. Reassemble the new gasket and strainer pot the same way the old one came off. Tighten the bolts evenly (so the new gasket compresses evenly) by gently securing one bolt, then the one opposite, then the last two. Continue tightening in this crisscross pattern until each bolt is hand tight. When dealing with plastic pumps, do not overtighten because the bolt will crack the pump components or strip out the female side.

Sometimes the bolts are designed to go through the opening in the pot and volute and are tightened with a nut and lock washer on the other side. Still, do not overtighten, because you will crack the pump components. The key to this simple procedure, as with virtually all other mechanical repair, is to carefully observe how the item comes apart. It will go back together the same way.

O-RINGS

RATING: EASY

When removing and replacing the strainer pot cover, be sure the O-ring and the top of the strainer pot are clean, because debris can cause gaps

TRICKS OF THE TRADE: O-RING EMERGENCIES

- **If no replacement is available, try turning the O-ring over. Sometimes the rubber is more flexible on the side facing the cover. Be careful to remove the O-ring gently. Too much stress will cause the rubber to stretch, making it too large to return to the groove in the cover.**
- **If the O-ring stretches, try soaking it in ice water for a few minutes to shrink it.**
- **Coat the O-ring liberally with silicone lube. This can take up some slack and complete the seal if the O-ring is not too worn out.**
- **Another emergency trick is to put Teflon tape around the O-ring to give it more bulk and make it seal. If you use this trick, be sure to wind the tape evenly and tightly around the O-ring, so loose or excess tape does not cause an even worse seal. If the O-ring has actually broken, it will almost always leak at that spot; however, I have used the Teflon tape trick successfully in these cases for a temporary repair when a new O-ring was not immediately available.**

in the seal. Sometimes these O-rings become too compressed or dried out and brittle and cannot seal the cover to the pot. In this case, replace the O-ring.

Changing a Seal

RATING: ADVANCED

As you have seen, all pumps have seals to prevent water from leaking out along the motor shaft. When these wear out (normally, or from overheating if the pump runs dry) they are easy to replace. The steps are shown in Fig. 4-11, based on the pump illustrated in Fig. 4-3. Before you start, turn off the electricity to the motor at the breaker.

1. **Unbolt** Figure 4-3 (item 8) shows a two-part seal. To access this seal for replacement, remove the four bolts that hold the pump halves together (item 5) (it is not necessary to remove the entire pump from the plumbing system).
2. **Disassemble Pump** Grasp the motor and pull it and the bracket away from the volute. Wiggle it slightly from side to side as you pull back to help break this joint. Do not wrestle with the equipment because you might bend the shaft—just be persistent.
3. **Remove Impeller** Take your pliers or ⅞-inch (22-millimeter) box wrench and hold the shaft extender to prevent it from turning. Unscrew the impeller (it unscrews counterclockwise as you face it, just like a bottle cap) from the shaft extender using an impeller wrench. You can also wrap a rag over the face of the impeller so you don't cut yourself and twist it off by hand.

   As a last resort, hold a large screwdriver against the impeller and tap it gently with a hammer. Use care not to damage the impeller. Use even more care that the screwdriver doesn't slip and damage you.
4. **Remove Bracket** Remove the four bolts (item 6) that hold the bracket on the motor. Use a hammer to gently tap the bracket away from the motor if needed.
5. **Remove Old Seal** Remove both halves of the old seal. Note how each half is installed so you get the new one back in the same way. One half is in the back of the impeller and is easily popped out with a flat-blade screwdriver. The other half is in the seal plate and motor bracket unit.

(Text continues on p. 130.)

FIGURE 4-11A Seal replacement. Step 1.

FIGURE 4-11B Step 2.

FIGURE 4-11C Step 3.

FIGURE 4-11D Step 4.

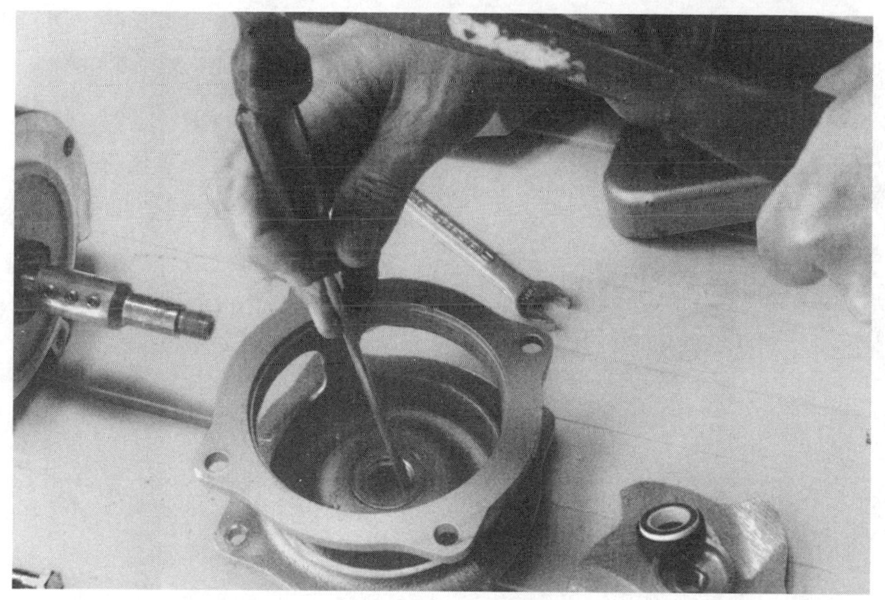

FIGURE 4-11E Step 5.

FIGURE 4-11F Step 6.

FIGURE 4-11F *(Continued)*

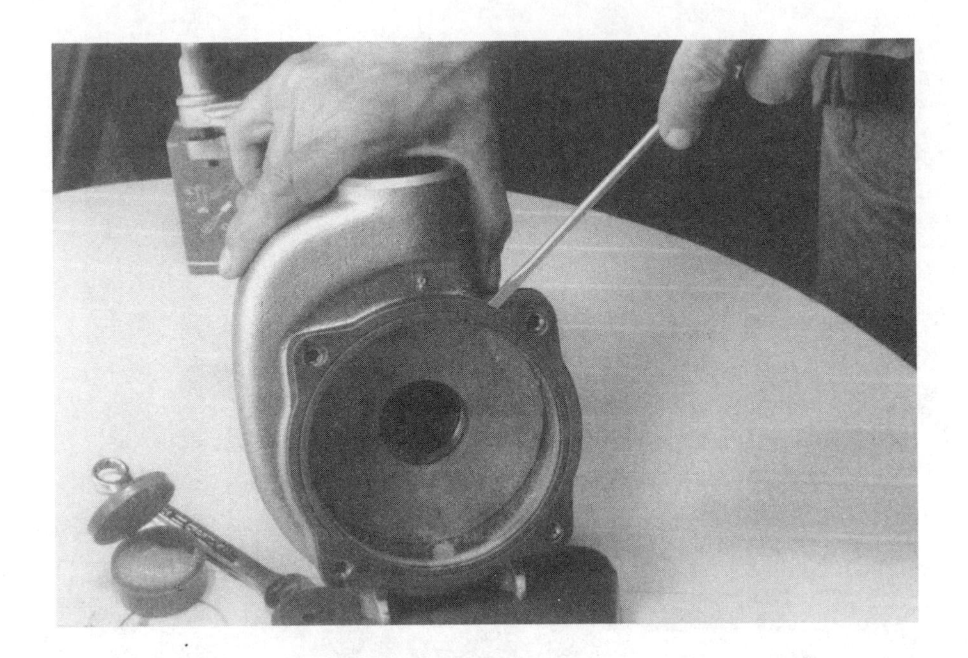

FIGURE 4-11G Step 7.

FIGURE 4-11H Step 8.

Lay the bracket on your workbench with the seal on the bottom. You will see the back of the seal through the hole in the seal plate. Use that trusty flat-blade screwdriver once again—put the tip on the back of the seal and tap it with a hammer. It will pop out easily.

6. **Install New Seal** First, look up your pump in the manufacturer's literature or supply house catalog to determine what model seal you need. Failing that, you can take the old one to the supply house so they can identify it for you. There are only three commonly used seals and only another six or so less common types used in pool and spa work. Clean out the seal plate and impeller where you have just removed the old seal. Use an emery cloth or a small wire brush and water. Dry each area and apply a small amount of silicone lubricant to help the new seal slide into place. Install each half of the seal the same way you removed the old one—white ceramic of one half facing the glazed carbon ridge of the other half.

   To ensure that the half that fits into the seal plate seats evenly, make a "seal driver" (shown in Fig. 4-11, step 6). A 1-inch (25-millimeter) PVC slip coupling, closed at one end with a 1-inch PVC plug, creates a unit that slips over the seal and allows you to tap it into place uniformly.

7. **Gaskets** (This information also applies to the gasket found on many pumps between the volute and strainer pot.) When you break apart a pump, the old gasket usually won't reseal (item 7). Clean all of the old gasket off of the seal plate and volute. Scrape it clean if needed with your trusty flat-blade screwdriver. Now reassemble the pump the same way you took it apart, placing a new gasket between the pump halves.

8. **Reassemble** Put the components back together the way they came apart. Notice that the setscrews for the shaft extender line up with a channel on the motor shaft. After you have resecured the impeller, check that it does not rub against the face of the seal plate (or extend so far away from it that the face of the impeller will rub against the inside of the volute). If it needs adjustment, loosen the three allen-head setscrews, position the impeller properly, then retighten the screws. Impeller gauges are made for some pump models to assist with this process. Be sure the impeller is tightly screwed onto the shaft extender. If not, when the motor is turned

on it will tighten and as it screws down it may jam against the seal plate.

9. **Check Your Work** Fire up the pump and watch for leaks. A new seal, improperly installed, might not leak for several minutes, so let the pump run awhile before deciding the job is done. A fresh paper gasket might leak for a few minutes until it becomes wet and swells to fill all the gaps, but it should stop leaking after a short time. If your job does leak, take it apart and go over each step again, making sure the seal halves are seated all the way and that there is no corrosion or debris left in the impeller or seal plate that might prevent the new seal from seating completely.

   If the pump is old and corroded, you might need two gaskets to seal the uneven gaps between the volute and seal plate. Soaking the gasket in warm water before reassembly sometimes helps as well. The best gasket is a rubber one, but even with these you might need two to seal an old, warped pump.

In Fig. 4-4, a Sta-Rite pump, you follow the same steps but the parts are slightly different. The clamp (item 12) is removed to disassemble the pump halves, and you must remove the diffuser (item 10) to get to the impeller.

To remove the impeller (Fig. 4-12A) you can grip it with your hand or a special wrench (Fig. 4-12C) and twist it off, but the trick with these units is to stop the shaft from spinning as you twist off the impeller. On some motors you can access the rear end of the shaft, which has been flattened on two sides to accommodate a 7⁄16-inch (11-millimeter) open-end wrench. Another way to secure the shaft is with a screwdriver. The proper method for securing the shaft will be determined by the configuration of the motor, which varies from one manufacturer or model (or year) to the next.

Instead of a gasket, the Sta-Rite pump uses an O-ring. Clean this and lubricate it with silicone before reassembly. If it has stretched and it seems like there is too much O-ring for the channel in the volute, try soaking the gasket in ice water for a few minutes to make it shrink a bit. Other than these few differences, the Sta-Rite seal replacement is the same as any other pump.

Figure 4-12B shows an impact tool for removing a stubborn shaft extender (top) and removing a stubborn AH-8 impeller (bottom).

FIGURE 4-12A Impeller removal.

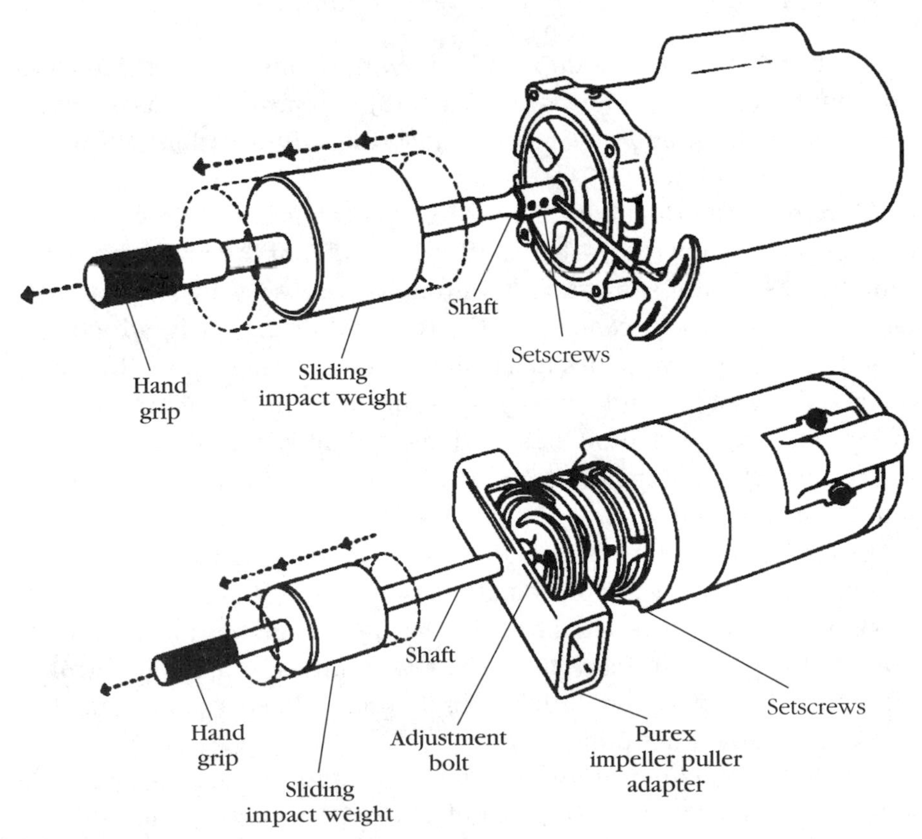

FIGURE 4-12B *Pool Tool Co., Ventura, Calif.*

One last problem. Some pumps use a plastic impeller with a housing that holds half the seal in place. If the pump has run dry and overheated the pot, this housing might be warped and the seal will not fit tightly. The only solution is to replace the impeller. This is a common problem with automatic cleaner pumps, which are not self-priming. These often run dry, for reasons discussed in a later chapter, warping the housing.

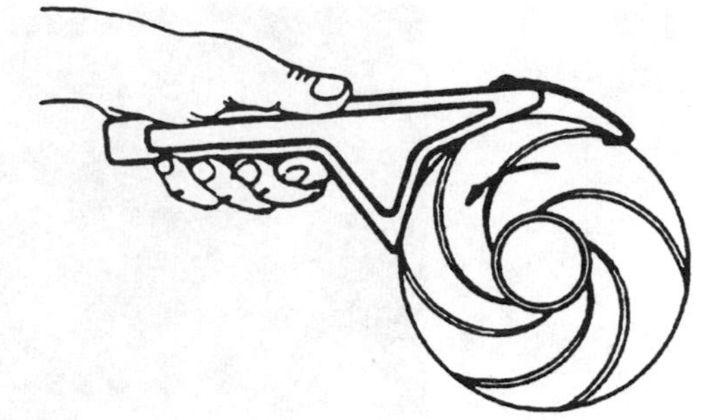

FIGURE 4-12C Closed-face impeller wrench. *Pool Tool Co., Ventura, Calif.*

If you master these few concepts, any pump seal will be easy for you.

Pump and/or Motor Removal and Reinstallation

RATING: EASY

Sometimes it is necessary to remove an entire pump and motor unit to take it apart or complete a repair. If the pump is damaged beyond your ability to repair it, you might want to take the entire unit to a motor repair shop. They can rebuild it as needed, and you can reinstall it at the job site. Your local pool and spa supply house can recommend a rebuilder, or you can consult the phone book.

Generally, to remove the pump and motor as a unit you will need to cut the plumbing on the suction and return side of the pump. Cut the pipe (Fig. 4-13) with enough remaining on each side of the cut to replumb it later. Ideally, a few inches on each side allows you to use a slip coupling to reglue the unit in place later (see the chapter on basic plumbing).

When installing or reinstalling the plumbing between the pump and filter take a look at the equipment area. Keep bends and turns to a minimum. Remember, each turn creates head (resistance) in the system. Also, don't locate the pump

TRICKS OF THE TRADE: SILICONE

- **Remember to use only nonhardening silicone lube on all pool and spa work. Vaseline or other lubricants are made of petroleum, which eats away some plastics and papers.**
- **Get silicone lube at your supply house or any scuba diving shop—it is used for scuba equipment repairs for the same reasons.**
- **Before reassembly, coat pump halves with silicone where they contact gaskets or O-rings. This helps to fill any tiny gaps that might still be present, especially on older pumps.**

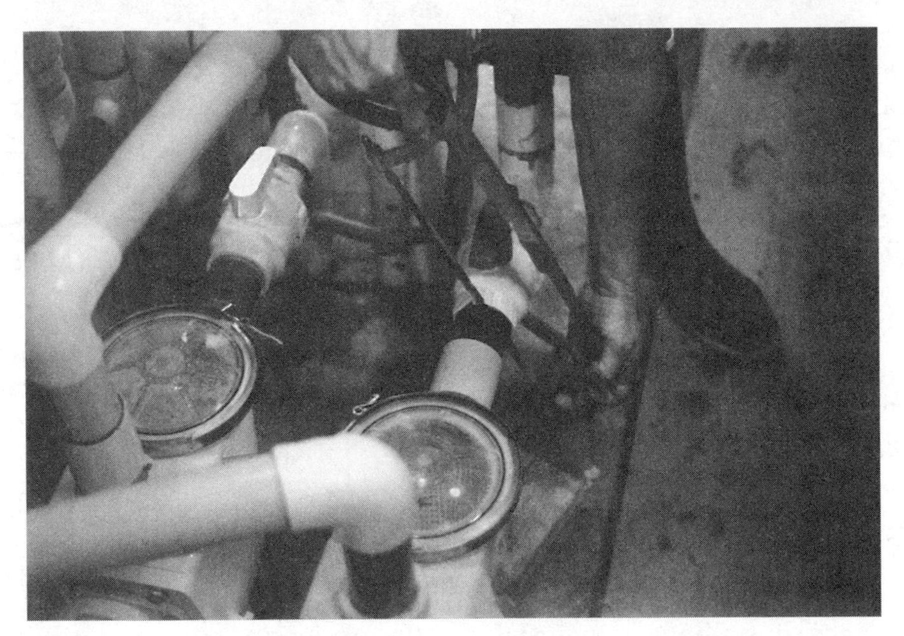

FIGURE 4-13 Pump and motor removal.

close to the base of the filter. When you open the filter for cleaning, water is sure to flood the motor. Lastly, try to keep motors at least 6 inches off the ground. The bracket of the pump does this in part, but heavy rains or flooding from broken pipes and filter cleanings can flood the motor if it is too close to the ground.

> **SAFETY FIRST**
>
> **Tape off the ends of the wires, even though the breaker is turned off, and put tape over the breaker switch itself. Leave a note on the breaker box to yourself, family members, or the customer to be sure no one accidentally turns the breaker back on while the pump and motor is away for service.**

When removing a pump and motor unit you have the opportunity to reinstall it on a raised surface for a greater margin of error. If you do this by adding a mounting block, don't use wood (it will deteriorate over time). A thick rubber mounting pad or large brick will work, but be sure to follow local building codes regarding bolting these to the deck and bolting the pump to the mounting material. All pumps should be bolted to the deck, but you will find many that are not. This is a good time to remedy such oversights.

The other component of pump and motor removal is the electrical connection. You have already turned off the breaker (right?). Now remove the access cover (Fig. 4-14, step 1) to the switch plate area of the motor, near the hole where the conduit enters the motor. Remove the three wires inside the motor and unscrew the conduit connector (Fig. 4-14, step 2) from the motor housing. Now you can pull the conduit and wiring away from the motor and the entire pump and motor should be free.

There might be an additional bonding wire (an insulated or bare copper wire that bonds/grounds all of the equipment together to a grounding system). This is easily removed by loosening the screw or clamp that holds it in place.

New Installation

RATING: ADVANCED

If you're lucky enough to install the pump and motor for the first time, do it right. I can't tell you how many installations look like they were done by someone who hated service technicians with everything plumbed together so tightly that later repairs were impossible knuckle-busters. You already know the plumbing and electrical techniques as discussed previously, so here are some tips on preparing for new installations.

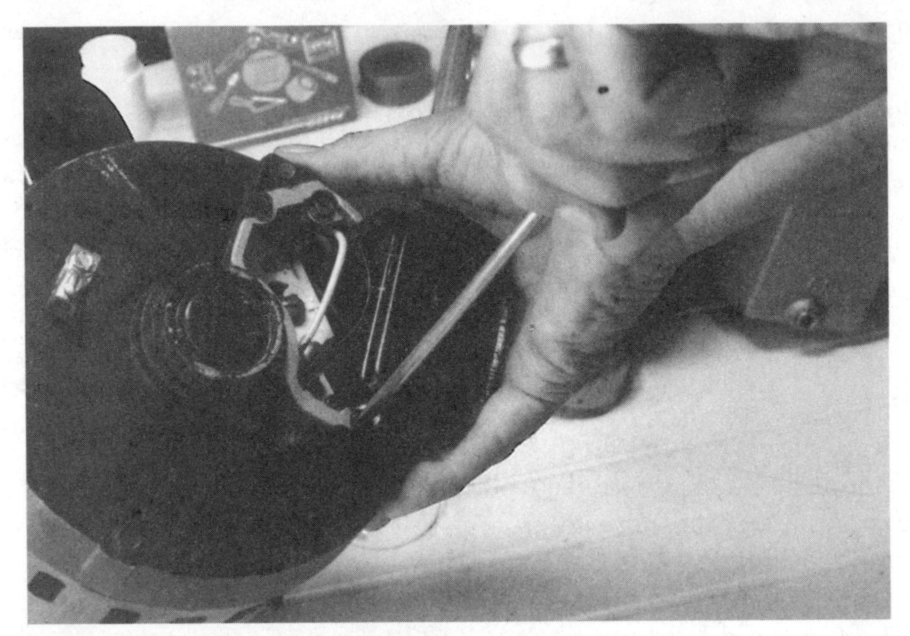

FIGURE 4-14A Motor electrical connections. Step 1.

FIGURE 4-14B Step 2.

Position the pump as close to the body of water and as near to water level as possible so it doesn't have to work so hard. Mount the unit on a solid, vibration-free base, not wood (which rots). Make sure there is adequate drainage in the area so that when it rains or if a pipe breaks the motor won't be drowned. Bolt or strap down the pump as required by local code.

Plumb in both suction and return lines with as few twists and bends as possible, to minimize head. A gate valve on both sides is advisable to isolate the pump when cleaning other components. A check valve is essential if the unit is well above water level. Plumb the unit far enough away from the filter that it won't get soaked when you take the filter apart.

Replacing a Pump or Motor

RATING: ADVANCED

Having learned how to remove and break down a pump and motor in the previous sections of this chapter, replacing any of the components is simply a matter of disassembling the pump down to the component that needs replacement, getting a replacement part, and reassembling the unit. Of course, if the entire pump and motor is to be replaced, you purchase the replacement as a unit and plumb it in as previously described.

Sometimes the motor will trip the circuit breaker when you try to start it. If this happens it is usually because there is something wrong with the motor; however, it could be a bad breaker or one that is simply undersized for the job and has finally worn out. The section on motor troubleshooting (later) and the chapter on basic electricity deal with checking the wiring, circuits, and breakers. However, to replace the motor depicted in Fig. 4-3, you follow the procedure of Fig. 4-15 as follows:

1. **Electrical** You can access the electrical connections through the switchplate cover in the front end bell.
2. **Disassembly** Remove the motor from the shaft extender by removing the allen-head setscrews and pulling the extender off the motor shaft. Sometimes this might need persuasion. Use your large flat-blade screwdriver to pry the extender away from the motor body. Sometimes corrosion will eat away at the setscrews and extender—if it is too tough to remove, replace it (it's only a few bucks).
3. **Preparation** Before sliding the shaft extender on the new motor, clean the motor shaft with a fine emery cloth such as you might

FIGURE 4-15 Replacing a motor.

have in your copper pipe solder kit. Apply a light coat of silicone lube to the shaft—no, the silicone won't make the extender slip loosely on the shaft. When you put the extender on the motor shaft, the setscrews go into a groove that runs along the shaft. This groove allows the screws to grip and not slide around the shaft.

4. **Reassembly** Secure the shaft extender by tightening the allen-head setscrews, but before doing so, be sure the impeller is properly positioned. The spring in the seal will push the impeller up against the inside of the volute when you loosen the setscrews, so now you must pull it back before securing the shaft extender. Insert your flat-blade screwdriver into the neck of the shaft extender (where it passes into the seal) and gently pry it back toward the motor, exerting pressure against the spring in the seal; then tighten the setscrews. If you pry the extender too far back, the impeller will rub against the seal plate. Pry the extender all the way back, then let up ⅛ inch (3 millimeters) or so to obtain the right setting. When you restart the pump, if you hear any scraping or if the impeller won't turn, you'll need to repeat this adjustment step. Reconnect the electrical connections.

Replacement of the motor for the Sta-Rite unit in Fig. 4-4 is the same process, but there is no shaft extender or adjustment to make when reassembling. All other pump and motor designs are variations on these themes and will be obvious once you have mastered these few steps.

Troubleshooting Motors

RATING: EASY

The first and most common motor problem is water. Motors get soaked in heavy rain, when you take the lid off the filter for cleaning, when a pipe breaks, or when you look at it wrong. In all cases, dry the motor and give it 24 hours to air dry before starting it up—moisture on the windings will short them out and short out your warranty as well. The basic problems beyond this are as follows.

MOTOR WON'T START

Check the electrical supply and breaker panel, and look for any loose connection of the wires to the motor. Sometimes one of the electrical supply wires connected to the motor switch plate becomes dirty. Dirt creates resistance that creates heat which ultimately melts the wire, breaking the connection. Similarly, if the supply wire is undersized for the load, it will overheat and melt. Check the proper wire size in Fig. 4-22, replacing the supply wiring if needed. Otherwise, clean dirty switch plate terminals and reconnect the wiring.

MOTOR HUMS BUT WON'T RUN

Either the capacitor is bad or the impeller is jammed. Spin the shaft. If it won't turn freely, open the pump and clear the obstruction. If it does spin, check the capacitor.

The best way to check a capacitor is to replace it with a new one. Check the capacitor (see Fig. 4-7) for white residue or liquid discharge. Either is a symptom of a bad capacitor. There is a screw or clamp bracket that holds it in place and two wires, connected to the capacitor with bayonet-type clips. When you see it, you'll realize that not much instruction is needed.

All of this assumes your internal motor switch connection is set for 120 volts, for incoming power of 120 volts, or set for 220 volts if the incoming is 220 volts. Check the wiring diagram and power supply.

A more rare condition that might cause the motor to hum but not run is that your line voltage is not what it should be. Your 120 volts, for example, might be coming in at only 100 volts because of a faulty breaker or a supply problem from your power company. Use your multimeter to test the actual voltage supply at the motor.

THE BREAKER TRIPS

Disconnect the motor and reset the breaker. Turn the motor switch (or time clock on switch) back on and if it trips again, the problem is either a bad breaker or, more likely, bad wiring between the breaker and motor. Be very careful with this test. Switching the power back on with no appliance connected means you are now dealing with bare, live wires. Be sure no one is touching them and that they are not touching the water, each other, or anything else.

If the breaker does not trip when conducting this little experiment, the motor is bad. This usually means there is a dead short in the windings and the motor needs to be replaced. Water can cause this.

TRICKS OF THE TRADE: NOISE CHECKLIST

Security

- **Is the pump properly secured to the deck or mounting block and is the mounting block secure?**
- **Are check valves rattling?**
- **Are pipes loose and vibrating? Grab sections of exposed pipe and see if the noise changes.**

Cavitation

- **Are suction and return line valves fully open or open too much?**
- **Is the suction-type automatic pool cleaner starving the pump?**
- **Undersized suction plumbing? Refer to the hydraulics section.**

Air

- **Is the pump strainer basket clean and the lid tightly fastened?**
- **Is the skimmer clogged or the water level low?**

Other troublemakers

- **Is the equipment located in a sound-magnifying environment, such as large concrete pad and masonry walls? Consider a vented "doghouse" cover.**
- **Is the heater "whining"? (See heater chapter.)**
- **Is the spa air blower loose or vibrating, or is the discharge restricted, producing a louder sound?**
- **Are loose filter grids rattling inside the filter canister?**

At Charles Bronson's house, I found a large lizard had crawled into the motor housing through the air vents and when the timer turned the system on, the poor reptile became the short across the winding wires, burning out the motor.

LOUD NOISES OR VIBRATIONS

This is most often caused by worn-out bearings. Take the pump apart and remove the load (impeller and water). If the motor still runs loud or vibrates, it is the bearings. Take it to a motor shop, or better still, replace the motor (unless the motor is relatively new or is still under warranty). This problem can also be caused by a bent shaft, although that is not common.

Not all noise is caused by the motor. Track down noises by a process of elimination, experimenting with various pieces of equipment (such as automatic pool cleaners, booster motors, automated valves, spa blowers, heaters) all turned off, then turned on one at a time. See the sidebar on p. 141 for a list that will help find other culprits.

Priming the Pump

Sometimes the most difficult step is getting water moving through the pump. *Priming* means getting water started, creating a vacuum so more will follow.

BASIC PRIMING

RATING: EASY

Let's go through the steps to prime most pools and spas.

1. **Water Level** Before starting a pump that you have had apart, always make sure there is enough water in the pool or spa to supply the pump. In taking equipment apart, water is usually lost in the process and there might not be enough to fill the skimmer. I have also encountered pools that seem to have enough water, but will not prime unless filled to the very top of the skimmer. Sometimes that extra inch or two is enough to change the hydraulics of the system and get it working. Factors such as distance of equipment from the pool and height above the pool also enter into the equation.
2. **Check the Water's Path** Often, priming problems are not related to the pump, but to some obstruction. Check the main drain and the skimmer throat for leaves, debris, or other obstructions. Next, open the strainer pot lid, remove the basket, and make sure there are no obstructions or clogs in the impeller. Last, make sure that once the pump is primed it has somewhere to deliver the water. In other words, be sure all valves are open and that there are no other restrictions in the plumbing or equipment after the pump. If all this checks out, proceed.

3. **Fill the Pump** Always fill the strainer pot with water and replace the lid tightly so air cannot leak in. Keep adding water until the pot overflows so you fill the pipe as well as the pot. Sometimes the pump is installed above the pool water level so you will never fill the pipe (unless a check valve is in the line as well). Just fill what you can and close the lid.
4. **Start Up** Start the motor and open the air relief valve on top of the filter. Give the pump up to two minutes to catch. Carroll O'Connor's pool in Malibu takes 4 minutes and 30 seconds to catch prime; you can set your watch by it (this is because it is about 4 feet higher than the pool level). Most pumps will catch prime sooner and you don't want to overheat a dry-running pump.

   Sometimes repeating this procedure two or three times will get the prime going. If there is a check valve in a long run of pipe, each successive filling of the pot pulls more and more water from the pool, which is held there each time by the check valve. Also, if it is warm outside, the air in the pipe might expand and create an airlock. The repeated procedures might finally dislodge the air.

THE BLOW BAG METHOD

RATING: ADVANCED

When basic priming fails, try a drain flush bag, also called a *blow bag* (Fig. 4-16). The drain flush is a canvas or rubber tube that screws onto the end of your garden hose. Slip this into the skimmer hole that feeds the pump and turn on the hose.

The water pressure makes the bag expand and seal the skimmer hole so the water from the hose cannot escape and must feed the pump. After running the hose a minute or two, turn on the pump. When air and water are visible returning to the pool, pull the drain flush bag out quickly, while the pump is running, so pool water will promptly replace the hose water.

This method is not effective if the skimmer has only one hole in the bottom. Remember from Chap. 2 that this hole is connected not only to the pump but also the main drain. The forced water from your drain flush bag will take the line of least resistance and flood through the main drain rather than up to the pump. In the two-hole skimmer, the

FIGURE 4-16 Using a drain flush bag.

hole farthest from the pool usually is plumbed directly to the pump. Your drain flush bag in this hole will give good results.

FILTER FILLING METHOD

RATING: ADVANCED

Another method is what I call *filter filling*. Open the strainer pot, turn on the motor, and feed the pot with a garden hose. Open the filter air relief valve and keep this going until the filter can is full (water will spit out of the air relief valve).

Close the air relief valve, turn off the motor and garden hose, and quickly close the strainer pot. Open the air relief valve. The filter water will flood back into the pump and the pipe that feeds the pump from the pool. When you think these are full of water, turn the motor back on. The pump should now prime.

DRASTIC MEASURES

RATING: PRO

Sometimes none of this works. Dick Clark's pool includes equipment that is about 10 feet (3 meters) above the pool water level and 40 feet

(12 meters) away. I have had to put a drain flush in the skimmer and a rubber plug in the main drain to get it primed. Obviously this meant diving to the main drain to plug it off. I got so tired of this I found a better way.

I cut the pipe from the pool to the pump, a foot (30 centimeters) in front of the pump. I plumbed in a T to rejoin the pipe. In the third opening of the T, I plumbed in a garden hose bib (faucet). When I need to prime the pump, I attach a garden hose to this faucet, turn on the pump and the hose, open the faucet, and the hose water starts a suction that starts pool water up the line. When I see water and air bubbling out of the pool return lines, I close the faucet and turn off the garden hose. In really difficult systems, try this method.

DETECTING AIR LEAKS

RATING: EASY

Okay, so none of this works. The problem might be that the pump is sucking air from somewhere, meaning it will not suck water (which is harder to suck).

Air leaks are usually in strainer pot lid O-rings, or the pot or lid itself has small cracks. The gasket between the pot and the volute might be dried out and leaking. Of course, plumbing leading into the pump might be cracked and leaking air.

If any of these components leak air in, they will also leak water out. When the area around the pump is dry, carefully fill the strainer pot with water and look for leaks out of the pot, volute, fittings, and pipes. Another way is to fill and close the pot, then listen for the sizzling sound of air being sucked in through a crack as the water drains back to the pool.

Sometimes there's just no easy answer—remove the pump and carefully inspect all the components, replace the gaskets and O-rings, and try again.

T-Handles

RATING: ADVANCED

Many pumps employ threaded T-shaped bolts that secure the lid to the strainer pot (Fig. 4-3). Sometimes these corrode and snap off, with part of the bolt in the pot and the other part in your fist.

If part of the broken bolt extends above or below the female part on the pot, try using pliers, especially Vise-Grips, to grasp the broken section and twist it out. You can buy new T-bolts at the supply house. You can also take a flat-blade screwdriver and place it on the broken bolt end, tap it with a hammer to create a slot in the end of the broken piece, and twist out the broken piece like removing a screw.

If this doesn't work, take your tap and die set or electric drill and tap a small hole inside the broken piece, then use your Phillips-head screwdriver to grip inside the hole and twist out the broken piece. If all else fails, remove the pot from the pump and take it to a machine shop to be tapped out and rethreaded.

A hint: If one of these handles breaks off on Friday afternoon, you can clamp that side of the lid on the pot with your Vise-Grips and run the system until Monday when you can get the parts to fix it.

Motor Covers

Protective covers are made to fit over motors. Some are plastic, some metal, some foam rubber. In all cases, they are designed to keep direct sunlight and rain off the motor housing. In fact, the motor is designed to do this itself. You will notice that the motors in the illustrations all have air vents on the underside. The greatest danger to a motor is flooding of the equipment area in heavy rain or when opening a filter, or allowing water from the ground to get up into the motor. This will short out the windings and void any warranty. Believe me, a warranty repair station or motor rebuilder will know if the motor was flooded.

Submersible Pumps and Motors

There are essentially two types of submersible pump and motor combinations that you will encounter.

High-Volume Pump-Out Units

Sometimes you need to drain a pool or spa. Several manufacturers make pump and motor units with long, waterproof electrical cords, that can be completely submerged (shown in Fig. 4-17). The suction side is at the bottom of the pump, as if a regular pump and motor unit were stood on end with the motor on top and the pump on the bottom.

FIGURE 4-17 Typical submersible pump.

The return line is sized to be attached to your vacuum hose. Smaller units are connected to a garden hose to feed water out of the pump.

Low-Volume Pumps and Motors

Fountains and small ponds use small submersible pump and motor units that contain all of the same components as their larger cousins, but range in size from no larger than a fist to about the size of a football. These have waterproof electrical cords so they can be submerged in the body of water.

Both high- and low-volume types contain the same components as the pumps and motors described earlier and are repaired in much the same ways. Submersibles have more crucial and tricky seals and gaskets, however, because leaks in these mean electricity in the body of water that can be fatal to both the motor and you.

I always leave repair of submersibles to a rebuilding shop, except for the wet-end parts, such as impellers and strainer housings, which are easily accessed and do not require breaking the seal that surrounds any electrical component.

Cost of Operation

As with any electrical appliance, you can easily calculate the cost of operation. I have heard many customers tell me their electric bills have gone up significantly and they can only attribute that to the pool or spa motors. Knowing how to calculate operating costs can help answer these questions—usually the customer is wrong.

Electricity is sold by the kilowatt-hour. This is 1000 watts of energy each hour. You know that volts × amps = watts, so you can look at the motor nameplate and see that the motor runs, for example, at 15 amps when supplied with 110-volt service, and 7 amps when supplied by 220-volt service.

Let's say the pump in our example is running on 220-volt service—220 volts × 7 amps = 1540 watts. Looking at an electric bill, you learn that you pay 15 cents per kilowatt-hour. As noted, a kilowatt is 1000 watts, so if you divide 1540 watts by 1000, you get 1.54. That is multiplied by your kilowatt rate (15 cents), equaling 23 cents for every hour you run the appliance. If you run the motor eight hours per day, that means 23 cents × 8 hours = $1.84/day. Over a month, that equals 30 × $1.84 = $55.20/month.

Booster Pumps and Motors for Spas

Pump and motor units that provide only jet action for spas (Fig. 4-18) generally are not equipped with a strainer pot and basket, otherwise they are the same as other units discussed previously. Some units are designed to perform two functions and therefore run at two speeds. They run at high speed (3450 rpm) to provide jet action, but also circulate, filter, and heat the water (and then, of course, would have a strainer pot and basket) at a low speed (1750 rpm).

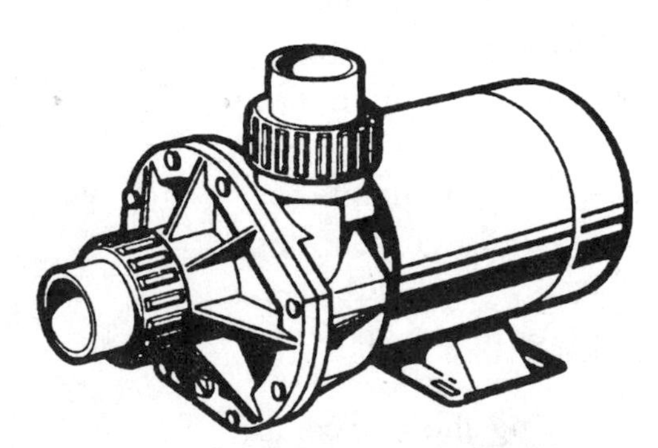

FIGURE 4-18 Spa booster motor (no strainer).

As a general rule of thumb, to operate efficiently, spa jets require 15 gpm (57 lpm) running through each one. Therefore, if you have a system that delivers 60 gpm (227 lpm), you can install up to four jets. As another generality, each jet requires ¼-hp from its pump and motor, so again, that four-jet spa would need at least a 1-hp unit.

Remember, this assumes the pump is doing no other work. If it is pushing water through the filter and heater before getting back to the jets, or if the equipment is more than 20 feet (6 meters) from the spa, then some power will be lost and you will need to calculate more than ¼ hp per jet.

When planning a system or replacing equipment, many codes require that a spa turn over completely two times per hour, so be sure the pump can handle that, especially when the filter gets dirty and head increases.

Basic Electricity

Because electricity powers motors, this is a good time to learn the basics. When troubleshooting short circuits or other specialized electrical problems, an electrician will repair it faster than you can, so call a professional.

Having said that, however, a little knowledge of electricity goes a long way for the do-it-yourselfer.

Electrical Terms

A comprehensive glossary of pool and spa terms appears at the back of this book, but to more easily understand the concepts of electricity, a few definitions are presented here.

- *Amperage* (*amps*) is the term used to describe the actual strength of the electric current. It represents the volume of current passing through a conductor in a given time. Amps = watts ÷ volts.
- *Arc* or *arcing* is the passage of electric current between two points without benefit of a conductor. For example, when a wire with current is located near a metal object, the electricity might arc (pass) between the two.
- *Circuit* is the path through which electricity flows.

- *Conductor* is any substance that carries electric current, such as a wire, metal, or the human body.
- *Current* refers to the rate of flow between two points.
- *Cycle* is a complete turn of alternating current (ac) from negative to positive and back again.
- *Gauge* refers to the size of an electric wire. Heavier loads can be carried on heavier-gauge wires; however, the numbering system of wire gauges works in reverse. A 10-gauge wire, for example, is thicker than a 14-gauge wire.
- *Line* refers to a wire conducting electricity.
- *Load* is an appliance that uses electricity.
- *Volts* is a basic unit of electric current measurement expressing the potential or pressure of the current. Volts = watts ÷ amps.
- *Watt* is a measurement of the power consumption of an appliance. One watt is equal to the volume of one amp delivered at the pressure of one volt. Watts = amps × volts.

Other terms are explained throughout the following text.

Electrical Theory

Figure 4-19 shows a simple circuit created with a battery and a small appliance—the light bulb. Without getting into more scientific detail than is needed, be aware that current flows from the negatively charged side of a battery to the positive, as shown. This diagram represents a *closed circuit*, because there is no interruption in the flow of electricity from the negative terminal, along the wire, through the appliance, and back along the wire to the positive terminal of the battery. If you were to cut the wire at any point and interrupt the flow of current, you would have what is called an *open circuit*.

In a simple 110-volt circuit, current travels along a wire, called the *hot* line, from the electrical panel to the appliance (called the *load*) and back to the panel. The current is then sent through a *neutral* line combined with other such return lines to ground. As long as the wires are insulated (covered with plastic or other nonconductive material) and the appliance is not damaged, the electricity will stay in this path. Electricity will take the line of least resistance, so if it finds a break in the insulation or can flow to a metal casing, it will find the ground by the shortest route, usually through whoever touches that spot.

This unintentional route to the ground without first returning to the electrical panel is called a *ground fault*. As noted, if you are part of the path to the ground, you will be shocked. This can also occur if you touch a hot line and a neutral or ground line, again becoming part of the circuit. These examples are called *short circuits*. To protect against ground faults and short circuits, grounding of appliances keeps you from being grounded, while bonding wires (heavy gauge wires that connect together all appliances and metal surfaces in an area) keep you from being part of the circuit.

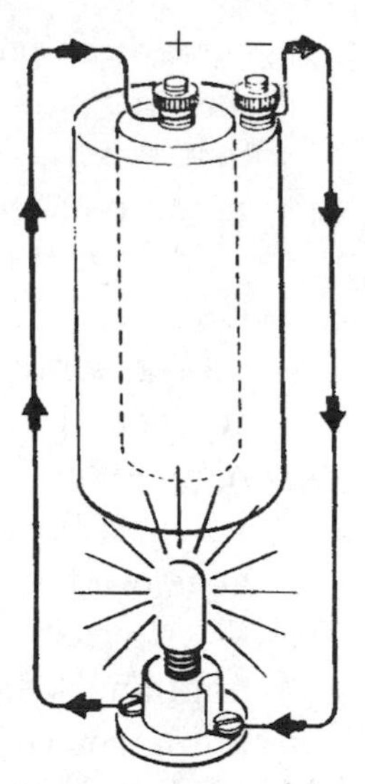

FIGURE 4-19 **A simple electrical circuit (closed).** *U.S. Government Printing Office.*

As noted above, conductors are any substance that allows the free movement of electric current. *Insulators*, on the other hand, are substances that do not conduct. Examples of each are

Conductors: silver, copper, aluminum, brass or bronze, iron or steel

Insulators: dry air, glass, rubber, plastic, ceramic

Before leaving general electrical theory and applying it to the world of pool and spa appliances, I need to briefly describe the type of current used. The battery in Fig. 4-19 provides direct current (dc). Touch the wire to it and it delivers its rated voltage without question. A battery is designed to deliver a certain voltage at all times until it is exhausted. For example, a radio battery of 9 volts will always deliver 9 volts. If the appliance uses very little power, it might draw that 9 volts slowly, say at 1 amp per hour. Another appliance might draw the 9 volts at 2 amps per hour, meaning the battery will be exhausted twice as fast. In these dc battery circuits, all current flows in one direction, from the negative side of the battery toward the positive. Therefore, the polarity of the appliance and battery must agree—the positive terminal of the battery must be connected to the positive side of the appliance. You see an example of this in the fact that batteries can only be inserted into a radio in one direction for the radio to function.

Alternating current (ac) travels in one direction then the other (alternating), so the appliance does not have to be connected to the power source in any special order. Unlike dc voltage, ac can be stepped up or down with a transformer, permitting the transmission of high voltage along municipal power lines that is transformed to lower voltages at each home or business. Because of this inherent versatility, ac is used in virtually all residential and commercial applications.

Alternating current is delivered to the home for consumption by appliances designed to accept it at either 110 volts or 220 volts (there are larger voltages in heavy-duty commercial applications, but those are best left to the electricians). Both designations are averages, since current supply varies slightly and operates most appliances in a range of 108 to 127 volts and 215 to 250 volts. Thus, you will sometimes see voltages expressed for appliances as 110, 115, 120 or 220, 230, 240.

Alternating current is also delivered at a certain rate. As noted, the alternating of the current one way, then the other, creates one complete cycle each time it reverses direction. The speed of that reversal can be controlled and makes a difference to appliances such as CD players or tape recorders that depend on a certain rate. In the United States, power is delivered at 60 cycles per second (60 hertz). In Europe and much of the rest of the world, it is delivered at 50 hertz. That is why you can take a voltage converter on vacation to step the voltage down from 220 to 110, but you can't operate appliances that require a certain cycle timing.

Electrical Panel

The municipal power supply enters the home or business as two (or three if there is heavy equipment use) lines of 110-volts ac and one neutral line in a protected metal box called the electrical panel. Figure 4-20 diagrams the concept of the electrical panel.

The power supply enters the panel and is connected to bars. Circuit breakers are attached to the bars. If the breaker is attached to one phase, it delivers 110 volts to anything that is connected to it. If the breaker is designed to be connected across both phases, it delivers 220 volts. All neutral lines returning to the panel are connected to the neutral bar, which is in turn connected to a ground. In this way, both 110- and 220-volt ac breakers are found in the same panel.

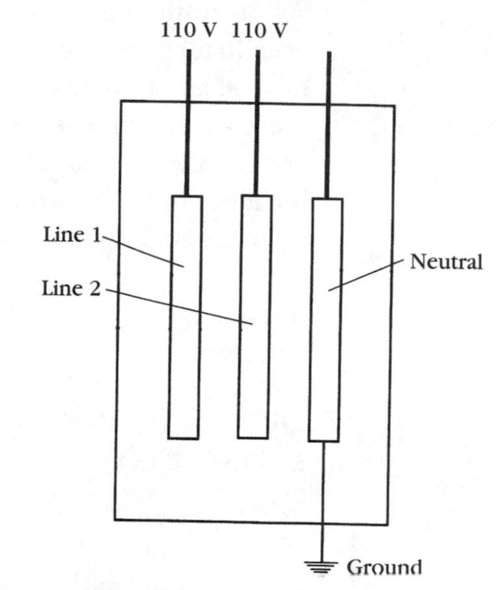

FIGURE 4-20 Diagram of the electrical panel.

Circuit Breakers

The supply lines are generally designed to carry 100 amps for the typical residential user. Each circuit breaker (Fig. 4-21) is designed to carry a specific load and break the circuit open when the load exceeds that value. Typical circuit breakers are 15, 20, 25, 30, and 50 amps, depending on the requirements of the appliances (or probable total of appliances on the same breaker). Wiring attached to the breaker leading to the appliances is sized in accordance with the amperage of the breaker.

When electrical volume exceeds the rating of the breaker, it opens the circuit and disconnects the power supply to the appliance or circuit in question. Such overload might occur as the result of an unintentional ground or short circuit at the appliance (or wiring to it).

Depending on the design of the breaker, resetting is accomplished in one of several ways. Sometimes it is not obvious which breaker has tripped. One style of breaker looks as if it is still on. You need to push the switch fully to off, then back to on to reset it. Another style pops halfway

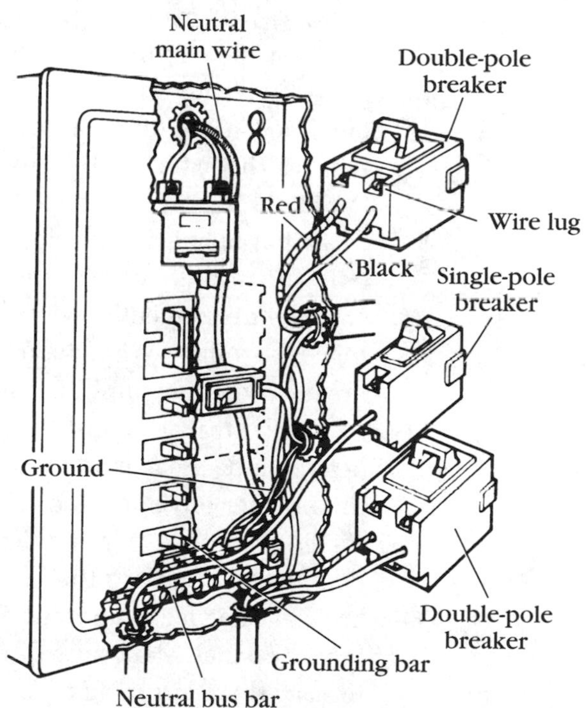

FIGURE 4-21 The electrical panel and circuit breakers. *Creative Homeowner Press, Upper Saddle River, N.J.*

between on and off, again requiring a hard push to off before going back to on. Another has a small window displaying a red flag when the breaker is off. Some of these require waiting up to 30 seconds before the breaker can be reset. Another type is off when a tab pops out and is reset by pushing the tab back in. In short, be aware that a tripped breaker might require some detective work.

It is also important to know that each manufacturer makes electrical panels to accommodate their breakers only. None that I know of are interchangeable, although some generic brands copy the major manufacturers.

TROUBLESHOOTING AND REPLACEMENT

RATING: EASY

When a breaker will not reset, it might mean that the breaker is faulty or the circuit is overloaded (demanding too much current). An overloaded circuit can be the result of an appliance that is faulty, an unintentional ground, or a short circuit in the wiring, or it might be that there are too many appliances on the same circuit (or one that is too large for the circuit).

Troubleshooting is simple. First, check the appliances on the circuit. Does their total amperage exceed the rating of the breaker? If so, remove the extra appliances or wire them to a circuit that can handle the load.

If that is not the problem, disconnect each appliance from the circuit one at a time, resetting the breaker after each disconnection. Be sure the disconnected wires are taped off and no bare wires are touching each other. When you have removed the faulty appliance, the breaker will stay on. You now know which appliance to repair.

If the breaker is still tripping, the problem might be in the wiring between the breaker and the appliance. Make a visual inspection (with the breaker off) of all the wiring that is accessible. If you don't find a frayed or broken wire or two bare wires touching each other, disconnect the wiring from the breaker. To do that, turn off the main service breaker that feeds the entire panel. Remove the faceplate from the breaker panel. Make sure the breaker in question is off (an added safety in case the main breaker is still on for any reason). Unscrew the wire lug screw at the base of the breaker (Fig. 4-21) and pull the load wires from the breaker. Turn the main service back on and reset the breaker in question. If it still pops off under this no-load condition, then the breaker itself is faulty and must be replaced.

Never try to repair a breaker. To replace a breaker, turn off the main service breaker. Place your flat-blade screwdriver on the front, top edge of the breaker and pry it out of the panel. Some breakers fit tightly, so apply firm, even pressure. If you have not disconnected the load wires, do so as described earlier. Look at the back of the breaker and the design of the hook connection that fits into the electric bar of the panel. When you have your replacement, reconnect the load wires to the new breaker, and return it to the panel reversing the steps taken to remove it. Put the panel faceplate back on and turn on the main service breaker.

If the breaker did not trip when you disconnected the load, the reason for the breaker tripping off must be in the wiring between the breaker and the appliance. Since you were unable to find a problem with the wiring during your visual inspection, you might need to replace the wiring. At this point, I recommend calling an electrician.

Sometimes electrical problems at the appliance or the tripping of a breaker is caused by a loose breaker. If you find that the breaker is loose when you first try to remove it, try pushing it back into the panel, and try your appliance again. If it won't seat firmly, replace the breaker.

You might think that it is easier to disconnect the load and check the breaker first, prior to following all of the other checks, but I present the steps in this order, beginning at the appliance, because it is usually here that you will find the problem.

Older homes might still have fuses. Fuses perform the same function as circuit breakers, but fuses must be replaced each time the overload breaks the circuit (blows the fuse). Fuses either clip or screw in place. As with breakers, always replace a fuse with one of the same amperage.

Whenever you approach a breaker panel, do so with great respect. Water, frayed wiring, or a poor previous service work might have created problems at the panel that you cannot anticipate. If you are planning to work on a panel, it's best to have a helper around to get help in case of electric shock. Other safety measures include wearing rubber gloves and boots, standing on a piece of dry wood to further insulate you from the ground, and leaving one hand in your pocket, so you can't inadvertently touch one hand to a live wire or panel and the other to a ground.

Wiring

Pulling new wires in a circuit or adding a circuit is a job best left to a professional electrician. But you might be called upon to replace the wires from an appliance to a junction box in the equipment area or the wiring of a heater connection, so it is valuable to know a few things about requirements.

Gauge and Type

The gauge of the wire refers to its thickness and, therefore, its ability to handle volume and pressure of current (amps and volts). AWG refers to American Wire Gauge and is the system described in Fig. 4-22. AWG wire is designed to operate under temperatures as high as 140°F (60°C).

Whenever you run wire for any reason, consult the chart in Fig. 4-22 to be sure you use the correct type. Remember, you can always use wire that is heavier (lower AWG number) than the breaker and appliance require, but never use wire that is thinner (higher AWG number) than required.

Wire is stranded or solid. There is less resistance in solid wire than stranded, so this should be your first choice. Wire is generally available in copper. During times of copper shortages and high prices, aluminum was used for wiring homes, but it has been the cause of overheating and fires and should be replaced whenever possible.

GENERAL WIRE SIZING CHART

Run (ft)	30	40	50	75	100	125	150	175	200
Amps	Minimum AWG size wire recommended								
5	14	14	14	14	12	12	12	12	12
10	14	14	14	12	12	12	10	10	10
15	12	12	12	10	10	10	8	8	8
20	10	10	10	10	10	8	8	8	8
25	10	10	10	10	8	8	6	6	6
30	8	8	8	8	6	6	6	6	4

Interpolate as needed between values given. Values given are based on 110/120 volts and are heavier than actually required with 220/240 volt equipment. However, using these values for either voltage will ensure safe, adequate installations.

FIGURE 4-22 General wire sizing chart.

Wires are sold in various colors. Green wire is always ground. Black and red are used for hot lines, white for neutral. When you need many hot lines to many circuits, they might be colored in any of the many other colors available. If you must use a wire color not in keeping with this code, tape the correct color tape over the wire or clearly label it. Never assume that the previous technician used the correct colored wire. Check everything as you go and try to leave wiring better than you found it.

When terminating wires to be attached to connections in appliances or at other terminal posts, use crimp connectors rather than simply wrapping the bare end of the wire around the post. Wrapping can come loose or be squeezed off the post. Figure 4-23 shows the correct way to make wrap connections if you have no other choice. Bend the wire in the same direction as you will tighten the screw, so when you tighten the screw it also tightens the wrap.

Figure 4-23 also shows a typical crimp connector. The connectors are available in various sizes and with various connection ends (called the *tongue*). The insulation is stripped off to accommodate the barrel of the connector. Using a crimping tool, secure the wire to the connector. Crimping sets are available for a few dollars at any hardware store.

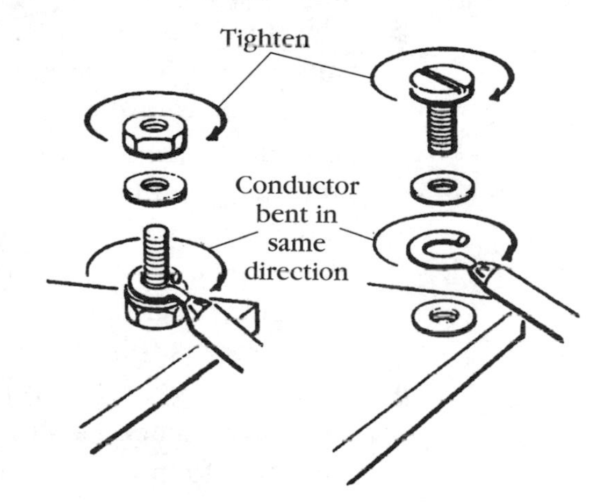

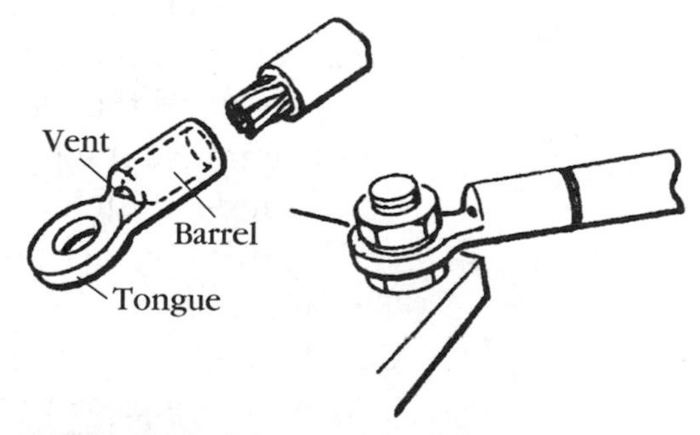

FIGURE 4-23 Wire wrapping and crimp-on connectors. *U.S. Government Printing Office.*

Ground Fault Interrupter (GFI)

When equipment or wiring fails it might draw more current than the appliance can use, burning out the appliance. The circuit breaker is designed to break the circuit when demand exceeds the rating of the breaker. As noted earlier, it takes so little current to kill a human that the typical breaker will deliver a lethal dose before breaking the circuit. In other words, circuit breakers are designed to protect equipment, not humans.

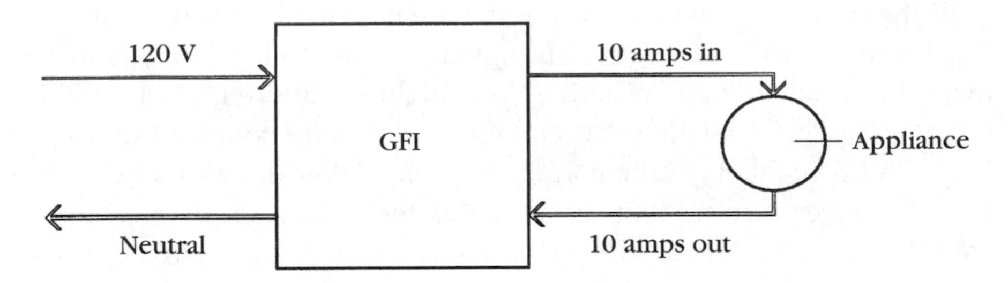

FIGURE 4-24 Ground fault interrupter (GFI).

The GFI is designed to protect humans. It is a circuit breaker that detects problems at a low enough level to protect you before lethal doses are delivered. It breaks a circuit when it detects a ground fault. Figure 4-24 diagrams the GFI concept. The GFI constantly measures the current going out of it (to the appliance) and coming back into it. If an inadvertent grounding takes place, such as if the metal case of an appliance were electrified, and you touch it, completing a pathway for current to the ground, the GFI detects the drop in the current it is receiving and breaks the circuit. The GFI detects variations as low as 0.005 amp, which is about half the lethal charge to a child and about one-sixth a lethal dose to you. The GFI cuts the circuit within one-fortieth of one second, so it is not only sensitive, it's quick.

There are three basic styles of GFI that you will likely encounter in pool and spa work. The first looks like a standard circuit breaker in the electrical panel, but it has a test button in the face of the breaker in addition to the on/off breaker switch. By pressing the test button, you are simulating an unbalanced current condition inside the breaker and thereby testing the efficiency of the GFI. The GFI breaker resets the same way a normal panel breaker does.

The second type of GFI is built into a wall outlet, such as the type you might install for plugging in a portable spa. It also contains a test button and a switch to reset the GFI.

The third type is a portable GFI, a unit that plugs into a wall outlet. The appliance is then plugged into the GFI, making the outlet a GFI outlet.

If a GFI keeps breaking the circuit, you troubleshoot the problem in the same manner as any other breaker.

Switches

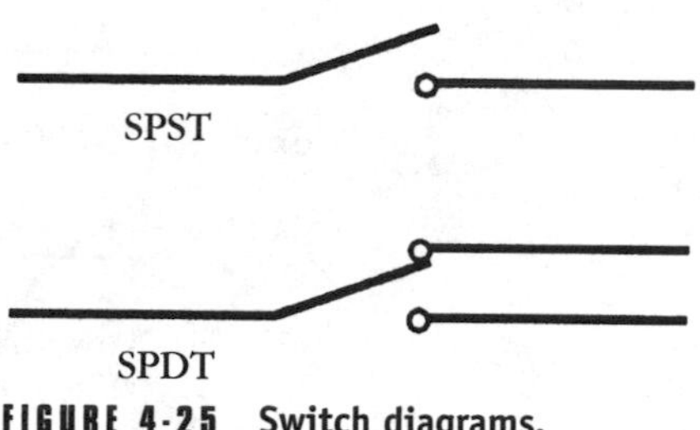

FIGURE 4-25 Switch diagrams.

Pool and spa equipment is not wired directly to circuit breakers in the electrical panel. Instead, these circuits are interrupted at some point by switches to control the operation of each appliance. A breaker should never be used as the on/off switch for an appliance because repeated switching will weaken the breaker.

Figure 4-25 depicts a basic switch, which is a break in the hot line of a circuit. This is the most basic on/off switch, called a single pole, single throw (SPST) switch. This switch handles one circuit (single pole) each time the switch is thrown. The second drawing depicts a single pole, double throw (SPDT) switch. In this case, there is still only one circuit of electric current, but when this switch is thrown one direction, it electrifies one appliance, and when it is thrown the other way, it electrifies another appliance. Depending on the appliance(s), you might use several variations of poles (circuits) and throws (destinations for the current). By understanding these basic concepts, you will recognize whatever type of switch you encounter.

A *relay* is a switching device on a circuit that controls current flow in another circuit. Figure 4-26 depicts a typical relay circuit. When the relay circuit is electrified, it energizes an electromagnet that pulls the two halves of the relay together. In doing so, the contacts of the controlled circuit are brought in contact, completing the circuit. Relays are normally used as safety devices. The purpose of this type of control is to use a low-voltage circuit (the relay circuit) to turn on or off a higher voltage circuit (controlled circuit). For example, a safe 12-volt circuit can be used near a pool or spa to control a dangerous 220-volt circuit that operates a pump motor or blower.

Safety

I will never forget grabbing an old light fixture that had been taken from the water and then lying on the deck and feeling a surge of 220 volts run through my body. Every muscle in my body contracted violently. I could see, but everything was reddish and hazy. I could hear perfectly, but could only force a gurgle out of my mouth. No matter

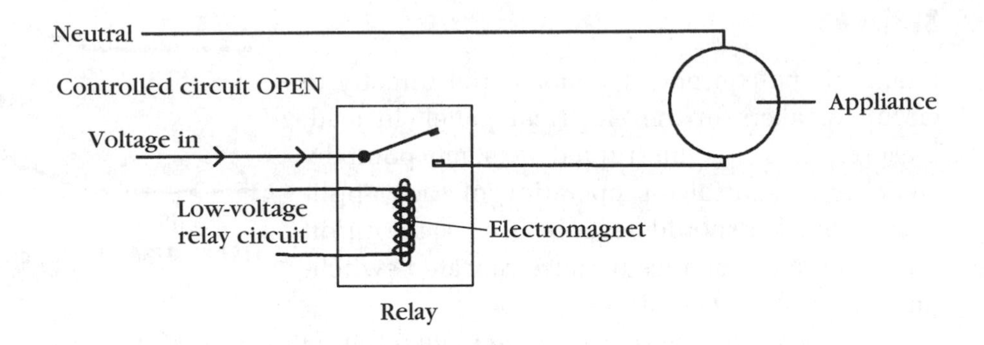

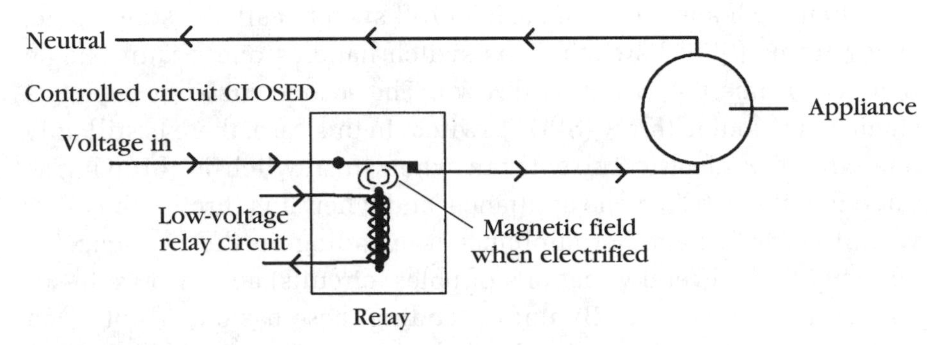

FIGURE 4-26 Relay.

what I tried or thought, I couldn't release the grip my hand had on that light fixture. If my helper had not been near the light switch, I wouldn't be writing this book today.

Safety with electricity is twofold. First, be aware. Never go into an equipment area and start grabbing wires, components, or disassembling things without turning off the power supply. Even once you have disconnected appliances, there might be current from another source, such as a parallel circuit or switch or a short circuit in another appliance that is connected by a bonding wire or water on the ground. In other words, always treat electrical circuits and appliances as if they were an animal that could bite you at any time.

Second, take precautions. Wear rubber-soled shoes and rubber gloves, and don't stand in water when you work on equipment. Have a helper around when you plan to do installations or electrical work. Take the extra time to walk over to the electrical panel and turn off the supply breaker in addition to any on/off switches that are closer.

Finally, never take shortcuts with wiring or installations, such as leaving off the bonding or ground wires, running circuits of different voltages in the same conduit, using wire of the wrong color because you are too lazy to make another trip to the hardware store, etc. Water and electricity are dangerous allies. Water is a very effective conductor, and an electrical fault in the equipment room, contacting a leak in an appliance, can charge the entire pool or spa and make it lethal. For your safety and that of your customers, not to mention the very existence of your business, be careful and take no shortcuts when working with electricity.

Testing

RATING: EASY

The last general area of basic electrical knowledge that is useful to the water technician is how to test for the presence and parameters of electric current.

As the name implies, the multimeter has multiple functions, testing circuit voltages, continuity, and resistance. Figure 4-27 shows a typical multimeter. It has a positive and a negative test lead and a switching device to set the meter for reading dc or ac (reading various ranges of each), resistance, or continuity. The meter is battery powered for continuity and resistance testing because you must send current into a line to test if it is continuous (unbroken) or broken and to test the amount of resistance in a conductor.

Testing for the presence of current at a connection or appliance is simple. It does require guessing what you expect to find because you need to set the tester for ac or dc and for the range of voltage you expect to find. For example, when testing the control circuit of a millivolt-controlled heater, you would set the meter for dc current in a voltage range of 0 to 1 volt (since you will be testing a circuit with up to 750 millivolts, which is equal to 0.75 volt). Similarly, if you are looking for the presence of current at your motor, you set the meter for ac in the voltage range of either 0 to 110 or 0 to 220 volts. Electronically controlled heater circuits operate on 25-volts ac, so you would set the meter for ac in a range of 0 to 50 volts. Generally you can't harm the meter by feeding it less current than the range you have chosen, but you can destroy

FIGURE 4-27 Using a multimeter.

it by feeding it more. Therefore, if you are uncertain about the voltage being tested, start with the 220-volt range and work down.

When testing dc circuits, remember that polarity (positive and negative) makes a difference. You must touch the positive meter lead to the positive contact of the appliance or switch and the negative lead to the negative contact. If you reverse these, you will see the meter register negative voltage. When testing ac voltage, the polarity doesn't matter, and you can touch either lead to either side of the circuit.

When testing 110-volts ac, touch one lead to the suspected hot line and one to a neutral line or to ground. When testing 220-volts ac, perform the same test on each of the two hot lines, then touch one lead to each hot line at the same time. If each line individually reads 110 volts, but when tested together they do not read 220 volts, it means the two hot lines are being supplied by the same pole of the power supply and therefore will not deliver 220 volts. This usually denotes a faulty breaker.

When buying a multimeter, make sure it can test millivoltage for working on millivolt heaters. Some meters won't accurately read less than 10 volts, and therefore are useless with millivoltage. Most electronic meters are pocket-size and can self-range, which is to say you need only dial in ac or dc and the meter will detect the voltage and adjust accordingly.

Something Better

"IntelliFlo" is the first programmable pump/motor (Fig. 4-28) that constantly monitors water flow and electrical current, making sure that the filtration system is operating at peak efficiency. It's an interesting idea that other manufacturers are rapidly copying, because it eliminates the need for pump curves and hydraulic calculations to determine the right pump for the job. Instead, it's a "one-size-fits-all" appliance that you program for your desired water turnover and the IntelliFlo pump adjusts accordingly. The manufacturer, Pentair, notes that this adjustable feature reduces energy cost by as much as 90% compared to conventional units.

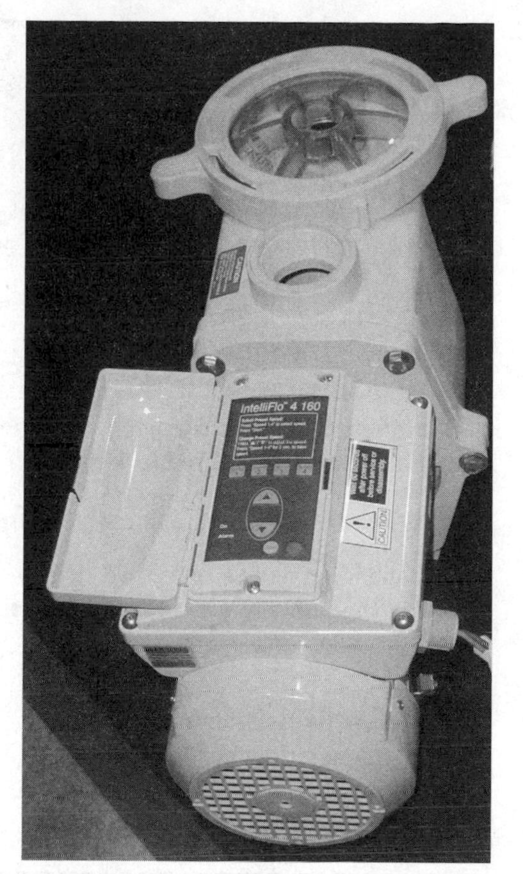

FIGURE 4-28 IntelliFlo pump.

FAQs: PUMPS AND MOTORS

Will My Pump Be Damaged if It Runs Dry?

- After several minutes of running dry, seals and plastic components will begin to warp, resulting in leaks later. Never run a pump dry for more than a minute.

Should the Pump Be Running When Swimmers Are in the Pool?

- Bather load will quickly turn the water cloudy, so circulation is a key to a clean pool. However, be careful if you have a shallow main drain and small children. Be sure all bathers know that suction at the skimmer and main drain can cause injury if hands, feet, or hair are allowed to block them off while the pump is running.

How Much Does It Cost to Run the Pump?

- Depending on the cost of electricity in your area and the number of hours you need to run the pump to keep your pool clean, it can cost pennies a day or up to a few dollars.

Is There Such a Thing as a Silent Pump/Motor?

- A properly functioning motor that is not laboring against blockages, such as a dirty filter or clogged skimmer, should not be objectionably loud. Loud motors are a sign of worn bearings or obstructions in the circulation system. You can also cover your motor with specially designed housings that reduce noise.

CHAPTER

5

Filters

As the name implies, the filter is the piece of pool or spa equipment that strains impurities out of water that is pumped through it. There are few moving parts (in fact no parts should be moving when the filter is in operation) and simple components.

Types

Three basic types of filter are in common use and each is preferred for various reasons. Each type is more efficient for various adaptations and you might also discover regional preferences around the world. For example, in regions where sand is plentiful but diatomaceous earth is scarce, the sand or cartridge filter will be more commonly in use than the DE filter. Setting aside such regional prejudice, however, as I review the types of filters, I will also discuss why each is preferred based on technical application.

Diatomaceous Earth (DE) Filters

In the diatomaceous earth type of filter, also called a DE filter (Fig. 5-1), the water passes into a metal or plastic tank, through a series of grids (also called filter elements) covered with fabric, and back out of the unit. The grids do not actually perform the filtration process, but instead are coated with a filter *media*, diatomaceous earth, that does the actual filtering work.

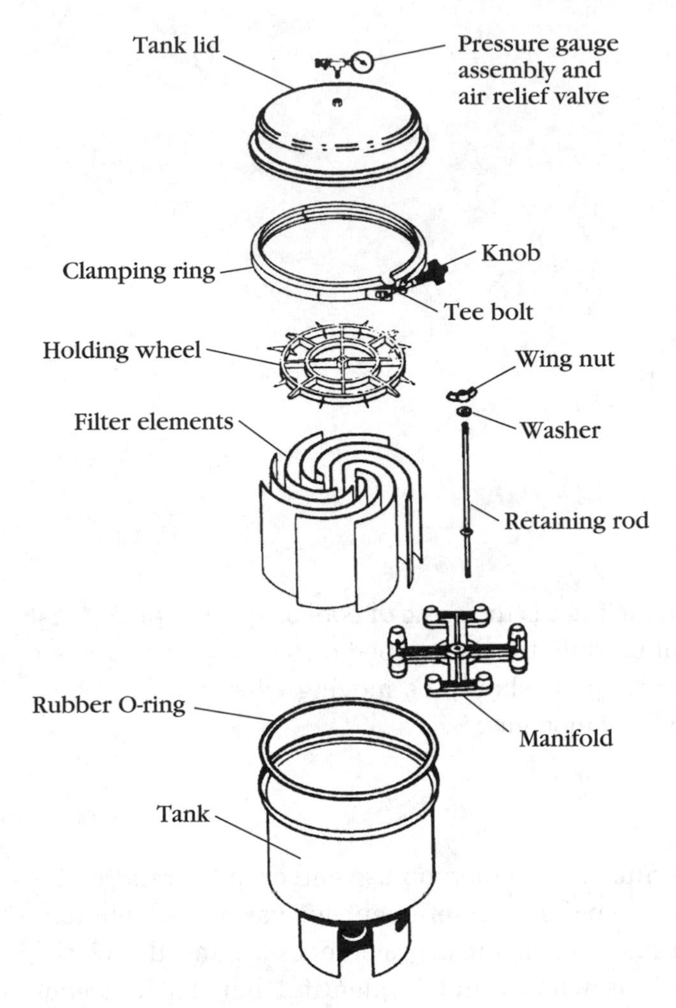

FIGURE 5-1A Diatomaceous earth (DE) filter. *Pentair Pool Products.*

DE is a white powdery substance found in the ground in large deposits. It is actually the skeletons of billions of microscopic organisms that were present on earth millions of years ago. So, in essence, this powder is akin to dinosaur bones. If you look at DE under a microscope, you will see what appears to be tiny sponges, thus the filtration ability becomes more apparent. Just like a sponge, water can pass through, but the impurities in the water can't. Because the DE particles are so fine, they can strain very small, actually microscopic, particles from the water as it passes through.

So why not just dump a few pounds of DE in a tank and pass the water through? Good question. Because the DE will collapse together and cake, making it impossible for even the water to get through. Therefore, the tank is equipped with free-standing grids (also called elements) that are coated with DE to accomplish the filtration process. DE filters are usually located on the return side of the pump (a pressure filter). There are basically two types of DE filter: the vertical grid and the spin type.

VERTICAL GRID DE FILTERS

As can be seen in Figs. 5-1A and B, the grids (elements) are mounted on a manifold and the resulting assembly fits into the tank. A retaining rod through the center screws into the base of the tank and a holding wheel keeps the grids firmly in place. The top of the tank is held in place by a clamping ring, and the two parts are sealed with a thick O-ring to prevent any leaking.

The water enters the tank at the bottom and flows up around the outside of the grid assembly. It must flow through the grids, down the stem of each grid, and into the hollow manifold, after which it is sent back out of the filter. This type of DE filter is called a vertical grid, for obvious reasons. Some DE filters place the manifold on top as depicted in Fig. 5-1C.

One way to clean this filter is called *backwashing*, a concept I will discuss in more detail later. As the term suggests, backwashing means the water is redirected through the filter in the opposite direction from normal filtration (accomplished with a backwash valve, also discussed later), thereby flushing old DE and dirt out of the filter.

Not all vertical DE filters are equipped with a backwash valve to allow backwash cleaning. These types must be disassembled each time for cleaning. Some are also equipped for *bumping* rather than backwashing. In this rather ridiculous process,

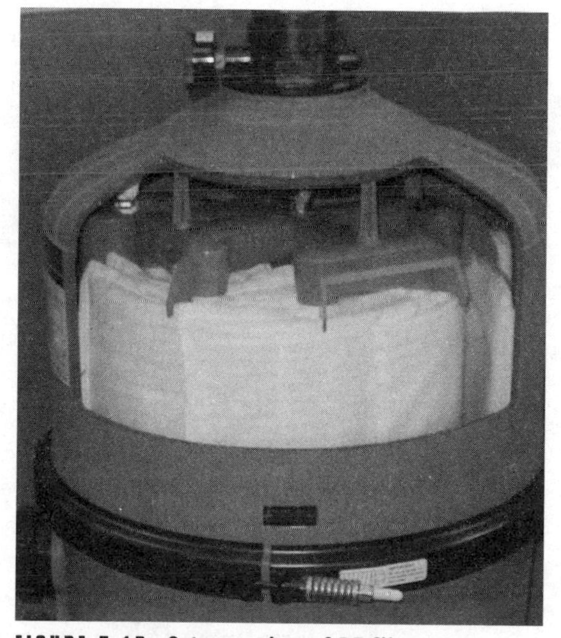

FIGURE 5-1B Cutaway view of DE filter.

FIGURE 5-1C Close-up of filter grid manifold (top mounted). *Pentair Pool Products.*

the dirty DE is bumped off the grids, mixed inside the tank so the dirt is evenly distributed within the DE material, then recoated onto the grids. As we will see later in this chapter, it is far better to disassemble the filter and clean it.

In some areas, out of concern that DE will clog pipes, local codes require that DE not be dumped into the sewer system. In this case, a *separation tank* is added next to the filter (Fig. 5-2). When draining or backwashing the filter, the dirty water is passed through a canvas strainer bag inside this small tank before going into sewer or storm drains. The canvas bag strains most of the DE out of the water so it can be disposed of elsewhere.

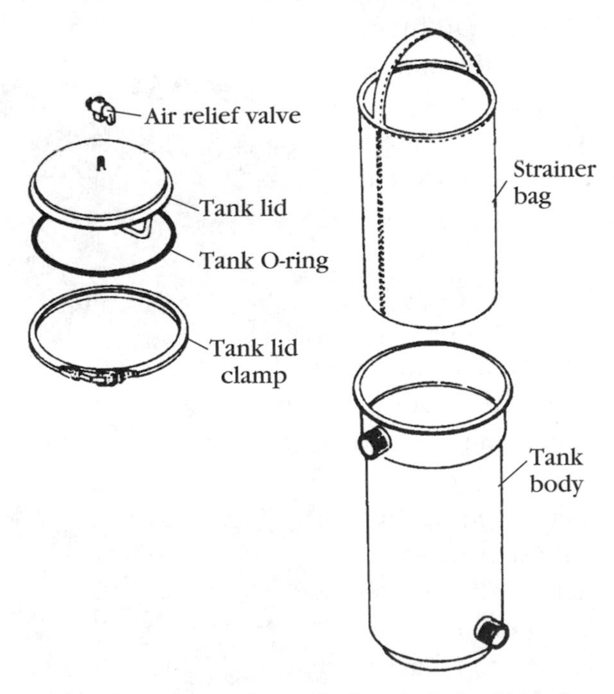

FIGURE 5-2 Separation tank. *Premier Spring Water, Inc.*

All filters are sized by the square footage of surface area of their filter media. In DE filters therefore, the total surface square footage of the grids would be the size of the filter. Typically there are eight grids in a filter totaling 24 to 72 square feet (2 to 7 square meters), designed into tanks that are 2 to 5 feet (60 to 150 centimeters) high by about 2 feet (60 centimeters) in diameter. Obviously the larger the filter, the greater capacity the filter will have to move water through it. Therefore, filters are also rated by how many gallons per minute can flow through them. More on this later.

SPIN DE FILTERS

There is also an obsolete, rather inefficient type of filter where the grids are wheel-shaped and lined up horizontally like a box of donuts. This design is called a spin DE filter.

The operating idea is the same as with vertical grid filters, but to clean this type you turn a crank on the tank that spins the grids, theoretically cleaning them. For obvious reasons, these are called *spin filters.* They don't work very well and you won't run across too many of them. If you do come across an old one in operation, you'll understand the mechanics if you first understand the basic vertical DE filter.

Sand Filters

If the filter media is not DE, it must be sand. Sand and gravel are the natural methods of filtering water. There are two types of sand filters in general use.

PRESSURE SAND AND GRAVEL FILTERS

These were once called *rapid sand* filters (older-style metal tanks designed for flow rates less than 3 gpm per square foot), but are now called *high-rate sand* filters (fiberglass or other composite materials designed for flow rates of 5 to 20 gpm per square foot or 20 to 75 liters per 0.1 square meter). The concept of the pressure sand and gravel filter is simple.

Water is passed through a layer of sand and gravel inside a tank, which strains impurities from the water before it leaves the tank. I designate these types of filters as *pressure* sand and gravel, because like their DE cousins, the water is under pressure inside the tank from the resistance created by trying to push it through the filter media. This differentiates it from another type of sand and gravel filter used especially in fish ponds where there is no such pressure (see free-flow filters).

In Fig. 5-3A, the water enters the tank through the valve (item 7) on top and sprays over the sand inside. The water then runs through the sand, impurities being caught by the sharp edges of the grains, and is

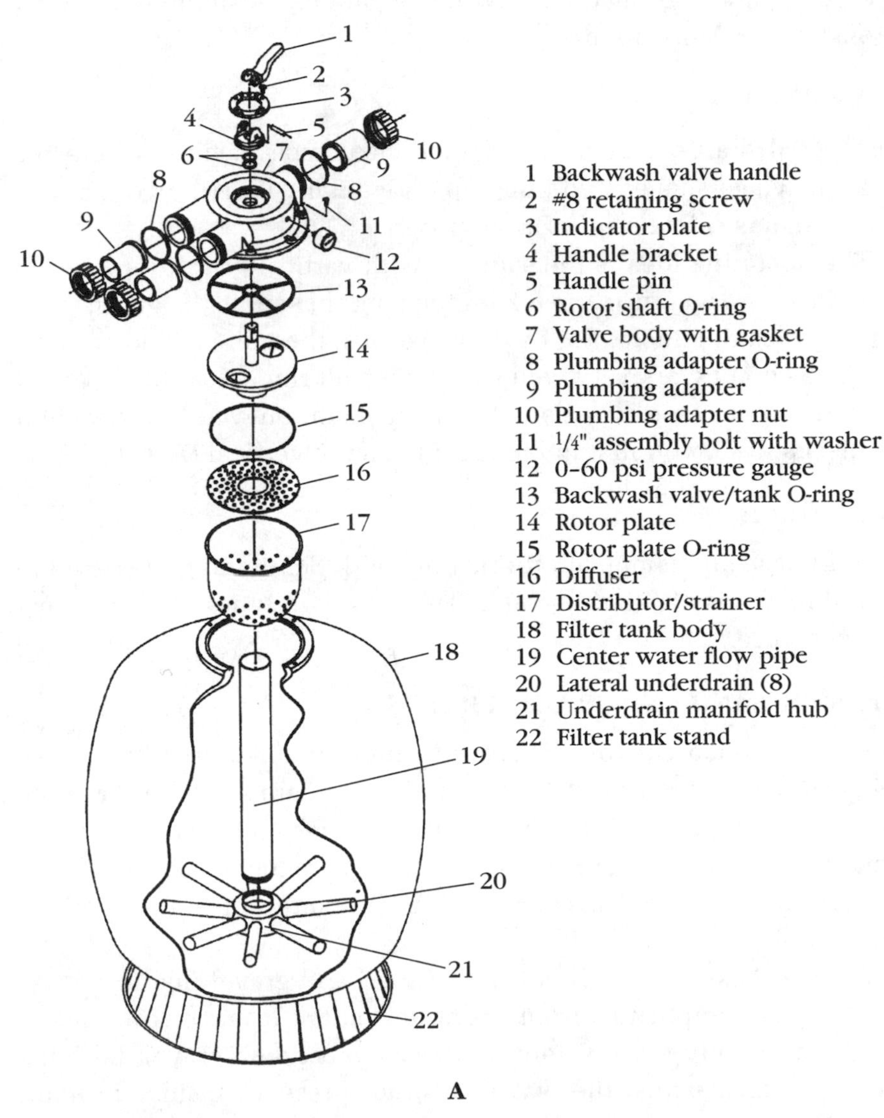

FIGURE 5-3 High-rate sand filters: (A) backwash valve is on top; (B) backwash valve is on side. *Pentair Pool Products.*

1 Air relief valve
2 Adapter
3 Pressure gauge
4 O-ring
5 Nut
6 Lid
7 O-ring
8 Strainer
9 Air relief tube
10 Diffuser assembly
11 Upper pipe assembly
12 Air relief tube connector
13 Lower pipe assembly
14 Tank and foot assembly
15 Laterals
16 Lateral hub
17 Drain spigot
18 Drain plug
19 Lid wrench
20 Lock nut
21 Spacer
22 O-ring
23 Spacer
24 Gasket
25 Bulkhead
26 O-ring
27 Washer
28 Basc
29 O-ring
30 2" threaded adapter kit
31 1.5" threaded adapter kit
32 (See 30, 31)
33 Closure kit
34 Fitting package
35 Spacer

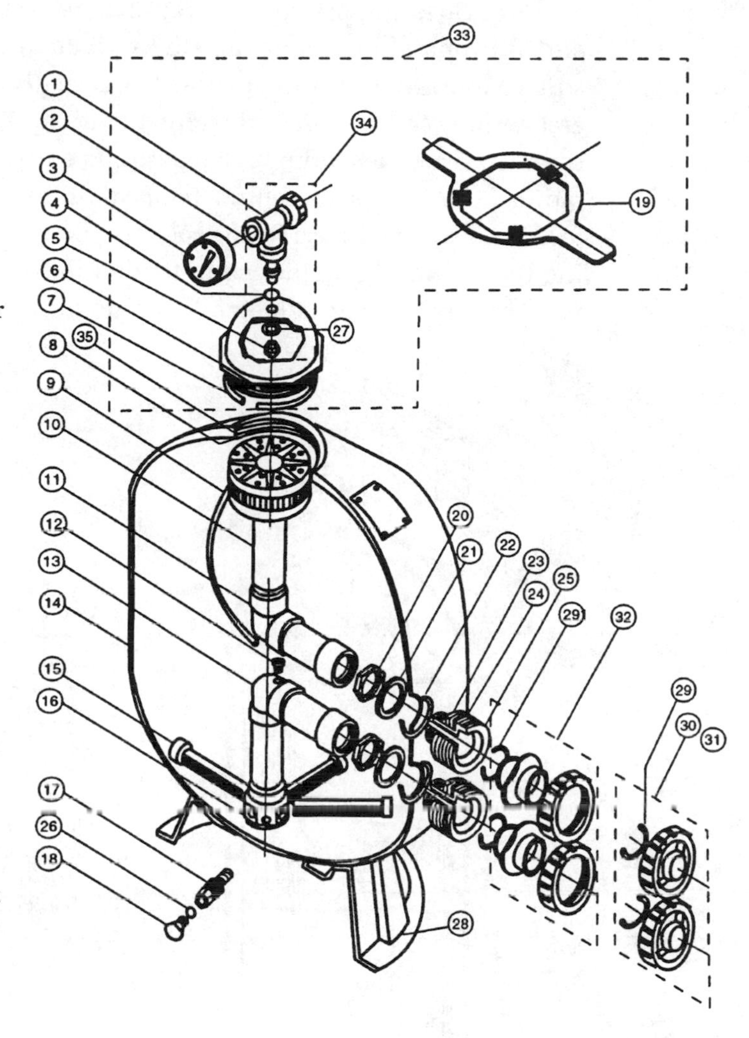

B

FIGURE 5-3 *(Continued)*

pushed through the manifold at the bottom where it is directed up through the pipe in the center and out of the filter through another port of the valve on top. The individual fingers of the drain manifold (item 20) are called *laterals* and the center pipe used to return clean water (item 19) is called a *stanchion pipe*. To drain the tank, a drain pipe is provided at the bottom as well.

Sand filters are also sized by square footage and gallons per minute. Does some engineer measure the surface area of the billions and billions of grains of sand in a given filter to arrive at the total square footage of filtration area? No, actually it is based on the surface area of the sand bed created inside the filter. Since most sand filters are round, a filter measuring 24 inches (60 centimeters) in diameter with a radius of 12 inches (30 centimeters) would have a sand bed of 3.1 square feet (0.29 square meter) (using the formula pi × radius squared). Knowing the volume of sand recommended for any given filter (expressed in cubic feet or cubic meters), the manufacturer arrives at a square

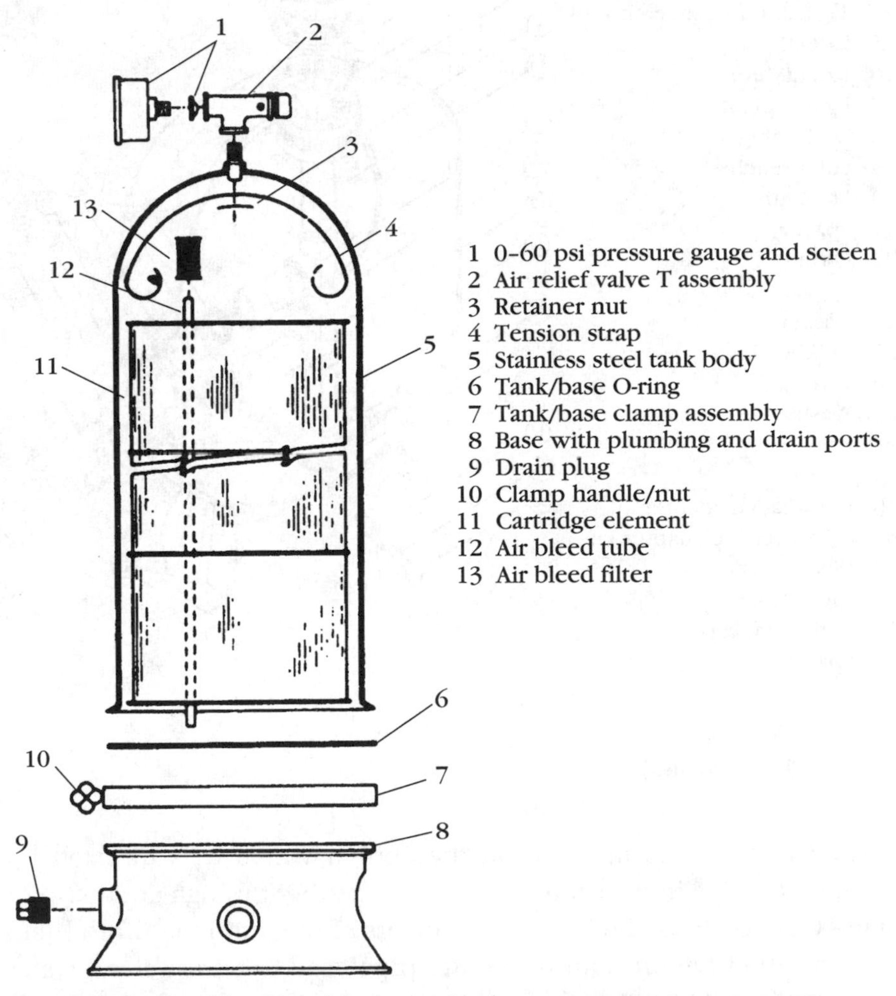

FIGURE 5-4A **Cartridge filter (single cartridge).** *Sta-Rite Industries, Delevan, Wis.*

footage value and a resulting gallons per minute (or liters per minute) rating. Sand filter tanks are usually large round balls, anywhere from 2 to 4 feet (60 to 120 centimeters) in diameter.

FREE-FLOW SAND AND GRAVEL FILTERS

Used mostly in decorative ponds or fish ponds, the free-flow filter is created as a natural filtration device. The main drain of the pond is covered with a layer of gravel, then small stones, then sand. As the water is sucked toward the drain for circulation, it must pass through these layers of sand and gravel, filtering out impurities in the process.

Over several weeks, the fish waste, decaying plant material, and algae form a biological "soup" in the filter media. Enzymes in this decaying matter further break down impurities that the sand and gravel have trapped, making them a very effective way of filtering fish ponds and other heavy-debris bodies of water. It is similar to nature's own way of purification and this type of filter is often referred to as a *biological* filter.

Cartridge Filters

A cartridge filter (Fig. 5-4A) is similar to a DE filter except there is no DE filter media. In Fig. 5-4B, water flows into a tank that houses one or more cylindrical cartridges of fine-mesh, pleated fabric (usually polyester). The extremely tight mesh of this fabric strains impurities out of the water.

Figure 5-5 also shows a cartridge filter, but instead of one large cartridge, this model uses 18 small ones. Why? To make life tough on the pool service technician, that's why. Well, okay, perhaps the manufacturer felt this design would filter better because water can circulate around 18 small cartridges more freely than around one large cartridge in the same size tank. If you've ever had to clean and reassemble one of these, however,

FIGURE 5-4B Cutaway view of a cartridge filter.

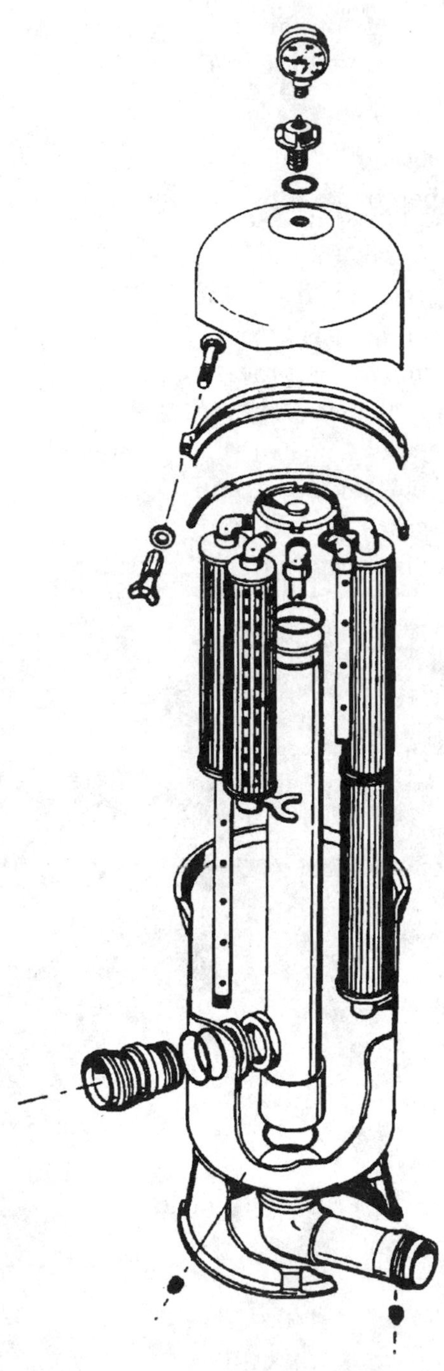

FIGURE 5-5 Cartridge filter (multicartridge).
Pentair Pool Products.

you'll agree with me that it was a designer who hated pool service technicians.

Cartridge filters are classified by square footage of filter surface as are DE and sand filters. By pleating the cartridge material, a lot of square footage can fit into a very small package. This would not be possible if DE were added, because the tight pleating would prevent DE from coating evenly and effectively. Typically, cartridge filters for residential use range from 20 to 120 square feet (2 to 12 square meters) in a tank not more than 4 feet (1.2 meters) high by 1½ feet (45 centimeters) in diameter.

Makes and Models

Now that you have had a brief introduction to typical pool and spa filter types, how do you make a selection and what size do you need for a given job? The intended use of the pool or spa and/or the local building and health codes will guide you in answering these questions.

Sizing and Selection

RATING: EASY

In many jurisdictions, the turnover rate (if you've forgotten turnover rate, see the section on hydraulics) requires a pool to be turned over in 6 hours, a wading pool in 1 hour, and a spa in 30 minutes. Using this information and knowing the total gallons in your pool or spa, you will know how many gallons per minute the filter must be able to handle.

As an example, my pool is 40,000 gallons. To turn that over every 6 hours I need a filter that can handle

40,000 gal (151,400 L) in 6 h
= 40,000 gal in 360 min
= 111.11 gal (420 L) per minute (40,000 ÷ 360)

Now, regardless of the manufacturer's claims, many building or health codes also determine the maximum number of gallons per minute permitted for every square foot of a filter's surface area. Just for the record, many codes also specify the minimum rate of flow during backwash. These might differ for residential and commercial pools, but many jurisdictions require the following:

Filter style	Maximum flow, gpm/sq. ft (L/1000 sq. cm)	Minimum backwash flow, gpm/sq. ft
High-rate sand	15 (57)	15
DE	2 (7.6)	2
Cartridge	0.375* (1.5)	No backwash

*Commercial installation.

Note that cartridge filters are rated at 1 gpm/sq. ft (3.8 L/1000 sq. cm) maximum flow rate on most residential applications and the more stringent 0.375 gpm/sq. ft on commercial installations. This is because cartridge filters are used primarily on spas where lots of bathers sit in a relatively small amount of water creating lots of bacteria, oil, and dirt for the filter to handle. The 0.375 rule simply means a commercial installation must have a larger filter than one at home where the bather load will probably be less.

So with this information, if you chose a sand filter, you divide the required gallons per minute for our sample pool, 111, by the maximum flow rate of the local code for sand filters, 15, giving you 7.4 square feet (0.68 square meter).

This means you need a sand filter of at least 7.4 square feet. Now the chances are that the manufacturer's claim of gallons per minute per square foot of filter area will exceed the local code requirement, but check to see that it does, at least, meet the code. In other words, the maximum flow rate permitted might be 15 gallons per minute per square foot of filter area, but if the maker says his 7.4 square foot filter can only handle 10 gallons per minute, then you will have to go to a larger size until you arrive at a filter that can do the job. Got it?

Using the same example, the 111-gpm rate with a DE filter is

$$111 \div 2 = 55.5 \text{ sq. ft } (5.1 \text{ sq. m})$$

So you need a DE filter of at least 55.5 square feet; again, checking the manufacturer's specifications.

Finally, the cartridge filter in the same application:

$$111 \div 0.375 = 296 \text{ sq. ft } (27.5 \text{ sq. m})$$

So you need a 296-square-foot cartridge filter to do the same job as the 7.4-square-foot sand filter or the 55.5-square-foot DE filter.

Another sizing and selection criteria is dirt. How dirty does the body of water get that this filter must service? You want to oversize the filter a bit so you don't have to clean it out every day, week, or hour (choose one). This time between filter cleanings is called the *filter run*. So you might as well get a filter 20, 50, or 100 percent larger than you actually need so it can hold more dirt and thereby leave more time between cleanings—right?

Well, not exactly. If the pump cannot deliver the gallons per minute for the square footage of this huge filter you have just installed, only part of the tank will fill with water, effectively giving you a smaller filter anyway. Also, the pump won't backwash the filter completely if it can't match the gallons per minute rating, which is why codes call for a minimum backwash flow rate. In the section on cleaning filters I tell you not to backwash DE filters anyway, so this might not be important to anyone except the building inspector (on new construction). More to follow. So the obvious question is, which do we use? Consider price, size, and efficiency.

Price is a factor of the current market and supplier availability in your area, so I'll let you check that one out for yourself. The size of the unit might be a deciding factor if the equipment area is small. To contain 7.4 square feet of filter area, a sand filter needs to hold about 8 cubic feet (226 liters) of sand. Wet or dry, that much sand weighs about 10 million pounds (well, okay, not quite that much, but you get the idea). In fact, for the filter to hold 8 cubic feet of sand it would need to be very strong and about 4 feet (1.2 meters) in diameter.

A cartridge filter with almost 300 square feet of filter area (28 square meters) is also rather large and will be very expensive. But a DE filter of around 60 square feet (5.6 square meters) is a fairly standard, serviceable tank in a modest price range.

If you're lucky enough to have a client like Barbra Streisand, and she just wants the filter that does the best job regardless of price, then how do you advise her which one to buy? The answer lies in the mighty micron. A *micron* is a unit of measurement equal to one millionth of a meter or 0.0000394 inch. Put another way, the human eye can detect objects as small as 35 microns; talcum powder granules are about 8 microns and table salt is about 100 microns.

So what? Well, each filter type has the ability to filter particles down to a particular size, as measured in (you guessed it) microns. Various references and manufacturers will give you different estimates and they might all be right as you will see in a minute. However, here's a rule of thumb:

- Sand filters strain particles down to about 60 microns
- Cartridge filters strain particles down to about 20 microns
- DE filters strain particles down to about 7 microns

I read an article recently (with data supplied by a manufacturer, of course) claiming sand filters at 25 microns, cartridges at 5 microns, and DE at 1 micron. Gentlemen, modesty, please!

Yet, as I said, both estimates are correct. Sand filter efficiency is based on the age of the sand. New sand is crystalline and many-faceted. These sharp-sided facets are what actually catch debris out of the water and filter it. As sand ages, the years of water passing over it cause erosion of those facets until each grain of sand becomes smooth and rounded, losing its ability to trap particles of the smallest size. So a new sand filter with fresh sand might strain particles of 25 microns, while a unit in service one year might be lucky to strain particles of 60 microns.

TRICKS OF THE TRADE: SAND FILTERS

- **To improve sand's ability to filter, add ½ cup or 125 milliliters of alum (aluminum sulfate) for every 3 square feet or 0.3 square meter of the filter's rated size. This acts like a DE coating on the grids of a DE filter and helps the sand strain finer particles.**
- **Choose a sand filter for fish ponds. Extreme amounts of dirt and waste will quickly clog a cartridge or DE filter and take more time to clean than the simple backwashing procedure for sand filters.**

Efficiency of a DE filter is affected by the cleanliness of the DE. As DE becomes dirtier, the filter will actually trap finer particles, but the flow rate will decrease, thus making it less efficient overall.

Because cartridges do not rely on organic material like sand or DE, they are not affected the same way. As long as the cartridge surface area remains clean, it will filter the same size particles (although extremely old cartridges that have been acid washed many times will stretch out somewhat, creating a mesh that is not so fine as when new). However, dirt does affect the cartridge filter, as it affects all filters, but with the cartridge it actually has a positive effect.

Dirt makes the mesh of the polyester cartridge material even tighter, essentially acting like DE. I know some service technicians who add a small handful of DE to a cartridge filter after cleaning it to start this process. Thus, with each successive pass of water through the cartridge filter, it becomes more efficient as the mesh gets finer and finer from the retained particles. This is also true with sand filters, up to a point.

Regardless of these variations, the one consistent fact you will note in the estimates is that sand filters are the least efficient filters in terms of how fine the particles are that they can strain from the water; cartridge filters appear to be the second best and DE the best. I say *appear to be* because a slightly dirty cartridge filter will strain particles of 5 to 10 microns, while a dirty DE filter simply clogs up and doesn't allow as much water to pass through as it should. Although it too is straining particles of 5 to 10 microns in size, if the flow is reduced because the DE is clogged, what good is the rated straining capability?

This is why cartridge filters are generally regarded as the best overall filtration method. They are easy to clean and require no added organic media. As you saw in the example, however, the size required for a cartridge filter might make its use impractical or too expensive. Generally you will find cartridge filters used on spas and small fountains, DE filters on pools and larger fountains, and sand filters on fish ponds, decorative ponds, and pools.

Remember too that each filter adds resistance to your system. (Remember the old villain *head*? If not, refer to the section on hydraulics.) A smaller filter will restrict water flow more than a larger one, so consider the amount of head you are adding to the system when replacing an old filter and be sure the pump can handle that value if it is significantly different from the value of the old filter.

Manufacturers' literature will tell you the amount of head their filter creates at various flow rates. You can read the pressure (in psi) from the filter pressure gauge and multiply by 2.31 to determine how many feet of head the filter is creating. Typically, filters run 12 to 20 psi, so let's use 15 as an average: $15 \times 2.31 = 34.65$ feet (10.6 meters) of head. A considerable amount. Remember too that as the filter gets dirty the pressure goes up. When it is only 10 psi higher, you are adding $10 \times 2.31 = 23.1$ more feet (7 meters) of head. A lot! Another reason to keep the filter clean.

Backwash Valves

As I noted previously, backwashing is a method of cleaning a DE or sand filter (cartridge filters do not backwash) by running water backwards through the filter, flushing the dirt out to a waste line or sewer line. There are basically two types of backwash valves—the piston and the rotary (the multiport being a variation of the rotary valve).

PISTON VALVE

Figure 5-6 shows a piston-type backwash valve. In the normal operating position, the water enters the valve (marked Pump Discharge) and, as you can see, the piston discs (item 7) only allow the water to go out to the filter tank inlet. The water is filtered through the sand or DE and returns at the top of the valve (marked Tank Outlet). Again, you can see that the piston discs (item 7) only allows the water to flow to the outlet of the valve (marked Return to Pool). Normal filtration has occurred.

To backwash the filter and flush out the dirt, the handle of the piston assembly (item 2) is raised. Figure 5-7 shows the same valve with the piston raised to the backwash position. Figure 5-6, item 3, shows two rolled pins that ensure exact placement of the piston inside the body. The outside pin acts as a stop when pushing the valve down, the inside pin acts as a stop when pulling it up.

Notice that the water still enters from the pump, but now the piston discs force the water to flow *in* to the filter tank through the *outlet* opening. The water flows backward through the filter, flushes the dirt out of the tank and *out* of the valve *inlet* opening. As you see, once inside the valve again, the dirty water is directed to the line marked Waste. O-rings on the piston discs ensure that water does not bypass the discs.

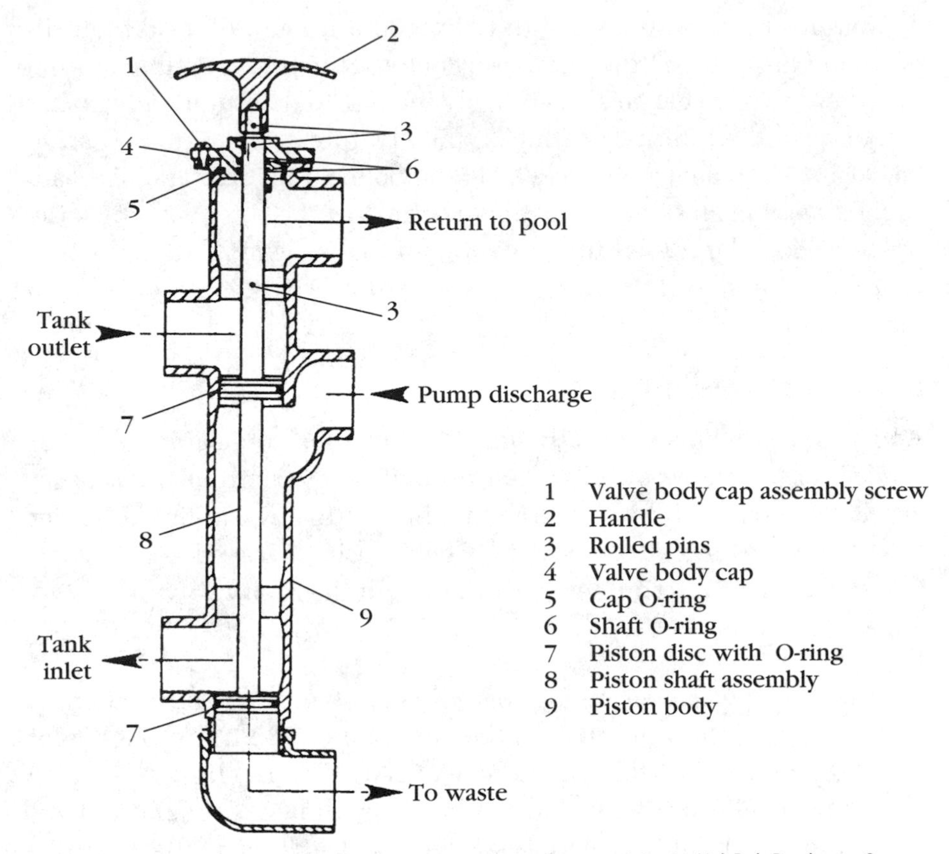

FIGURE 5-6 Piston backwash valve (normal operating position). *Val-Pak Products, Canyon Country, Calif.*

By the way, never change the piston position when the pump is running, because this puts excessive strain on the pump and motor and the O-rings of the valve, causing it to leak. The piston-type backwash valve is usually plumbed onto the side of the filter tank.

ROTARY VALVE

The rotary backwash valve, used only on vertical DE filters, does the same thing as the piston valve and in pretty much the same way. Figure 5-8 shows the typical rotary valve. Changing the direction of water flow is done by rotating the interior rotor (item 11 for the bronze, or item 11A for the plastic version). A rotor gasket seal (item 3 or 3A) or O-rings (item 4) keep water from leaking into the wrong chamber.

This type of valve is mounted underneath the filter tank. The tank has a hole in the bottom to accommodate the unit. The valve body (item 1) is held under the tank while a compression retaining ring (item 13) is placed inside the tank and the two halves are bolted together with cap screws (item 14). The handle underneath allows you to rotate the interior rotor to align it with the openings of the valve body as desired.

In normal filtration, water comes into the valve through the opening marked From Pump Discharge and flows up into the filter through the large opening on top of the rotor marked Normal Water Flow. After passing through the DE-coated grids, the water flows back through the inside of the grids, into the manifold (see the DE filter section), down through the center of the rotor (marked Return Water), then out the effluent opening and on to the heater or back to the pool.

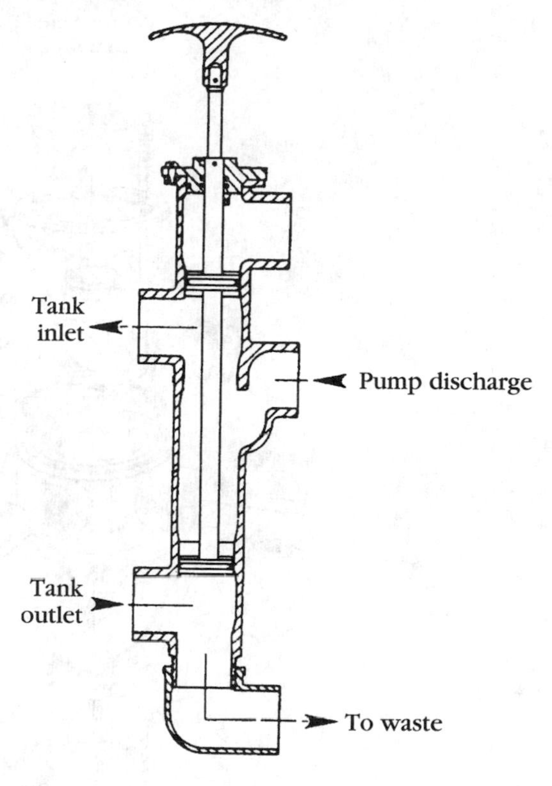

FIGURE 5-7 Piston backwash valve (backwash position). *Val-Pak Products, Canyon Country, Calif., modified by author.*

When the valve is rotated, water is sent up through the middle of the rotor, up inside the grids. The dirt and DE are pushed off the grids and the water flows from inside the grids to the outside. This is flushed back through the rotor, opposite the normal flow, and directed to the opening marked Backwash. This line is connected to a waste or sewer line. As with all backwash valves, do not operate it while the pump is running or you might damage the valve and the pump and motor.

MULTIPORT VALVE

The multiport backwash valve (as shown in Fig. 5-3) is used on sand filters and looks just like a rotary valve when disassembled, except that there is one more choice for water flow. A rinse is added, so that after the water has backwashed through the filter, clean water from the pump can be directed to clean out the pipes before returning to normal

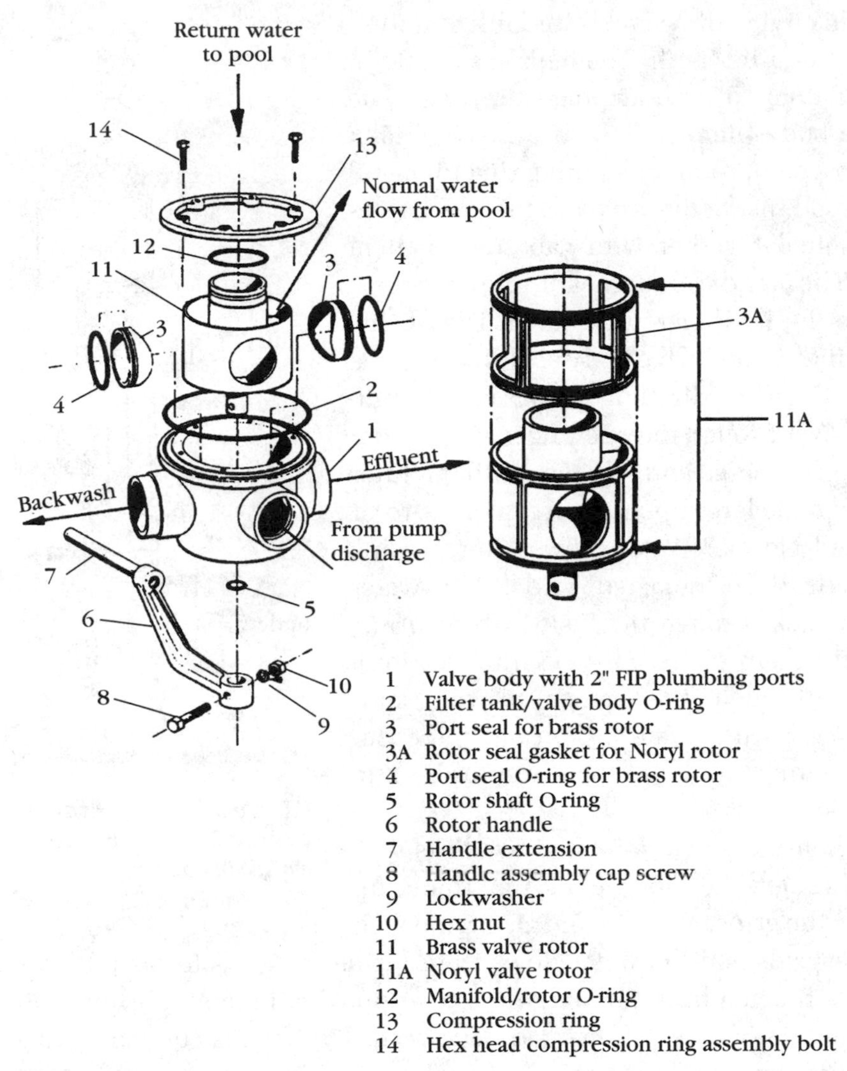

FIGURE 5-8 Rotary backwash valve. *Pentair Pool Products.*

filtration. This prevents dirt in the lines from going back to the pool after backwashing. If you understand piston and rotary valves, you will have no trouble with a multiport valve when you encounter it in the field.

Backwash Hoses

To accomplish the noble purpose of backwashing, the dirty water has to go somewhere. Some waste or backwash openings are plumbed directly into a pipe that sends the water into a nearby deck drain and on toward the sewer. Many are not hard-plumbed and you must connect a hose to direct the waste water wherever you want it to go.

By the way, dirty filter water is an excellent fertilizer for lawns because it is usually rich in biological nutrients, algae, decaying matter, and DE (which, as you now know, is a natural material). You can run your backwash hose on the lawn or garden, providing the water chlorine residual level is not above 3 ppm (see Chap. 8). Chlorine levels higher than that might burn the grass.

A backwash hose can be your pool vacuum hose, clamped onto the waste line of the backwash valve with a hose clamp. Normally, however, a cheap, blue, collapsible plastic hose is attached to the waste opening with a hose clamp. This 1½- or 2-inch-diameter (40- or 50-millimeter) hose is intentionally made flimsy because the water is not under much pressure when draining to waste and also to allow it to be easily rolled up and stored near the filter. This ease of rolling up is an advantage because backwash hoses normally come in lengths of 20 to 200 feet (6 to 60 meters)—you might have to route waste water into a street storm drain, so very long lengths are common.

TRICKS OF THE TRADE: THE POOL BLOOD PRESSURE MONITOR

- **When a filter is newly put into service or has just been cleaned, I make it a habit to note the normal operating pressure (in fact, I carry a waterproof felt marking pen and write that pressure on the top of the filter). Most manufacturers tell you that when the pressure goes more than 10 pounds (700 millibars) over this normal operating pressure it is time to clean the filter.**
- **The other value of the pressure gauge is to quickly spot operating problems in the system. If the pressure is much lower than normal, something is obstructing the water coming into the filter (if it can't get enough water, it can't build up normal pressure). If the gauge reads unusually high, either the filter is dirty or there is some obstruction in the flow of water after the filter.**
- **When the pressure fluctuates while the pump is operating, the pool or spa might be low on water or have some obstruction at the skimmer—when the water flows in, the pressure builds, then as the pump sucks the skimmer dry, the pressure drops off again. This cycling will repeat or the pressure will simply drop altogether, indicating the pump has finally lost prime.**

Pressure Gauges and Air Relief Valves

Most filters are fitted with a pressure gauge, mounted on top of the filter (Fig. 5-4A, item 1). Sometimes the gauge is mounted on the multiport valve (Fig. 5-3A, item 12). These gauges read 0 to 60 psi (0 to 4000 millibars) and are useful in several ways.

Some codes, especially for commercial pools or spas, require a pressure gauge be mounted on the incoming pipe and on the outgoing pipe to compare the differential. Obviously, as the water tries to push from the pump into the filter, the pressure is greater than as it leaves the filter. As you saw in the hydraulics section, even a clean filter creates this back pressure.

One way for a health department inspector to check the cleanliness of your filter system is to compare these two pressure readings and when that difference exceeds a certain amount, they know the filter is dirtier than whatever standard they have set. Normally the differential should be 2 to 4 psi (140 to 280 millibars); when it reaches 10 psi (700 millibars), the filter needs cleaning.

Mounted on a yoke or T fitting along with the pressure gauge, you will normally find an air relief valve (Fig. 5-4A, item 2). It is simply a threaded plug that when loosened allows air to escape from the filter until water has fully filled the tank. When a system first starts up, particularly after cleaning or if the pump has lost prime, there is a lot of air present in the filter. If it is left like that, the filter might be operating at only half its capacity.

Figure 5-4B shows an inside view of a cartridge filter. The area above the cartridge is called the *freeboard*. This empty area is present above the filter media of all filters. For some of this to be air rather than water is okay, but if the air in the tank were down to, for example, the halfway mark of the cartridge and water were flowing around only the lower half of the cartridge, you would effectively cut the filter square footage in half.

TOOLS OF THE TRADE: FILTERS

- **Heavy flat-blade screwdriver**
- **Hacksaw**
- **PVC glue**
- **PVC primer**
- **Pipe wrench**
- **Teflon tape**
- **Silicone lube**
- **Needle-nose pliers**
- **Hammer**
- **Emery cloth or fine sandpaper**
- **Vise-Grips**

FIGURE 5-9 **Sight glass.** *George Spelvin.*

So it is important to let the air out of the filter at every service call. Again, some air is normal, so don't be obsessive about it; just be sure that the filter media itself is covered with water, not air. Newer filters now have automatic air relief valves, leaving one less thing to chance or checklist.

Sight Glasses

A *sight glass* is simply a clear section of pipe (Fig. 5-9). The sight glass is actually composed of a metal frame with a glass interior, constructed to the same diameter as a standard pipe (1½ or 2 inches or 40 to 50 millimeters). A sight glass is installed anywhere in a line of pipe where you want to see the quality of the water.

A sight glass is used to see the effectiveness of your backwashing. Sight glasses are normally installed on the backwash line coming out of the backwash valve (going to the waste or sewer or going to a separation tank). When the dirty water starts to look clean, you know you have backwashed enough. A sight glass is handy if your waste pipe goes into a deck drain or if you use a long backwash hose where you can't see the clarity of the water being flushed out.

Repair and Maintenance

As noted previously, there are no moving parts on a filter when it operates and few when it is at rest. Therefore, there's not much to break down and, when they do, filters are easy to repair.

Installation

RATING: EASY

Replacing a filter is one of the easiest repair jobs you can get. There is no electricity or gas to hook up and normally you are dealing with only three pipes.

1. **Shutdown** Turn off the pump and switch off the circuit breaker. This way you can make sure it won't come back on (from the time clock, for example) until you're ready.
2. **Drain** Drain the old filter tank by opening the drain plug or backwash valve. If you don't do this first, you'll take a bath when you cut the plumbing.

TRICKS OF THE TRADE: FILTER INSTALLATION

- On a filter with threaded openings in a plastic base (like most cartridge filters) or a plastic multiport valve (like most sand filters), be careful not to overtighten—you'll crack the plastic. On a vertical grid DE filter with the rotary valve on the bottom, lay the filter down and screw the MIPs into place. Some manufacturers make this a knuckle-busting experience. Be sure you have the fittings tight—if they leak, you'll have to cut out all your work to fix it.
- Some manufacturers still provide filters with bronze plumbing openings on the backwash valve for sweating copper pipe. If you encounter this type of plumbing, remove any parts that might suffer from heat when you sweat the fittings in place. Plastic grids and manifolds will melt if they are not removed, and with the lid on the filter, the heat inside will build up quickly. Just as the backwash valve conducts water efficiently, so too will it conduct heat to those plastic parts.
- If a check valve was not a part of the plumbing between the filter and the heater, it is not a bad idea to add one now. If there is any chance of hot water flowing back into the filter, you run the risk of melting the grid or cartridge materials. Also, you don't want the heater to sit without water in the heat exchanger (see the heater chapter).

3. **Cut Out** Cut the pipe between the pump and filter in a location that makes connecting the new plumbing easiest. There's no rule of thumb here, just common sense. If the original installation has more bends and turns in the pipe than needed, now is a good time to cut all that out and start over. Eliminating unnecessary elbows increases flow and reduces system pressure. Cut the pipe between the old filter and the heater using the same guidelines. Last, cut the waste pipe if there is one plumbed into a drain.
4. **Remove** Remove the old filter. Even without water, the old filter will be heavy, so you might want to disassemble it for removal. Sand filters are the worst, and you will have to scoop out the heavy, wet sand before you will budge the tank. Save any useful parts. Old but still working grids, valves, gauges, air relief valves, lids, lid O-rings, cartridges, and other components make great emergency spares to carry in your truck. There are some companies that recondition old filters or parts, so when you make the deal with the homeowner to replace the filter, make it like a car battery or tire sale—you get the trade-in. The customer doesn't want it anyway and their trash service probably will refuse to pick it up for disposal, so agree to do the hauling away in exchange for any usable parts.
5. **Installation** Set up the new filter. After removing it from the box, make sure all the pieces are there—grids, pressure gauge, etc. Most new filters come with instructions and it really pays to read these. In fact, if the homeowner doesn't want the booklet, add it to your library for reference. While the unit is out in the open and easy to work on, screw into place the appropriately sized MIP fittings [1½ or 2 inch (40 or 50 millimeter)], after applying a liberal coating of Teflon tape or pipe dope (see the plumbing section). Some manufacturers include their preference of tape or pipe dope right in the box.
6. **Plan the Plumbing** Place the new filter in the location by the pump (Fig. 5-10A) and figure out the plumbing between the pump and filter, between the filter and heater, and between the filter and waste line (if appropriate). Some creativity and planning here will save many service headaches later. Basically you want to avoid elbows and you want to leave enough room between the

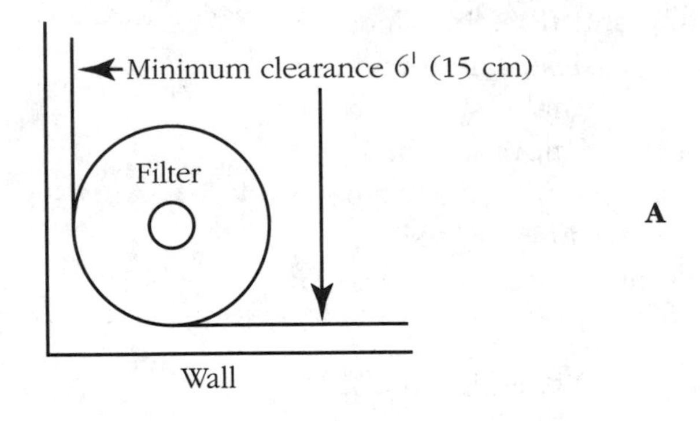

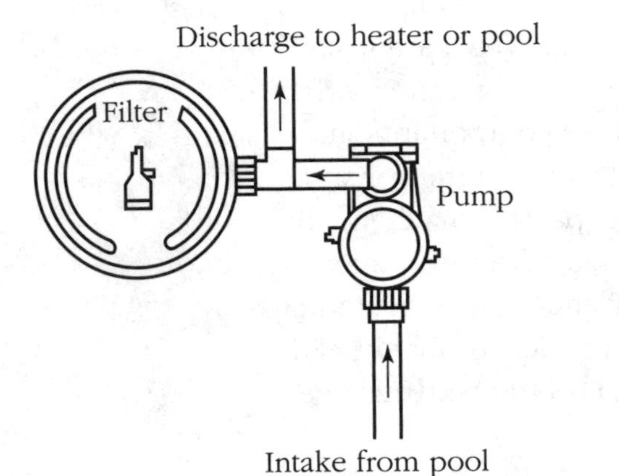

FIGURE 5-10 (A) Placement of a filter, (B) plumbing a filter.

pump, filter, heater, and pipes for service access (Fig. 5-10B). Remember, you'll have to clean this filter someday. Can you easily access the lid? Will water flowing out of the tank as you clean it flood the pump? Can you access the backwash valve and outlet as needed?

7. **Plumb** Plumb it in. Use the plumbing instructions in the chapter on basic plumbing and make careful connections. Who was it that said, "There never seems to be enough time to do it right, but there's always enough time to do it over when it leaks"?

Starting up the newly installed filter is just like restarting after cleaning, so read further for startup procedures and hints.

Filter Cleaning and Media Replacement

The most important thing you can do for a pool is to keep the filter clean. This is also the simplest way to ensure the other components work up to their specifications and you end up with satisfied clients. Let's review the process for cleaning each type of filter.

DE FILTERS

I do not believe backwashing a DE filter is of any value as a regular practice and, in fact, I know it can be harmful to the filter and pool cleanliness. Obviously the makers of backwash valves and those who have bought into their technology over the years will disagree with me, but here's what I believe based on years of experience. When you backwash, some dirt and some DE are flushed from the filter. The remainder drops off the grids and falls to the bottom of the filter in clumps. The manufacturers say that after backwashing 70 percent of

QUICK START GUIDE: FILTER CLEANING (ALL TYPES)

Rating: Easy

1. Prep

- **Shut off pump and tape over switch or breaker, so no one can turn it on before you finish.**
- **Isolate equipment by closing valves at suction line (at skimmer and/or main drain connection before pump) and return line (at pool discharge outlet).**

2. Disassemble

- **Remove filter top or lid.**
- **DE FILTER: Carefully remove retainer, grids, manifold (Fig. 5-11A).**
- **CARTRIDGE: Remove cartridge.**
- **SAND: Remove debris basket or other components to expose sand inside filter.**

3. Clean

- **DE or CARTRIDGE: Hose off grids or cartridges thoroughly (Fig. 5-11B). Soak in cleaning solution, if needed, and rinse thoroughly. Rinse interior of filter and any other components.**
- **SAND: Insert hose and flush dirt from sand, stirring it with a broom handle as you flush. Rinse until water flows out clean.**

4. Reassemble

- **Rebuild components, tops, and lids in reverse order of disassembly procedure.**

5. Restart

- **Reopen pool plumbing valves.**
- **Start pump, purge air, and check for leaks.**
- **DE ONLY: Add DE at skimmer slowly, watching for any that passes back into pool (if DE enters pool after first 20 seconds of circulation, repeat steps 1–4).**

the DE has been removed, so you need to replace that amount. I have opened up filters that were just backwashed and have seen as little as 10 percent of the DE washed away, while others had virtually 100 percent washed away.

If you don't know how much went out, how do you know how much to put back in? If you add too little, the filter grids will quickly clog with dirt and the pressure will build right back up, even stopping the flow of water completely. If you add too much, you will get the same effect by jamming the tank with DE.

Moreover, backwashing cannot remove oils from the grids, which get there from body oil, oil in leaves, and suntan lotions. You can backwash for hours and when you open up the filter, you will find the grids clogged with oils and a layer of DE and dirt that sticks to these oils.

Finally, backwashing wastes water. If you break down the filter and clean it completely, you will use some water to wash the grids and tank, but you will not have to clean it again for weeks or even months.

When you backwash you really are not cleaning the filter thoroughly. You'll be backwashing again in a few days or weeks and when you get tired of that you will break down and clean the filter anyway.

Okay, I've had my say! Now I'll tell you the one time when backwashing a DE filter is useful. You have a pool that has been trashed by winds, mudslides, algae, or other heavy debris. You start to vacuum it and quickly the filter can't hold any more dirt. To save a lot of time you backwash, add a little fresh DE, and get on with the job. You repeat this process until the big mess is cleaned up, then you break down the filter and clean it properly.

The other time you might backwash is when you're vacuuming a normally dirty pool, but the filter hasn't been cleaned in awhile and is just about full of dirt. You're getting no suction because the filter is clogged. It's Friday night at 4:45 p.m. and you have a hot date. Okay, okay. Backwash, add some fresh DE, finish cleaning the pool, and make a note to do a breakdown (clean the filter) on Monday.

Another important fact about backwashing. Since the water is going inside the grid and flowing outward, any debris in the water from the pool will clog the inside of the grids (or laterals on sand filters), rendering them useless. On a new pool startup where a lot of plaster dust or gunite debris might be in the water, don't backwash. If you must, open the strainer pot and turn on the pump. Flood the pot with water from a hose and backwash as needed that way. Obviously, never vacuum a pool with the filter on backwash because the dirt and debris you vacuum will flow directly inside the grids (or laterals).

Cleaning (Rating: Easy): So how do you properly break down (or tear down) and clean a DE filter? I'm going to describe a common style of vertical grid tank DE filter (Fig. 5-1). They are common in the field and if you can do these, you can do them all.

1. **Shutdown** Turn off the pump and switch off the circuit breaker.
2. **Lid Removal** Remove the lid of the filter. That might sound easy, but depending on the design, it can be real work. On some filters, it is as easy as removing the clamping ring and applying light pressure under the lid with a screwdriver (be careful not to gouge the lid or O-ring—if you do, leaks will develop in these spots).
3. **Grid Removal** Open the tank drain and let the water run out. Remove the retainer's wing nut and remove the retainer (also called the holding wheel). Now gently remove the grids (elements). One design flaw of many grids lies in the fact that they are made like small aircraft wings—large, curving units—but they are set into the manifold on stubby little nipples. Applying a reasonable amount of

TRICKS OF THE TRADE: FILTER CLEANING

- **Some filters make such a tight seal with the O-ring and lid that a nuclear bomb will not remove them. Draining the tank first won't help—as the water drains out, it sucks the lid on even tighter. What I do is, after removing the clamping ring, turn on the pump for a few seconds. The pressure from the incoming water will pop the lid off. When it does, grab it quickly, otherwise when the pump is shut off and the water recedes, it will suck the lid right back onto the tank.**
- **Some manufacturers do not approve of this procedure. I have "popped" literally thousands of filter lids and never had one pop more than a few inches off the tank. I have also never observed damage to the equipment by this technique. The pressure is applied evenly as the lid pops off, so the tank or lid doesn't warp or bend. Be sure not to grasp the lid by the gauge assembly—they snap off easily.**
- **The only caveat to this practice is to be sure that you don't stand in the water that will inevitably flow out of the tank as the lid pops while you are holding onto the pump/motor switch. Water and electricity can be deadly. Also, if the motor is installed directly adjacent to the filter tank, the water flowing out might flood the motor. In that case, get a screwdriver and crowbar to remove the lid, or wrap the motor in a plastic bag.**

force on the rather large wing part of the grid won't hurt it, but the resulting torque on the flimsy nipple will snap it right off. Therefore, to remove the grids, wiggle them gently from side to side as you pull them straight up and out (Fig. 5-11A).

Tennis star John McEnroe fired his pool man of many years and called my company. I went over and found the pool very dirty and the filter pressure almost off the scale. I popped the lid and found so much dirt and DE in the tank that there was literally no room for water. This condition is called bridging because the DE and dirt bridge the normal gaps between the grids, clogging and effectively reducing the amount of filter area. It was so packed that it took me three hours of archaeologist-like excavation with a small spade, stick, and lots of water to finally get the grids free. This, by the way, was the result of a pool guy who backwashed and added fresh DE about once a month—for 6 years. He never once opened the filter.

The point of telling you this is that to some lesser degree you might find the same condition when you open a filter. Be prepared to hose out the tank while the grids are still in place (if the drain hole isn't also clogged with DE) or patiently excavate the dirt and DE until you can free the grids.

4. **Rod Removal** Remove the retaining rod. It threads into the base of the rotary valve like a screw, so just unscrew it. Sometimes it is corroded in place, so have pliers (Vise-Grips work best) handy to grip the rod and unscrew it. A word of caution—the rod might be corroded enough that if you force it with your pliers, it snaps off at the bottom, leaving the threaded end in the rotary valve.

   If you find too much resistance to your effort to remove the rod, leave it in place and clean the tank as best as you can. If it has broken off and you don't have time to disassemble the entire tank and rotary valve to get out the stub, lay a brick on top of the holding wheel when you put the unit back together to keep it and the grids in place. Then come back and fix it when you have time.

5. **Manifold Removal** Reach in the tank and remove the manifold. It rests just inside the rim of the rotary valve; it is not threaded in place.

6. **Cleaning** Hose out the inside of the tank, the manifold, and the holding wheel. Hose off the grids (Fig. 5-11B). You might need to

scrub them lightly with a soft bristle brush to loosen the grime. If the grids are still dirty, soak them in a garbage can of water, trisodium phosphate (1 cup per 5 gallons of water or 250 milliliters per 19 liters), and muriatic acid (1 cup per 5 gallons) (acid alone will not clean grids because it does not affect the oil). After 30 minutes, try scrubbing them clean again. Don't use soap—you won't get it all out, no matter how well you rinse the grids, and when you start up the circulation again you'll have soap suds in the pool.

7. **Reassemble Rod and Manifold** Inspect the manifold for chips or cracks. DE and dirt will go through such openings and back into your pool. Cracks can be glued. If chunks of plastic are missing, buy a new manifold—they're only about $30. Particularly inspect the joint between the top and bottom halves of the manifold. Where these two parts are glued together, they often start to separate. Replace the manifold as you took it out. Reinstall the center rod.

8. **Reassemble Grids and Retainer** Carefully inspect the grids before putting them back inside. Look for worn or torn fabric, cracked necks on the nipples, or grids where the plastic frame has collapsed inside the fabric. Replace any severely damaged grids. When you reinstall the grids, notice that inside each hole in the manifold is a small nipple and on the outside of each grid nipple is a small notch. By lining up the nipple and notch as you reinsert each grid, the grids will go back as intended.

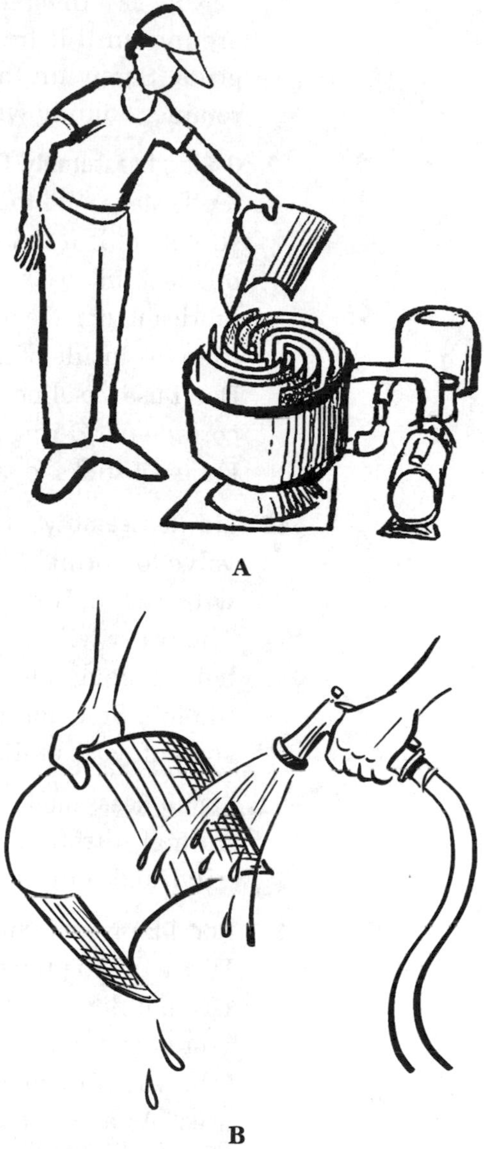

A

B

FIGURE 5-11 Key steps of filter cleaning.

Now lay the retainer over the tops of the grids and spin it around until it finds its place holding down and separating the grids. Screw on the wing nut and washer that holds down the retainer holding wheel.

9. **O-Ring Reassembly** Getting the lid back on can be as tough as getting it off. Make sure the O-ring on the tank is free of gouges and has not stretched. If it is loose, soak it for 15 minutes in ice water and it might shrink back to a good fit. If not, replace it. Apply tile soap as a lubricant to make it slide on easier (or silicone lube if you can afford it) to the inside of the lid around the edge that will meet the O-ring. Don't use Vaseline or petroleum-based lubricants because these will corrode the O-ring material. Don't use that green slime called Aqua Lube—it sticks to everything and comes off of nothing.

10. **Lid Reassembly** Now close the tank drain, turn the backwash valve to normal filtration, and turn on the pump. Let the tank fill with water. Turn off the pump and turn the valve to backwash. The water will drain out, sucking the lid down. Don't be afraid to help it along by getting on top of the lid. Your weight will finish the job. Be careful not to hit the pressure gauge assembly—they snap off very easily.

11. **Clamp Reassembly** Replace the clamping ring, return the valve to normal filtration, and start the pump/motor. Open the air relief valve and purge the air until water spurts out the valve.

12. **Add DE** Never run a DE filter without DE, even for a short time. Dirt will clog the bare grids. Remember, it's not the grids, but the DE that does the actual filtering. The label on the filter will tell you how much DE to add, or refer to the table on the bag of DE. It tells you how many pounds of DE to add per square foot of filter area. As a convenient scoop, use a 1-pound (½-kilogram) coffee can; but remember that because DE is so light and powdery, a 1-pound coffee can holds only ½ pound (226 grams) of DE. One pound of DE covers 10 square feet (1 square meter) of grids.

    DE is added to the system through the skimmer. Do not dump it in all at once. It will form in clumps at the first restricted area, like a plumbing elbow or the inlet on the filter tank. Sprinkle in one can of DE at a time, mixing it in the skimmer water with your hand. It should appear to be dissolving in the water. In fact, it is not dissolving, just freely suspending itself in the water, but this will keep it from clumping.

Add one can and wait about a minute. If you have any gaps in the manifold or holes in the grids or you didn't assemble the unit correctly, DE will get through these areas and flow back into the pool. If that happens, it is better to have one can of DE flowing back into the pool rather than all 10 or 15. On most startups you will see a little milky residual entering the pool from any DE or dirt that settled in the pipes during cleaning. This is normal, but if you see great clouds of DE returning to the pool, shut the system down and take the filter apart. You missed something.

I must emphasize this problem of adding DE too fast (or using too much—follow the DE package directions). Comedian Rich Little's system was sluggish and the filter pressure unusually low. I cleaned the pump strainer, skimmers, blew water through the suction lines with a drain-flush bag, and cleaned the filter twice . . . and no luck. I took apart the pump to make sure the impeller was clean and operating properly. I checked the power supply, thinking a bad breaker was perhaps delivering low voltage and making the motor work too slowly. No luck.

Finally, after hours of hunting, I emptied the filter tank and fed hose water directly into the pump and watched it trickle into the tank. I knew the obstruction was in the plumbing between the pump and the rotary valve. I cut open the plumbing and, sure enough, at one of the 90-degree elbows I found clumps of DE. The

TRICKS OF THE TRADE: DE

- **A good way to add DE to prevent clumping is in a slurry. On large commercial installations that require large amounts of DE, there is actually a slurry pit, but you can use a bucket. The concept is that you thoroughly mix the DE in water (achieving that suspension I spoke of) before pouring the solution into the skimmer.**
- **Most pools have skimmers where you can add DE. But what if there isn't one or if a hot tub or other body of water that has no skimmer uses a DE filter? Again, make a slurry in a bucket. After cleaning the filter, take the lid off the pump strainer pot and turn on the pump. Add the slurry to the strainer pot, followed by clear water (have a hose handy) to make sure all the DE gets to the filter and completely and evenly coats the grids. Turn off the pump, fill the pot with water, replace the lid, and reprime the system. Again, remember to let the air out of the tank.**

pressure had crystallized it and made it rock hard. I had to replace that section of plumbing and later in the shop I took a hammer and chisel to the DE to see just how hard it had become.

SAND FILTERS

Sand filters are designed to use #20 silica sand, a specific size and quality of sand. Larger sand will not filter fine particles from the water and finer sand is pushed through the slits in the laterals, clogging them.

Sand filters do need regular backwashing and, unlike DE backwashing, it is effective. Although wet sand is heavy and lays on the bottom, when the tank is full of circulating water the sand is suspended in the tank. In fact, you can reach into a tank of sand and water and get your hand all the way to the bottom of the tank. Try that without water in the tank. Your hand will push about 6 inches into the thick sand.

Backwashing (Rating: Easy): Most rotary valves have the steps printed right on them, and they are very simple.

1. **Prepare** Turn off the pump. Rotate the valve to Backwash. Roll out your backwash hose or make sure the waste drain is open.
2. **Flush** Turn on the pump and watch the outgoing water through the sight glass. It will appear clean, then dirty, then very dirty, then it will slowly clear. When it is reasonably clear, turn off the pump and rotate the valve to Rinse.
3. **Rinse** Turn the pump back on and run the rinse cycle for about 30 seconds to clear any dirt from the plumbing. Turn off the pump, rotate the valve back to Filter, and restart the pump for normal filtration.

Backwash as often as necessary. When the filter gauge reads 10 psi (700 millibars) more than when the filter is clean, it is usually time to backwash. A better clue is when dirt is returning to the pool or when vacuuming suction is poor.

When backwashing, be sure there is enough water in the pool to supply the volume that will end up down the drain. It is usually a good idea to add water to the pool or spa each time you backwash.

Teardown: Twice per year, I recommend opening the filter. Sand under pressure and with the constant use of pool chemicals or dissolving pool plaster will calcify, clump, and become rock-like over time. Passages are created through or around these clumps, but less and less water is actually filtering through the sand and more is passing around it. This is called *channeling*. To correct or avoid this problem, regular teardown is the answer.

1. **Shutdown** Turn off the pump. Disconnect the multiport valve plumbing by backing off the threaded union collars. Some valves are threaded into the body of the tank, others are bolted on. Remove the valve.
2. **Flush** Some sand filters have a large basket just inside the tank. Remove this and clean it out. The sand is now exposed. Push a garden hose into the tank and flush the sand. As noted previously, it will float and suspend in the water. Use a broom handle to bust up clumps. As the water fills the tank, it will overflow, flushing out dirt and debris. Be careful not to hit the laterals on the bottom of the tank because they are fragile and break easily.
3. **Reassemble** When the sand is completely free and suspended in the water, not clumped, turn off the water and replace the basket (if any), multiport valve, and plumbing. Backwash briefly to remove any dirt that was dislodged by this process but not yet flushed out.

This teardown process also allows you to check to see if the regular backwashing has flushed out too much sand. You might need to add some fresh sand. Most sand filters need to be filled about two-thirds with sand and have one-third freeboard. Backwash after adding any new sand to remove dust and impurities from the new sand.

Replacing Sand (Rating: Easy): Every few years you need to replace the sand completely because erosion from years of water passing over each grain makes them round instead of faceted and rough. Smooth sand does not catch and trap dirt as efficiently and it slowly erodes to a smaller size than the original #20 silica, allowing it to clog laterals and pass into the pool. Some manufacturers suggest adding a few inches of gravel over the laterals first. This keeps the sand separated from the laterals so the sand cannot clog them. To replace sand, or add sand to a new installation:

> **TRICKS OF THE TRADE: ALUM**
>
> **If channeling is a problem because of hard water or pool chemistry (which speeds up calcification of the sand), introduce aluminum sulfate (alum) through the skimmer just like you would add DE to help prevent this problem. Use the amounts recommended on the bag, but usually about ½ cup per 3 square feet (125 milliliters per 0.3 square meter) of filter size.**

1. **Remove** Open the filter as described previously. Remove the old sand by scooping it out with your hands or with a sand-vac (Fig. 5-12).
2. **Add Water** Fill the bottom third of the tank with water to cushion the impact of the sand on the laterals.
3. **Add Sand** Slowly pour the sand into the filter, being careful of the laterals. Fill sand to about two-thirds of the tank. Reassemble the filter parts and backwash to remove dust and impurities from the new sand, then filter as normal.

CARTRIDGE FILTERS

RATING: EASY

Cleaning a cartridge filter is perhaps easiest of all.

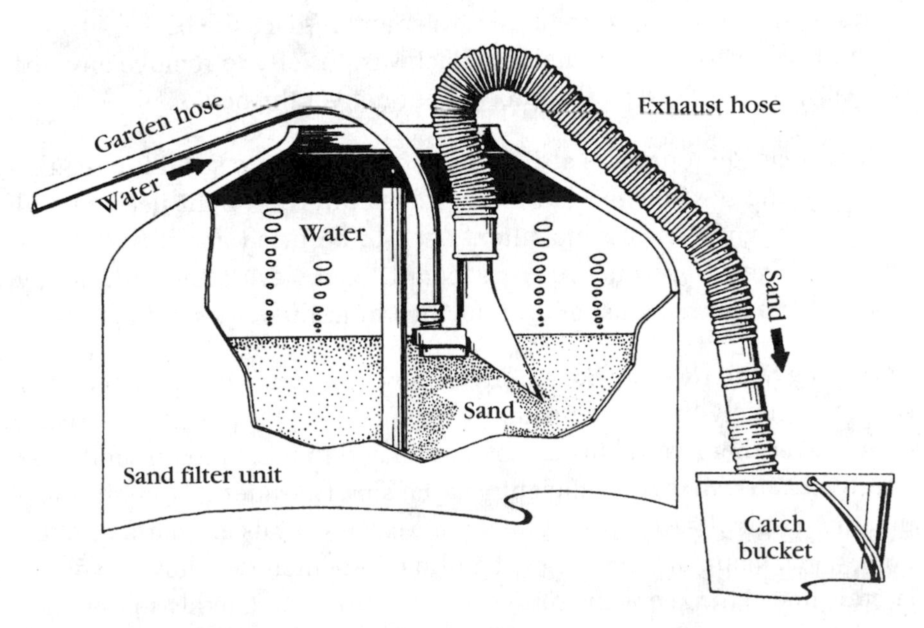

FIGURE 5-12 Sand filter vacuum. *Lass Enterprises, Altamonte Springs, Fla.*

1. **Shutdown** Turn off the pump. Remove the retaining band (Figs. 5-13A and B) and lift the filter tank or lid from the base. Remove the cartridge.
2. **Clean Cartridge** Light debris can simply be hosed off (Figs. 15-13C and D), but examine inside the pleats of the cartridge. Dirt and oil have a way of accumulating between these pleats. Never acid wash a cartridge. Acid alone can cause organic material to harden in the web of the fabric, effectively making it impervious to water. Soak the cartridge in a garbage can of water with trisodium phosphate (1 cup per 5 gallons or 250 milliliters per 19 liters) and muriatic acid (1 cup per 5 gallons). About an hour should do it. Remove the cartridge and scrub it clean in fresh water. Don't use soap. No matter how well you rinse, some residue will remain and you will end up with suds in your water.
3. **Reassemble** Reassemble the filter and resume normal circulation.

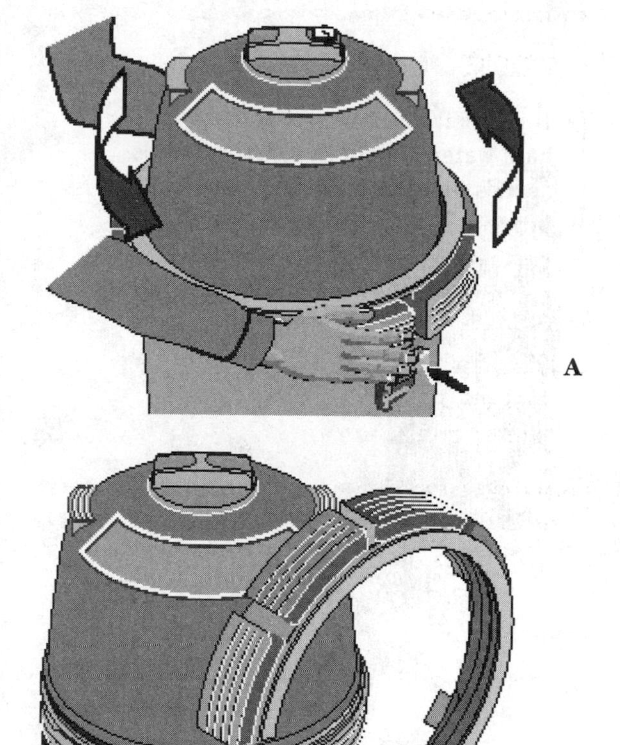

FIGURE 5-13 Cleaning a cartridge filter. *Sta-Rite Industries, Delevan, WI.*

Leaks

Filters, being the rather simple creatures they are, don't have many repair or maintenance problems beyond cleaning as discussed. The biggest general complaint, however, relates to leaks of various kinds.

BACKWASH VALVES

Backwash valves leak in two ways: internally and externally. Internally, O-rings deteriorate and allow water and DE or dirt to pass into areas not intended.

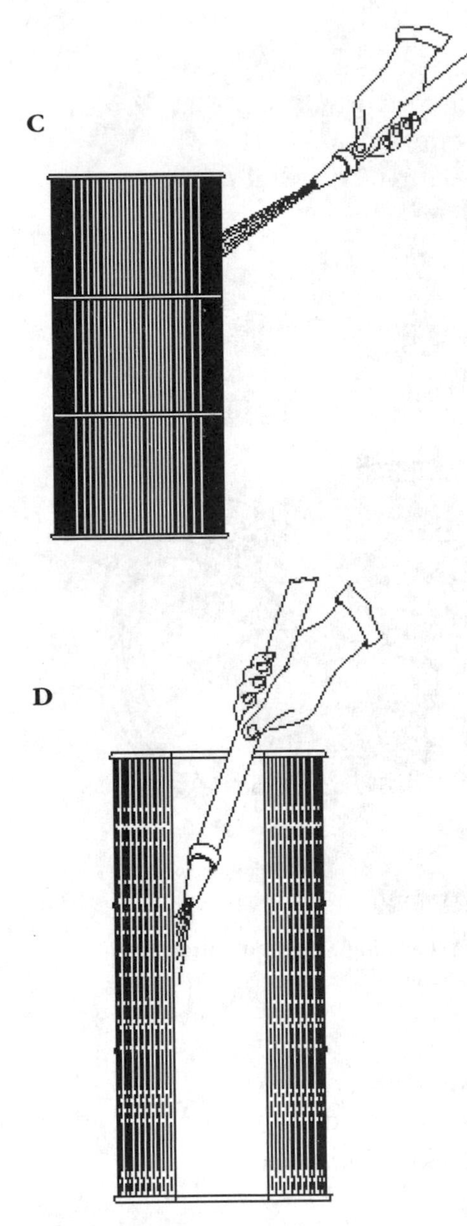

FIGURE 5-13 *(Continued)*

Piston Backwash Valves (Rating: Easy): The valve in Fig. 5-6 has piston discs equipped with O-rings (item 7). As these wear out, water or dirt bypasses the intended direction. Similarly, the O-rings on the shaft (just under the handle) wear out from regular repeated use. If you suspect the disc O-rings or see water leaking from the top of the shaft, tear down the valve as follows:

1. **Teardown** Turn off the pump. Remove the screws on top of the valve cap. Pull the handle up as if you were going to backwash, but keep pulling straight up to remove the entire piston assembly.

   Replace the O-rings on each disc. They pull off like rubber bands and the new ones go on the same way. Apply silicone lube to the O-rings.
2. **Shaft O-Rings** Remove the handle from the piston stem. It is held in place by setscrews or allen-head screws. This also allows you to slide thc cap off the stem. Look inside the cap. You will find two small O-rings. Pull these out with the tip of a screwdriver and replace them. Apply silicone lube.
3. **Rebuild** Clean (as needed) the stem and disc assembly and flush out the inside of the valve body. Grit or sand can create leaks or cause your new O-rings to wear out sooner than necessary. Reassemble the unit the same way you took it apart.

Rotary or Multiport Backwash Valves (Rating: Advanced): Rotary and multiport valves are similar in construction, so if you understand one you will not have trouble with the numerous other designs. Figure 5-8 shows a rotary valve normally mounted under a vertical grid DE filter. As with piston-type units, these leak either externally or within the chambers of the unit itself. If water appears under the filter, use a flashlight to inspect underneath as carefully as possible. If you can see or feel a leak where the plumbing enters the valve openings, you can repair that without disassembling the entire filter. If the leak appears to be at the joint of the valve and filter tank, or if the problem is DE and dirt bypassing the normal flow and getting back into the pool, you will need to tear down the filter and valve.

Another typical symptom of an internal leak is drips coming from the backwash outlet even though the valve is turned completely to the normal filtration position. Figure 5-8 employs a rotor seal (item 3A) that can compress or wear out. When the body gasket wears out (and it will wear out prematurely by rotating the valve with the pump on) and water bypasses the normal flow, some leakage gets to the backwash side and appears as a leak under the filter. If the backwash outlet is plumbed directly into a waste or sewer drain, this leak might not be visible.

Remember this when you're looking for a pool leak—sometimes the problem is not in the pool or spa itself, but in some hidden area within the system plumbing. If the gasket completely wears out, the leaking can be substantial and, as noted previously, if the plumbing prevents you from seeing it you will never know. Such a hidden problem can also cause the system to lose prime overnight when the pump is off. The leak drains the water from the filter tank, then siphons the water out of the pump.

On startup the next day, the pump has no prime. If the pump runs dry for several hours, overheats, loosens or melts the plumbing fittings, you will attribute the loss of prime to the damaged plumbing. You repair the plumbing and the same problem occurs the next day.

The moral of the story is that it pays to have a sight glass on the backwash outflow line so you can see any leaks and/or have a shutoff gate valve on that line that stays closed when the valve is in the normal filtration position.

To tear down this type of valve (Fig. 5-8), use the following procedure.

1. **Filter Teardown** Cut the plumbing to isolate the filter, and take the unit apart as described previously.

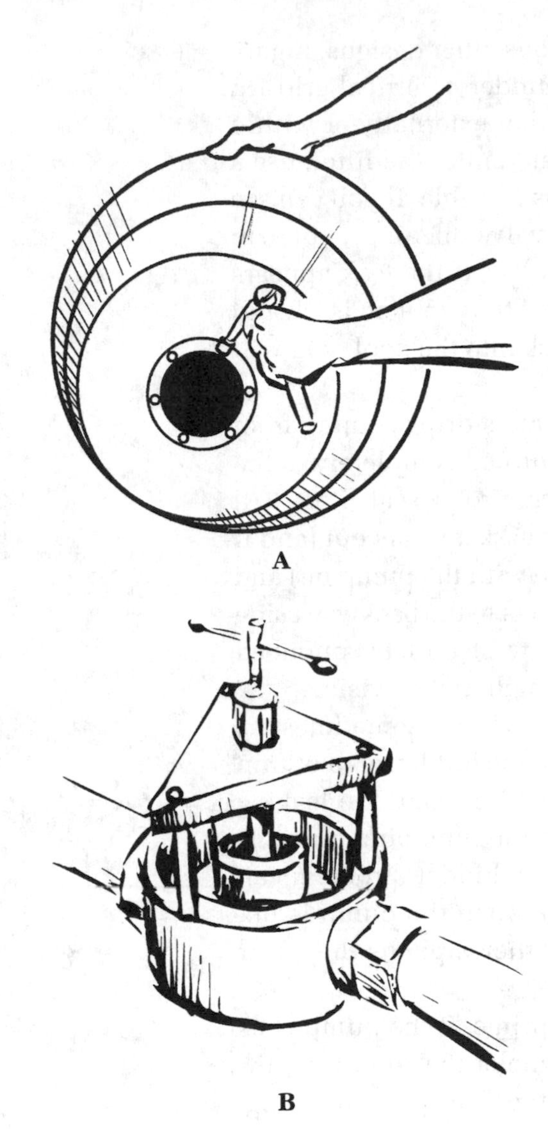

FIGURE 5-14 Rotary valve teardown.

2. **Disassemble** Reach inside the bottom of the filter (Fig. 5-14A) and remove the bolts (Fig. 5-8, item 14, usually ¼-inch or 6-millimeter hex-head bolts) that hold the compression ring (item 13) with a nut driver. This ring holds the valve in place as well, so the valve will now fall away from the filter tank.
3. **Rotor Teardown** You now have the valve body (item 1) with the rotor inside. Remove the handle (items 6 and 7) on the underside of the valve by removing the bolt assembly (items 8, 9, and 10) that holds it on the rotor shaft and slide it off the shaft. Pull the rotor out of the body. Bronze rotors are very hard to remove and you might have to take the valve to a pump rebuilding shop. Most shops have special rotor pulling tools (Fig. 5-14B) or they can carefully heat the valve body so it expands and releases the rotor. I wouldn't try to do this yourself. Without a lot of experience, you will probably warp or destroy the components.
4. **Gasket** Pull the old rotor seal gasket (item 3A) from the rotor with needle-nose pliers. Clean the rotor and inside the valve body. Put a new gasket on the rotor, being careful not to overstretch the new gasket.
5. **O-Rings** Lube the gasket with silicone lube and replace it in the valve body. On bronze rotors, each port has an O-ring instead of one body gasket seal as you will find on the plastic versions. Before reassembling the filter, replace the O-ring (item 2) that sits between the tank and valve and the O-ring (item 5) that seals the shaft as it passes through the valve body to the handle. Also

replace the O-ring (item 12) on the neck of the rotor. The grid manifold sits on this neck and the O-ring seals that joint, so to prevent dirt from bypassing the correct direction of flow, you need a good seal here. Lube all O-rings with silicone lube.

6. **Rebuild** Reassemble the valve and tank the way you took it apart. Be sure the tank itself is clean (dirt or sand will prevent the O-ring from sealing tightly) and that the opening in the bottom shows no rust or cracks. If it does, you should clean it thoroughly and have the cracks welded. Such weak spots will come back to haunt you, so it might be time to suggest to your customer a new tank (or new filter). Replumb and restart the filter as described previously.

LIDS AND GAUGE ASSEMBLIES

RATING: EASY

Lids on filters leak in two places: the O-ring that seals them to the tank and/or the pressure gauge air relief valve assembly. The lid O-ring can sometimes be removed, cleaned, turned over (or inside out), and reused. I have not recommended that for other O-rings on the filter, such as in the backwash valve, because if they are too worn or compressed to reseal, you must tear down the entire filter again to replace them. Hardly worth the price of an O-ring. But the lid O-ring is thick and expensive and easy to remove and replace. So try the cleanup/turnover method and if you still have leaks, then replace it.

Some filters will crack on the rim of either the lid or the tank where the O-ring is seated. Obviously, the problem in this case is not a bad O-ring, but a bad lid or tank. Inspect these stress areas carefully for hairline cracks that might be the source of the leak.

Air relief valves (Fig. 5-15) sometimes leak if they become dirty or they simply wear out. Some are fitted with an external spring that applies tension to create the seal, and when the spring goes, so does the watertight seal. Others have a small O-ring on the tip of the part that actually screws in to create the seal. Unscrew this type of valve all the way. The screw part will come out to reveal the O-ring on the tip that makes the seal, and you can easily replace that. Air relief valves themselves simply screw out of the T assembly. Apply Teflon tape or pipe dope to the new one and screw it back in place.

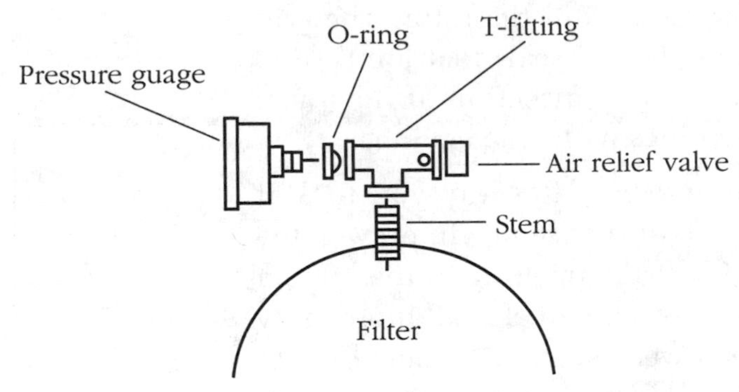

FIGURE 5-15 Typical gauge and air relief valve assembly.

The pressure gauge also threads into the T assembly. If you have a leak there, unscrew the gauge, apply Teflon tape or pipe dope to the threads, and screw it back into place. If the gauge doesn't register or seems to register low, take it out and clean out the hole in the bottom of the gauge. Dirt or DE can clog this small hole, preventing water from getting into the gauge.

Remember, when removing an air relief valve or pressure gauge, you must secure the T with pliers or a wrench while removing the component. The T assembly can easily snap off the filter lid or come loose if you fail to hold it securely when removing or replacing a valve or gauge.

The T assembly itself can come loose and create a leak where the close nipple passes through the hole in the lid. In this case you must remove the lid and tighten the nut from the underside of the lid. Some makes of filters have a nipple welded to the lid, so you won't have this problem unless you crack the weld.

DIRT PASSING BACK INTO POOL

RATING: EASY

I have reviewed most of the ways dirt or DE gets through the filter and back into the pool and the methods to make repairs. Just as a summary, however, if you see this condition when vacuuming, check the following:

- Damaged grids, laterals, or cartridges;
- Backwash valves with bad gaskets or O-rings; and
- Broken manifolds or retainers.

Prevention is always the best cure, so when you feel a backwash valve getting hard to turn, do a teardown and lubrication before the leaks occur. Examine grids, laterals, cartridges, and manifolds carefully each time you break down a filter for cleaning, and always take your time when reassembling. Sloppy reassembly after cleaning is the cause of more leaks than anything else in filters.

FAQs: FILTERS

Which Is Better—Sand, Cartridge, or DE Filters?

- Each filter type will keep a pool clean. The key to success is proper sizing and regular cleaning. The best type is the one you are most likely to keep clean, so cartridge filters may be the best choice because of their ease of maintenance. That said, if you have a large pool, a cartridge filter may not be practical, so choose the filter that fits your pool—and clean it often!

If I Backwash, Do I Also Need to Tear Down and Clean My DE or Sand Filter?

- Backwashing of DE filters should be used as a temporary cleaning measure when complete teardown and cleaning is impractical. Sand filters respond well to backwashing, and because there is no DE to add back, there is no potential for errors that may lead to a dirty pool.

How Often Should I Backwash?

- The schedule for backwashing and teardown of a filter is a factor of how much the pool is used and how dirty it gets in normal service. Typically, DE filters should be torn down and cleaned fully at least six times per year. Unless your pool gets very dirty, you won't need to backwash it. Sand filters can be backwashed once a month and torn down twice a year.

When Replacing a Filter, Should I Buy a Bigger One?

- Bigger is not always better. If you find your filter needs cleaning more than once a month, it may be undersized. Consult a pool professional to get a new filter of the proper size—a filter that's too large for your pump will not fill with water and, therefore, will not give you the improved results you expected.

CHAPTER 6

Heaters

Let's begin the discussion of pool and spa heaters with a point that seems obvious to the technician but which serves to create confusion with many homeowners. The pool and spa heater doesn't work like the water heater in your home. Customers have asked me countless times why the water coming out of the return line isn't hot (like tap water in the home). The assumption in that question is that the pool heater holds a large reservoir of preheated water. They don't realize that the pool and spa heater heats the water as it passes through copper coils, creating a mixture of the heated water with cold water so that the resulting flow out of the heater is not more than 10° to 25°F (5° to 10°C) warmer than the water that originally went in.

The basic principle of the pool and spa heater (Fig. 6-1) is simple. A gas burner tray creates heat. Heat rises through the cabinet of the heater, raising the temperature of the water that is passing through the serpentine coils above.

Gas-Fueled Heaters

Figure 6-2 shows a typical gas-fueled heater (those that use natural or propane gas as the heating fuel). The water passes in one port of the front water header (item 23), then through the nine heat exchanger tubes (item 45). The water reaches the rear header (item 24) and is

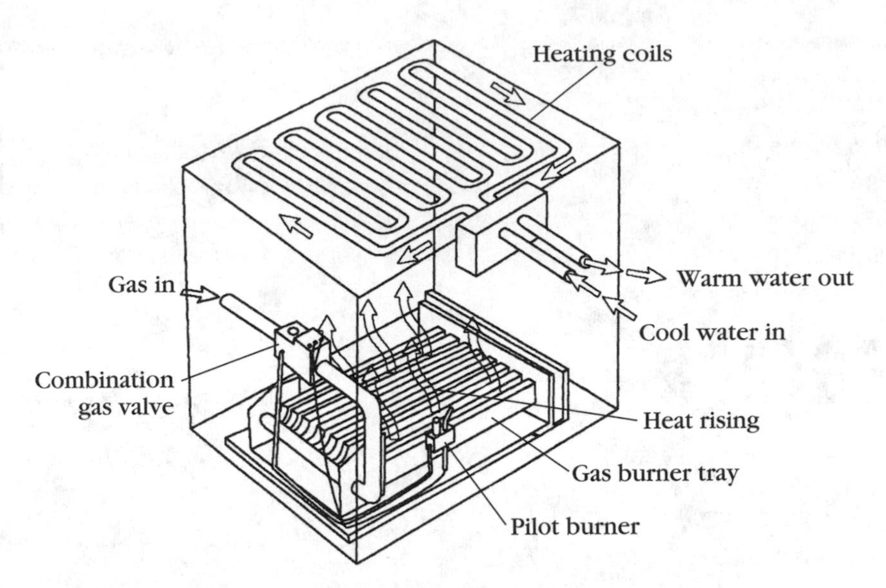

FIGURE 6-1 The concept of the pool and spa heater.

returned through other exchanger tubes to the front header and out the other port.

Most modern exchangers are four-pass units, meaning the water goes through at least four of the tubes, picking up 6° to 9°F on each pass, before exiting the heater. Generally, these are self-cleaning unless extreme calcium (scale) is present in the water.

The heat exchanger tubes are made of copper which conducts heat very efficiently. The tubes have fins (about eight per inch or 2.5 centimeters) to absorb heat even more efficiently and are topped with sheet metal baffles (item 53) to retain the heat. The heat rising from the burner tray (item 55) is effectively transferred to the water in the exchanger because of the excellent conductivity of copper; however, improper water chemistry can easily attack this soft metal and dissolve it into the water. More on that later.

Notice that there is a flow control assembly on the front header (items 38–42). This spring-loaded valve is pressure sensitive, designed to mix cool incoming water with hot outgoing water to keep the temperature in the exchanger from becoming excessive. This design keeps the outgoing water no more than 10° to 25°F (5° to 10°C) (depending on manufacturer) above the temperature of the incoming water to prevent condensation and other problems that greater differentials would

TRICKS OF THE TRADE: HEATER SAFETY

Heaters are unquestionably the most potentially dangerous component of the pool or spa equipment group. They combine water under pressure and heat, gas or other combustible fuel, and electricity. The point is simply that whatever care you exercise normally must be doubled when working with heaters. Therefore, I have a simple safety checklist for working around heaters.

- **Never bypass a safety control and walk away. Jumping controls (discussed below) is a good way to troubleshoot, but do not operate the unit this way. Always remove your jumpers after troubleshooting.**
- **Never repair a safety control or combination gas valve. Replace it. You will notice that your supply house doesn't even sell parts for gas valves. They should never be repaired, because future failure could be catastrophic.**
- **Never hit a gas valve—it might come on, but it might stay on.**
- **Keep wiring away from hot areas and the sharp metal edges of the heater.**
- **Communicate. Tell your customer about heater part failures and repairs. Disable the heater and tape a shutdown notice on the unit until repairs are made. *You can be held liable* if you are the last person to work on a heater and it causes damage or injury by firing incorrectly or before repairs have been made.**
- **When jumping a safety control or otherwise trying to fire a heater that will not come on, keep your face and body away from the burner tray, where flashback might occur. It might be awkward to squat alongside a heater and jump a control through the opening in the front, but awkward is better than burned. Double that warning with LP-fueled units.**

create. Temperature control is achieved by flow regulation rather than direct temperature regulation. Maintaining a constant flow through the heat exchanger results in a constant water temperature.

When reaching temperatures over 115°F (46°C), water breaks down, allowing minerals suspended in it to deposit in the heat exchanger. Also, water is designed to flow through the unit at no more than 100 gpm (378 liters per minute) with 1½-inch (40-millimeter) plumbing or 125 gpm (473 liters per minute) with 2-inch (50-millimeter) plumbing. Above that, a manual bypass valve is installed.

The other major component of the gas-fueled heater is the burner tray. This entire assembly can be disconnected from the cabinet and pulled out for maintenance or inspection. Depending on the size of the

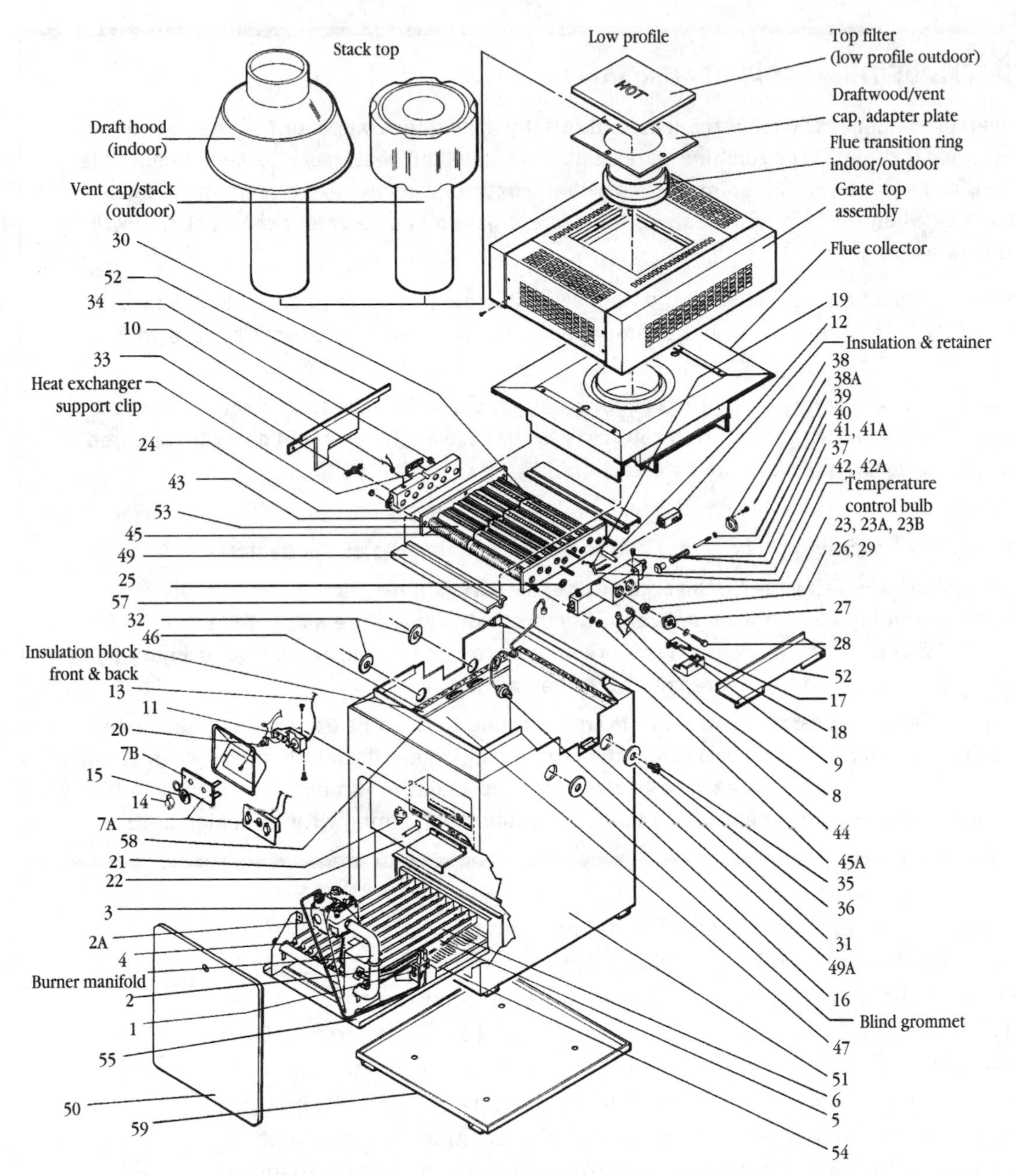

FIGURE 6-2A Detail of a typical millivolt, standing pilot heater. *Jandy-Teledyne Laars, Moorpark, Calif.*

Key No.
1 Pilot generator assembly
2 Visoflame lighter tube
2A Pilot tube
3 Automatic gas valve
4 Burner orifice
5 Burner w/pilot bracket
6 Burners
7A Plate assembly
7B Thermostat dial
8 High-limit switch (135°F.)
9 High-limit switch (150°F.)
10 Redundant limit
11 Temperature control
12 Protective sleeve, bulb
13 Wire harness
14 Thermostat knob
15 Temp-lok
16 Pressure switch
17 High-limit switch retainer clip
18 High-limit switch cover
19 O-ring
20 On-off switch
21 Fusible link
22 Fusible link bracket
23, 23A, 23B Front water header
24 Rear water header
25 Header gasket
26, 29 Flange packing collar w/copper sleeve
27 Water header flange
28 Water header flange bolt
30 Heat exchanger baffle retainer
31 Drain grommet
32 Drain grommets
33 Drain valve
34 Drain plug
35 Drain valve
36 Bushing, drain valve
37 Brass plug
38 Flow control cap
38A, 43 Bolt & washer
39 Flow control gasket
40 Flow control shaft
41, 41A Front control spring
42, 42A Flow control disc
43 Bolt, front & rear header
44 Header nut, 3/8" hex
45 Heat exchanger
45A Syphon loop
46 Fiberglass blanket
47 Insulation block, side
49 Insulation block cover, front & back
49A Insulation block cover, end
50 Door
51 Jacket assembly
52 Gap closure
53 Heat exchanger baffle
54 Burner tray shelf
55 Burner tray assembly
57 Rear deflector
58 Lower deflector
59 Noncombustible floor base (optional)

FIGURE 6-2A *(Continued)*

heater, there will be 6 to 16 burners (item 6), the last one on the right having a pilot (item 5) mounted on it. Individual burners can be removed for replacement. The combination gas valve (item 3) regulates the flow of gas to the burner tray and pilot and is itself regulated by the control circuit.

Gas-fueled heaters are divided into two categories based on the method of ignition.

The Millivolt or Standing Pilot Heater

As the name implies, the standing pilot system of ignition uses a pilot light (burner) that is always burning. The heat of the pilot is converted into a small amount of electricity (0.75 volt or 750 millivolts) by a thermocouple which in turn powers the control circuit. The positive and negative wires of the thermocouple (also called the pilot generator) are connected to a circuit board on the main gas valve.

FIGURE 6-2B Typical pool heater installed.

When lighting the pilot, it is necessary to hold down the gas control knob to maintain a flow of gas to the pilot. When the heat has generated enough electricity (usually a minimum of 200 millivolts) the pilot will remain lit without holding the gas control knob down. The positive side of the thermocouple also begins the electrical flow for the control circuit. When electricity has passed through the entire control circuit, the main gas valve opens and floods the burner tray with gas which is ignited by the pilot.

The Control Circuit

The control circuit is a series of safety switches—devices that test for various conditions in the heater to be correct before allowing the electrical current to pass on to the main gas valve and fire up the unit. Figure 6-3 shows a millivolt control circuit; Fig. 6-4, an electronic control

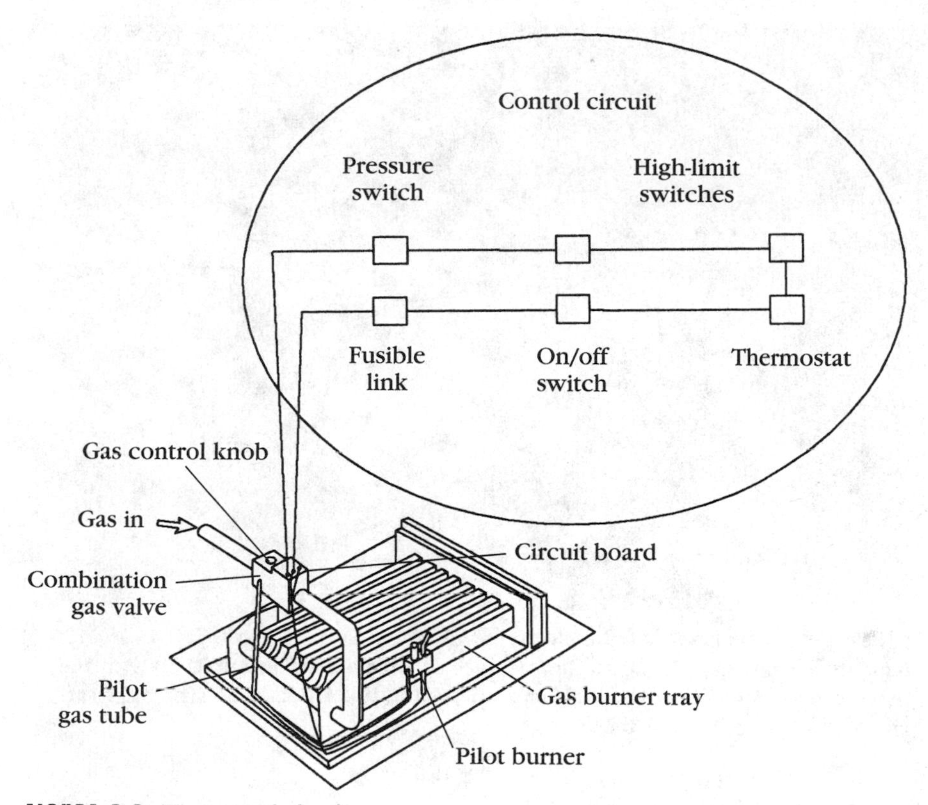

FIGURE 6-3 The control circuit.

circuit. Following the flow of electricity (not all manufacturers follow the same routing of their control devices, but they all include the same devices), a control circuit includes the following items.

FUSIBLE LINK

The fusible, or fuse, link (Fig. 6-2A, item 21) is a simple heat-sensitive device located on a ceramic holder near the front of the gas burner tray.

If the heat becomes too intense, the link melts and the circuit is broken. This would most commonly occur when debris (such as a rodent's nest or leaves) is burning on the tray or if part of a burner has rusted out causing high flames. Other causes are improper venting (allows excessive heat buildup in the tray area), extremely windy conditions, or low gas pressure causing the burner tray flame to roll out toward the link.

Figure 6-5 diagrams a fusible link. A wax pellet, designed to melt at certain high temperatures, melts and allows the spring to break its nor-

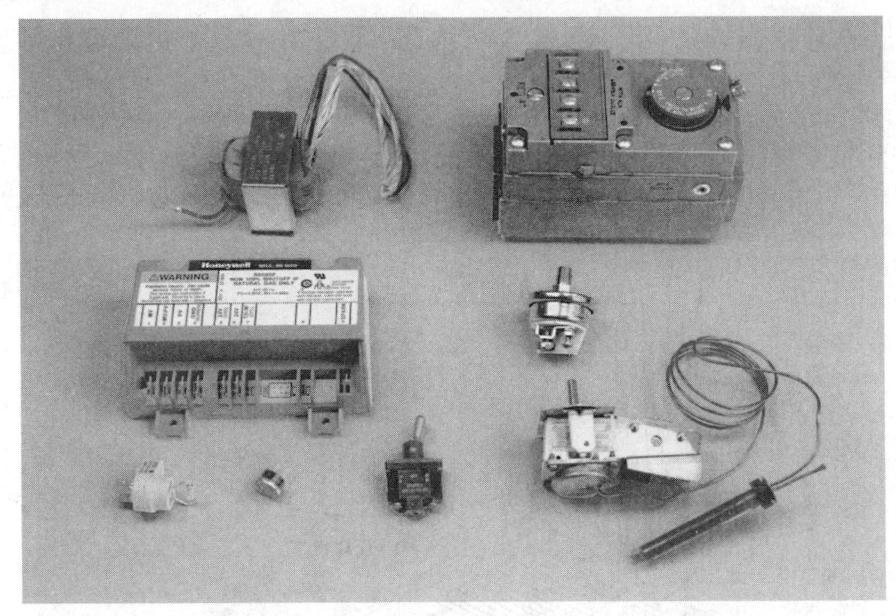

FIGURE 6-4 Components of the electronic control circuit. Top row (left to right): transformer, automatic combination gas valve. Middle row: intermittent ignition device (IID), pressure switch. Bottom row: fusible link, high-limit switch, on/off switch, mechanical thermostat.

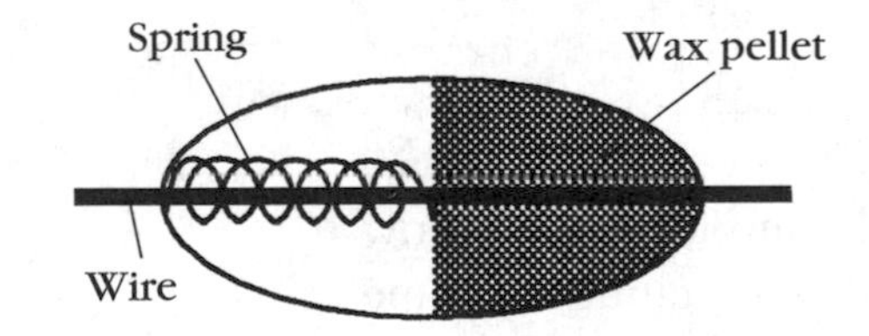

FIGURE 6-5 Detail of the fusible link.

mal contact, thus cutting power in the circuit. My early experience was that these links often need to be replaced even when the heater was new or operating normally; however, current engineering has overcome the oversensitivity of these devices. When the fusible link burns out, it pays to examine the other components of the heater or the installation to find the cause.

Improper venting is not the only cause of overheating. Rusted components, improper installation, low gas pressure, and insect or rodent nests more frequently cause the problem, so the fusible link is one more safety device in a component of your pool and spa equipment that needs a lot of safety.

Not all manufacturers included a fusible link in their control circuit when they were first introduced by Teledyne Laars (now Jandy). Today, virtually all heaters use one.

ON/OFF SWITCH

As the name implies, the on/off switch (Fig. 6-2A, item 20) is usually a simple, small toggle-type switch on the face of the heater next to the thermostat control. Sometimes, particularly on older Teledyne Laars models, the switch is located on the side of the heater in a separate metal box that also contains the thermostat. In older Raypak models, the switch and thermostat might be located on the side, mounted directly through the sheet metal side of the heater cabinet.

Often the switch is remotely located so the user can switch the unit on and off from a more convenient location than where the equipment itself is located. Manufacturers recommend that a remote on/off switch for a millivolt heater be located no more than 20 to 25 feet (6 to 7 meters) from the heater. This is because with less than 0.75 volt passing through the control circuit, any loss of voltage from running along extended wiring means that there might not be enough electricity left to power the gas valve when the circuit is completed.

Also, as the thermocouple wears out and the initial electricity generated decreases, the chance that there won't be enough power becomes very real. Therefore, I suggest from experience that remote switches be located no more than 10 feet (3 meters) from the heater and that they be run through heavily insulated wiring to avoid heat loss. Better yet, run a remote switch off a relay so the control circuit wiring does not have to be extended at all (see the chapter on basic electricity).

If the heater has two thermostats, the switch has three positions: high, off, and low. Why would you have more than one thermostat? Let's say you have a pool and spa, both operated by the same equipment. You can set one thermostat for your desired pool temperature and the other for your desired spa temperature. Then, instead of having to reset the thermostat when you run the spa, you simply flick the switch to read the second thermostat. Many manufacturers now sell only dual thermostat heaters, because it costs supply houses too much to stock heaters of both kinds when the difference is only a few bucks for one additional thermostat.

It is also worth noting that if the switch has been remotely located or duplicated in a remote on/off system, the factory-installed unit might be left in place but not be operative. Factory technicians will place a sticker above such a nonfunctioning switch to alert future users or technicians. However, they are often lost or not used by pool builders or other technicians and, as a result, you might be confused when troubleshooting the heater.

THERMOSTATS

Thermostats (Fig. 6-2A, item 11), also called temperature controls, fall into two categories: mechanical and electronic. The mechanical thermostat is a rheostat dial connected to a metal tube that ends in a slender metal bulb. The tube is filled with oil and the bulb is inserted in either a wet or dry location where it can sense the temperature of the water entering the heater. These thermostats are precisely calibrated, but many installation factors affect the temperature results. In other words, setting the dial at a certain point might result in 80°F (26°C) water in one pool while the exact same setting might result in 85°F (29°C) water in another pool. Therefore, pool heater thermostats generally are color-coded around the face of the dial, showing blue at one end for cool and red at the other end for hot. Settings in between are by trial and error to achieve desired results.

Usually, as shipped from the factory, thermostats will not allow water in the pool or spa to exceed 103° to 105°F (39° to 40°C), although they can be set higher. Also, they do not generally register water cooler than 60°F (15°C), so if the water is cooler than that you might turn the thermostat all the way down and the heater will continue to burn. Therefore, the only way to be sure a heater is off is to use the on/off switch (or turn off the gas).

The electronic thermostat uses an electronic temperature sensor that feeds information to a solid-state control board. These are more precise than mechanical types; however, because of the same factors noted previously they are also not given specific temperatures, but rather the cool to hot, blue to red graduated dials for settings. Some manufacturers of spa controls make specifically calibrated digital thermostats, but my experience is that no matter what the readout says, the actual temperature will vary greatly.

HIGH-LIMIT SWITCHES

High-limit switches (Fig. 6-2A, items 8, 9, and 10) are small, bimetal switches designed to maintain a connection in the circuit as long as their temperature does not exceed a predesigned limit, usually 120° to 150°F (49° to 65°C). The protection value is similar to the fusible link and often two are installed in the circuit, one after the other, for safety and to keep the heater performing as designed.

The first high-limit switch is usually a 135°F (57°C) switch, and the other is a 150°F (65°C) switch. Where the fusible link detects excessive air temperatures, the high-limit switch detects excessive water temperatures. They are mounted in dry wells in the heat exchanger header. Sometimes a third switch, called the *redundant high-limit*, is mounted on the opposite side of the heat exchanger for added safety.

PRESSURE SWITCH

The pressure switch (Fig. 6-2A, items 16 and 45A) is a simple switching device at the end of a hollow metal tube (siphon loop). The tube is connected to the header so that water flows to the switch. If there is inadequate water flow in the header there will not be enough resulting pressure to close the switch. Thus, the circuit will be broken and the heater will shut down.

Although preset by the factory (usually for 2 psi or 138 millibars), most pressure switches can be adjusted to compensate for abnormal pressures caused by the heater being located unusually high above or below the water level of the pool or spa.

AUTOMATIC GAS VALVE

The automatic gas valve (Fig. 6-2A, item 3) is often called the combination gas valve because it combines a separately activated pilot gas valve with a main burner tray gas valve (and sometimes a separate pilot-lighting gas line combined with the pilot gas valve). After the circuit is complete, the electricity activates the main gas valve which opens, flooding the burner tray. The gas is ignited by the pilot and the heater burns until the control circuit is broken at any point, such as when the desired temperature is reached and the thermostat switch opens, if the on/off switch is turned to off, if the pressure drops (such as when the time clock turns off the pump/motor) and the pressure switch opens and breaks the circuit.

There are several different designs of automatic gas valve, but all have aspects in common. Figure 6-6 shows various valve designs. For millivolt-powered units, the terminal board will have three terminals (or four, where terminals 2 and 3 are connected by a common connection or fusible link). Terminals 1 and 2 are the neutral and electric hot

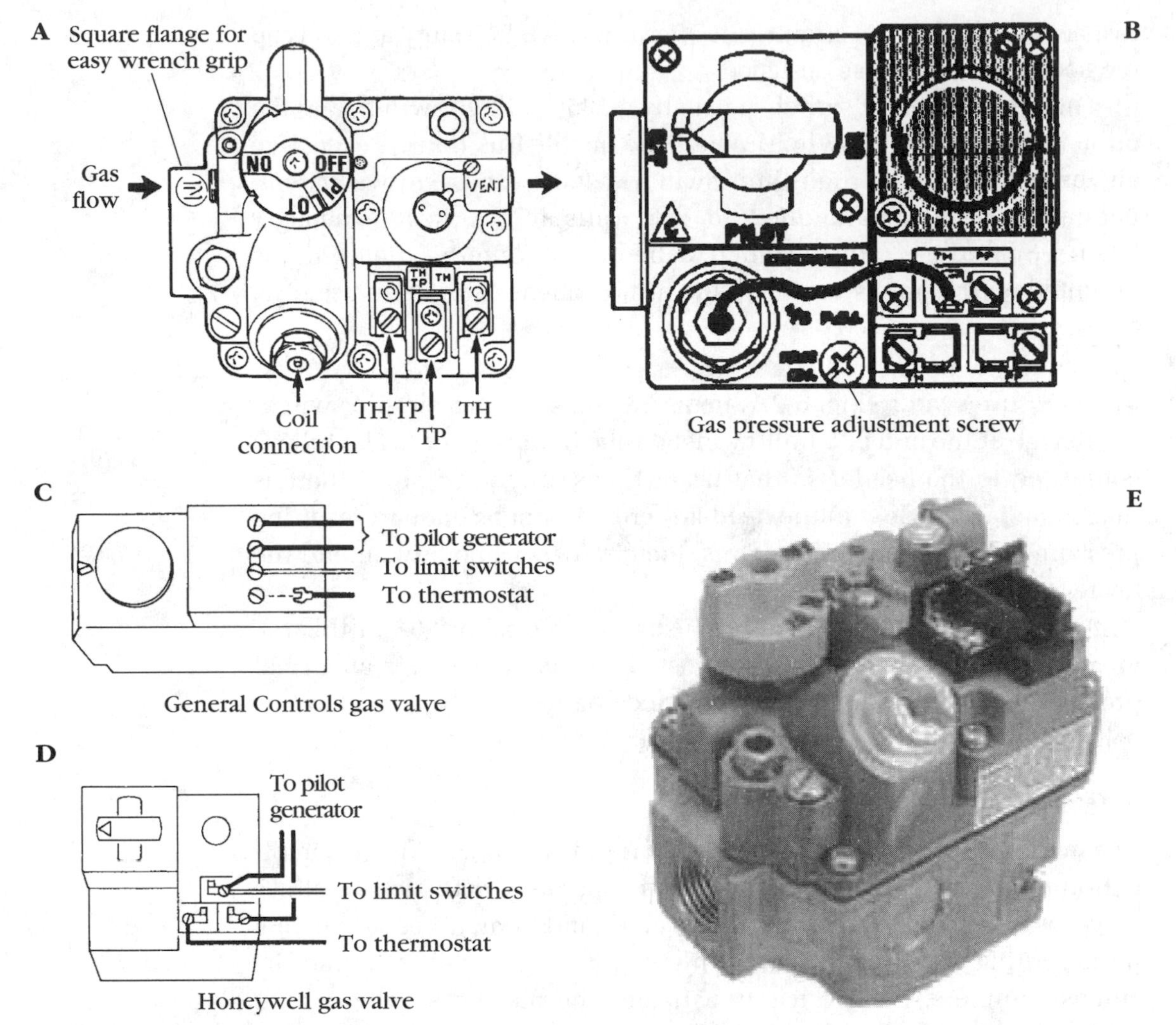

FIGURE 6-6 The automatic (combination) gas valve. *A, B: Jandy-Teledyne Laars, Moorpark, Calif.; C, D: Raypak, Inc., Westlake Village, Calif.*

(positive) lines from the pilot generator to start the control circuit. When the circuit is complete, power arrives at terminal 3 (or 4) which opens the main gas valve.

On 25-volt units, there is a pair of terminals to power open the pilot valve and to return the current to the common (or neutral) line of the intermittent ignition device (IID) (detailed later). Another pair of terminals power open the main gas valve and return the current to the common (or neutral) line of the IID.

If you are unsure how a valve should be wired, look for markings near each terminal. They are often marked *PP* meaning powerpile (the pilot generator) or *TP* meaning thermopile (which are the same thing and, just to confuse you, they're also called thermocouple, but I have never seen *TC* on a gas valve terminal); *TH* meaning thermostat or other connection to the control circuit; and *TR* meaning transformer.

The gas plumbing of the automatic gas valve is self-explanatory. The large opening (½ or ¾ inch or 13 or 19 millimeters) on one end, with an arrow pointing inward, is the gas supply from the meter. Note that it has a small screen to filter out impurities in the gas, like rust flakes from the pipe. The hole on the opposite end feeds gas to the main burner. The small threaded opening is for the pilot tube and a similar hole is for testing gas pressure. These are clearly marked. Teledyne Laars heaters employ an additional small tube to assist in lighting the pilot (see visoflame tube).

Automatic gas valves are clearly marked with their electrical specifications, model numbers, and most important, Natural Gas or Propane. Black components or markings usually indicate Propane.

All combination gas valves have on/off knobs. On 25-volt units, the knob is only on or off. With standing pilot units, there is an added position for *pilot* when lighting the pilot. As a positive safety measure in most, you are required to push the knob down while turning.

Honeywell, General Controls, and Robertshaw make most of the combination gas valves, pilot assemblies, and IIDs in use today.

ELECTRONIC IGNITION HEATERS

When a heater (Fig. 6-7) with electronic ignition is turned on, an electronic spark ignites the pilot which in turn ignites the gas burner tray in the same manner as described previously. In all other respects, these heaters operate the same way as those already discussed. Where the control circuit on the standing pilot heater is powered by millivolts, the electronic ignition heater is controlled by the same kind of circuit but is powered by 25 -olts ac (Fig. 6-8). Regular line current (at 120 or 240 volts) is brought into the heater and connected to a transformer that reduces the current to 25 volts.

This voltage is first routed into an electronic switching device called the IID (intermittent ignition device), which acts as a pathway to

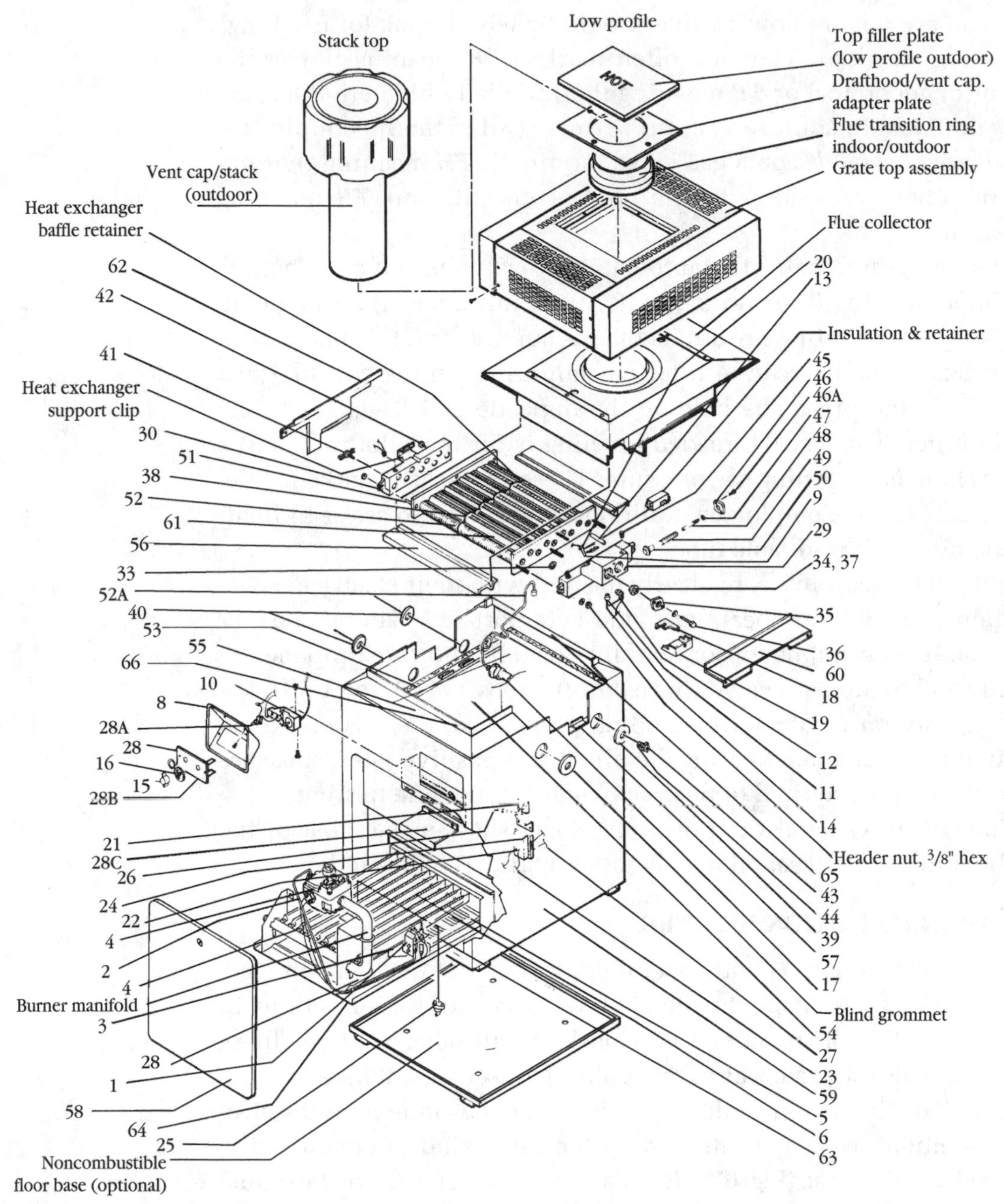

FIGURE 6-7 Detail of a typical electronic ignition heater. *Jandy-Teledyne Laars, Moorpark, Calif.*

Key No.
1 Pilot burner electric assembly
2 Pilot tube
3 Ceramic insulator assembly
4 Automatic gas valve
5 Burner w/pilot bracket
6 Burners
7 Burner orifice
8 Gasket, temperature control
9 Temperature control bulb
10 Temperature control (type EPC)
11 High-limit switch (135°F.)
12 High-limit switch (150°F.)
13 Protective sleeve, bulb
14 Wire harness temp. control
15 Thermostat knob
16 Temp-lok
17 Pressure switch
18 High-limit switch retainer clip
19 High-limit switch cover
20 O-ring
21 Transformer
22 Ignition control
23 High voltage lead
24 Fuse pack
25 Fusible link assembly
26 Electrical fuse assembly
27 Wire harness
28 Fusible link bracket
28A On-off switch
28B Plate assembly
28C Rain guard
29 Front water header
30 Rear water header
33 Header gasket
34, 37 Flange packing collar w/copper sleeve
35 Water header flange
36 Flange bolt
38 Clip for tube baffles
40 Drain grommets
41 Drain valve
42 Drain plug
43 Drain valve
44 Bushing, drain valve
45 Brass plug
46 Flow control cap
46A Bolt
47 Flow control gasket
48 Flow control shaft
49 Flow control spring
50 Flow control disc
51 Bolt, front & rear header
52 Heat exchanger
52A Syphon loop
53 Fiberglass blanket
54 Insulation block, side
55 Insulation block front & back
56 Insulation block cover, front & back
57 Insulation block cover side
58 Door
59 Jacket
60 Gap closure
61 Heat exchange baffles
62 Gap closure
63 Burner tray shell
64 Burner tray assembly
65 Rear deflector
66 Lower deflector

FIGURE 6-7 ***(Continued)***

and from the control circuit. From here the current follows the same path through the same control circuit switches as described previously.

When the circuit is completed the current returns to the IID, which sends a charge along a special wire to the pilot ignition electrode creating a spark that ignites the pilot flame. The IID simultaneously sends current to the gas valve to open the pilot gas line.

When the pilot is lit, it is sensed by the IID through the pilot ignition wire. This information allows the IID to open the gas line to the burner tray, which is flooded with gas ignited by the pilot.

Natural versus Propane Gas

The differences between heaters using natural gas and those using propane gas are nominal. Most manufacturers make propane heaters

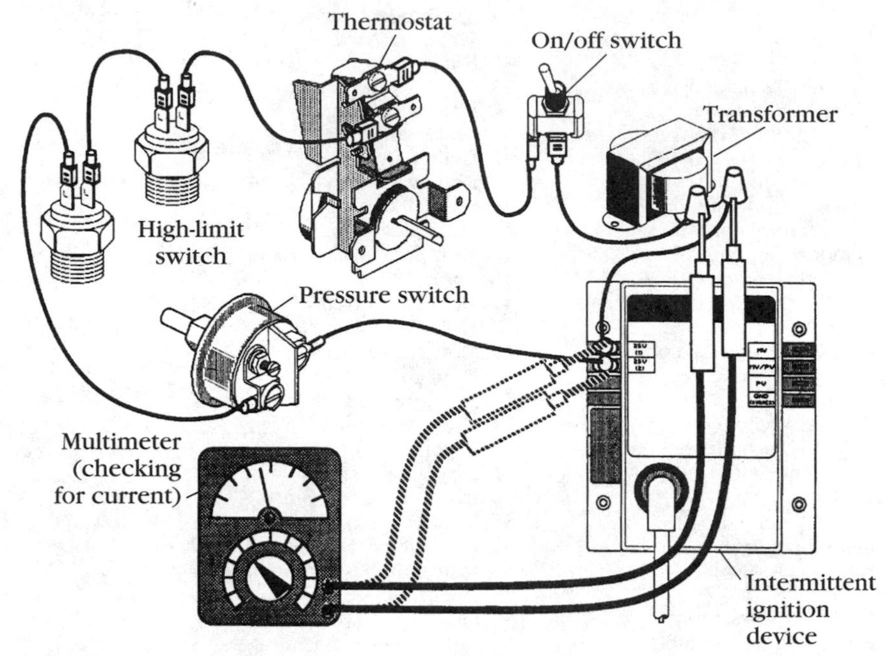

FIGURE 6-8 Electronic ignition heater control circuit (with manual thermostat).
Raypac, Inc., Westlake Village, Calif.

in standing pilot/millivolt models only. Because of different operating pressures, the gas valve is slightly different (although it looks the same as a natural gas model), as are the pilot light and the burner tray orifices. The gas valve is clearly labeled *Propane*. The heater case, control circuit, and heat exchanger are all the same as for a natural gas model.

Natural gas is lighter than air and will dissipate somewhat if the burner tray is flooded with gas but not ignited for some reason. Similarly, the odor added to natural gas will be detected if you are working nearby as the gas floats out and upward. Make no mistake, this is still a serious situation and explosions can occur.

Electric-Fueled Heaters

I have thus far only discussed gas as a fuel for heating. Electricity is also used in some small heaters, usually for spa applications. Because of the cost of operation, the slower recovery and heating time, and the high amps required with the corresponding heavy wiring and electrical supply, electric heaters are useful only where gas is unavailable. They are also used in small portable spas where gas hookups would be impractical.

Figure 6-9 shows a typical electric spa heater. The components are similar to gas heaters except the heat is derived from an electric coil that is immersed in the water flowing through the unit. This is also true of the small in-line electric heaters used in small spas. Often these in-line units have no control circuits or they might have only a thermostat control because the other controls are built into the spa control panel itself.

Several sizes of electric heater are manufactured, rated by the kilowatts consumed and, therefore, the Btus produced. Here is a comparison of the energy use and output of the most common models:

- 1.5-kilowatt (1500-watt) heater = 5119 Btu
- 5.5-kilowatt (5500-watt) heater = 18,750 Btu
- 11.5-kilowatt (11,500-watt) heater = 37,500 Btu

Each of these generally consume about one-third more power to start up than to run at the designated wattage.

TRICKS OF THE TRADE: PROPANE SAFETY

Propane gas is heavier than air and if it floods the burner tray without being ignited it tends to sit on the bottom of the heater. Because it remains undissipated and because you are less likely to smell it because it is not floating out and upward, if it does suddenly ignite it will do so with violent, explosive force. Rarely is the heater itself damaged—the explosion takes the line of least resistance, which is out through the open front panel. Never position your face in front of the opening to try to learn why the heater hasn't fired. Remember to follow your safety checklist, and treat propane with great respect.

Solar-Fueled Heaters

Figures 6-10 and 3-7 show a typical solar installation. The concept to understand here is that the water should go through the solar panels before it passes through the heater. In this way, whatever heat can be gained from the sun is obtained first, then the gas heater adds additional heat if desired. Sensors detect if the panels are warm enough to heat the water, and if so, open motorized valves to divert the water to the panels before it gets to the heater. When the panels are cold, the normal flow is to bypass the solar panels and go directly to the heater.

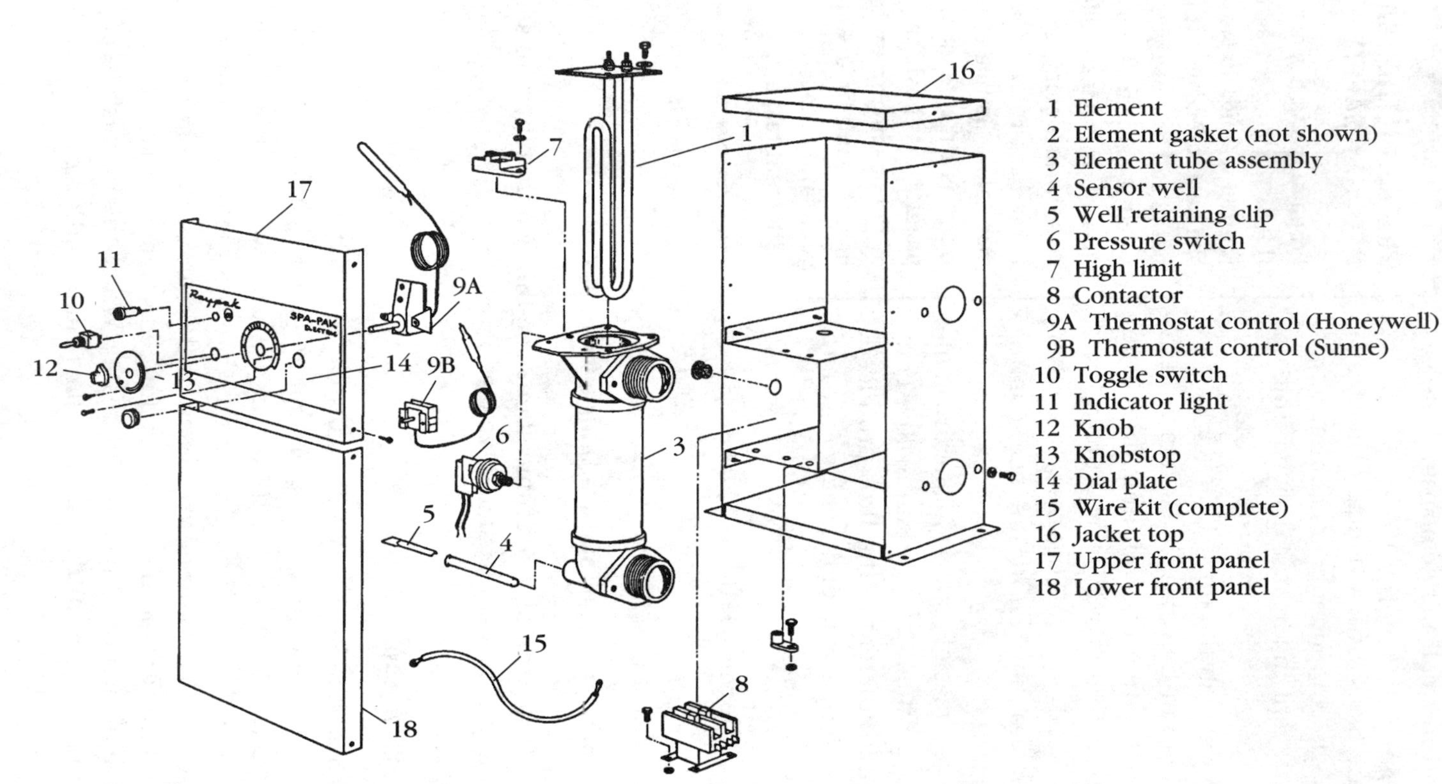

FIGURE 6-9 Detail of a typical electric heater. *Raypak, Inc., Westlake Village, Calif.*

FIGURE 6-10 Rooftop solar installation. *Suntrek Industries.*

Solar heating systems are controlled by time clocks and/or thermostats because, in summer, the panels might add too much warmth to the water and some means of regulation is needed. Also, they typically have simple on/off toggle switches to completely disable the system.

Solar heating systems are becoming more user friendly. Panels are made of lighter materials than just a few years ago, including some made from doormat-like rubber that simply tacks onto a roof or hillside in large, flexible sheets. Most manufacturers of this technology or traditional panels sell complete kits—systems composed of the necessary panels, controls, and installation instructions.

Because solar heating is essentially a plumbing job, it is described in more detail in Chap. 3.

Heat Pumps

An old technology, used in refrigeration and air conditioning, is also employed in pool and spa heating. It is the heat pump. The heat pump does not pump any more heat than any other design of pool and spa heater. Like the others, pool and spa water circulates through the unit (Fig. 6-11) and heat is transferred to that water. Instead of using gas,

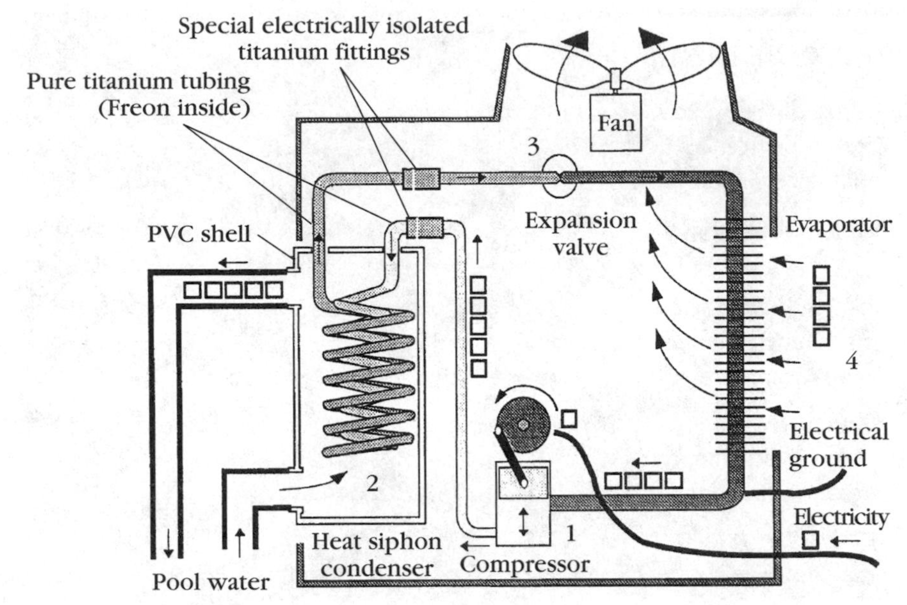

FIGURE 6-11 Detail of a typical heat pump. *Heat Siphon® Swimming Pool Heat Pumps, Latrobe, Pa.*

electricity, or solar heat as a fuel, the heat pump takes warmth out of the air that is created by compressing a gas.

As you all remember from high school physics (you don't?), when you compress a gas, it increases in temperature. A compressor in the unit exerts pressure on a gas (usually Freon) and heat is generated (Fig. 6-11, #1). The water is circulated through a heat exchanger (#2) that is warmed by contact with the hot gas. The gas cools (#3) from contact with the water and is recompressed and heated to start the cycle all over again.

By the way, Freon belongs to a family of chemicals you might have heard about—fluorocarbons, a combination of fluorine and carbon. It is a nonflammable, noncorrosive gas, which makes it suited to this application. Some fluorocarbons contain chlorine as well and are called chlorofluorocarbons, which are in part responsible for depletion of ozone in our atmosphere. The Freon used in heat pumps does not contain the chlorine component that makes it environmentally hazardous.

Heat pumps are energy efficient and last a long time. For every kilowatt of electricity used to run the compressor, you gain the equivalent of 5 to 7 kilowatts of energy (heat) in return. This is known as the heat pump's coefficient of performance and is one fac-

TRICKS OF THE TRADE: HEAT PUMPS

- **Heat pumps take longer to heat the water, but whether you heat with a heat pump or other device, 75 percent of the heat lost by your pool will be from the water's surface, so use a cover.**
- **If you are replacing another type of heater with a heat pump, try to locate the heater closer to the water (or wrap the pipes) to prevent heat loss along the way.**
- **Follow installation instructions closely, especially regarding proper sizing of the electrical supply wiring and breakers. Follow manufacturer's recommendations for inspection of fans, compressor, and refrigerant levels.**
- **Consider adding a heat pump to your existing system. Since heat pumps are more efficient over time, but fossil fuels are quicker to heat the pool or spa, an investment in a heat pump added to the system (plumbed in before the fossil fuel heater) will pay for itself over time, but you still enjoy the benefits of "speed warming."**
- **Be cool—heat pumps can also work in reverse and cool the water in hot climates.**

tor used to compare products. Heat pumps can repay their high initial cost after years of use, especially where the heater is used regularly. They are not effective spa heaters because they take a long time to do the job. Because they rely in part on taking warmth from the air (#4), the hotter the surrounding (ambient) temperature, the better and quicker they work.

Unlike gas-fueled heaters, heat pumps are rated like air conditioners, expressed in tons. In this rating, a *ton* is the amount of energy required to keep one ton of ice at 32°F for 24 hours. As a rough rule of thumb, one ton equals 15,000 Btu (described in the section on sizing).

Oil-Fueled Heaters

Less common in residential applications are oil-fueled heaters. These heaters are designed identically to gas-fueled units, but they burn #2 diesel fuel instead of natural or propane gas. Because these are not very common, I will not describe much about their operation, but the plumbing, electrical, control circuits, and many of the actual components of the heater are identical to those of gas-fueled units. So if you encounter one you should have no problem troubleshooting or repairing it if you understand the other lessons in this chapter.

Makes and Models

Five firms dominate the pool and spa heater market—Raypak, Teledyne Laars (now owned by Jandy), Pentair (which includes Purex/Triton, PacFab, and old Hydrotech models), Sta-Rite, and Hayward—with products that are remarkably similar. Several smaller companies market spa heaters. A.O. Smith formerly made pool and spa heaters and still provides replacement parts.

You'll learn more about the sizes of heaters and their functions in the next section. For now you need to know that heater models are based on their size as expressed in output of heat (measured in Btus). Each manufacturer produces models of similar size; for example, 50,000, 125,000, 175,000, 250,000, 325,000, and 400,000 Btu. After this range, you enter the realm of commercial heaters.

Selection

Beyond manufacturer preferences or price, there are two basic parameters to consider in selecting a heater: sizing and cost of operation.

Sizing

RATING: EASY

Sizing of heaters is fairly easy, particularly if you follow a simple guideline. Starting out with a heater that is not large enough for the job is the first mistake, which is quickly made worse by other factors. Is the pool located in an extremely windy area which causes rapid cooling of the water? Is the surface area of the pool very large (the greater the surface exposure, the faster the heat loss)? How much water are you heating? What average temperature will you start at and how much temperature rise is needed? All of these factors, and changes over the years in them, will determine how large a heater you need.

Although each manufacturer supplies heating capabilities in their literature, here's some guidelines to help make that determination. One

Btu is the amount of heat needed to raise the temperature of 1 pound of water 1°F. There are roughly 8 pounds of water in every gallon.

Let's say your pool is 15 feet by 30 feet, with an average depth of 4 feet. Remember, there are 7.5 gallons in each cubic foot of water (see Chap. 1). Let's also say the average temperature of the pool is 70°F (21°C) and you want to get it to 80°F (27°C)—a 10°F (6°C) increase.

15 ft × 30 ft × 4 ft = 1800 cu. ft (50 cu. m)
1800 cu. ft × 7.5 gal/cu. ft = 13,500 gal (51,000 L)
13,500 gal × 8 lb/gal = 108,000 lb (49,000 kg) of water
108,000 lb × 10 = 1,080,000 Btu needed

If you are planning to circulate the pool 8 hours per day, and you have just calculated that you need 1,080,000 Btu in the entire day, then your heater output needed each hour is 1,080,000 ÷ 8, or 135,000 Btu per hour. Therefore, you need a heater rated at least 135,000 Btu.

Rated? Here's a simple trap in heater sizing—there is a difference between the input and output ratings of a heater. Obviously some heat is lost up the venting, so you need to know how much is actually going into the water. Look at the heater label or manufacturer's literature to determine each. Typically with gas- or oil-fueled heaters, the output (the amount that actually heats into the water) is 70 to 80 percent of the input.

So the point is that when calculating Btu heating needs, you will arrive at an output amount. When buying the heater, you must add 20 to 30 percent to arrive at the input amount, which is the nominal (or advertised) rating of the heater. In the example, where you wanted 135,000 Btu *output* to your water, you would need to buy at least a 175,000 *input* Btu rated heater.

Before going on I must point out that the example of a 10°F (5°C) increase might be true in California in summer (or Florida at any time), where the water temperature might average 70°F (21°C) without help. But when sizing a heater, consider the coldest average temperature you might start with and the hottest average temperature you might want to reach. For example, let's say you want to swim in spring or fall when the average water temperature is 50°F (10°C). You might want to heat the pool to 85°F (29°C). That's a 35°F (19°C) increase, far above our 10°F (5°C) example. Keep overall heater use in mind when selecting a heater.

As I said, estimating heater size is more art than science because of all of the factors that can affect the job. So ballpark estimates are generally good enough because you will want to up-rate your choice anyway. My examination of manufacturers' heater sizing charts reveals that there is a constant rate used in the calculations. This method of calculation tells you how many Btus per hour, rather than the method just described, which gives you total Btus in 24 hours. Here it is.

You need 100 Btu every hour for every 10 square feet (1 square meter) of pool surface for every 1°F temperature rise desired. So, using the previous example:

15 ft × 30 ft = 450 sq. ft (42 sq. m)
450 sq. ft ÷ 10 sq. ft = 45
45 × 100 Btu/1°F = 4500 Btu for each degree rise desired
4500 Btu × 10° rise = 45,000 Btu needed per hour to hold temperature
45,000 Btu × 24 h/day = 1,080,000 Btu total
1,080,000 Btu ÷ 8 h circulation/day = 135,000-Btu heater

So either way you get the same answer: that a minimum 135,000 output Btu heater is required, regardless of the heating fuel used. It was noted that about 15,000 Btu equals 1 ton of heat pump rating, so 135,000 ÷ 15,000 = 9-ton heat pump.

Now remember, this calculation shows the heater needed under ideal conditions. One engineer told me that a 10-mph wind would require you to double the heater size to get the same results. Suddenly your 135,000-Btu heater is a 270,000-Btu unit if you are in a consistently windy area. That is what I mean about always going up in size when recommending a heater.

Let's say you already have a heater and you want to know how quickly it will heat your pool or spa. Here's a formula that works, not including any wind or other cooling conditions:

Heater Btu output/(gallons × 8.33) = degrees temperature rise/hour

So going back to the sample pool of 13,500 gallons, let's say you have a 175,000 input Btu heater already installed and want to know its expected abilities. First, you must determine the true Btu output. Using a 75 percent efficiency, that would be

$$175{,}000 \text{ Btu} \times 0.75 = 131{,}250 \text{ actual Btu output}$$
$$131{,}250 \div (13{,}500 \times 8.33) = 1.16°\text{F/h}$$

Therefore, in 8 hours of circulation you might expect a 9°F rise in temperature (8 hours × 1.16).

To calculate heater size for spas you use the same calculations. But with much less water in a typical spa than in a typical pool, a much smaller heater will do the trick. Or will it?

There might be less water, but if you have a hot date (pun intended) and a cold spa, do you want to wait eight hours to raise the temperature 10°F? Oh, that's right, there's less water, but you want it much hotter and much faster. My spa is 6 feet round and 4 feet deep. What heater do I need to get it from its usual 70° to 100°F in 1 hour? (Refer to Chap. 1 if you've forgotten how to measure the surface area of a circle.)

$$28 \text{ sq. ft} \times 4 \text{ ft} = 112 \text{ cu. ft}$$
$$112 \text{ cu. ft} \times 7.5 \text{ gal/cu. ft} \times 8 \text{ lb/gal} = 6720 \text{ lb of water}$$
$$6720 \text{ lb} \times 30°\text{F rise} = 201{,}600 \text{ Btu total}$$

Remember, heaters are rated in their Btu capacity per one hour, and since I want that spa hot in one hour, then I need at least a 201,600 output Btu heater for this little spa. Many customers don't want to wait an hour either, more like half that, meaning they need twice that size when selecting a heater.

For the record, the American Red Cross suggests pool temperatures of 78° to 82°F (26° to 28°C) and spas not more than 104°F (40°C). Most building and safety codes follow these guidelines.

Now that you can calculate the proper size of heater for any job, you can test your estimates against the charts on p. 232. Sizing pool heaters is generally based on a desired temperature outcome in the normal filtration cycle. Sizing spa heaters is generally based on how fast the spa needs to be heated to the desired temperature. See the sidebar on p. 233 for help in making those ballpark estimates.

Pools

	10°F/5°C	15°F/7°C	20°F/10°C	25°F/13°C	30°F/16°C
200sf/18sm	21,000 Btu	32,000 Btu	42,000 Btu	53,000 Btu	63,000 Btu
400sf/36sm	42,000 Btu	63,000 Btu	84,000 Btu	105,000 Btu	126,000 Btu
600sf/54sm	63,000 Btu	95,000 Btu	126,000 Btu	157,000 Btu	189,000 Btu
800sf/72sm	84,000 Btu	126,000 Btu	168,000 Btu	210,000 Btu	252,000 Btu
1000sf/90sm	105,000 Btu	157,000 Btu	210,000 Btu	263,000 Btu	315,000 Btu

Spas (minutes required for every 30°F/16°C temperature rise desired)

	Heater size				
Spa size	**125,000 Btu**	**175,000 Btu**	**250,000 Btu**	**325,000 Btu**	**400,000 Btu**
200 gal/757 L	30 min	20	15	12	<10
400 gal/1514 L	60	45	30	25	20
600 gal/2271 L	90	65	45	35	30
800 gal/3028 L	120	85	60	45	40
1000 gal/3785 L	150	110	75	60	47

Cost of Operation

The cost of operating a heater is simple to figure out if you know what you pay for a therm of gas. A *therm*, the unit of measurement you read on the gas bill, is 100,000 Btu/hour of heat. My last gas bill showed I pay about 50 cents per therm. The heater model tells you how many Btus per hour your heater uses. Divide that by 100,000 to tell you how many therms per hour it is. Next, determine how many hours of operation are needed to bring the temperature up to the desired level.

Let's use the example from before. You need at least a 135,000 output Btu heater, but to allow for the other factors noted, you buy a 250,000 input Btu model. You decided it would run eight hours a day, remember? Well, using the facts you have, here is your operating cost:

$$250{,}000 \text{ Btu} \div 100{,}000 \text{ Btu/h} = 2.5 \text{ therms/h}$$

$$2.5 \text{ therms/h} \times 8 \text{ h/day} = 20 \text{ therms/day used to run the heater}$$

$$20 \text{ therms/day} \times 50¢/\text{therm} = \$10/\text{day}$$

In the spa example, let's say I use the same 250,000-Btu heater to heat my spa and I figured it would take an hour to do so. Using the same calculations, it costs $1.25 per day to heat the spa. Of course, the heater will run while the spa is in use and the heat loss will be considerable as jets and blowers stir up the water.

Once you have reached your goal temperature, if you want to know what it will cost to keep the spa at that goal you must know how much heat the water is losing. In an hour without the heater, if the spa lost 10°F, you would use the same calculations using a 10°F heat rise to arrive at the cost. Obviously it will not be as much per hour as the original 30-degree rise in one hour.

Someone will ask you one day how much it costs to keep a standing pilot burning. Well, it uses between 1200 and 1800 Btu per hour. Figure it out from there. By the way, the temperature of that little flame is over 1100°F, so when removing a pilot assembly for repair, don't grab one that has recently been lit.

TRICKS OF THE TRADE: HEATER GENERAL SIZING CHARTS

Based on the square footage (sf) or square meters (sm) of the pool surface...

...and the desired rise in temperature (shown in deg. Fahrenheit or Celcius)...

- **Based on an 8-hour filter cycle for pools**
- **Based on the desired speed for raising the temperature for spas**

Select a heater with a Btu rating as indicated (or higher).

Interpolate as needed for values between those shown.

Installation, Repairs, and Maintenance

Heater troubleshooting is both art and science. After years in the field, most good service technicians can sense what is wrong by the customer's description or by a look at not just the heater, but the overall pool and spa and the equipment area. Since nothing replaces experience in this field, I will limit my comments to the most common heater failures and how to repair them. Beyond that, refer to the reference section to get and absorb as much information as you can from manufacturers' seminars, literature, pool and spa seminars, trade publications, and other service technicians.

QUICK START GUIDE: INSTALLING A HEATER

Rating: Pro

1. Location

- Indoor: Place the heater on a bare concrete slab. Follow the manufacturer's guidelines for clearance distances from walls. Pay close attention to air circulation and venting guidelines.
- Outdoor: Place the heater on a bare concrete slab or other level, solid, nonflammable foundation. Follow the manufacturer's guidelines for clearance distances from structures. Avoid areas with overhanging trees, roofs, or other potential fire hazards.

2. Plumbing

- Use compression fittings or threaded connections (both are provided on most heaters), following basic plumbing guidelines in Chap. 2.
- Solar panel pipes are plumbed BEFORE the heater. All other suction lines (automatic cleaner pumps, chlorinators, and so forth) are plumbed AFTER the heater.

3. Gas

- Be sure the gas supply is shut off before removing the old heater and installing the new one.
- Replace any corroded gas pipes. Make threaded connections leakproof with pipe dope and test for leaks when finished.

4. Electrical

- Be sure the electricity is turned off before starting.
- Review the manufacturer's wiring diagram and specifications.
- Connect properly sized wires from heater to time clock power supply.

5. Testing

- Double check water, gas, and electrical connections before test-firing the heater.
- Watch heater operation for several minutes to ensure that all systems work properly under full heat and pressure.
- Turn the pump on and off several times to ensure that the heater goes off when the water circulation stops.

Installation

RATING: PRO

Before worrying about how to fix a heater, you will want to know how to install one. A lot of heater failures result from improper installation, so you need to understand what's proper in the first place. Any heater installation is composed of four basic steps: location (including ventilation), plumbing, gas, and electrical.

LOCATION

Hot air rises. Very hot air rises very quickly in large volumes, requiring replacement by adjacent cooler air. Burning fossil fuel, such as gas or oil, results in by-products such as carbon monoxide, which is deadly. These simple concepts are at the heart of your decision about where to locate a heater.

Most residential heaters are designed to be installed along with the pump, motor, and filter at an outdoor or indoor location. When purchasing the heater, you ask for a *stackless* heater (Fig. 6-2 shows both the stack and stackless tops) for an outdoor installation, meaning there will be a draft hood on top of the heater, but no additional vent pipe or stack to remove excess heat and the products of combustion (carbon monoxide).

If your installation is to be indoors, ask for a *stack* heater, which comes with a vent hood for attachment to the stack pipe. Heaters used to come with no top and you would be given one designed for your setup when you bought the heater. Now, most manufacturers' tops are adjustable to either indoor or outdoor installation. Never use a stackless top on an indoor installation. Burning gas produces carbon monoxide, and even if the heater is small and the indoor location large and well ventilated, this is a deadly gas.

Take the heater out of the box and look for instructions. **Read them.** I have installed over 500 heaters, but constantly changing designs mean I still have to read those instructions before working with any new unit.

Open the front panel on the heater. Usually packed in here is a plastic bag with the hardware you will need. Set the heater near the filter, but remember you will need to do future service work on both the filter and heater, so leave enough space to work. Other spacing

TRICKS OF THE TRADE: GOLDEN RULES OF HEATER REPAIR

- **Always turn the heater off when making repairs. Preferably turn the pump off as well and disconnect or turn off any source of electricity. If you don't, you might complete the repair or touch some wiring together causing the heater to restart when you don't really want it to. By shutting everything down, you control the entire process—you check your work and control the test-firing when you think you're done. Otherwise the heater controls you.**
- **It is generally better to replace components rather than repair them and, if it's a well-worn heater with other parts that will soon give out, replace the entire heater instead of adding new parts every month ad infinitum. Associated with this point is the fact that each part in a heater fails for a reason, rarely old age. When you replace or repair something, find out why it failed in the first place or it will happen again.**
- **Most heater repairs are not the heater at all. The majority of heater failures are the result of dirty filters (and add to that low or obstructed water flow from the pool). In short, nothing to do with the heater at all. Moral of the story? Look around at the entire installation before starting on what has been called in as a heater problem.**
- **If the heater has been running prior to any repair you are making, watch out for hot components. Pilots generate over 1100°F, so they stay hot for a long time after they've gone off. Cabinets and other metal parts get hot too, so watch what you grab.**

guidelines are provided by the American National Standards Institute (ANSI) and are designed to keep carbon monoxide from entering the closed living spaces of your home and to allow enough draft area around the heater. Remember, it needs a large supply of air, which it will heat, and enough distance from flammable walls or fences so that the hot air and hot metal cabinet don't burn the house down.

Refer to Fig. 6-12 for clearances, but here are the highlights.

- The rear and nonplumbed sides must have at least 6 inches (15 centimeters) of clearance [12 inches (30 centimeters) is better]; the plumbed side, 18 inches (45 centimeters).
- The front must have 24 inches (60 centimeters) of clearance. If that is all you allow, you might be safe with ANSI, but the next service technician who comes to work on the unit will curse you. To effectively work on a heater, which requires squatting

and lying in front of it and sometimes pulling out the burner tray, I would allow at least 4 feet (120 centimeters). Obviously on a small heater you might allow less, but I would say never less than the depth of the heater cabinet (measured front to back).

- Leave at least 3 feet (90 centimeters) from the top of the heater to the underside of any overhang and at least 5 feet (150 centimeters) below any window (or 4 feet or 120 centimeters away from such a window, door, or vent). Again, I think ANSI is shy about this. For example, trees that overhang by only 3 feet (90 centimeters) above a large residential heater that operates for long periods will dry and burn. I think this clearance should be at least 6 feet (2 meters).

The rest of the clearances, including vent stack requirements, are shown in the diagram. The stack pipe should have no 90-degree angles, and the stack itself should not be altered. The vent stack is an integral part of the design that each manufacturer has engineered for compensation between indoor and outdoor atmospheric pressure differentials.

Set the heater on a solid, level, noncombustible base. Although heat rises, the metal cabinet will get quite hot underneath as well, so a concrete slab over a wood floor is insufficient protection. Heaters have their own feet or runners to hold them off the surface, but even these get warm and would not be compatible with a wooden floor or wall to wall carpet. The best flooring is a concrete pad, hopefully the pad all of your equipment is resting on.

Alternatively, you might use a precast concrete slab available at your supply house or an air-conditioning supply house. These come in slabs of 3 or 4 feet

TOOLS OF THE TRADE: HEATERS

- **Flat-blade screwdriver**
- **Phillips screwdriver**
- **Hacksaw**
- **PVC glue**
- **PVC primer**
- **Pipe wrench**
- **Teflon tape**
- **Silicone lube**
- **Needle-nose pliers**
- **Hammer**
- **Emery cloth or fine sandpaper**
- **Pipe wrench**
- **Multimeter (and millivolt tester if separate)**
- **Nut driver set**
- **Electric drill with reaming brush**
- **Channel lock-type pliers**
- **Manometer**
- **Knee pads**

square (¼ or ½ square meter) by 1½ to 2 inches (40 to 50 millimeters) thick. All you need to do is level out the ground, lay down the slab, and add the heater.

The last choice would be brick or cinder block, with the holes lined up for ventilation, with a piece of sheet metal over the top to keep the heat from cracking them. In any event, refer to local building codes for what is required in your area.

REVERSIBLE WATER CONNECTION

Most heaters are right-handed; that is, as you face the heater, the plumbing is on the right side of the unit. What if your installation would be more convenient if the plumbing were on the left? Good question.

Answer—just reverse the heat exchanger on the heater so the plumbing (and gas and electrical) connections are on the left side of the cabinet. Each make of heater is slightly different, but if you disassemble one, the others will be easy to figure out. Each heater comes with a booklet that describes the exact manufacturer's procedure for reversal. It will take you no more than 30 minutes the first time you try it and about 15 minutes each time thereafter.

Flipping the heat exchanger from right to left is simple and it will help you understand many important components of the heater at the same time. Flipping is actually a misnomer; you don't actually turn the heat exchanger over, you turn it around.

1. **Plan** As with any disassembly, make careful note of how you remove components so you can put them back together the same way. In this procedure, it is especially true of the wiring. Note how it is routed to keep it away from heat or sharp metal edges. Refer to Fig. 6-2 to identify the components discussed in this procedure.
2. **Disassemble Facia** Remove the front faceplate (access door) of the heater. Remove the vent top (unscrew the retaining screws) and flue collectors underneath (simply lift straight up and off). Remove the faceplates around the plumbing header (unscrew the retaining screws) and the corresponding plate on the left side of the heater.
3. **Disassemble Wires and Bolt-ons** Remove the thermostat tube or sensor from the drywell in the header (some lift out, some are held in

place by a single screw and small bracket). Disconnect the wires to the high-limit switches. Unscrew and remove the tube feeding water to the pressure switch.

4. **Disassemble Plumbing** Remove the heat exchanger drain plugs or air relief valves. Most designs have these passing through the heater cabinet, so they must be removed to free the exchanger unit. Lift out the heat exchanger unit and rotate it so the plumbing connections are on the left side of the cabinet. Again, remember you are not flipping the unit over like a pancake. You simply rotate the unit 180 degrees.
5. **Inspect** While the exchanger unit is out, you have a great opportunity to examine the interior of the heater. Replace any broken firebrick, and examine the inside of the cabinet and top of the burner tray for soot (if this is not a new heater). You will see now how simple the heater really is.
6. **Reassemble** Replace the components in the reverse order of their removal. The pressure switch will need to be repositioned on the left side of the heater so the water tube feeding it will reach. Reroute the wiring as needed for the pressure switch, high-limit switches, and thermostat sensor (or bulb and tube). Make sure that wiring is not routed over hot or sharp metal surfaces or edges. The only other difference in assembly from right to left is that the faceplates that were on the right side of the heater (around the plumbing connections) now move to the left, while the plates from the left move to cover the right side. Hole patterns are drilled by the factory to accommodate this right/left switch.

I mention reversible connections at this point because before you plan your installation, it pays to know that you can hook up from either side and, of course, you would do it now before setting up the heater, venting, etc.

VENTING

As noted, outdoor installations require no additional venting, they are vented sufficiently within their stackless tops. In windy areas, however, you might want to consult the manufacturer's recommendations and add a short stack and cap (about 3 feet [1 meter] total) to cut down on excess drafting (see vent/cap stack in Fig. 6-2). Refer to each manufacturer's specifications on this.

For indoor installations, follow the guidelines in the heater booklet. Working with stack pipe (made of flexible sheet aluminum), which you get at your supply house, is no tougher than working with PVC, except there is no glue, they just snap together. Vent pipe comes in lengths of 2 to 6 feet (60 to 180 centimeters) and diameters of 3 to 12 inches (7 to 30 centimeters). You also buy preformed 45- and 90-degree elbows and connector sleeves just like PVC plumbing.

Going through walls or a roof to get outside, however, is trickier because you need to use doublewall pipe, flashing, and waterproofing to make the passage safe and rainproof. Some local codes require a permit and a licensed building contractor to handle this part. If your indoor installation is a replacement of a heater that otherwise worked well (and your replacement is the same Btu output), you really only need to connect your new heater venting to the old vent that goes up and out of the building. But if it is a new vent stack that means cutting walls and a roof, I'd hire a contractor to handle that. It is well worth the few extra bucks.

In either case, it is your responsibility to make sure the clearances meet the ANSI code so the heater works properly. Familiarize yourself with the guidelines in Fig. 6-12 and supervise the contractor's work. One simple way to perform a rough test of the venting is to light a match and hold it under the draft hood—the smoke should be drawn up into the vent system and out of the building.

Some howling, whistling, or other ventilation harmonic noise is normal and acceptable. As mentioned, a very large volume of air is rushing up through the heater and as it passes over vent fins it might howl. Notice what is normal for the heater and then compare that to any future noises. Changes in the sound might denote problems.

A few minutes after the heater fires, a knocking noise that might actually rock the heater is not normal. This denotes overheating for some reason and is caused by the superheated water expanding and trying to escape. Some high-pitched whining is caused by debris in the gas line. It should be disassembled, cleaned, and reassembled.

PLUMBING CONNECTIONS

Modern heaters offer basically two kinds of plumbing connections: compression fittings and threaded. No other aspect of installation could be easier, as there is one pipe in and one out. The manifold (the

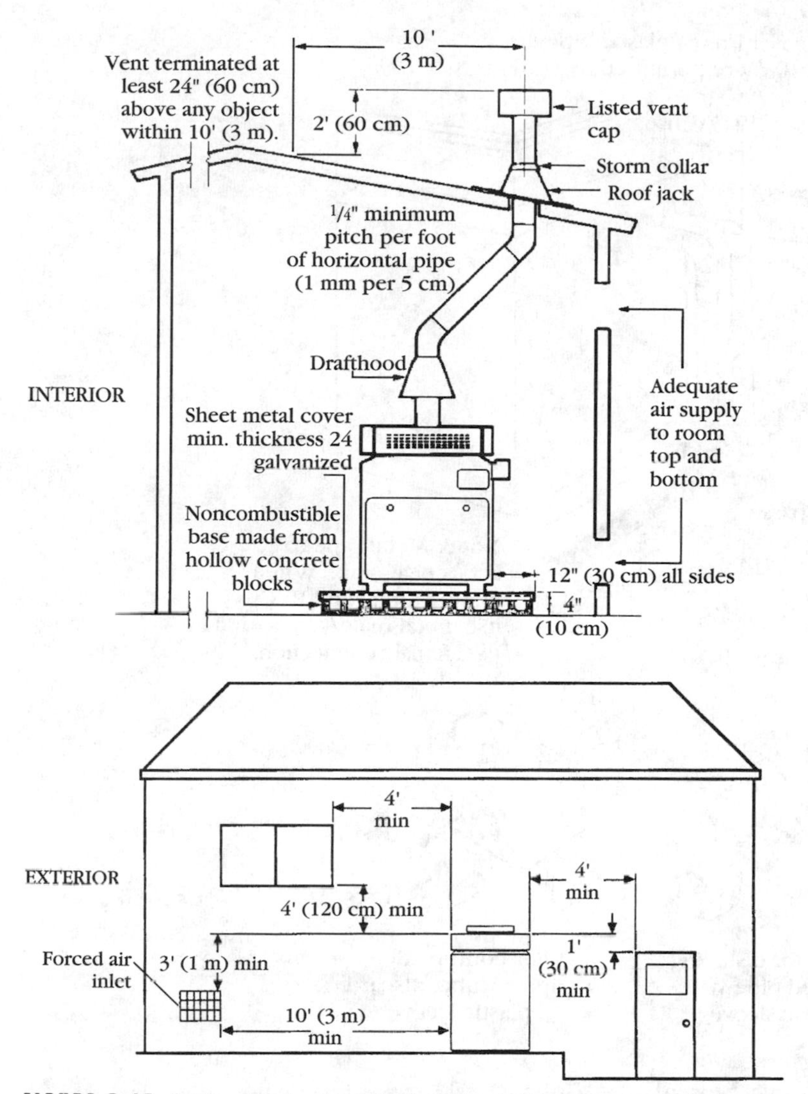

FIGURE 6-12 Indoor heater installation. *Jandy-Teledyne Laars, Moorpark, Calif.; Raypak, Inc., Westlake Village, Calif.*

end where the plumbing enters the heat exchanger) is set up for 1½- or 2-inch (40- or 50-millimeter) pipe, sometimes allowing either.

Compression Fittings: The heater comes with two flange gaskets (Fig. 6-2A, items 26 and 29; the flange is item 27). Very simply, the pipe slides through the flange and gasket and into the header as far as it will travel (Fig. 6-13). The flange is then bolted to the header, in the

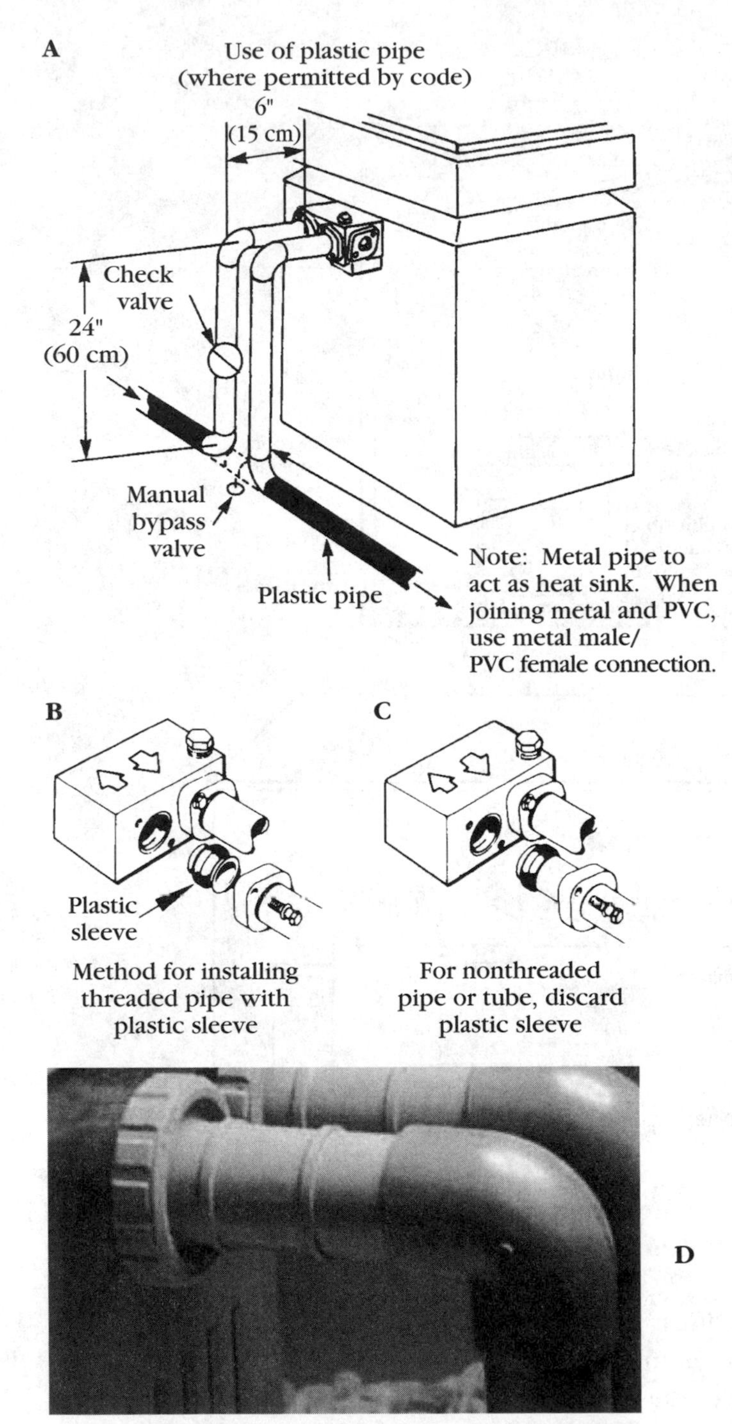

FIGURE 6-13 Heater plumbing installation. *Jandy-Teledyne Laars, Moorpark, Calif.*

process applying pressure to the gasket which tightens around the pipe, sealing it.

As noted, some heaters supply gaskets for 1½- and 2-inch (40- and 50-millimeter) pipe, your choice, while larger heaters might require 2-inch (50-millimeter) pipe only.

Threaded Fittings: To screw a threaded pipe directly into the flange, you first assemble the flange gaskets with the plastic or metal sleeves provided. The sleeves act like the pipe in the compression fitting (otherwise when you bolt down the flange, the gasket will collapse and leak or restrict water flow). With the gaskets in place, bolt the flange to the header (Figs. 6-13B and C). The flange is female threaded, usually for 2-inch (50-millimeter) pipe, so you can directly thread fittings or threaded pipe into the flange.

Remember, a check valve should be installed between the filter and heater so that when the pump shuts down, the extremely hot water does not backflow into the filter and damage the plastic grids or cartridge fabric. If your system includes a chlorinator (which is always installed in the plumbing after the heater), a check valve between the heater and chlorinator is also a good idea so that when the pump shuts down there cannot be backflow of the highly chlorinated water into the heater where it might corrode the soft copper of the heat exchanger.

Suction lines for automatic pool cleaner pumps should also be installed in the plumbing after the heater. To put them before the heater could result in too much water being pulled away from the plumbing before it enters the heater, resulting in overheating or low-pressure shutdown of the heater.

Heaters require a minimum water flow to operate efficiently. Sometimes, however, the system provides a flow that is too fast. If the flow is more than 125 gpm (473 lpm), a manual bypass should be installed (as shown in Fig. 6-13A). This is a simple gate valve that allows some of the flow to bypass the heater.

Some codes require that a pressure safety valve be installed on the heater, just like on your household water heater. On the front water header where the plumbing is attached, you will find a ¾-inch (19-millimeter) threaded plug that can be replaced with a ¾-inch pressure relief valve available at your supply house.

GAS CONNECTIONS

When replacing a heater, you will generally hook the new heater up to the old gas line. Gas plumbing must follow rigid guidelines because a gas leak is a greater hazard than a water leak. Having said that, the procedures are the same as described in the chapter on basic plumbing and require no special skills or tools, just lots of care.

Each manufacturer specifies the required size of pipe to ensure an adequate supply of gas, depending on the size of the heater and the distance from the source of the gas (natural gas meter or propane tank). Generally, these will be ¾- to 1½-inch-diameter (19- to 38-millimeter) pipes.

Gas plumbing can be steel or special PVC pipe. Some building codes permit special PVC gas pipe for underground runs (a heavy-duty PVC pipe material that is green-tinted to denote gas). It can be argued that especially with underground pipe, PVC might be better because it doesn't rust. The disadvantage is that anyone digging in the yard can more easily rupture a PVC pipe than a metal one. If you do run PVC gas pipe, it must be accompanied in the ground by a 16-gauge tracer wire so the gas line can later be found by a metal detector if needed. Most codes also require that PVC gas lines be buried at least 18 inches (45 centimeters) below ground [12 inches (30 centimeters) under concrete]. Finally, any risers (lines coming up out of the ground) or lines above ground must be metal, so you must transfer from PVC to metal before leaving the trench.

Metal gas pipe must be 12 inches (30 centimeters) below the ground or 6 inches (15 centimeters) below concrete. Metal gas pipe is painted green (to denote gas) and any underground portions must be wrapped with waterproof 10-millimeter-thick gas tape to at least 6 inches (15 centimeters) above the ground. Lines should be run as close as possible to the meter at one end and the heater at the other. Flexible gas lines are not permitted in any building code I know of.

Gas hookups must include a shutoff valve (Fig. 6-14) just before the heater, followed by a sediment trap (just a drop of pipe off of a T that can collect impurities in the gas before it gets to the heater), followed by a threaded union. When you need to service the heater, you can close the valve, break the union, and remove any parts of the system without shutting off the gas to the entire house. Similarly, if a leak or other emergency develops in the heater you can quickly shut off the supply of gas in a convenient location. Because of this, I recommend an easy shutoff valve with a handle rather than a style that uses a nipple on the valve that requires a wrench to close.

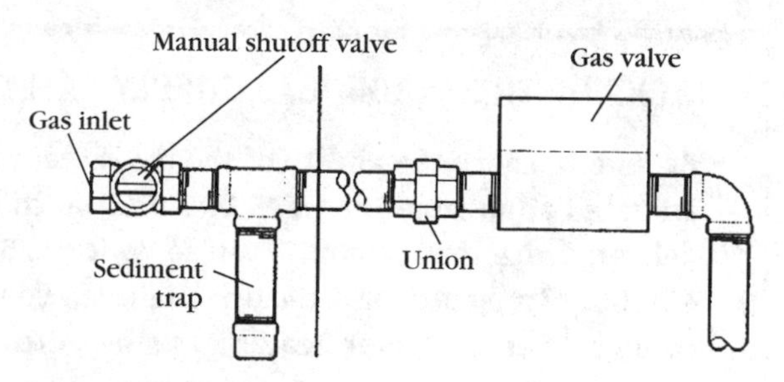

FIGURE 6-14 **Gas connection plumbing.** *Raypak, Inc., Westlake Village, Calif.*

ELECTRICAL CONNECTIONS

If the heater has an electronic ignition, you must supply the electricity. Wire should be 14-gauge copper, run in its own waterproof conduit (not shared with wiring for other purposes). For either 120- or 240-volt supply, run three wires. One wire, the ground wire, is green. With a 120-volt supply, the other two wires should be black or red (hot line) and white (neutral). With a 240-volt supply, both wires might be black or red.

Electronic ignition heaters are designed to operate on either 120 or 240 volts. The heater control circuit works on 25 volts regardless of the voltage supplied. Carefully study the wiring diagram provided by the manufacturer to make sure you are connecting 120 volts to 120-volt connections or 240 volts to 240-volt connections. If you set up the heater's transformer wires to 120 volts and supply 240 volts, you will be eating fried transformer.

Where should the electrical supply come from? I prefer to run the wiring from the system time clock so that the heater can't operate unless power is also being supplied to the pump and motor. This doesn't guarantee that the heater will only come on when water is flowing, but it is one more safety measure. The ground should be attached to the ground lug or bar in the clock and to the ground lug inside the heater cabinet (you'll see the lug near the transformer wiring, usually with a big sticker that says Ground).

As you will see in the chapter on time clocks, there is a line side and a load side. Always connect the heater (and pump and motor) to the load side, otherwise the wiring to the heater will always be hot.

Some manufacturers suggest including a *fireman's switch*. This switch is mounted in the time clock and is designed to turn the heater off about 20 minutes before the time clock turns off the pump. The idea is that by running cool water through the heater for 20 minutes, you cool the heater components and extend their life. Most time clocks

TRICKS OF THE TRADE: GAS SUPPLY SAFETY

- **Be sure to check the ability of the gas meter to supply the required amount of gas. Many are rated at no more than 175,000 Btu, which the gas company will tell you actually delivers twice that amount. Even so, twice 175,000 is only 350,000 Btu, not enough for a 400,000-Btu heater or a smaller one when your house is simultaneously using a gas clothes dryer, gas water heater, gas stove, etc.**
- **Gas lines must be pressure tested. Building codes determine how much pressure the line must hold and special gauges are available at the supply house, but I suggest hiring a licensed plumber to conduct this test for you. They have the tools and knowledge to conduct this important safety check.**
- **When putting a heater into service or when connecting a replacement heater to existing lines, you can conduct a simple test of your own. Turn on the gas supply to the heater and wipe liquid soap over any joints or unions (ammonia in a spray bottle also works well). Escaping gas makes the soap or ammonia bubble so you can detect any leaks quickly. Never leak test with a match—you might end up in the next county.**

have predrilled holes for standard fireman's switch installation (and directions are provided in the switch package).

Electric-fueled heaters need heavier gauge wire because they consume large amounts of electricity as the heating fuel. Follow the manufacturer's guidelines for these installations or bring in an electrician to make the hookup.

Repairs

Now that I have reviewed the most common problems that can occur, let's turn to more detailed troubleshooting and repair. For simplicity, I will go down the list in the same order as the troubleshooting checklist. Items that are self-explanatory or which are detailed in another chapter are not repeated here.

IS THE WATER FLOW ADEQUATE?
RATING: PRO

If you can answer yes to all of the water flow questions posed in the checklist, yet the flow of water is still too low, the problem might be inside the heat exchanger.

QUICK START GUIDE: HEATER STARTUP

Rating: Easy

When you first fire up your heater, a few pointers are worth noting.

1. **BLEED** If the gas supply line is new, bleed the air out of the pipe by opening the line at the union near the heater and opening the shutoff valve. When you smell gas instead of air coming out of the pipe, shut off the valve and reconnect the union. Open the gas shutoff valve.
2. **PUMP** Make sure the heater on/off switch is off. Turn on the pump and motor and make sure the air is out of the water system. Check for leaks.
3. **PILOT** Light the pilot (refer to the instructions that follow or those provided with the heater) if the unit is a standing pilot type. Never try to light an electronic ignition pilot with a match or other fire source. After lighting the pilot, turn on the on/off switch and turn up the thermostat as needed to fire the heater.

   If it is an electronic ignition unit, turn the valve on the combination gas valve inside the heater to on. Turn on the on/off switch and turn up the thermostat as needed to fire up the heater. It might take a minute or two to bleed the air out of the system and replace it with gas, so don't worry if it seems to take awhile for the pilot to ignite, and the unit to fire the first time.
4. **FIRE-UP** When the heater first fires, the heat will burn off the oil that is applied by the factory to the heat exchanger as a rust preventive. Light smoke for a few minutes is normal. Also normal is some moisture condensation as very cold water runs into the very hot heat exchanger. The condensation will drip down onto the burner tray and sizzle—a little of this is normal too.
5. **LOOK** Observe the heater for the first 10 minutes. Make sure the smoke and condensation stop and that there are no leaks. Use your eyes and nose (gas leaks?). Turn the heater off and on a few times to be sure it operates properly. Do this from the on/off switch and the thermostat a few times. Remember, if the water is colder than about 65°F, the thermostat won't turn the unit off because it doesn't register that low. Now turn the pump off. The heater should shut off within 5 seconds. If it doesn't, the pressure switch needs adjustment.
6. **FEEL** With the pump running, touch the inlet pipe to the heater and then the outlet. The temperature differential should not be more than 10°F. If it is much more than that, refer to the discussion of bypass valves that follows.

QUICK START GUIDE: TERRY'S HEATER TROUBLESHOOTING TIPS

Rating: Advanced

This is the only section of the book I named after myself, because after years of troubleshooting heaters, I have developed a method that seems to work for me (I think it will work for you, too). Although some of the troubleshooting tips refer to gas-fueled heaters, the majority of the technology and diagnostic tips will apply to any heater.

Check the Water System

- Is the pump primed and running without interruption?
  - Enough water in the spa?
  - Air purged from system?
  - Skimmer/main drain clear?
- Pump strainer pot and impeller clear?
  - Filter clean?
  - ALL valves open?
- Is the water chemistry correct?
  - High pH could mean scale in heater.
  - Low pH could be causing leaks.
- Any visible leaks?
  - At exterior plumbing connections.
  - At interior heat exchanger components.

Check the Gas System

- Is gas getting to the heater?
  - Look for pilot flame.
  - Combination gas valve turned to "on."
  - Gas supply lines adequate/unobstructed?
  - Propane—tank full?
- If pilot and/or burner are working . . .
  - Is flame 2″–4″ and steady?
  - Is flame steady and blue?
  - Are all burners fully lit?
    - Does pilot ignite within 5 seconds?

- Does burner ignite within 10 seconds after pilot?
- Does tray ignite without "flash" or loud boom?

- Is ventilation adequate?
  - Adequate air supply to heater?
  - Adequate venting of hot air away from heater?
- Smell for leaks!
  - Sniff around outside connections/unions.
  - Sniff around combination gas valve/joints.

Check the Electrical System

- If electric-fueled heater, is the circuit breaker "on" for the unit?
  - Is "reset" breaker button tripped?
- Is the "on/off" switch on? Is the remote switch (if any) on?
- Is the thermostat turned up high enough?
- On electronic ignition gas heaters, are 25 volts coming out of the transformer?
  - If not, is 120/240 volts coming into transformer?
  - Is heater grounded properly?
  - Are all connections tight and clean?
- Check the pilot:
  - Is the pilot clear of rust/dampness or insects?
  - Has the electronic ignition fired the pilot and is there a strong, blue flame?
- Check the control circuit—follow the path of electricity:
  - Power from pilot generator or transformer?
  - Power to each switch?
    - On/off.
    - Thermostat.
    - Fusible link.
    - Pressure switch.
    - Hi-limit (2).
    - Inline fuse (some models).
    - Fireman's switch.
- Power to gas valve?

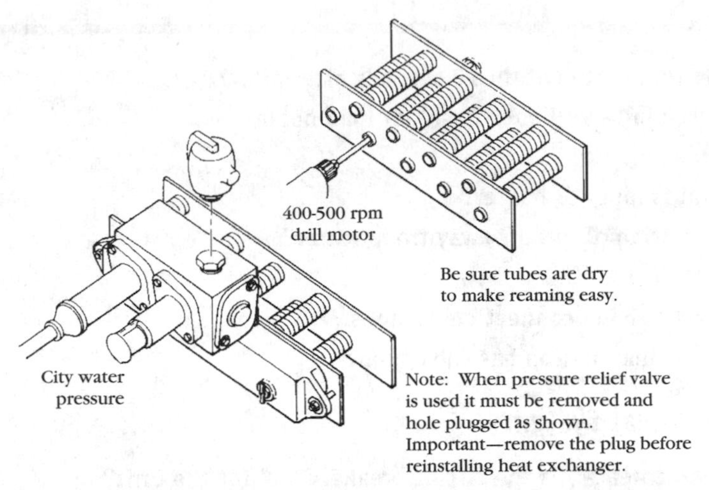

FIGURE 6-15 Heat exchanger reaming (cleaning). *Jandy-Teledyne Laars, Moorpark, Calif.*

Over time and with improper water chemistry, scale (calcium, liming) can build up inside the tubes of the heat exchanger slowly closing the passages, somewhat like fat closing the arteries in your body. To correct this, proceed as follows (Fig. 6-15):

1. **Look** Make a visual inspection to verify the problem. On one or both sides of the heater, depending on the make, you will find heat exchanger drain plugs. Some are fitted with air relief valves similar to those on a filter. Remove the drain plugs and make a visual inspection inside. Scale formation will be obvious.
2. **Disassemble** Remove the heat exchanger assembly from the heater as described in the installation section. Remove the headers from the exchanger by removing the bolts on the top and bottom at each end.
3. **Ream** The tubes of the exchanger can be cleaned out with a special reaming tool available at your supply house. The reamer is attached to your power drill. Withdraw the reamer frequently to clear lime deposits from the tool and prevent binding. For best results, work with a dry heat exchanger. When finished, flush each tube with water to be sure it is thoroughly cleaned.
4. **Reassemble** Install new gaskets and replace the headers on the heat exchanger. Tighten the bolts from the center first, working toward the outside, alternately tightening bolts on top and bottom to

ensure even compression of the gaskets and a solid seal. Working the reverse of the disassembly procedure, reinstall the exchanger and the components removed.

IS THE INTERNAL FLOW CONTROL VALVE OPERATING?

RATING: PRO

As noted, if the temperature differential between the incoming and outgoing pipes exceeds 10°F (5°C), the internal flow control is not operating correctly. I have also seen water flow blocked when the plunger of one of these units has broken apart and lodged in the header. Age and heat can destroy the stem or plunger on these valves.

Figure 6-16 shows the simple disassembly of these units for inspection. The Raypak Unitherm Governor acts like your car thermostat, opening and closing in response to a specific temperature. The spring-loaded type responds to water pressure. In either case, unless the heater is several years old, these units should not appear to be disintegrating or failing. Failure indicates another problem—poor water chemistry, unit overheating from some other cause, or scale buildup. Check further.

IS THE WATER CHEMISTRY CORRECT?

RATING: EASY

Improper water chemistry can cause heater problems. Refer to the chemistry chapter for more details; however, heater-friendly water should be

- pH 7.4 to 7.8.
- Alkalinity 120 to 150 ppm.
- Total dissolved solids of no more than 2500 ppm (pools) or 1500 ppm (spas).

IS GAS GETTING TO THE HEATER?

RATING: EASY

If this is not apparent (from a working pilot, for example), turn off the gas valve and open the union near the heater. Make sure there are no open flames or sparks nearby that might ignite the gas, then turn the gas valve back on and listen for gas flow (you will hear the hissing before you will smell the gas). If there is no flow, go to the source—the

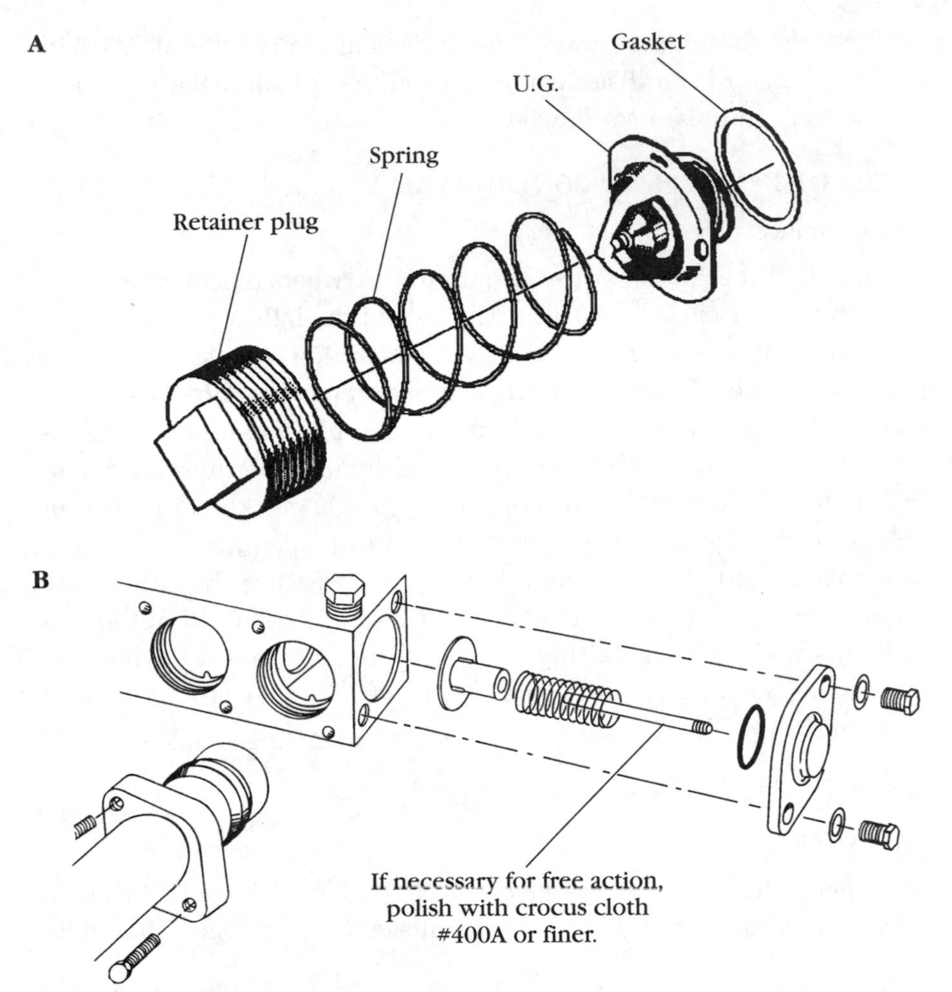

FIGURE 6-16 Flow control valves. *(A) Raypak, Inc., Westlake Village, Calif.; (B) Jandy-Teledyne Laars, Moorpark, Calif.*

main meter or propane tank. Notice if the gas supply pipe to the heater is dedicated to the heater alone, not also servicing a nearby gas barbecue or household water heater.

REPAIR OR REPLACE BURNER TRAY COMPONENTS

RATING: PRO

Rusted-out burners or tray components will be obvious to the naked eye. Following Fig. 6-2, remove the burner tray assembly and replace any defective burners (Fig. 6-18). The burner tray assembly

TRICKS OF THE TRADE: LIGHTING (OR RELIGHTING) THE PILOT

Rating: Easy

This procedure applies to standing pilot units only, because the electronic ignition is automatic unless something is broken. Instructions are almost always printed on the heater itself. Look for these and follow them. If the directions for your particular heater are obscured or missing, the following procedure is most common:

1. SHUT OFF Turn the gas valve control to Off and wait five minutes for the gas in the burner tray or around the pilot to dissipate (safety consideration). Turn the on/off switch to Off.
2. LISTEN Turn the gas valve control to Pilot and depress. If the area is quiet, you should hear a strong hissing sound as the gas escapes from the end of the pilot. If not, it might be clogged by rust or insects.
3. LIGHTING Light the pilot and continue to depress the control. If the heater is equipped with a pilot lighting tube (Raypak), you should be able to hold your match in front of the tube and ignite the gas from the pilot (Fig. 6-17). Check to see that it is clear of obstructions.

   If the heater is equipped with a visoflame tube (Teledyne), you should be able to hold your match over the mini-burner of this tube which ignites its own gas supply and in turn ignites the gas supplied to the pilot (Fig. 6-17). Visoflame tubes sometimes flare up because they have their own gas supply, so keep your hands and face as far back as practicable and light the tube promptly after depressing the control.
4. WAIT Hold the control down for at least 60 seconds. This allows heat from the pilot to generate electricity in the thermocouple to power the gas valve (which will then electronically hold the pilot gas valve open and power the control circuit which ultimately opens the main burner gas valve). Release the control and the pilot should remain lit.
5. VERIFY This step is not usually included in instructions printed on the heater—that the pilot flame is healthy; that is, look to see that a strong-burning blue flame of 2 to 3 inches (5 to 7 centimeters) extends toward the burner tray and a secondary flame of equal value is heating the thermocouple. Make sure the pilot is securely in place; it might be burning properly, but rusted fasteners or poor installation from a previous repair might have left it dangling near the burner tray, but not in close contact with it.

   As mentioned previously, burner tray design leaves a lot to be desired when servicing the pilot assembly of most heaters. A common problem is that a lazy service technician previously left the pilot assembly hanging in the general area by wire, tape, a screw not fully in place, etc. The result is that the pilot burns just fine, but the burner tray needs to be flooded with excessive amounts of gas before any reaches the pilot flame to be ignited. The resulting ignition is explosive. Therefore, always make this visual inspection.

6. FIRE UP Turn the gas valve control to On. Stand back from the heater (and to one side—not in front of the open front of the heater) in case of flashback (the explosive ignition described previously). Make sure the pump is operating and water is flowing freely, turn the on/off switch to On, and turn the thermostat up. The heater should fire normally.

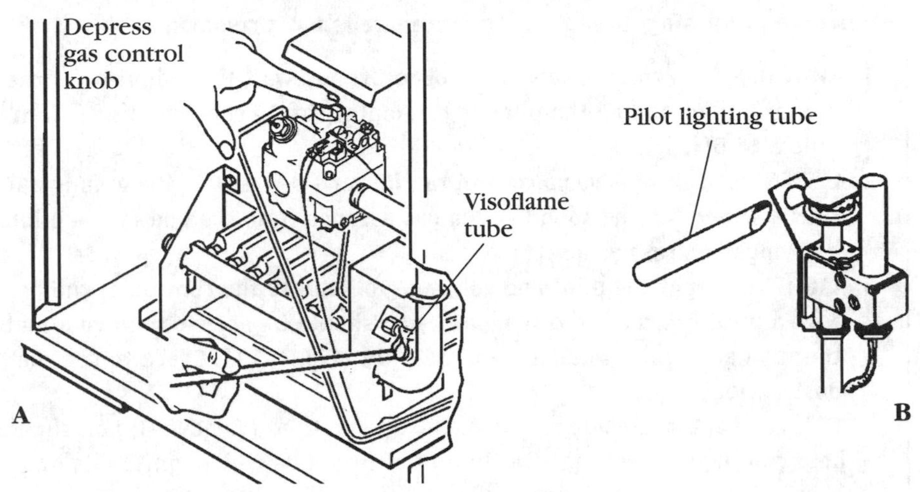

FIGURE 6-17 Pilot lighting systems. *(A) Jandy-Teledyne Laars, Moorpark, Calif.; (B) Raypak, Inc., Westlake Village, Calif.*

includes the combination gas valve, burner tray, and pilot assembly. Figure 6-2 shows a standing pilot type heater, but the electronic ignition unit looks the same except you must also disconnect the high-tension wire that serves the pilot. You also need to disconnect the three wires connected to the IID and the ground wire in order to completely remove the tray.

If the pilot fails to light it might be clogged. If the heater "booms" when it lights, it means the pilot is not igniting the gas in the tray soon enough. Is the flame too small because of an obstruction in the pilot? Is the pilot bent, rusted, or not positioned close enough to the burner tray? Take it apart.

There are several types of pilots used in various heaters, but once the tray is removed, the disassembly should be obvious. Let's review typical installations and problems with standing pilot units and electronic ignition units.

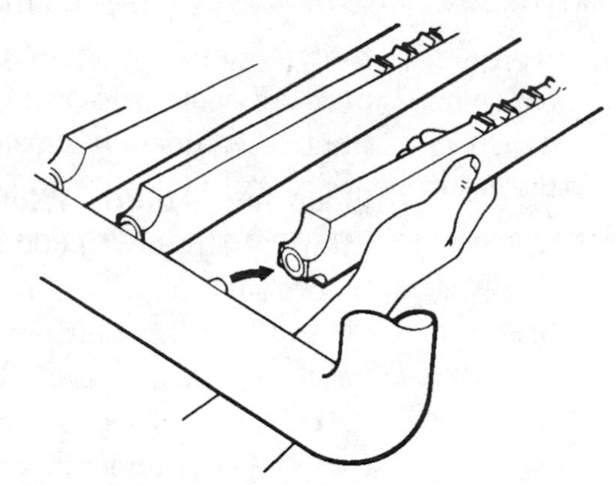

FIGURE 6-18 Replacing burners in the burner tray. *Jandy-Teledyne Laars, Moorpark, Calif.*

Standing Pilot Units: Disconnect the wires of the pilot generator from the combination gas valve. Remove the pilot gas tube from the combination gas valve and remove the entire pilot assembly from the tray. The pilot generator either clips in place or is held by a threaded ring (Fig. 6-19). Remove the pilot generator. If it is rusted or swollen, replace it.

The pilot itself will further disassemble into two or three sections, depending on the make. Examine each part for obstruction or rust. Blow through one end of the pilot gas tube and be sure it is unobstructed. The pilot can be cleaned by soaking it in muriatic acid for 30 to 60 seconds, then cleaning thoroughly with fresh water and blowing dry (do not leave acid or water on metal parts—they'll rust).

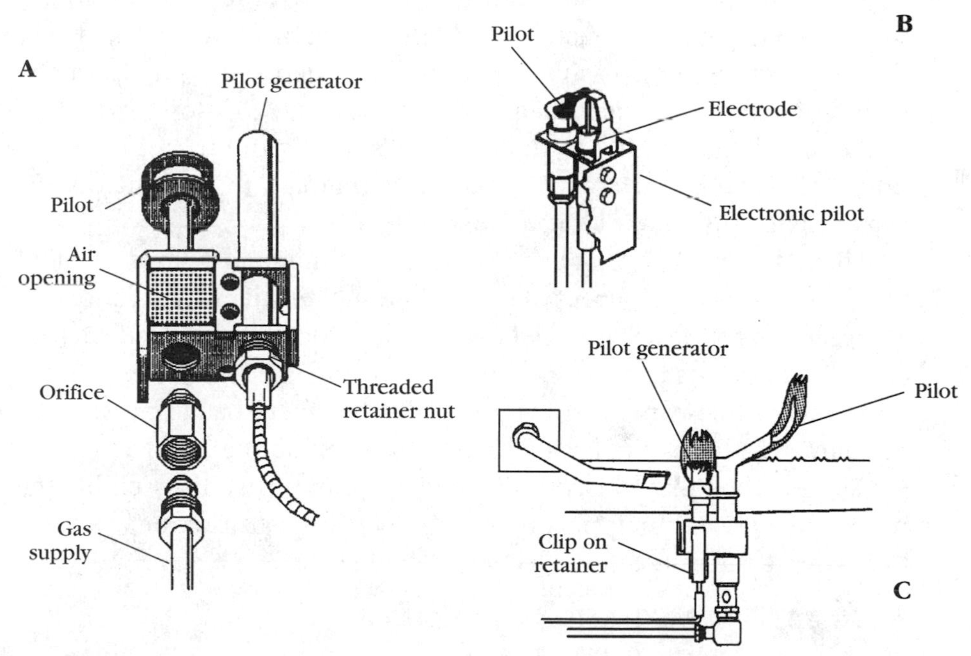

FIGURE 6-19 Types of standing pilot assemblies. *(A) Raypak, Inc., Westlake Village, Calif.; (B, C) Jandy-Teledyne Laars, Moorpark, Calif.*

Natural gas is actually odorless, and to make leaks more detectable, the gas supplier adds a perfume. This perfume has been found to attract insects who build nests in pilot assemblies. Insects and rust are the two main enemies of the pilot.

Electronic Ignition Pilot Units: The ignition electricity feeding the pilot is between 10,000 and 20,000 volts, so be sure the power is off before servicing. Remove the high-tension wire that supplies electricity to the pilot. Remove the pilot gas tube from the combination gas valve and remove the pilot assembly from the burner tray.

On some units, the orange insulated wire terminates at the burner tray and is connected to a bare wire which itself is held in place by a ceramic insulator in a bracket. This wire then attaches to the pilot electrode with a threaded end and two nuts. Sometimes, this bare wire can get too close to the metal burner tray and the electricity jumps or shorts out to the tray instead of getting to the electrode. This is solved by bending the bare wire (with the power off) so it remains clear of any other metal. Be careful when you replace the burner tray, because this wire can easily be bent back into contact with the burner tray.

The end of this wire on the electrode is also subject to failure. Working with installations on the Malibu beach, I have noticed that this connection rusts easily and is often coated with salt from the moisture in the air of the ocean. This is not enough to stop the flow of the high voltage coming to the pilot for spark ignition, but once the pilot lights the connection, it is corroded enough to prevent the IID from sensing that the pilot has indeed been lit.

The IID senses the pilot by the electricity generated from the heat of the flame—a reverse current of 0.00002 amp (2 milliamps) is generated along the line. It is so sensitive that any electrical obstruction such as corrosion prevents this signal from reaching the IID. I have cleaned and reassembled these units and successfully put them back into service, but they usually soon fail again. In this case, replace the pilot assembly.

As described previously with standing pilot units, disassemble the pilot assembly and supply tube and clear any obstructions or rust. Reassemble the opposite of the way it came apart.

AUTOMATIC COMBINATION GAS VALVES

There are no in-the-field repairs that can be made to the automatic combination gas valve. If you determine that it is failing, replace it.

They are sometimes rebuilt at the factory, but the cost of doing so is greater than the replacement cost ($70 to $120).

The plumbing of the valve is female threaded and screws directly onto the gas pipe of the burner tray. When replacing, apply Teflon tape to the male gas pipe of the burner tray. I do not like pipe dope in this case because it can easily squeeze off the threads and into the opening of the gas valve, obstructing gas flow.

Note that on the end of the valve, a flat ridge has been provided (Fig. 6-6) around the female threaded opening so a standard box wrench can be used to tighten or loosen the valve from the gas pipe of the burner tray. This is easier than trying to get a huge pipe wrench around the entire combination gas valve unit. When tightening, be careful to support the pipe so it does not bend or snap off of the burner tray.

The gas tube that supplies the pilot is prebent by the factory. Trying to bend these in other ways usually leads to crimping and leaks. Your supply house now carries a universal supply pipe in a thinner, more flexible pipe style, which is coiled in sufficient length to serve any make of burner tray. Buy these for replacements.

Having said you cannot service a gas valve in the field, there is one adjustment that can be made—gas pressure to the pilot or main burner. Figure 6-6 shows the typical location for the small screw that adjusts these pressures, sometimes beneath a threaded cap which must first be removed (black caps denote propane). Clockwise turns increase pressure; counterclockwise decrease pressure.

IS VENTILATION ADEQUATE?

RATING: ADVANCED

You would be surprised how many problems are created by improper ventilation. You need only place your hand about 2 feet over the top of a gas-fueled heater to feel the power and volume of hot air moving through the ventilation system to understand why any restrictions under such heat result in damage of some kind.

Sooting, which is black carbon buildup on the heat exchanger, is a symptom of improper ventilation. Carbon starts as a dirty black coating and builds up to the point where chunks of coal burn and break off, falling on the burner tray below. The burners become clogged, shutting down the heater's full capacity. The heater smokes when it operates.

Here's a perfect example of a problem that is easy to repair, but the repair will be of no value if you do not also address the cause. Refer to the installation section and to the manufacturer's guidelines about proper ventilation. Correct any problems in this area first.

The remainder is easy. Remove the burner tray as described previously, cleaning all the components with soap and water, and drying out everything before reinstallation. While the burner tray is out, remove the draft hood and any vent stack to reveal the heat exchanger. Using a stiff bristle brush and soapy water, clean the exchanger. Be sure to get inside the cabinet to clean the underside of the exchanger as well (it might be easier on small heaters to remove the heat exchanger for cleaning).

Be careful not to soak electrical components or the firebrick. After reassembling the heater, small amounts of carbon that you might have missed will burn away over time, providing that you have corrected the ventilation problem and no new sooting is occurring.

Do not use a wire brush for soot removal. It can cause sparks that might ignite the carbon. Use a stiff natural or plastic brush.

CHECK THE CONTROL CIRCUIT

RATING: ADVANCED

Replacing the various control circuit switches as needed is very easy—the on/off switch and thermostat (mechanical or electronic) come off the cabinet with the removal of two screws; the high-limits are easily removed from the header; the pressure switch simply unscrews from the end of the water tube feeding it (Fig. 6-20); the fusible link is held by a ceramic holder that is held in place by a screw or two; and the wires to each switch simply unclip or come off by loosening a screw.

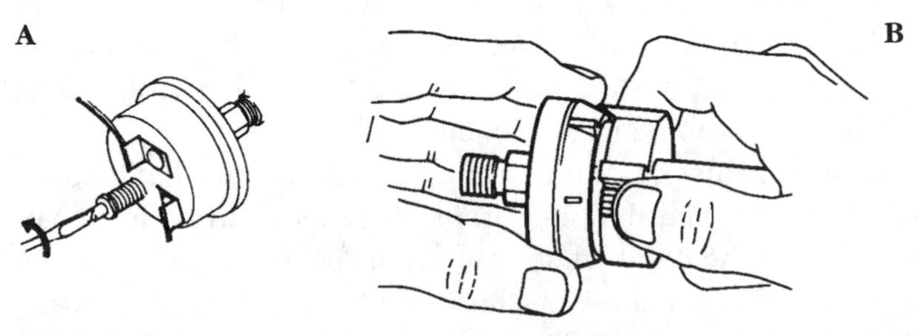

FIGURE 6-20 Pressure switch adjustment. *Jandy-Teledyne Laars, Moorpark, Calif.*

No, the problem is not in replacing a faulty control circuit switch, the problem is determining which switch is faulty . . . and why.

Whether the circuit is powered by 750 millivolts or 25 volts, the simple troubleshooting procedure is to follow the path of the electricity. Using your multimeter (discussed in more detail in the chapter on basic electricity), you check to see if electricity is getting to the switch (see Fig. 6-8) and, if so, is it getting out? If not, then that is the faulty switch.

On electronic ignition heaters, check the transformer to see that 25 volts is leaving the transformer (Fig. 6-22). Set your meter to the appropriate scale (on mine, 50-volts ac).

To save time, check if there is 25 volts at the end of the circuit (Fig. 6-22, item 2). Leave one lead touching the neutral side of the transformer and move only the hot lead to test each part of the circuit. If there is current, then obviously all of the switches in the circuit are closed and you need to look elsewhere for failures. If there is no power at item 2, then one of the switches is open for some reason.

On standing pilot heaters, use the same technique. Check the beginning of the circuit (Fig. 6-21) to determine if the pilot generator is delivering 400 to 700 millivolts. As discussed previously, set the meter to the correct setting (on mine, 1-volt dc) and touch the negative lead of your meter to the negative terminal on the gas valve and the positive lead to the positive terminal (item 1).

The heater will work on as little as 200 millivolts, but the electricity required to power the circuit and the combination gas valve will

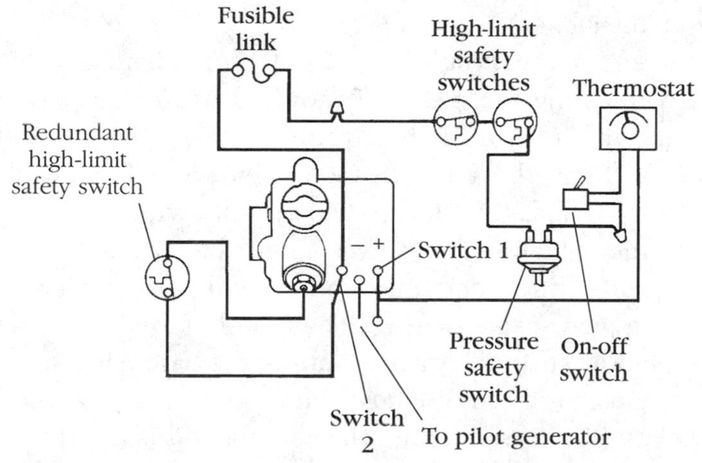

FIGURE 6-21 Millivolt wiring schematic. *Jandy-Teledyne Laars, Moorpark, Calif.*

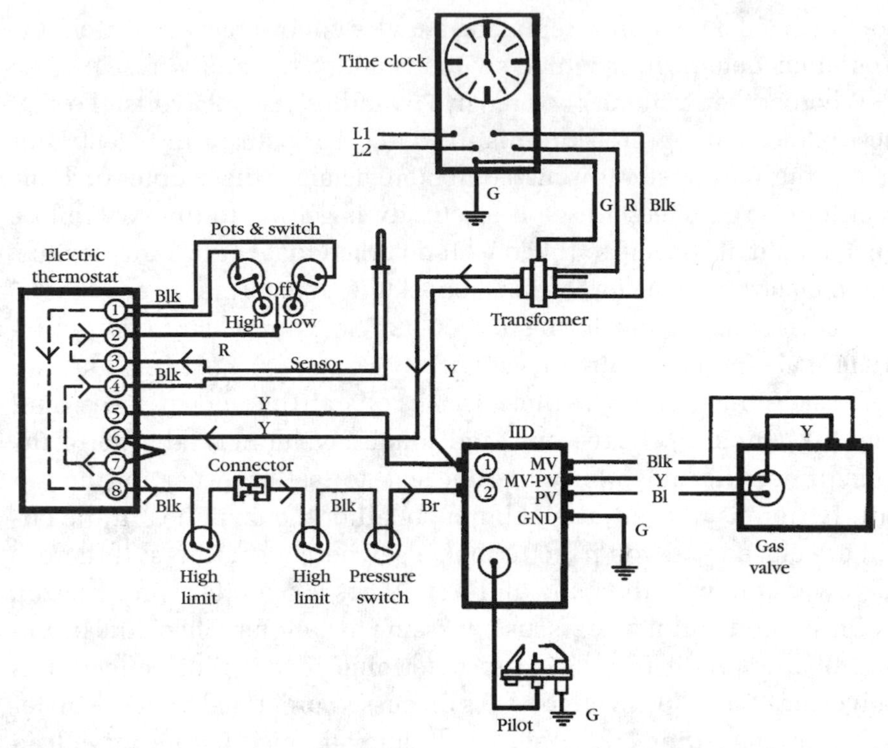

FIGURE 6-22 **Electronic ignition heater control circuit (with electronic thermostat).**
Raypak, Inc., Westlake Village, Calif.

use almost all of that, so the heater might fire, but it will soon shut down. A healthy pilot generator delivering over 400 millivolts will withstand the 200-millivolt operational drop and still retain enough power to continue the job.

Now check the end of the circuit (item 2). As noted before, leave the negative lead touching the negative terminal on the gas valve and move the positive lead around the circuit. Again, if there is 200 millivolts or more there, the circuit is complete and the heater should fire. If not, start checking the control circuit switches.

Be aware that the switches themselves might be fine but there could be a short in the wiring. Sometimes rodents will nest in a heater and for some reason gnaw at the wiring. They rarely cut all the way through the wire, but by stripping the insulation, the bare wire sometimes comes in contact with the metal cabinet and creates a dead short—the electricity just flows through the cabinet and is dissipated or is sent through the ground wire.

On a millivolt heater, the symptom of this is usually that even though the pilot generator is producing 400 to 700 millivolts, the pilot goes out when you let the gas valve control knob up after lighting. To test if this is the problem, remove both wires of the control circuit from the gas valve. If the pilot now stays lit, you have a short in the wiring. If it still goes out, the gas valve is bad and should be replaced. On a 24-volt system, there is no simple way to test for bad wires, except visual inspection.

For both millivolt or 24-volt systems, the control circuit switches are similar. As noted previously, follow the path of the electricity and test for voltage into and out of each switch until you find the one that is open. Also as noted, to test a terminal or switch connection, keep the negative lead of the meter on the negative of the transformer and the positive lead on the wire or connection being tested.

Pressure Switch Failure: Most pressure switch failure is caused by obstructions in the circulation—low water in the pool or spa, a clogged skimmer basket or main drain, a clogged pump strainer basket, or a dirty filter. Check all of these before touching the pressure switch.

Sometimes the obstruction is in the water supply to the pressure switch. Remove the wires to the switch and unscrew the switch from the tube (Fig. 6-20). Turn on the pump. Water should flow vigorously from the tube. If it does, look in the hole at the end of the pressure switch to be sure it is clear of obstruction.

If no water is coming out of the tube, remove the heater vent top, front header faceplate, and flue collectors to expose the end of the tube in the header. Unscrew the tube and remove it. Blow through the tube to clear any obstructions. Turn on the pump. Water should shoot out of the hole in the header if there is no obstruction. I recall clearing lime from this location on Dick Clark's spa heater more than once (the spa heater always seemed to fail when he had houseguests and I was always the villain). Finally I corrected the water chemistry and reamed out the heat exchanger and had no such problems thereafter.

If the problem is not in the water supply, adjusting the pressure switch to be more or less sensitive is done with the screw or knob on the switch (Fig. 6-20). On the Hobbs-type unit, loosen the nut that holds the screw in place (some units have a spring on the screw to provide tension and keep the adjustment). Tighten the screw to

> **TRICKS OF THE TRADE: HEATER TROUBLESHOOTING**
>
> **Another way to test a switch is to jump it. Take a short wire and connect the two terminals of a switch, in essence, completing the circuit by bypassing the switch itself. If the heater fires, then obviously the switch is open.**
>
> **As a time-saver, you might want to test the last one in the circuit first and work back to the first—when you find a switch with power to it, you know everything prior to that one is working and it might save you from checking several switches unnecessarily. Another time-saver is to start with the most accessible switches first, because you might find the problem without digging into the cabinet (some high-limit switches and thermostats require faceplate removal to test). You will develop your own preferences, but I start with the switches that are most accessible (and most often the problem), as described starting on p. 261.**

make the switch more sensitive (the switch closes and the heater comes on with less pressure), or loosen it to make it less sensitive (needs more pressure to come on) if the heater burns with no water circulating.

When adjusting either way, turn the screw only a one-quarter turn at a time because these switches are very sensitive. Check the operation after each quarter turn until the heater operates correctly. The heater should fire and stay lit when the circulation is running; it should shut down within three seconds when the circulation stops.

Remember, the pump might be off but the circulation might still run if the heater is substantially above or below the pool or spa water level (or if solar panels are part of the system). Therefore, a lag in the heater shutoff might be several seconds after the pump is off. Retighten the nut to keep the screw at the adjustment made. The other type of pressure switch is adjusted with a knob in the same manner as the setscrew on the Hobbs-type unit.

If the switch can no longer be adjusted or is rusted out, replace it as follows:

1. Turn off the pump.
2. Remove the two wires and unscrew the unit from the line.
3. Reverse these steps to install the new unit.
4. Test the new unit by turning on the pump and verifying that the heater will fire. Leave the heater switched on and the thermostat on High and turn off the pump. The heater should go out promptly. If it continues to run, refer to the previous section and adjust the new switch accordingly.

Fuses and Fusible Links: As described previously, some 24-volt heater designs include an inline fuse, like a car fuse, in a bracket on the positive wire coming out of the transformer. Make a visual inspection or use your multimeter to test for current at a point after the fuse. The fusible link is easy to test with your multimeter, following the procedures outlined previously. If current is not passing through the link, it must be replaced, but you want to know why.

Is the venting adequate or is there an overheating condition? Is there soot or debris on the burner tray causing overheating somewhere? Is the gas pressure low, causing *lazy flame* (yellow flames that seem to lick out of the burner instead of strong, blue flames burning straight up)? Replace the link, but also address the cause of the failure.

High-Limit Switches: High-limit switches are slightly different in each make of heater, but the principles are the same. In Teledyne heaters, two high-limits are located side by side beneath a small protected cover (as seen in Fig. 6-2). Remove the cover (usually one screw) and pull the switches with their retainer bracket from the header area. This operation can be done with the pump on or off. Pull the switches from the bracket, noting how this setup is assembled (pay attention or it can be a jigsaw puzzle to put back together). Pull the switch from its socket and replace it.

I usually replace both. If one has worn out, the other one will usually fail shortly and the customer will not be understanding about two labor charges for what appears to be the same repair within a few days or weeks of the first job. Start the pump or turn the heater back on and verify that the heater operates.

On/Off Switch and Thermostat: The mechanical thermostat can be tested for current just like the others. The electronic version looks more complex, but is actually easier to diagnose because it is connected through a terminal block where all the connections are in one place for easy testing.

To avoid heat-related injuries, manufacturers limit their thermostats to 104°F (40°C) (maximum temperature recommended for spas by most local codes and health laws). Because the mechanical thermostat has an accuracy of +/−3°F, these units are calibrated for 101°F (38°C) maximum to allow for the extra 3°F. Of course that means it might be 3°F the other way and you might only get your spa to 98°F (37°C). Now add bubbling water and a cool evening breeze and the water only goes to the low 90s. This is a common complaint with mechanical thermostats, so be prepared to educate your customer.

Electronic thermostats are calibrated to go no higher than 104°F (40°C), but their accuracy rate is within 0.5°F. Still, on a cool, windy night with the jets and blower going full bore, the water might not get higher than the high 90s and your customer might complain of insufficient heat. Be aware that the problem might be in the design and not in the components or the heater itself.

Figure 6-22 shows the terminal block of a typical electronic thermostat. Twenty-five volts travel from the transformer to terminals 6 and 7. Check here to verify current supply.

From here the current goes to terminal 4 and into the temperature sensor (called a *thermistor*), then back to terminal 3. If no current is present at terminal 3, the sensor is bad.

From terminal 3, the current goes to terminal 2 and into the on/off switch. The switch sends the current into one of two thermostats (also called potentiometers or pots), and the thermostat sends current to terminal 1. If no current is present at terminal 1, the switch or thermostat is bad.

From terminal 1, the current runs to terminal 8 which feeds the high-limits, fireman's switch, and pressure switch. As noted earlier, a time-saver is to look for current at terminal 8 first. If 24 volts is there, everything in the line before it is good.

You can also test the terminal block by jumping connections. If you jump terminals 1 and 2 and the heater fires, you know the problem is in the switch or thermostats. If you jump terminal 8 to the IID's terminal 2 and the heater fires, the problem is in the high-limits, fireman's switch, or pressure switch. Examine the diagrams and see if you can find several other jumps that will quickly tell you without a meter the condition of various switches.

With new models or different manufacturers, the terminal block might not be set up exactly like the one in Fig. 6-22, but if you understand this concept you can read the wiring diagram for the heater you are working on and follow the same procedures. The wiring diagram can be found inside the cabinet door or printed on a panel inside the heater. If it is missing or obscured, refer to the booklet that comes with the heater.

Replacing a mechanical thermostat requires removing the draft hood top of the heater to expose the oil-filled bulb and tube that are attached to the thermostat dial. Remove this along with the thermostat and be sure the replacement is the same model, not one with a shorter tube that might not reach the dry well.

Replacing the electronic type means replacing the unit on the panel in the front of the heater, unless you also suspect the thermistor. If so, the thermistor is found in the same location as the tube or bulb in the header.

Fireman's Switch: A fireman's switch as previously explained is simply an on/off switch attached to the time clock that shuts off the heater 20 minutes before the pump. The result is a cooldown of the heater before the water circulation stops, prolonging component life. You can troubleshoot this switch by jumping across the electrical wires leading to it at any convenient place. If it is defective, replacement is self-explanatory.

Intermittent Ignition Device: The IID is actually a sophisticated switching device, not an electronic brain as some call it. There are no hidden computer chips in this device.

All IIDs have the same terminals, though perhaps not in the same location on the box. Figure 6-22 shows that power is supplied to the device by the transformer after passing through the control circuit. The IID sends power to the pilot to spark ignition and simultaneously opens the pilot gas valve by sending power (from the pv terminal) to the appropriate terminal on the combination gas valve. When the pilot lights, the heat creates a voltage that is detected by the IID and acts as a signal to send power to the main gas valve on the combination gas valve, opening it to send gas to the burner tray.

The IID must be grounded to operate properly, so that is a good place to start your troubleshooting. If you have 25 volts at terminal 2 on the IID but no spark on the pilot, check the pilot. If it looks good, test the orange high-tension cord. The easiest way to do this is to remove the end of the cord from the IID and hold it about ⅛ inch (3 millimeters) away from its terminal on the IID. Turn the heater on and if you get a spark jumping from the cord to the terminal, the cord is probably good. If you get no spark, it is faulty.

If you get a spark here but not at the pilot, the IID or high-tension cord is faulty. If you get a spark but no gas to the pilot, the problem might be in the high-tension cord (it might be sparking the pilot, but not sensing that the pilot has lit). If replacing the cord doesn't solve the problem, the fault is in the IID.

Finally, if you have a spark and a lighted pilot but no voltage at the main valve (MV) terminal of the IID, the IID is faulty. I always carry a spare high-tension cord and IID. If I suspect either component, it is easy to replace them to locate the problem. If it turns out that the component is okay, then my spare goes back in the toolbox and I put the original back in the heater.

IIDs, like combination gas valves, can be repaired at the factory but not by you. However, the cost of a new one ($70 to $100) is less than the labor, parts, and trouble of taking it to the factory for repair.

Electric-Fueled Heater Reset Buttons: The small red button found on electric-fueled heaters is a safety circuit breaker. Some are activated by an overheating condition, such as when the unit is allowed to run dry. Some are only sensitive to excess amperage, like the circuit breakers in your house. After they cool (in a minute or so), they can be pressed back in to reset them. However, as with other troubleshooting, you must determine the cause of the circuit breaking.

Usually the reset breaker is tripped by an electric heater that is overheating from low (or no) water flow. Check your circulation before resetting, and afterwards observe the unit in operation for several minutes to make sure it doesn't pop again. Of course, sometimes these little breakers simply wear out from old age and need to be replaced, but this is not common.

Solid-State Controls: Although the majority of heaters still use mechanical controls, there is a growing trend toward touch panel solid-state controls. These may seem tougher to troubleshoot and repair, but in fact they are generally more reliable and actually easy to service. Because each manufacturer uses a different panel and configuration with various display abbreviations, there is no "one fits all" troubleshooting guide. However, since many elements are similar, if you understand one system you will probably be confident to work on other systems, especially if the owner's manual for the heater is readily at hand. Since solid-state controls are more common in spas (controlling the heater, blower, pump, and other elements), refer to the troubleshooting guide in Chap. 11.

Preventive Maintenance

RATING: EASY

The best preventive maintenance is to use the heater regularly. As noted previously, corrosion, insects, nesting rodents, and wind-blown dirt create many heater problems that can be eliminated by regular use. The heat helps to dry any airborne moisture that might otherwise rust the components. It discourages insects and rodents before they get too comfortable. It keeps electricity flowing through the circuits, preventing corrosion that creates resistance that might ultimately break the circuit completely. It burns off the odd leaf or debris that lands inside the top vents before a greater accumulation builds up, which when the

FIGURE 6-23 Routine preventative maintenance.

heater comes on might start a fire or send flying embers into the air. In short, running at least a few minutes a day is great therapy for a heater.

Otherwise heaters need only be visually inspected from time to time. A look around will detect sooting, gas or water leaks, or other problems before they begin (Fig. 6-23). Keep leaves and debris off the top of the heater. Look at the pilot and burner flames. Are they strong, blue, and burning straight up at least 2 to 4 inches (5 to 10 centimeters)? Open the drain plug on the heat exchanger and look for scale buildup.

FAQs: HEATERS

Do I Need Both a Heater and a Cover?

- **No, but a cover and a solar heater will insulate and warm your pool significantly, so a gas heater doesn't need to work as hard (or cost as much for fuel). The cover also keeps the pool clean and reduces evaporation, which is exacerbated by heating in the first place. In order of value and importance, you should consider a cover, solar heating panels, and, finally, a gas heater.**

Will a Pool Heater Work on Electricity, Propane, or Natural Gas?

- **Yes, but an electrically fueled heater is slow to heat the pool and the fuel costs are considerably more than natural gas or propane. Natural gas or propane will heat the pool about the same, but the heater needs slightly different burner- and gas-regulating components, so be sure to use the heater that matches the fuel in your area.**

How Long Will It Take to Heat My Pool?

- **You will feel warm water coming from the return outlet immediately when heating with solar or gas heaters. Depending on the size of the pool and the capacity of the heater, you should be able to take the chill off the pool and make it swimmable within a day, especially if insulated by a cover.**

How Much Does It Cost to Run a Heater?

- **Depending on the size of your pool and how warm you want the water, you can heat your pool for less than $5/day in the coolest months of spring to make the water comfortable for swimming. As the weather warms and especially if you have a cover, that cost will drop significantly.**

CHAPTER

7

Additional Equipment

We have now examined the major components of water circulation, filtration, and heating. There are several pieces of equipment that complete the mechanical resources available to the pool, spa, or water feature.

Time Clocks

A key component of any water maintenance system is a time clock. Time clocks make sure that water is circulated through the filter each day, that it is heated sufficiently and/or at the correct time of day, that fountains or decorative lights come on at the right time, that automatic cleaners operate daily, and a host of other useful functions.

Electromechanical Timers

Figure 7-1 shows an exploded view of a typical 120-volt time clock, and Fig. 7-2 shows the same product assembled. It allows various on or off settings throughout each 24-hour period. Some models also include a seven-day feature, allowing you to determine which days of the week you want the clock to control the system. The only difference between a 110-volt and a 240-volt clock is that the 240-volt unit does not have a neutral line, but includes two incoming lines and two outgoing loads for the equipment. The neutral location is a ground. For

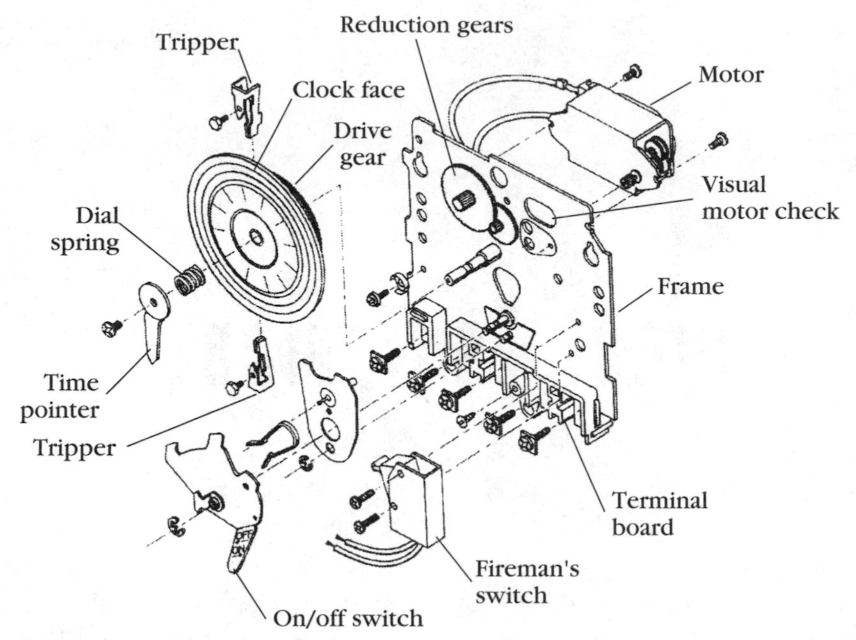

FIGURE 7-1 Typical electromechanical time clock. *Intermatic, Inc.*

explanation purposes, however, this diagram shows the basic parts found in any time clock.

The small electric motor runs on the same push-pull concept of opposing magnets as the motors used for pumps. A winding is electrified that creates an electromagnetic field and sets up the rotation, which happens at a predictable 400 rpm.

The 24-hour dial would not operate very well at that speed, so a series of reduction gears in the motor housing result in the actual drive gear turning at a speed that rotates the dial just once every 24 hours. In this way, you can tell the time of day by the clock face and use it to set on or off switching of the system.

Most ordinary watches or clocks have a face that remains fixed, while the hands of the clock rotate to give the time. The time clock operates on the opposite concept—it has only one hand, which remains stable, and a face that rotates.

Note the on and off trippers (the words on and off are etched on the trippers) (Fig. 7-2B). By setting these where you want the system to go on or off, the clock controls the flow of electricity to your system accord-

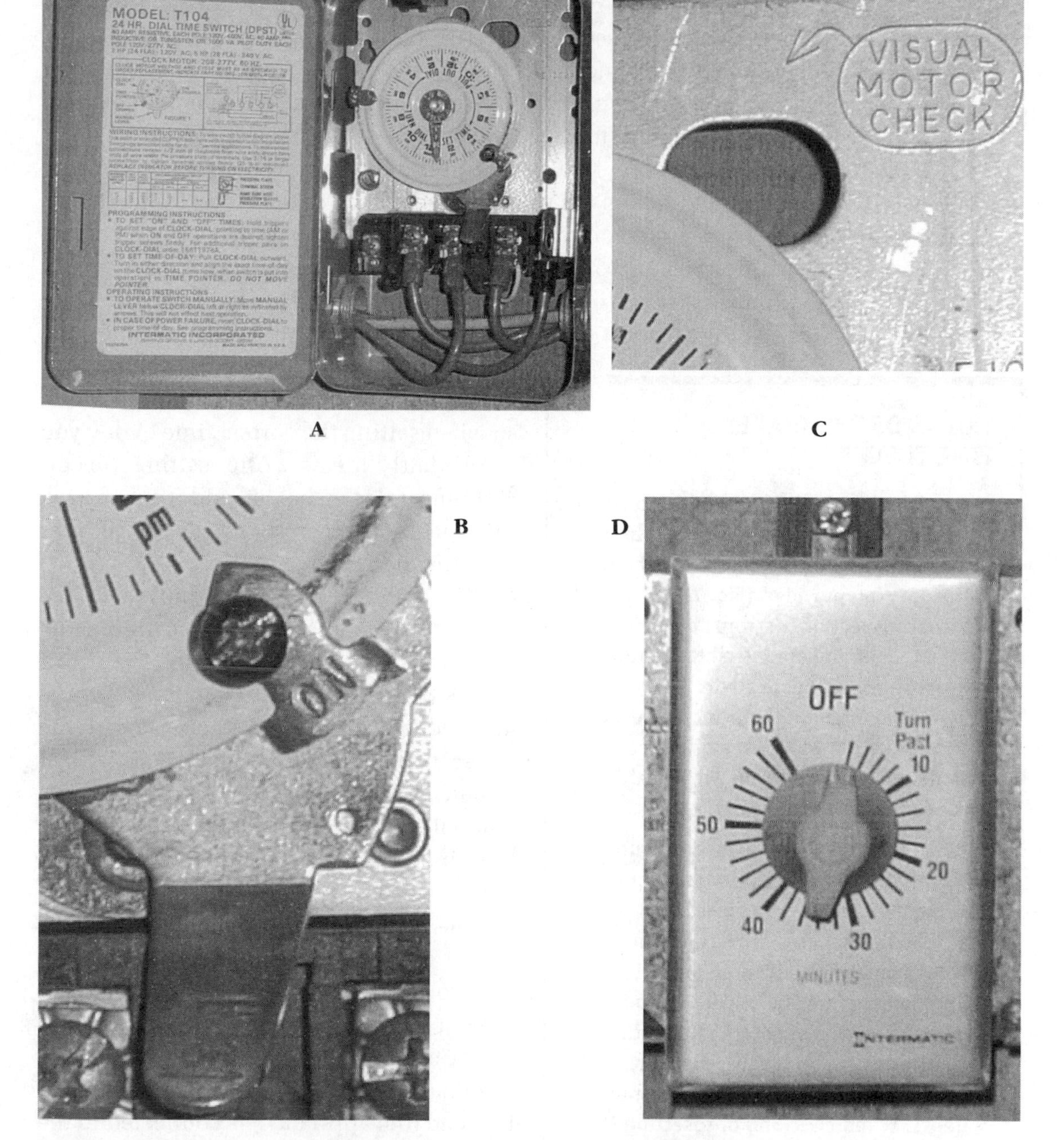

FIGURE 7-2 (A) Typical electromechanical time clock; (B) on/off tripper; (C) visual inspection port on time clock; (D) typical twist timer.

ingly. More than one set of trippers (also called *dogs*) can be affixed to the dial, so the system can go on and off several times during a 24-hour period. As you can see on the top tripper, the nipple on the bottom of the tripper is located almost midway along the edge. The nipple on the lower tripper is located at the end. These nipples engage the cam on the on/off switch of the clock, turning it on or off as the clock face rotates.

To set the clock, you simply pull the face toward you and rotate it until the number on the dial, corresponding to the correct time of day, is under the time pointer. By pulling it forward, its drive gear is disengaged from the motor drive gear and it rotates freely. As noted earlier, the time pointer is the only hand on the clock. The dial is divided into each of the 24 hours of the day (differentiating between a.m. and p.m., some clocks say Day and Night) and each hour is further divided into quarter hours, so when setting the correct time of day you can be fairly precise. After setting the correct time, release the dial and it snaps back into position, reengaging its drive gear with the motor drive gear.

When the switch lever moves to the right, it lifts the contacts apart, breaking the electrical circuit. When the lever moves back to the left, the contact arm is allowed to drop back into contact with the lower contact, completing the electrical circuit. Simple, eh? The lever also allows manual operation, so regardless of where the trippers are set you can manually operate the system.

The contacts are attached to screw terminals for attaching the wires of the appliance to be controlled. On a 120-volt model, the hot 120-volt wire is attached to the terminal marked Line and the neutral to the terminal marked Neutral. The hot wire for the load (the appliance) is connected to the load terminal and its neutral is connected to the neutral terminal. Some clocks provide separate neutral terminals for the line

TRICKS OF THE TRADE: TIME CLOCKS

- **A quick way to tell if power is getting to the clock (assuming the clock motor is in working condition) is to look through the opening provided (Fig. 7-2C) marked Visual Inspection, or words to that effect. Look in here to see the motor gears spinning. Some clocks have the motor mounted on the front of the clock and the center hub of the drive gear is visible. Close inspection will reveal if the hub is spinning or not.**
- **When a second clock is used to run an automatic pool cleaner booster pump, make sure it is set to come on at least one hour after the circulation pump comes on and is set to go off at least one hour before the circulation pump goes off. These boosters rely on the circulation pump for water supply, and without it will burn out their components. When checking or resetting the system clock, be sure to follow this guideline with the booster pump clock.**

and load, but they both join to the same wire that returns to the circuit breaker. As noted, the 240-volt version includes two line terminals for the two hot lines coming in, and two load terminals for the supply to the appliance.

The wires feeding electricity to the clock are wrapped on the line and neutral terminals (with a 240-volt clock, the one lead would be attached to each line terminal). This way there is a constant supply of electricity to the clock.

Waterproof boxes in metal or plastic are available to house time clocks. The unit shown in Figs. 7-1 and 7-2 simply snaps into clips built into the box. Other versions have built-in brackets that are designed to align with the screw holes on the clock plate, and these are held in place by two machine screws. The boxes are built with knock-out holes to accommodate wiring conduit of various sizes and have predrilled holes in the back for mounting the box to a wall.

Twist Timers

Twist timers (Fig. 7-2D) are used mostly with spa equipment such as booster motors or blowers when you want a limited amount of appliance operation as the user demands. A twist timer is built to fit in a typical light switch box and contains no user serviceable parts. This unit has a faceplate showing 15, 30, 45, or 60 minutes. The knob attached to the shaft that comes through the faceplate is twisted until its pointer or arrow aligns with the desired number of minutes. The circuit is completed in any position except off and the mechanical timer is spring-loaded to unwind for the number of minutes selected. When the spring is unwound, the circuit is broken and the appliance shuts off.

Twist timers are available in 120 and 240 volts and are used in place of a simple on/off switch, usually where users might forget to turn off such a switch. This also functions for safety. You might want to use a twist timer on a spa so a user cannot turn on the spa and remain in the hot water too long. When the system shuts down, the user is forced to leave the spa and reset it, perhaps making them realize that they've had enough. Another good use for a twist timer is for lights in a public area, where the user must set the desired time and the lights will go off even if the user forgets.

In the repairs section later, I don't discuss twist timers because, as noted, there are no user serviceable parts. If they fail, replace them. It is no more difficult than replacing a light switch and all you need is a screwdriver. When you buy the replacement, follow the instructions in the box if you don't find it self-explanatory. As with any electrical repair, be sure the electricity is disconnected at the breaker panel.

Electronic Timers

Do you own a VCR? Are you one of those people who never sets the time of day on it, but lets it flash 12:00 all the time? Well, when you learn about electronic timers for water system controls, you will also know how to set your VCR.

Electronic timers are used in many spa control packages that also include electronic thermostats and other sophisticated controls. Some pool and spa heaters are now manufactured using electronic timers or other controls. There are many makes and models, but a general troubleshooting guide is presented in Chap. 11.

Setting these units, however, is usually easy and sometimes self-directed (the prompts come up on the screen to tell you what to do). There are so many types and manufacturers that I could not outline typical instructions. If you are servicing a unit with no obvious instructions, call the maker and ask for a duplicate set. It pays to do this because customer complaints of malfunction are often related to electronic timers that are simply not set correctly. For instance, sometimes the timer functions have been inadvertently set to override manual functions, so the spa shuts down in the middle of use. I have also found that after a power failure some systems require resetting, and customer complaints of malfunctioning spa equipment can be traced to this simple procedure.

The advantages of electronic controls are precision and low voltage. Some units are designed with such low voltage that you are permitted to install them next to a body of water (like a spa) and not worry about electrocuting your customer. Most have digital readouts so you can precisely set the time and temperature desired, as well as programming a host of other features.

Lastly, most electronic devices have small backup batteries so that when main power is interrupted, they don't lose the time of day or previous programming.

TOOLS OF THE TRADE: TIME CLOCKS AND REMOTE CONTROLS

- **Screwdriver set**
- **Loose-joint pliers**
- **Needle-nose pliers**
- **Spray lubricant**
- **Multimeter electrical circuit tester**

Again, system malfunctions can sometimes be traced to batteries that need to be replaced.

Repairs

There's not much that goes wrong with electromechanical time clocks and it's easy to service most of them.

REPLACEMENT

RATING: EASY

When a time clock rusts or the gears wear out and the unit needs to be replaced, it takes longer to buy the new unit than to perform the replacement. When buying the replacement, be sure to get the same voltage clock as you have in the existing installation. Also, buy the same make of clock, because a different manufacturer's clock probably won't fit in the existing box.

1. **Power** Turn off the power supply at the breaker.
2. **Disconnect** Remove the line and load wires from the old clock and any ground wire. Remember (or mark) which wire is which.
3. **Replace** Unscrew the holding screws or unclip the old clock to remove it from the box. Snap or screw the new one in place.
4. **Connect** Reconnect the wires to the appropriate line or load terminals. Replacement clocks might not be arranged the same as the old one, so be sure you are getting the line wires (two hots or one hot and one neutral) attached to the line terminals and the two appliance wires attached to the load terminals. If it is not clearly marked, the simple way to tell which terminal is which is to notice where the clock motor is connected—the two clock wires are always attached to the line terminals.
5. **Test** Turn the power back on and test the clock operation by turning the manual on/off lever to On.

Most manufacturers sell just the clock motor for replacement. I have found that if the motor is old enough to burn out, the clock mechanics are probably wearing out too, so it pays to replace the entire mechanism. Besides, considering the $20 or so price difference between the motor and the entire clock unit, it's not worth messing with motor replacement only. In fact, it takes longer to replace the motor only than it does the entire clock.

By the way, if your installation is a new one, follow the electrical guidelines discussed in Chap. 6 and remember to supply power directly from a breaker. If the power supply goes to an on/off wall switch or other device first, the power might be interrupted, meaning the clock cannot keep accurate time. Keep in mind that the clock motor itself must always have a source of power.

CLEANING

RATING: EASY

Sometimes insects will nest in your time clock. I have found countless times when a clock fails to turn a system on that it is clogged with nesting ants. The cure is simple: turn off the power and remove and clean the contacts. When finished, apply a liberal amount of insecticide to the interior of the housing before reinstalling the mechanism. Don't spray the mechanism itself. You might create unintended electrical contact through the liquid.

Other than that, corrosion will occasionally bind up a clock. If you know it is getting power but cannot see the gears turning, take the tip of your screwdriver and gently force the gear (the one visible through the inspection hole) in a clockwise direction. Often that will get it started and it will continue to run fine after that. If it happens once, it will probably happen again unless you give the gears a good general lubrication.

To lubricate a time clock, turn off the power source and remove the clock from the box (wires still attached) to expose the rear of the clock. Apply some spray lubricant liberally around the gears. Do the same on the front to lube the gears behind the dial face. As noted, be careful not to wet the electrical contacts. Put the clock back in the box and turn the power back on. Turn the clock on and off a few times to work in the oil. If the clock fails again, replace it.

MECHANICAL FAILURES

RATING: EASY

When you pull out the dial face of the clock to set the time, take care that when you release it you get a true reengagement of the dial with the motor drive gear. Try setting different times on the clock and you will note that sometimes the two gears don't mesh, but rather the dial gear sits on top of the motor gear. Obviously, in this case the clock won't work.

The answer is to wiggle the dial face as you release it. As you release it, twist the dial back and forth very slightly in your hand to make sure the gears mesh. With a few practice settings you'll feel the difference between a dial that has gone back into place completely and one that is slightly hung up. Often, clocks that don't work are a result of this setting problem, so make this one of the first things you check when you suspect time clock failure.

The second mechanical problem of time clocks is with the trippers. If the screw is not twisted tightly on the face of the clock, they come loose and rotate around the dial, pushed by the control lever instead of doing the pushing themselves. Check the trippers regularly because they can come loose over time or from system vibration.

Also, if you are trying to set a time that happens to be close to a tripper setting, the dial will not engage with the motor gears. You need to move the tripper a little to make room for the dial gear. Finally, trippers can wear out, so if the clock is keeping good time but not turning the system on or off, try a new pair of trippers.

SETTINGS

RATING: EASY

When setting on/off trippers, you can place them on the dial face side by side to what appears to be about 30 minutes between them. I have found that when they are too close, they won't operate the lever. Generally, trippers must be at least an hour apart to operate.

Make frequent checks of your time clocks. Power outages, someone working at the house who shuts off all the power, the twice yearly daylight savings time changes, and any number of other household situations can interrupt power to the time clock. Each time this happens, the clock stops and needs to be reset when the power returns in order to reflect the correct time.

You might also find that you set the clock to run the filter eight hours a day and the customer resets it to run two hours a day. Very often a cloudy or green pool is the result of someone fooling with the time clock.

Remote Controls

Remote control devices allow you to operate pumps, heaters, lights, blowers, and other devices without actually going to the pool or spa.

Such devices are also available with switches at or in the water, so you can control appliances without getting out.

Remote control devices fall into two categories: those operated by pneumatic (air) switches and those operated by electronic wireless switches. Low-voltage controls are usually found mounted right at the water's edge rather than remotely located (see Chap. 11 for a discussion of these control systems which are typically found in spas).

Air Switches

You have probably driven into a gas station over the black rubber hose and heard the bell in the station. Pool and spa air switches (Fig. 7-3A) operate on the same idea—by compressing air in a hose, you force a switch to go on, off, or to a new operating position.

Fortunately a man named Len Gordon pioneered these devices with push buttons for pool and spa use so it isn't necessary to use a car each time you want to compress the air in the hose.

Figure 7-3B shows typical air switch buttons. By depressing the plunger in the middle, air is forced into a flexible plastic tube (usually ¼-inch or 6-millimeter diameter or less). The force is transmitted along the air tube until it reaches a simple, bellows-actuated electromechanical switch at the other end, turning it on, off, or to another position. You can mount this type of button near or even in the spa and not worry about electricity near the water. Air switches operate up to 200 feet (60 meters) from the button.

The most basic of these are simple on/off switches. Some activate a rotating, ratcheted wheel that in turn activates up to four switches. You push the button once to turn on the circulation pump and heater, a second push to add a blower, a third push to add a booster, and a fourth push to turn it all off. Most of these four-function units allow combinations that vary the pattern, depending on your use preference or the additional equipment you have.

The electrical part of the air switch is usually located in a waterproof box, housing the wiring connections, the actual switches (usually microswitches), and the terminals for attaching line and load just like a time clock. In some units there is a small electromechanical time clock that can operate one or more appliances at preset times in addition to the activation by the air switch button. Some units are very compact for use with a single appliance (Fig. 7-3C).

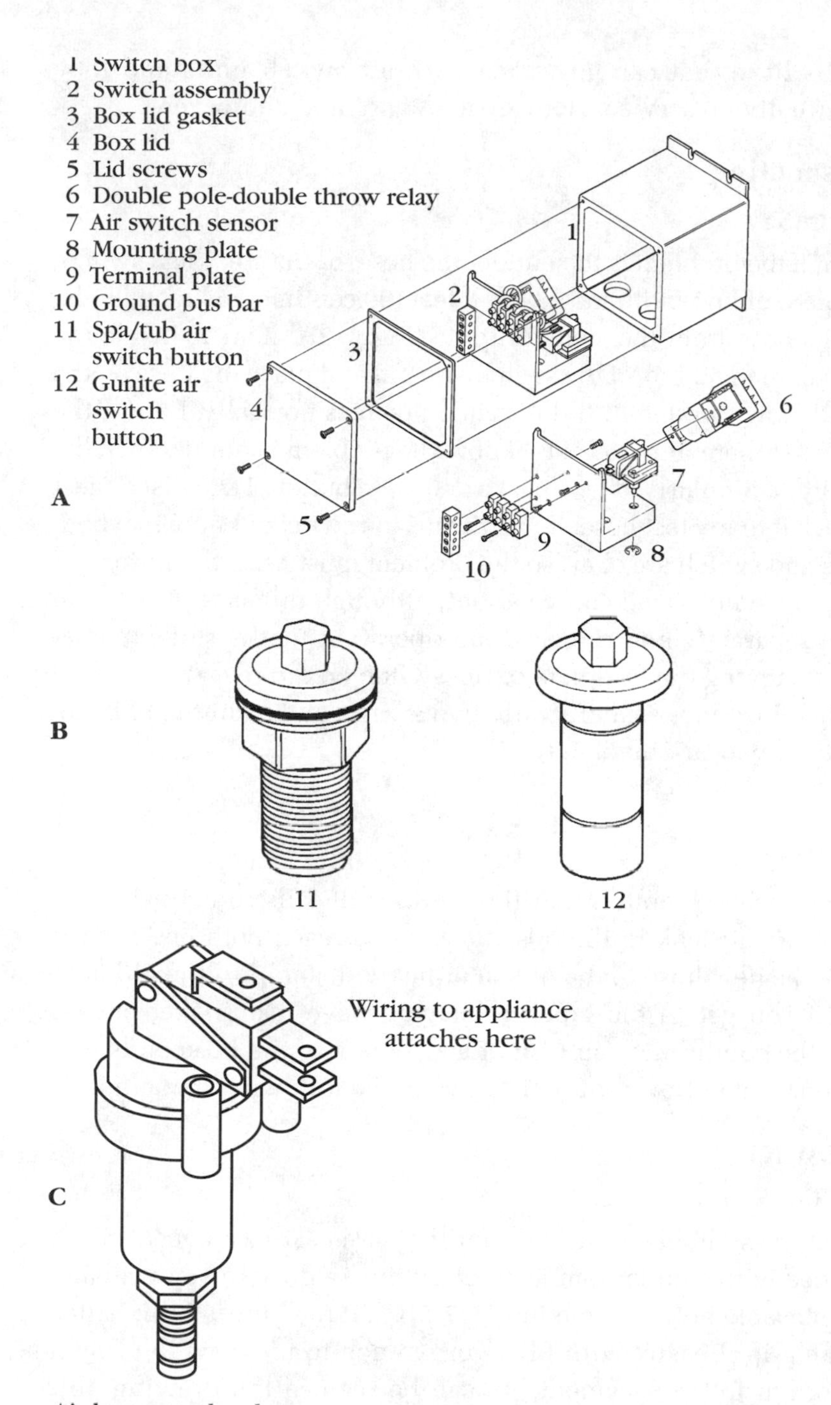

FIGURE 7-3 (A, B) Air switch components; (C) compact air switch unit. *(A) Len Gordon Company.*

There is little that can go wrong with air switch units and they require virtually no service. Here are a few pointers, however.

Troubleshooting

RATING: EASY

To find out if the problem is the button, the hose, or the electrical switch, I carry a piece of hose with me [about 3 feet (90 centimeters) long]. I disconnect the hose from the switch nipple (Fig. 7-3A, item 7, inside the electrical component box of the switch system) and place my hose on the nipple and blow through it. If the switch operates normally, I know the problem is not there. If it doesn't, I know the problem is not likely in the hose or button. Similarly, if I suspect a defective button, I can disconnect it and attach it to my test hose. If it activates the switch, I know that both the button and switch are okay, so the problem must be in the hose.

The microswitch itself can wear out, although this is rare, and can be bought separately and replaced (an operation that is self-explanatory after removing the faceplate of the switch box to reveal the actual microswitch location—usually only two screws and a couple of bayonet wire terminals are involved).

AIR LEAKS

RATING: EASY

When you push the button and the equipment fails to activate, you might have an air leak in the hose. If you can reach both ends of the hose, tape the new hose to the old, at either end, and pull the old hose out. When you get to the taped joint, you have just snaked a new hose into the conduit. If you cannot access one of the hose ends, follow the replacement procedure detailed in the following paragraph.

REPLACEMENT

RATING: EASY

Sometimes air switches wear out, but they are easily replaced. Some are designed with a collar that is mounted into a deck or spa wall and have a removable button. Note in Fig. 7-3B that the button is six-sided so you can grip it easily with pliers or wrench to unscrew the center button portion for replacement. As can be seen in the drawing, this unit is fitted with a large nut on the underside so you can mount it on a deck and tighten the nut to hold the collar in place.

When you remove the center button portion, if there is no slack in the air hose, the hose might come off the end of the button and might be hard to reach for attachment to the new button. Try pushing the hose from the other end (the end attached to the electrical switch box) toward the button to force it back out of the conduit. If there just isn't enough hose, pull out the old hose and run an electrician's snake through the conduit. When the end comes out, tape the new hose to it and pull the snake back through. Now you can attach the hose ends to the button and the switch.

NEW INSTALLATION

RATING: ADVANCED

Original installation of an air switch system is not difficult, but you might want to hire an electrician. If you try it yourself, follow the directions that come with whatever system you choose. To save time and money, you might want to buy the system, mount the electrical box, run the hose, and mount the button, then call in the electrician to do the electrical connections.

You will find that not all air hoses are run in conduit. Because they have no water or electricity in them, there are no standards or rules about running a length of hose. When making a new installation, I strongly recommend running the hose through a ½-inch (13-millimeter) PVC electrical conduit. This protects the hose from the elements and keeps it supple. It also keeps out rodents or dogs that might want to chew the hose.

Use PVC electrical conduit and fittings rather than plumbing PVC. If you have to use a snake to put in a new line, the tape will get hung up on the sharp angles of the plumbing elbows and connections, whereas electrical connections are swept gradually into the elbow or angle to make that less likely.

Wireless Remote Control

In the past few years, wireless remotes (Fig. 7-4) have gained popularity. These are composed of a battery-powered wireless sending unit with anything from 4 to 24 buttons in a waterproof case (low voltage for safe water-side use). The buttons send a signal to a receiver that might be as much as 1500 feet (450 meters) away (usually mounted in the pool or spa equipment area). The receiver interprets which button has been pressed and activates a switch that turns on or off that piece

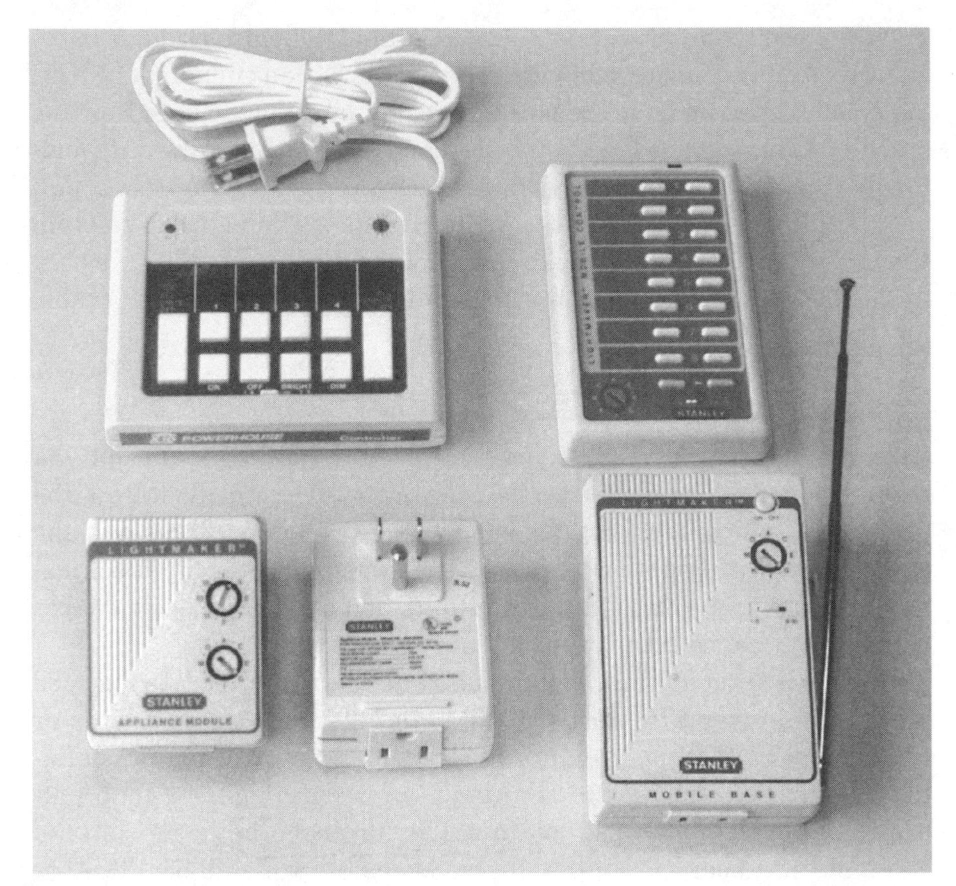

FIGURE 7-4 Wireless remote controls. Top row (left to right): 110-volt control transmitter, battery-powered control transmitter. Bottom row: 110-volt receiver module (front and rear view), remote receiver module.

of equipment. In essence, it works just like the air switch, except the signal is sent by radio signal instead of compressed air.

Some units do not have batteries, but plug into any household outlet. When you press the button, the signal is sent along the household wiring to the receiver, wherever it is located, which is powered by the same household current. Obviously, because these sending units are powered by 120-volt household current, they cannot be located near the body of water. The value of these is that they are not subject to battery failure or weak radio signals that sometimes fail to penetrate thick walls or long distances.

The 120-volt remote control is usually installed where the customer plugs the sending unit into an electrical outlet in the house so he is able to turn on and heat up a spa that might be located out in the yard. The disadvantage, obviously, is that to turn appliances on or off while using them you must get out of the spa.

There are too many makes and models of these wireless remotes to detail the wiring or technical data for installation or repair. As with a new air switch system, you might want to include your friendly neighborhood electrician in the price of installing a new system. Many are sold in modules that allow you to configure any system arrangement you wish, adding modules to existing wiring and equipment.

Hardwired Remote Control

The same remote control components can also be hardwired from sender to receiver to appliance. A hardwired remote can also be as simple as a standard on/off wall switch in the home that turns the spa pump and heater on or off. When more control features are needed, or when more pieces of equipment are being controlled, most pools and spas today will employ sophisticated low-voltage automated units which control relay switches to power each function.

Each make and model of hardwired remote controls will differ in setup, nomenclature, features, appearance, and repair. But all systems have basic elements in common, which means that if you can install, operate, troubleshoot, and repair one, you will probably be successful with any other system. The common features of automated control systems include the following:

Hardwired remote controls typically command

- Filter pump
- Jet pump
- Air blower
- Lights
- Heater (and temperature)
- Other auxiliary equipment

Standard 110-volt or 220-volt power feeds relays that actually energize each piece of equipment.

Relays are controlled by low-voltage (less than 24-volt) switches that are remotely activated or hardwired.

Low-voltage switches are operated manually or by sophisticated computer controls.

Display panels allow for programming of function, timers, temperatures, and troubleshooting.

Today's automated controls are more complex than the ones of just a few years ago, but the good news is that they offer many more features and are organized like familiar computer programs. Widespread computer literacy has also led the spa industry to create owner's manuals that are more user-friendly, including easy-to-follow "quick start" menus and more "plug and play" control components that can be easily added or simply unplugged from the system and replaced when they fail.

Figure 7-5 shows the power center (with the safety panel removed to show the relays and wiring) of a popular automated control system, the Aqualink, made by Jandy. The control panel, also called the printed-circuit board (item 1); relays (item 2); and 110/220-volt circuit breakers (item 3) are housed in a weatherproof metal box (item 4) with the wiring diagram (item 5) pasted inside the door. The power center is programmed to operate the spa equipment automatically or manually and is therefore the heart of any automated system.

The power center's control panel is the true brain of the system, constantly scanning the system to determine if an action is required. For example, it senses a change in water or air temperature and turns the heater on or off. It senses that a manual button has been depressed at the spa and activates the specified piece of equipment.

Figure 7-6A shows the same power center with the control panel removed. Now you can see how the electricity moves from the breakers through the relays to the spa appliances. A relay is simply a switch which allows household current (110- or 220-volt) to power an appliance. The relay switch closes, completing the circuit, when it is energized by a low-voltage magnetic coil. In this way, safer low-voltage controls can be used to turn high-voltage appliances on or off.

Figure 7-6B shows the functions of the control panel and the wiring diagram that is pasted inside the power center door. While the control panel can be programmed to automate all functions, the

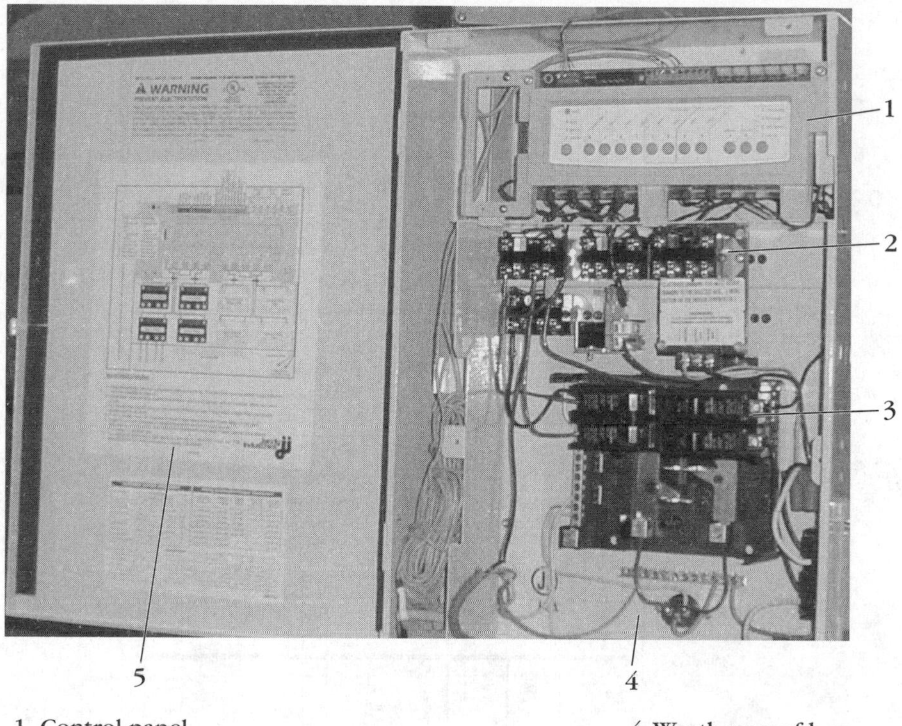

1 Control panel
2 Relays
3 110/220-volt circuit breakers
4 Weatherproof box
5 Wiring diagram

FIGURE 7-5 Automated control power center.

power center is usually located in an equipment area some distance from the pool or spa.

Figure 7-7 shows a hardwired spa-side remote control, an in-home remote control, and a wireless remote control. All these are designed to work with the power center shown in Fig. 7-5. The hardwired units are designed as modules, typically four pin connectors, that plug into the power center and require no special connection, soldering, or other wiring. The spa-side unit is water-resistant, although not designed to be mounted underwater. There are also simple push-button models that can be mounted spa-side. Remotes are powered by 10 volts of direct current (DC).

Like many automated systems, the one shown in Fig. 7-5 is designed with several important features:

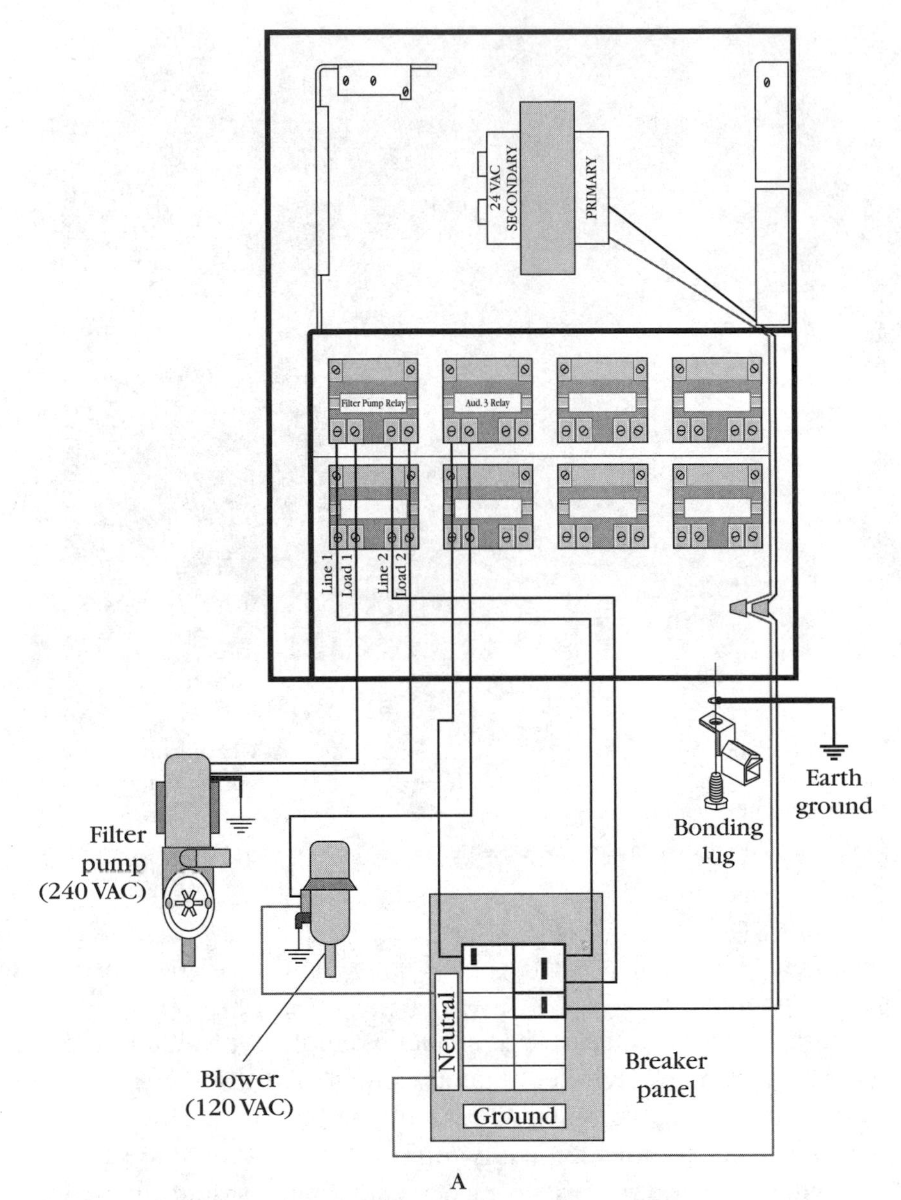

FIGURE 7-6 (A) Wiring of power center. (B) Close-up of power center control panel and wiring diagram. *Images courtesy of Jandy.*

- The programming is done from the remote control panel in the home. A service module is sold separately, which is the same control unit but one that can be plugged into the power center directly when you are performing service or repairs. When plugged into

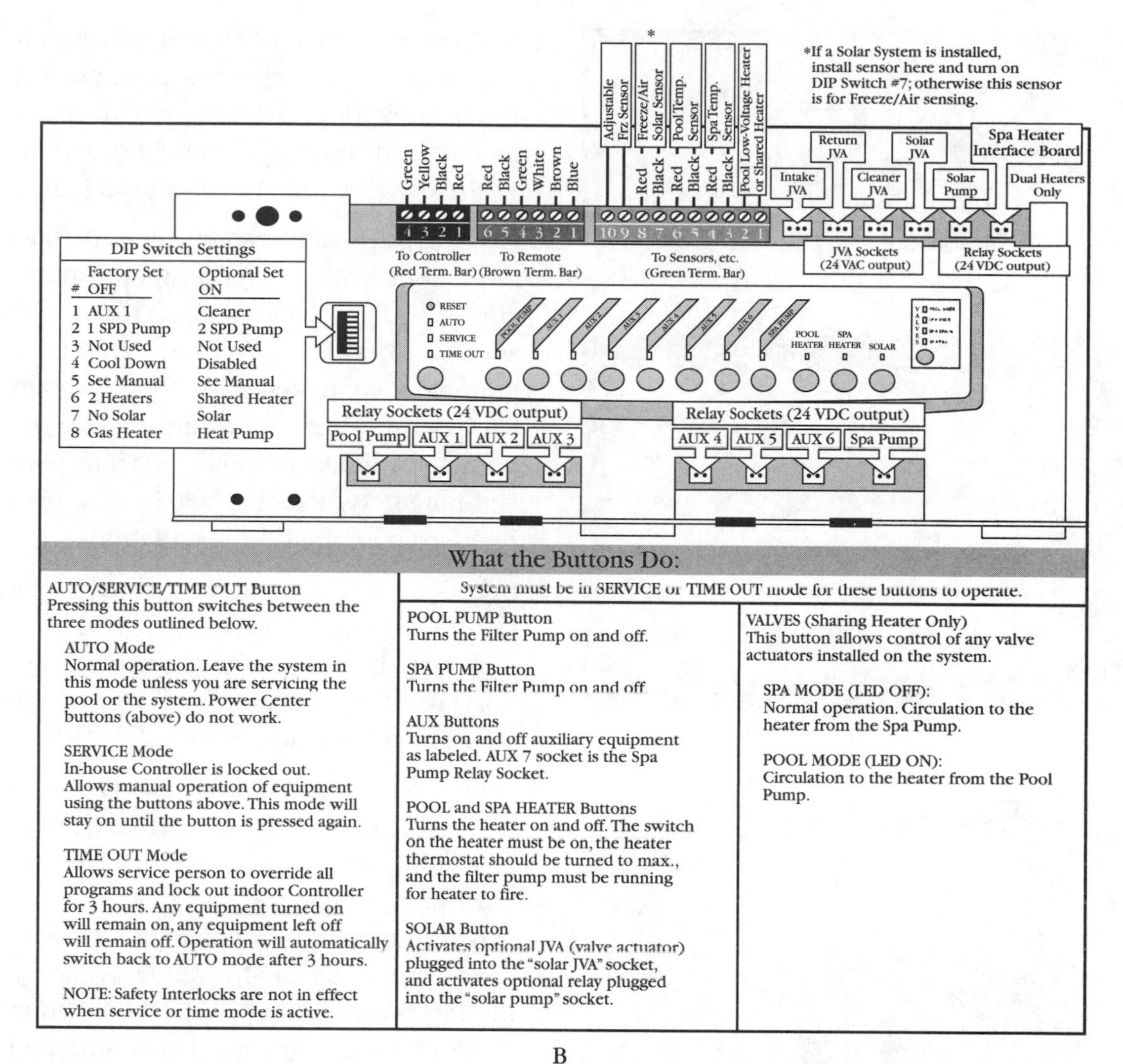

FIGURE 7-6 *(Continued)*

the power center, the service module can lock out control from other remote locations to ensure that only one set of commands is being given to the programmable circuit board.

- The control panel in the power center contains the memory of all programming and is powered by the incoming electricity from the breakers. This unit also employs a backup 9-volt battery to prevent memory loss in case of power outages.

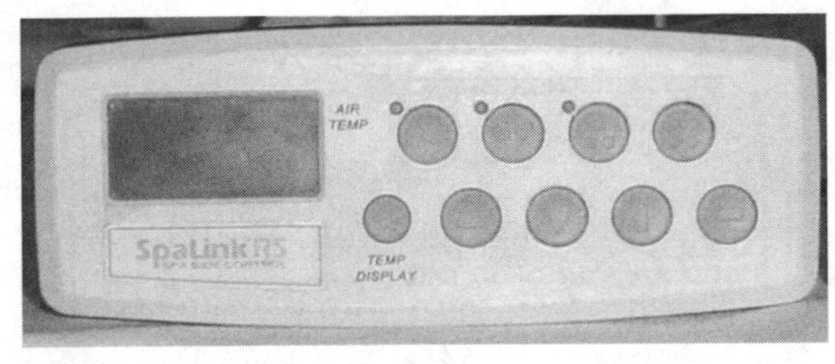

A

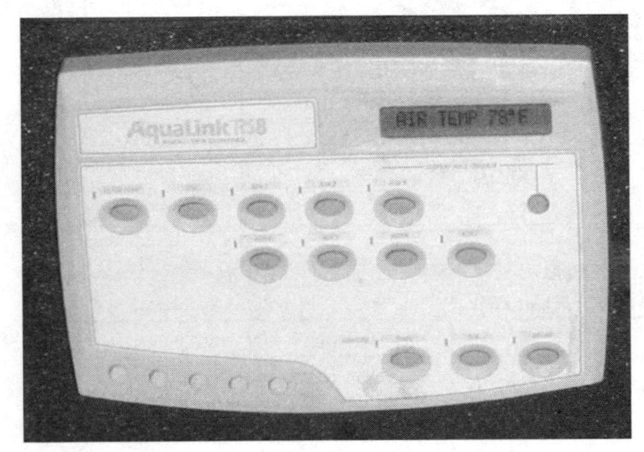

B

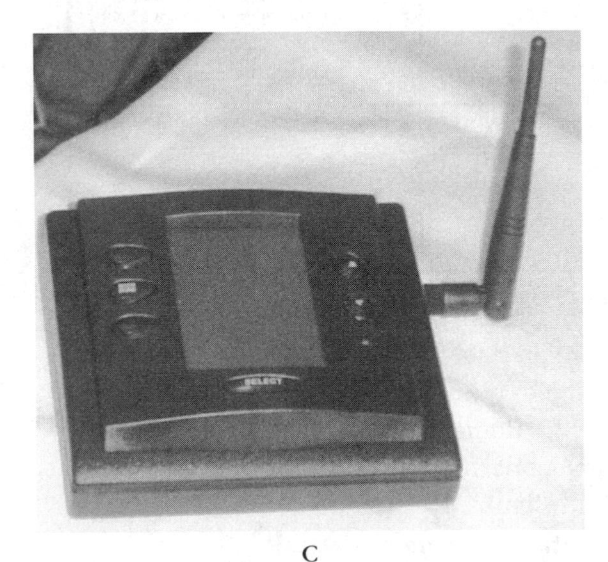

C

FIGURE 7-7 **(A) Spa-side remote control; (B) in-home remote control; and (C) wireless remote control.**

- The power center is also served by air and water temperature sensors, called *thermistors* (Fig. 7-8), which measure temperature and send the information via hardwiring back to the control panel.
- Control panel and remote control buttons, along with the relays they control, are typically labeled *Filter, Heat,* and several *Auxiliary* buttons. Auxiliary buttons are connected to other optional equipment, such as an air blower, jet pump, spa lights, outdoor lights, or pool equipment (when the spa is part of a pool/spa combination installation).
- Spa-side remote controls include a small heating element to drive out moisture, but the units themselves are not meant to be mounted below the waterline of the spa.
- Automated control systems are also designed to operate motorized valves when a spa shares equipment with a pool. For example, the system can switch the valves from pool filtration and heating to spa filtration and heating and can change the thermostat setting of the heater from the desired pool temperature to the desired spa temperature automatically.

Remote controls increasingly use computer-style programming. The remote control shown in Fig. 7-7 has a button for each function, while the unit in Fig. 7-7B and the wireless unit in Fig. 7-7C have a small screen and programming keys. There are also in-home remote control panels, called *one-touch* units, that have computer-style menus and programming. These are easy to follow and program, thanks to on-screen prompts and self-explanatory menu options.

If you are installing a new spa or adding an automated system to your existing spa, you will follow the installation instructions that come with each make and model. Typically, installation requires basic carpentry skills to mount the remote control units in the home or at spa side. You may want to have a professional electrician assist with the wiring and mounting of the power center, but other than that, installation of these units is quite simple.

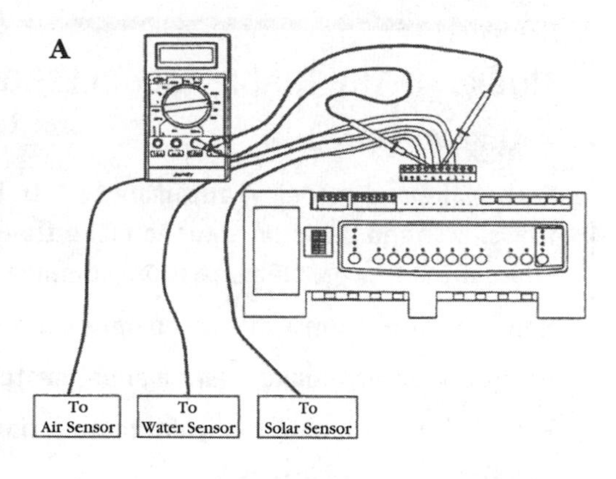

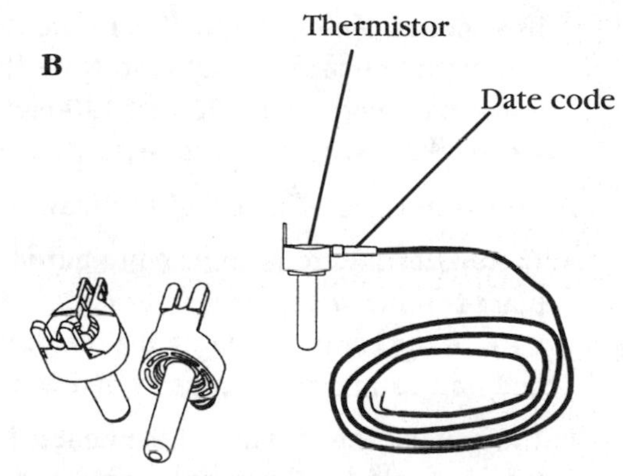

FIGURE 7-8 **(A) Testing remote sensors; (B) thermistors.** *Image courtesy of Jandy.*

Flow Meters

As the name implies, flow meters record the flow rate in a particular pipe or piece of equipment. Most look like clear Plexiglas thermometers with a graded scale (expressed in gallons or liters per minute) over a liquid-filled tube (Fig. 7-9). Inside, a small lead weight rises and falls depending on the flow of water in the line. One end of the flow meter is male threaded for easy installation into a female threaded opening in a pipe or, for example, on the drain plug opening of a heater.

Flow meters are generally used in commercial installations where health department or building inspectors test the system by the rate and volume of water flow. Some codes require that one flow meter be installed somewhere in the system. Only one is required because the flow rate is the same anywhere in the system. Remember, pressure might change where restrictions occur, but the rate of flow (in gallons or liters per minute) is the same anywhere in the system.

When installing a flow meter, locate it away from the inflow end of a 90-degree elbow. Most codes say installation must be at least ten times the diameter of the pipe away from the inflowing side of a 90, and two times the diameter of the pipe on the outflow side. To illus-

TRICKS OF THE TRADE: TROUBLESHOOTING AUTOMATED CONTROLS

Rating: Advanced

Each make and model of automated control comes with an owner's manual that provides flowcharts and decision trees to make troubleshooting easy. Start by consulting these charts. There are a few problems that are common to most makes and models of automated controls.

Remote control unit fails to operate the selected appliance.

- Check circuit breakers and manual switches at power center to ensure all are on.
- Check programming to ensure that no timer or other overrides are controlling the supply of power to the appliance.
- Attempt to operate appliance from the power center control panel. If appliance operates from power center, the fault is in the remote control unit or wiring between the power center and remote control unit. Note that most hardwired remote control units do not operate at more than 300 feet (90 meters) from the power center because of voltage loss along longer wire runs (especially on cold days).

Heater is not operating or fails to maintain desired temperature.

- Thermistor failure is common to automated systems. Thermistors are sensors that are placed in the water stream, usually in the plumbing after the heater, and send signals back to the control panel in the power center. Your owner's manual will show you where and how to test these sensors, using a multimeter electrical tester (Fig. 7-8).
- The manual thermostat on the heater itself must be set at the highest temperature setting possible for the remote control to operate properly. Otherwise, this mechanical switch will shut the heater off when its setting is reached, rather than waiting for the automated control temperature setting to be achieved.

There are unusual or no displays on remote control units.

- Most automated controls have a Reset button in the power center, and many have in-line fuses in the remote controls themselves. Try pressing the Reset button, and if that does not reboot the system, look for a blown in-line fuse.
- Automated controls also typically have a 9-volt backup battery to maintain programming during power failures. Malfunctioning spa equipment can sometimes be traced to backup batteries that simply need to be replaced.

In general, a thorough reading of the owner's manual will solve any problem you are likely to have with your automated controls. If the manual is lost, most manufacturers now provide them on their websites for easy reference and downloading. Some manufacturers also offer online troubleshooting assistance.

trate this point, assume you have a 2-inch-diameter (50-millimeter) pipe. Your flow meter would have to be 20 inches (50 centimeters) before the next 90-degree elbow in the path of the water (10 × 2 inches = 20 inches) and 4 inches (10 centimeters) past the last 90-degree elbow (2 × 2 inches = 4 inches).

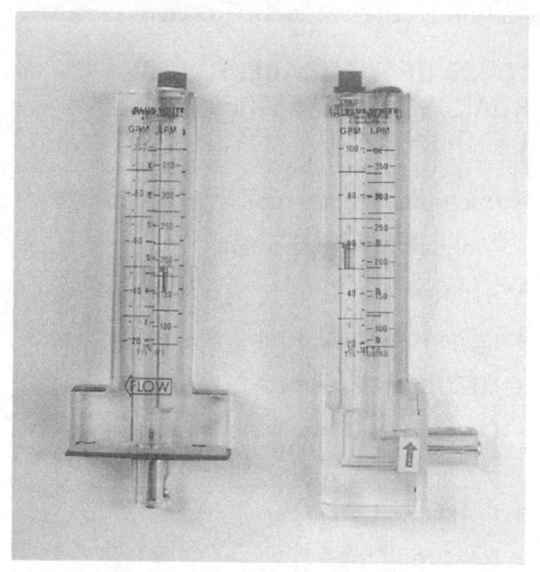

FIGURE 7-9 Flow meters, top and side mount.

Some flow meters are threaded. To install this type, cut the pipe and plumb in a T fitting to reconnect the pipe. Two ends of the T are used in reconnecting the pipe, and the third opening of the T is plumbed with a female threaded reducer to make the opening the same size as the male threaded end of the meter, which is then screwed into place. Apply Teflon tape to the male end to prevent leaks. As with pressure gauges, I don't use pipe dope for these installations because any excess dope can clog the opening of the meter.

Another method is to drill a hole in the pipe that is slightly smaller than the diameter of the male threaded end of the flow meter. Because the PVC is softer than the metal end of the meter, when you apply pressure and screw it into the pipe it will self-tap. These installations usually leak, so I don't recommend this method unless the plumbing or equipment area is just too tight to cut the pipe and add the T fitting. You can use this method with copper plumbing because although the copper is soft, it is less likely to strip when you screw in the flow meter. Some meters fit loosely in the hole (sealed with a gasket) and are held in place with pipe clamps.

Diving Boards, Slides, Ladders, and Rails

A good source of added income for a pool technician is installation of new diving boards, slides, or ladders. Here are a few guidelines.

Diving Boards

Diving boards are not as popular as they once were because pools today are built smaller and shallower as people try to save water and

TOOLS OF THE TRADE: DIVING BOARDS, RAILS, SLIDES, AND LADDERS

- Open-end wrench set
- Channel lock–type pliers
- Hammer
- Large flat-blade screwdriver
- Electric drill and bits
- Concrete drill (rental for new installations)
- Marker
- Measuring tape
- Two 5-gallon buckets
- Quick-set concrete, mixing bucket, spatula, cleaning rags/sponge

energy and because more people use pools for swimming exercise laps rather than for jumping into deep water. Still, there are many boards out there, and as they age they need repair or replacement. Diving boards are wood, for flexibility, covered with fiberglass, for waterproofing, with a nonskid tread applied on the top.

Diving boards come in simple fulcrum models and spring-assisted jump models for extra bouncing action. Figure 7-10 shows several designs. Residential boards are from 6 to 12 feet (1.8 to 3.6 meters) in length and 18 inches (45 centimeters) wide. Commercial boards are usually 20 inches (50 centimeters) wide and up to 16 feet (5 meters) in length.

Some building and safety codes specify pool dimensions for a diving board to ensure that there is adequate water area and depth for safe diving. Generally, to accommodate the smallest board, a pool must be at least 7½ feet (2.3 meters) deep and at least 15 feet (4.6 meters) wide by 28 feet (8.5 meters) long. These suggestions go up for longer boards or those with more springing action. Check local codes before adding a diving board to a pool. Remember, the customer depends on you for this type of advice and you might be held liable for installing a product they can't use safely.

REPAIR AND REPLACEMENT

RATING: ADVANCED

Most diving boards you encounter will need repair or replacement, not new installation. I have repaired or replaced hundreds of boards, but in the past 10 years I have not been asked to install one new board. Also, although replacement boards are not cheap ($200 to $500 wholesale, including hardware), I suggest replacement rather than repainting or repairing.

Because of the wood interior, if the fiberglass has cracked or started to delaminate, the wood underneath might be wet or have dry rot. If you cosmetically repair a board by patching cracks with fiberglass resin and adding new nonskid, then the board snaps in half two weeks after you worked on it, you are asking for a lawsuit.

Figure 7-10 shows the hardware used to attach a board. Here's how you proceed.

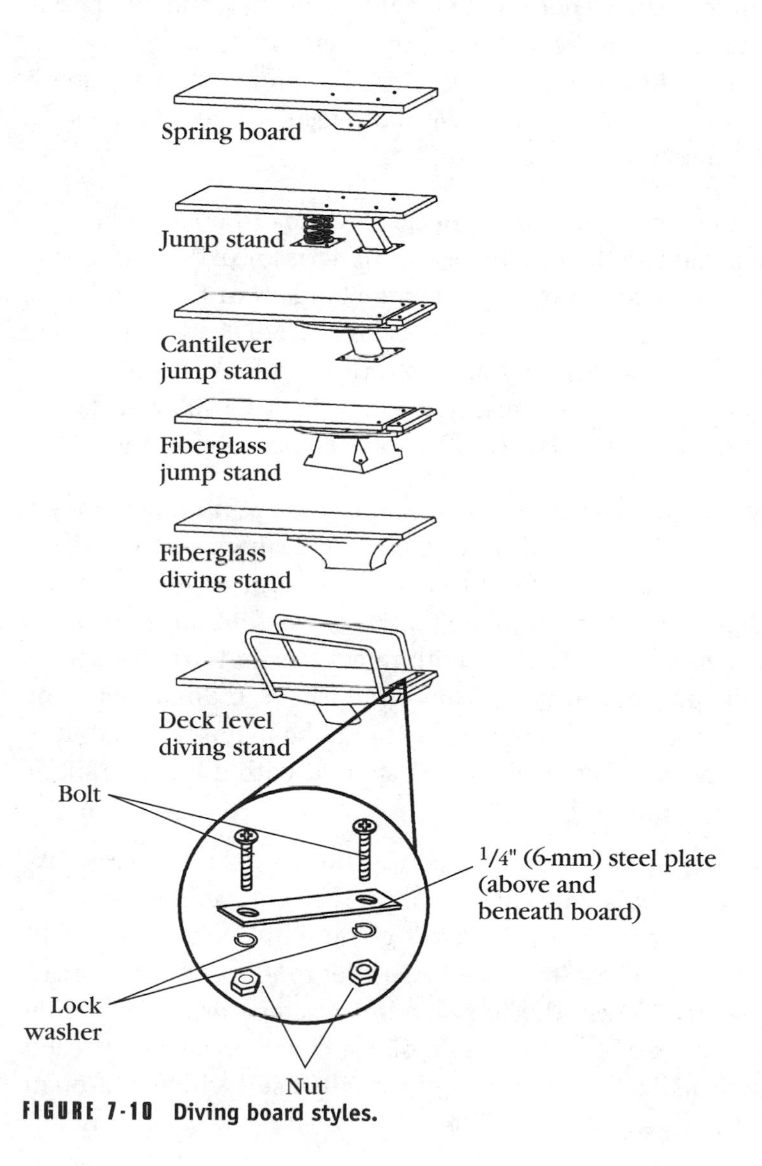

FIGURE 7-10 Diving board styles.

1. **Unbolt** Looking underneath the board, remove the nuts, lockwashers, flat washers, and/or boltplate from the mounting hardware that holds the board to the stand. As you remove the last one, remember that the board might tip forward into the pool, so have someone hold it or put some weight on the end of the board, away from the pool side, so you can control the removal.
2. **Disassemble** Pull out (or push up from the bottom) the bolts that hold the board on the stand. There are two bolts at the very end and possibly two more in the middle at the fulcrum. Remove the bolt plate and gaskets. Remove the board from the stand. Be careful and have some help nearby—boards can be very heavy, especially if the wood inside the board has become waterlogged.
3. **Set Up** Lay the new board on the deck. Lay the old board on top of the new one and use the old holes as a pattern for drilling the new holes. Drill the needed holes. On some stands, the fulcrum has a rubber cushion that might need to be replaced. A new cushion comes with most replacement boards. Lift off the old cushion and push the new one in place. Most stands also have a rubber gasket or cushion on the bolted end as well. Apply the new gasket there.
4. **Install** Lay the new board in place. Again, remember that it might tip into the pool if you don't have someone help you hold it. You can also fill two 5-gallon (20-liter) buckets with water to use as counterweights. Lay the bolt plate on the new board and insert the bolts. Replacement boards come with new bolts, but even if the old ones look good, replace them. They might have hidden cracks or rust and after years of pressure are bound to be metal-fatigued. It is no good to have a strong new board installed with bolts that might snap off at any minute.
5. **Rebolt** Working underneath, replace the nuts with whatever lockwashers, flat washers, and/or bolt plate that are included for the underside. If the replacement doesn't come with lockwashers, add them so that the nuts can't vibrate loose over time, causing failure of the entire board. Never add nuts without washers or an underside bolt plate. The stress on the bottom of the board is enormous each time a jump is made and, over time, the nut by itself will tear through the board. The washers or bolt plate distribute that stress evenly.

6. **Test** When the new board is bolted on, walk out on the length of the board and jump up and down. I have always felt it was better to have something fail under me, when I'm expecting it, than under the customer who is not. Be sure your bolts are tight, but don't be crazy and torque them so far that the board cracks or weakens at that point.

For legal and safety reasons, do not replace a board with a longer one. Use the same type and length.

Having said earlier that I wouldn't upgrade a board cosmetically, the one exception might be to replace the tread. Sometimes, on a board that is only a few years old, heavy use and sunlight can destroy the nonskid material on the top. This can be bought in self-adhesive rolls (called *tread kits*) at the supply house with instructions for application. Replacement springs are sold for each make of jump-type board with instructions for replacement, and again, although fiberglass patch kits are available, I wouldn't use them.

Slides

Slides are generally made of fiberglass with metal frames and steps. A straight slide will be 8 to 13 feet (2.5 to 4 meters) long, requiring a lot of deck space. If deck space is limited, left-handed or right-handed curved slides are available (Fig. 7-11). As with diving boards, safety is

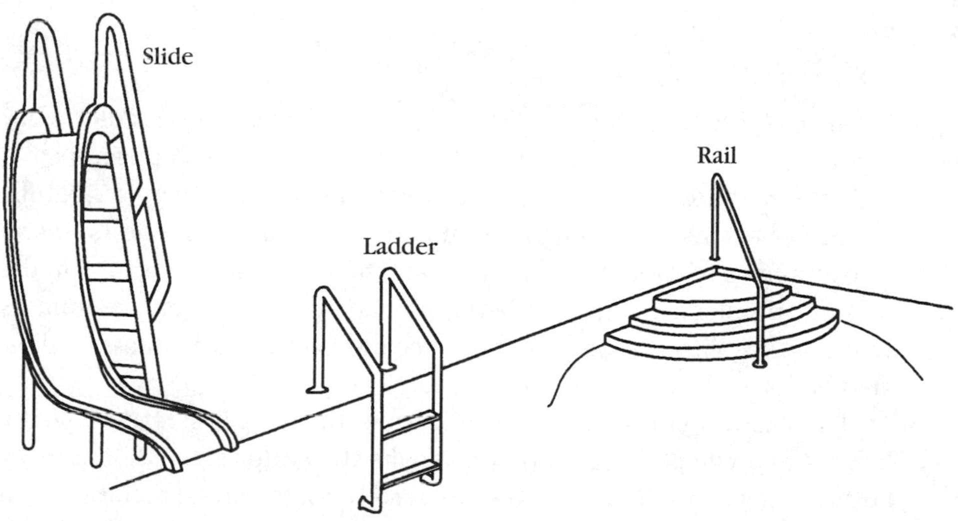

FIGURE 7-11 Slides, ladders, and rails.

a key consideration—the higher the slide, the deeper the pool needs to be.

INSTALLATION

RATING: ADVANCED

Slide installation is relatively simple and anchor kits are available at your supply house with instructions. Some installations provide brackets that you bolt to the deck with special concrete fasteners. Others require drilling 1½- or 2-inch (40- or 50-millimeter) holes for each of four legs and securing the installation with quick-set concrete. Following the instructions in the kit, anyone can install a slide.

Some slides are plumbed to provide a sheet of water along the slide area to make them more slippery. These can be plumbed into a water supply line, but prolonged use will lead to an overflowing pool. A better installation is to plumb the slide to a return line from the pool's circulation system. This might require running a long and complicated pipe from the equipment area to the slide, but the supply pipe is usually ¼-inch (6-millimeter) diameter, so concealing it in gardens or along building edges is not difficult. Always fit the line with a shutoff valve so the slide can be turned off when it is not in use. The water running over the large surface area of the slide, which is usually quite hot, will evaporate rapidly, slowly draining the pool.

MAINTENANCE

RATING: EASY

Because of constant exposure to the elements, slides often discolor. Your supply house sells a glaze/polish kit that can restore the appearance to almost new. I don't recommend painting slides, even with epoxy paints, because the constant sliding friction and exposure to sun and temperature extremes will quickly make the painted surface look worse than the original problem. Painted surfaces often oxidize and your swimmers will end up with powdery streaks over the backs and bottoms of their swimsuits.

The only other maintenance concern with slides is safety. Make it a habit when you service the pool to shake the slide to detect loosening bolts. Because water on the deck can lead to rusted bolts, make a visual inspection and replace bolts with signs of rust. Finally, check the water supply line (if the system has one) for leaks that might deplete the pool water level, leading to equipment that runs dry and expensive repairs.

Ladders and Rails

Ladders and rails are available in a wide variety of configurations to meet the design needs of any pool or spa. Some are anchored in the deck only, simply resting on the bottom or side of the pool, while others are anchored in the deck and pool.

INSTALLATION

RATING: PRO

Installation of a rail or ladder is easier than the manufacturers try to make it, but does take some care and skill. Most installation kits have you drill holes in the concrete deck and/or pool floor and fit the holes with bronze cups that are equipped with bolts and wedges. You mount the cup in the hole, secured by some quick-set concrete, then put the legs of the rail or ladder into the cups and tighten the wedges. I do not use this method because the bolts and wedges ultimately come loose, allowing the ladder or rail to fail under someone's weight. These cups are designed to allow easy replacement of the rail or ladder in the future, but I have found that if the material of the rail or ladder has discolored or rusted, so has the metal cup and bolt system. My installation is more secure and if you need to replace the item, you can cut the old unit out with a hacksaw and drill out the deck holes with a concrete drill, just like creating holes for a new installation. In the long run, safety is the key and my method is much safer.

1. **Plan** Fit the pool or spa for the desired rail or ladder by deciding where you want the item and measuring the available space. In the case of a ladder (Fig. 7-11), it needs 36 inches (90 centimeters) of deck width and 24 inches (60 centimeters) depth below the waterline. With a rail, you measure the horizontal distance from the deck to the end of the steps. With these measurements, your supply house can provide the appropriate rail or ladder.
2. **Tools** Rent a concrete drill from your local tool rental yard with a 2-inch (50-millimeter) bit. Most rails and ladders are made from 1½-inch-diameter (40-millimeter) stainless steel pipe.

3. **Drill** Position the item on the deck as needed and mark the spots where the legs touch the deck and pool. In the case of the rail, which also needs to be anchored in the pool bottom, the pool must be empty for this procedure. Drill out the holes in the deck and pool bottom to a depth of 3 to 6 inches (8 to 15 centimeters), depending on how high you want the final installation to be.
4. **Fit** Dry fit your ladder or rail. Slide the escutcheon plates into place (the decorative rings that conceal the holes in the deck or pool). Be sure to put them on facing the correct direction. After the item is set in the concrete, you can't remove the escutcheon plate. Set the item into the holes and be sure you have the desired fit and height. Be sure the item is level—that it doesn't lean to one side or the other. If you drilled your holes to the same depth, the item should also be horizontally level. Now remove the rail or ladder.
5. **Install** Mix a batch of quick-set concrete (available at any hardware store). Fill the holes with the mixture and set the ladder or rail into the holes as deeply as it will go. The concrete should fill and surround the hollow pipe. Keep it in position until the concrete sets enough to hold the rail or ladder by itself.

   Let the concrete dry overnight before refilling the pool or applying any pressure to the item. Slide the escutcheon plates into place to conceal the installation holes and test the rail or ladder. Apply pressures and forces consistent with normal pool or spa use to test your installation.

Safety Barriers

Many jurisdictions now require some type of fence, solid cover, or other safety barrier around a pool or spa to prevent drowning ("barrier codes"). If more people paid attention to this commonsense requirement, many needless deaths and injuries could be prevented every year.

Generally, barrier codes are written for "swimming pools" but are meant for both pools and spas (including hot tubs) and include at least these requirements:

- The pool must have a fence on all four sides at least 4 feet (1.2 meters) high.

- Gates must be self-closing and latching.
- Spas must have a "safety cover" (as defined by various certifying agencies).
- Where the home itself forms one or more "walls" of the fence, doors must have locks or other provisions to prevent children from accessing the pool.
- Sometimes an alarm can be substituted for certain barrier requirements (check on the type of alarm permitted in your area).

Remember, many drownings of children occur even when barriers are provided if gates are left open or covers left off. Safety barriers are only as good as their actual use.

Automatic Pool Cleaners

There are numerous designs of automatic pool cleaners available today. Some of them are no longer manufactured but you might encounter them in existing installations.

There are two categories of automatic pool cleaner in common use today and three other less effective technologies that might still be lurking around a pool you are servicing. Let's dispense with those three first.

Electric Robot

The electric robot type of automatic pool cleaner sounds futuristic, but is nothing more than a battery-powered (some are 120 volt, some transformed down to 12 volt) vacuum cleaner with a bag that catches debris as the unit patrols the pool bottom.

Once reserved for commercial pools, robotic pool cleaners are now quite effective and affordable for residential pools, too. Figure 7-12 shows a robotic pool cleaner on its way up the wall of a demonstration pool. This model operates on 24 volts supplied from a cable that plugs into a household electrical outlet on the pool deck (household current is transformed down to the operating voltage).

Debris is captured in an internal bag and a filter strains fine particles and dirt. Computer chips direct the unit around even the most complicated obstacle course in any pool, ensuring thorough cleaning.

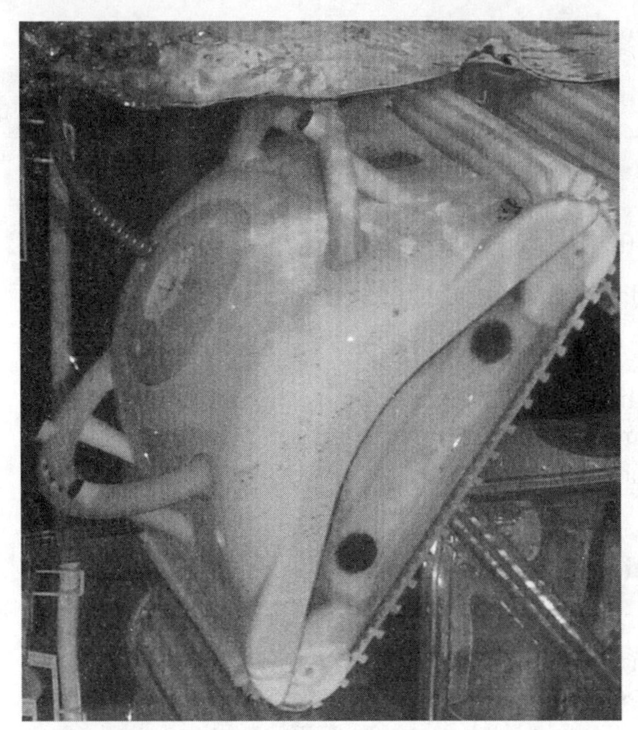

FIGURE 7-12 Robotic pool cleaner.

Although the electrical components of these units are watertight, I do not recommend leaving them in the pool when swimming. Damaged cables can result in electrocution and, although unlikely, it's better to be safe than sorry.

Booster Pump Systems

Booster pump systems take water after the filter and heater, which is already on its way back to the pool, turbocharge or pressurize it by running it through a separate pump and motor, then send this high-pressure water stream through flexible hoses into a cleaner that patrols the pool bottom.

In one style, called a *vacuum head* type, the cleaning device has its own catch bag for collecting debris, much like a vacuum cleaner. In another variation, the *sweep head* type floats on top of the water with long flexible arms that swirl along the walls and bottom, stirring up the debris. A special basket is fitted over the main drain so that the stirred-up debris is caught in either the main drain or the skimmer and any fine dirt is filtered out normally. The sweephead type is obsolete nowadays, but many popular models of the vacuum head style are available.

VACUUM HEAD TYPE

Although there are several manufacturers, Polaris Vac Sweep is the leading maker of this style and will be the example. As with other pool and spa equipment, if you understand the leading manufacturer's equipment, you will easily comprehend the operating concepts of the others.

Figure 7-13 shows a common vacuum unit with catch bag. Pressurized water from the booster pump enters the unit through the stalk and some is immediately jetted out the tail. This water pressure causes the tail to sweep back and forth behind the unit to brush loose any fine dirt on the bottom that is then filtered out by the pool circulation system. The remainder of the water powers a turbine (item 20) that has a horizontal shaft with gear teeth to engage comparable gear teeth on the inside of the single left-side wheel (item 21) and the front right-side wheel. A small right-side drive wheel (item 2) transfers power to the trailing right-side wheel as the unit moves forward.

Some jetted water is diverted to the thrust jet (item 26) which can be adjusted up or down to help keep the unit from moving nose-up. The head float (item 31) also serves this function and keeps the unit upright.

Installation (Rating: Advanced): Vac Sweeps are available as preplumbed units (where the supply pipe from the equipment to the pool area is plumbed into the original pool plumbing) or as over-deck models, which require a garden hose be run from the equipment area over the deck to the pool's edge. The booster pump and vacuum unit are identical, only the plumbing between the two are different with these two models. A complete installation guide is provided with either unit when purchased, so I will outline here only the general steps to give you an overview of what you will be doing.

1. **Install the Booster Pump** The Polaris booster pump is a ¾-hp pump and 120/240-volt motor. The electrical connections are the same as described in the pumps and motors chapter, normally powered through a time clock.

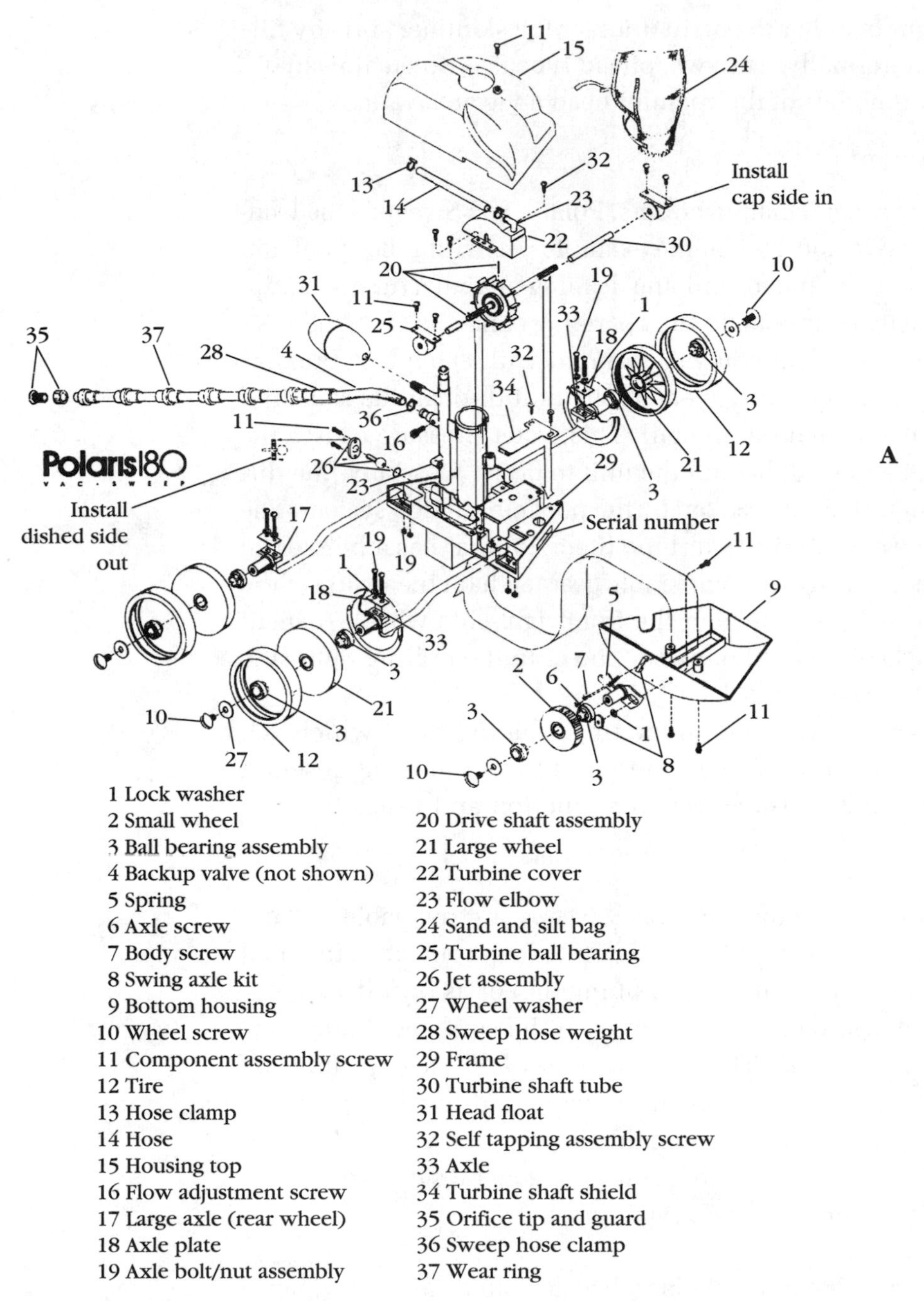

FIGURE 7-13 **(A, B) The vacuum head type of automatic pool cleaner.**

B

FIGURE 7-13 *(Continued)*

The pump's suction plumbing is taken from the return line of the pool after any other equipment (after the heater and solar panels) but before an automatic chlorinator, if any. Where the circulation system supplies both pool and spa and is diverted to one or the other by a three-way valve, you must install the Polaris plumbing after the three-way valve on the return line to the pool only. If you added it before the three-way valve, when you circulate the spa water, some water will be returned to the pool through the booster pump line, slowly draining the spa.

Follow procedures outlined in the pumps and motors chapter for setting, plumbing, and electrifying the pump and motor.

2. **Plumbing** Plumb from the booster pump to the pool. If an automatic cleaner line was included in the pool construction, as it probably was if the pool was built in the past 15 years, this step has already been completed. If you are adding over-deck plumbing, either hard-pipe or use a garden hose to run the water supply from the pump to the pool edge. Garden hoses tend to fail, so I strongly

recommend ¾-inch (19-millimeter) PVC pipe for this installation, and in either case, run the line through garden areas or along building edges as much as possible to avoid the line being run over, kinked, or kicked. Make the shortest open-space run possible. Bring the line to the middle of the pool's longest side.

3. **Feeder Hose** Add the feeder hose. A long plastic hose [30 feet (9 meters) provided with the unit, and additional 10-foot (3-meter) lengths available] is supplied to run from the booster pump supply pipe to the vacuum unit itself. Every 10 feet (3 meters) a swivel connection is provided so the hose doesn't coil up during use. Lay the hose out to its full length, preferably on a hot deck. Let it lay in the sun for an hour or so. The plastic has a memory and wants to recoil itself back into the form it was as packed in the box. This layout procedure helps the plastic hose relax.
4. **Connect** One end of the hose is fitted with a female snap connector to attach to the male end that is plumbed to the end of the supply pipe from the booster pump. Connect this fitting together. The other end fits over the top of the stalk on the vacuum unit itself. A securing nut is provided (with reverse threads) to attach the hose to the stalk. When these break or wear out, I use a 1-inch (25-millimeter) stainless steel hose clamp instead.
5. **Layout** Walk the supply hose and vacuum unit around the pool deck. If you can reach all around the pool, you have enough hose. Allow enough for the depth of the pool at the deepest end. Now cut off any excess hose length or add more sections if it's too short. After taking the vacuum head unit out of the box, you might need to attach the tail, fitting the open end over the tail jet on the stalk and securing it with the hose clamp provided. Add the catch bag provided, secured by a Velcro strip. Set the unit on the deck and turn on the pool circulation pump.
6. **Start Up** After the air has purged from the lines, you should see water squirting from the tail, the thrust jet, and out through the catch bag on top. Lower the unit into the pool and turn on the booster pump (it is educational to try this once while the unit is still on the deck so you can observe the difference between running the water with and without the booster in operation). The unit will now patrol the bottom.

7. **Adjustments** Each pool has design quirks that make adjustments to the vacuum head unit necessary.
   - Is the tail sweeping vigorously, but not so hard that it flies off the pool bottom? The tail jet flow can be adjusted by tightening or loosening the screw (item 16) that acts like a valve to open or close the flow to the tail.
   - Is the vacuum unit running nose up? Does the unit go in left or right circles? If so, loosen the holding screws (item 11) of the thrust jet and adjust the aim of the thrust jet to compensate for the misdirection. A few experiments with the thrust jet position will teach you this technique.
   - Is the unit hanging up on steps, ladders, or deep corners? You might need to shorten or lengthen the supply hose to allow the unit to turn around more easily when it reaches a corner. If the hose is too long, the unit might just circle in one area.

After each adjustment, watch the unit for several minutes. Some problems work themselves out and others might not be immediately apparent, so take your time with this procedure to get the best results. Follow additional directions in the manufacturer's instruction booklet.

Operation (Rating: Easy): A few operational guidelines will help you keep the vacuum head automatic cleaner cleaning the pool efficiently. Never operate the booster pump without the circulation pump also working. The booster is not self-priming, it relies on the system circulation pump to provide water. If it runs dry, the plastic pump will overheat and warp the seal plate and burn out the seal.

Set the booster time clock to come on at least one hour after the circulation pump and to go off at least one hour before the circulation pump does. This allows for slight time differences between the clocks. The vacuum head will probably cover as much of the pool as it's going to cover in about three hours. More than that and you are just wearing out components. In windy or very dirty locations, you might want to set it to operate for two hours in the morning and two hours in the afternoon.

Before putting the catch bag in place, slip the end of an old stocking over the opening and secure it with a rubber band. Then install the catch bag. The fine mesh of the stocking will capture fine dirt and sand, saving you from having to vacuum the pool so often. Some supply houses sell small stockings made for this purpose, which might be easier than constantly raiding a nearby panty drawer.

Empty the catch bag and/or sock as needed, but don't be obsessive about it. The bag actually works better with some leaves or dirt in it because the debris helps filter as well, catching fine particles in the leaves. Make sure the openings on the bottom and through the center of the unit are not clogged with large leaves so there is always a clear path for the debris to get into the bag.

Watch the unit perform for a few minutes each time you service the pool. If it runs listlessly or fails to run in a pattern that will clean the entire pool, follow the adjustment procedures outlined in the installation section.

Repairs (Rating: Easy): Perhaps the simplest way to explain the few repairs needed by these cleaners is to list the symptoms of the problems you might encounter.

- Water is not flowing out one or more of the jets in the vacuum unit. Because the jets inside the unit are small, grains of sand can clog them. A fine mesh strainer is provided that is installed at the connection of the plumbing to the feeder hose to catch any small particles or DE that might get past the filter system. Dirt or sand can, however, be picked up by the unit and clog any of the internal jets. If this happens, there is probably sand or dirt in other parts of the unit as well, creating excessive wear and abrasion on the drive system, tires, and other components.

  Using Fig. 7-13, disassemble the unit and clean each part thoroughly. As with every other disassembly and repair procedure in pool and spa work, carefully note how the unit comes apart so that you will know exactly how to put it back together. Use a thin wire to clear out the jets. Follow the path of the water and simply clean it all out.

- Wheels are not turning. If the jets are all clear, there are two other reasons for sluggish performance. First, if the booster pump is not getting enough water because of restrictions in the main circulation system, the vacuum unit will not perform well.

Clean the filter or address whatever other problem is creating poor circulation in the entire system and you will usually find that the automatic pool cleaner works better as well.

Second, over time the metal drive gear wears out the plastic drive gear inside the wheels. Check to make sure the gears are meshing and that there are enough teeth on the inside of each wheel. If the wheels are sloppy, they will also fail to properly engage with the drive gear.

Replace the wheel bearings (item 3), which simply pop in place like a pump seal. Last, as the tires (item 12) wear out, they do not properly engage on the drive wheel, so replace them. They simply stretch in place over the wheel like a large rubber band. Polaris provides a tire gauge to help you judge the thickness of the tire all around its circumference to determine when you should replace them.

- Vacuum unit falls over. Remove the head float by gently but firmly pulling it off of the stalk (be careful—the stalk it is mounted on can easily break). If it is full of water, it is not floating the unit upright. Replace it.
- Stalk breaks. Polaris sells a mast repair kit. Using a hacksaw, you cut off the mast just above the tail jet. The replacement glues in place with PVC glue and includes a new headfloat stem. Simple, detailed instructions are included with the kit.
- The wheel screw breaks. The screws that secure the wheels are plastic and overtightening will snap them off. If this happens, you can gently tap a flat-blade screwdriver into the broken stem to unscrew, remove, and replace the screw.
- Tail wear. The tail assembly will be the first thing to wear out because it is constantly sweeping the pool bottom and sides. Water will squirt out of parts of the hose where it shouldn't, making the tail swing wildly. To help prevent this, the tail is fitted with rubber rings that absorb the wear, so as you see these donuts wearing down, replace them before the tail goes.
- Wheels seize up. Sometimes the drive wheel (item 2) gets hung up and actually prevents the wheels from turning. Tension for the drive wheel is spring loaded (item 8 assembly), and as the spring

wears out the tension will be too much or too little. (The new model 360 uses an internal belt-driven propulsion instead of this cumbersome assembly.) The folks at Polaris might not agree, but I have actually removed this drive wheel completely from the units I service and had great operational results. The turbine powers the front wheel on the right side and the single wheel on the left side with the rear right-side wheel just trailing behind. The unit works fine and the wheels never seize up. Try it.

- Unit gets caught in ladder, corner, or steps. The irregular-shaped pools that are popular today are the automatic pool cleaner's nightmare. If all adjustments and hose lengths are correct but you still have problems, a backup valve is the answer.

  Installed about halfway down the feeder hose, this valve shuts off the water supply to the vacuum unit about every five minutes, shooting the water out of the valve to act as a jet to pull the unit backwards. It sounds complex, but it is actually amazingly simple. Read the directions that come with the backup valve for installation and servicing instructions. They work very well.

- Unit runs too fast, just skipping over the dirt. On some pool systems where the return water pressure is very strong, the vacuum head pressure is too great for normal operation. Polaris makes a pressure tester, available for your own assembly from a kit. Follow the simple instructions provided and test the pressure at poolside to determine if pressure-reducing washers are needed. If so, this simple reduction technique employs a washer with a smaller diameter than the plumbing, thus restricting the amount of water that can flow to the vacuum head.

This pressure tester is a valuable tool to use when you suspect inadequate pres-

TRICKS OF THE TRADE: TRAINING SEMINARS

Polaris (and most other manufacturers) offers terrific seminars all over the country (often with dinner included) that give you hands-on training in teardown and rebuilding, troubleshooting tips, adjustment procedures, etc. These seminars will not only give you a good foundation in automatic pool cleaning technology, but they will keep you up to date on new features and products. For these reasons, it is worth attending one each year, but also because the first time you might not comprehend the subtleties or ask the important questions. After some field experience with automatic cleaners, you will get more out of each seminar. Attend one.

sure might be the cause of sluggish operation. Pressure values and test techniques are explained in the installation booklet or test kit instructions.

Suction-Side Systems

Suction-side automatic pool cleaners work off of the suction at the pool's skimmer. It is like vacuuming the pool without a pole or pool person. In this design, a standard vacuum hose of 1½-inch (40-millimeter) diameter is connected between the skimmer suction opening at one end and a vacuum head that patrols the pool bottom at the other end (Fig. 7-14).

The problem with these systems is that the skimmer was designed to skim dirt and debris from the surface of the pool before it sinks to the bottom. By using the skimmer's suction for another purpose, everything sinks to the bottom. Special valves can be installed on the skimmer to allow some suction to continue the skimming action while the remainder operates the vacuum head, but few pumps have enough suction to do both efficiently, so you end up with two systems that work poorly. Some valves alternate the suction between the skimmer and the vacuum head, but then you are back to the original problem that when the vacuum head is running, the skimmer is not.

The second major problem with these systems is that the debris collected is sucked directly into the pump strainer, often overloading the basket and cutting off circulation, making the entire system of filtration and heating inefficient. Canisters have been designed to float in-line along the suction hose, but when these clog, you are still slowing the circulation of your pool which means it is not filtering or heating effectively.

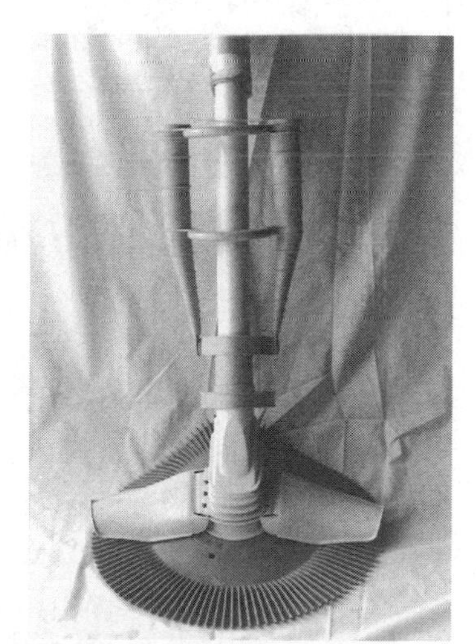

FIGURE 7-14A A suction-side automatic pool cleaner. *Kreepy Krauly (PacFab, Inc.).*

Lighting

External lighting of your pool, spa, or water feature is possible with a variety of high- and low-voltage systems. For installation and maintenance of external systems, I suggest you consult your local electrician. For lighting the water from within, however, there are also a variety of options.

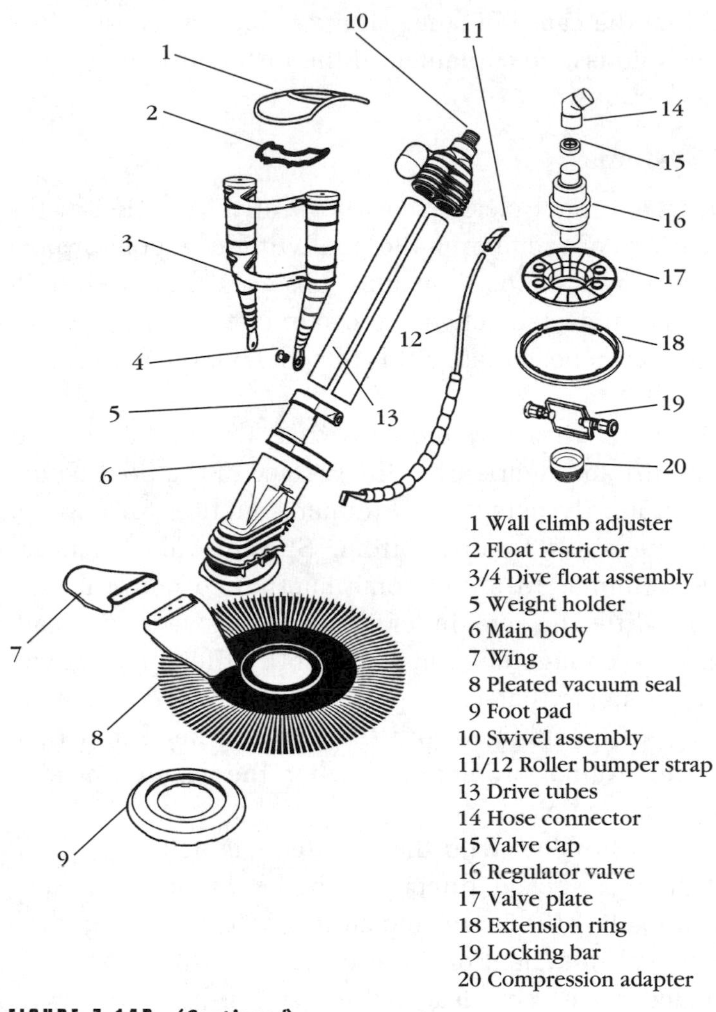

FIGURE 7-14B *(Continued)*

Standard 120/240-Volt Lighting

Figure 7-15 shows a typical pool or spa light housed in a stainless steel conical-shaped fixture, about 8 inches (20 centimeters) in diameter by 6 to 10 inches (15 to 25 centimeters) deep. The fixture is mounted in the wall of the pool or spa in a container called a *niche*.

Like the lights in your house, pool and spa light fixtures have a standard, screw-in socket for a bulb (in fact, your household 100-watt bulb will work in a 120-volt pool or spa fixture). The cord that sup-

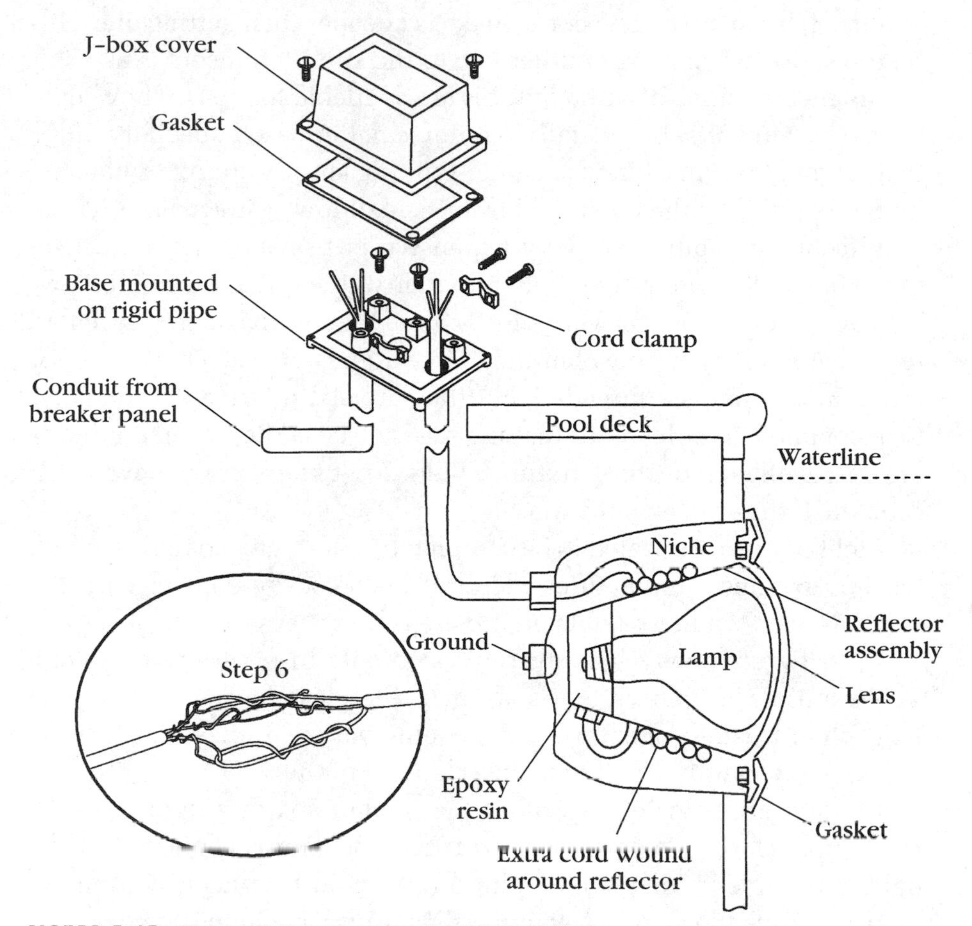

FIGURE 7-15 The typical pool or spa light installation.

plies electricity to the fixture is waterproof and enters the unit through a waterproof seal. There are no user serviceable parts in the cord, seal, or fixture except for changing the bulb. Although the correct term is *lamp*, for clear understanding I will use the more common expression *bulb*. Fixtures and bulbs are available in 120 or 240 volts, but 120 volts is most common for residential use. Bulbs generally run from 300 to 500 watts.

Fixtures are sold with cords of 10 to 100 feet (3 to 30 meters), so you need to know the distance from the light niche to the junction box to determine what you need when purchasing a replacement fixture. A simple rule of thumb is to buy the next longer length than you think you need, because the few extra bucks is cheaper than getting one a little too short and making another trip to the supply house.

Smaller versions of these fixtures are available for spas. They come in 120 or 240 volts, but usually employ smaller-based specialty bulbs that might screw in or have bayonet-type sockets. A quartz or halogen bulb will give you the most light from a small low-wattage bulb. These new technology bulbs are very expensive but save energy and need replacement less often than hotter, higher wattage standard bulbs.

You might encounter very old fixtures that are basically designed the same as modern ones, but they have a *mogul* base. The socket for these are also female threaded, but they are about twice the diameter to accommodate bulbs with that mogul-style base. Replacement bulbs are still available for these fixtures, but I don't know anyone who still sells the fixtures (or would want to).

Light fixtures are sealed to be completely watertight, so the air inside reaches extreme temperatures unless it is cooled by contact with the water. Never turn these lights on if there is no water in the pool.

Most fixtures have lens overlays available in a variety of colors. These are plastic lenses that snap over the glass lenses to make the light blue, green, or whatever color. Being very thin plastic, these too will quickly melt if the light is operated out of the water.

Fixtures are held in the wall of the pool or spa in a light niche. A niche is a metal can large enough to contain the fixture, cemented watertight to the side of the pool. A waterproof conduit leads away from the niche to above the water level out of the ground to a junction box.

INSTALLATION

RATING: ADVANCED

Since you probably won't be installing too many light niches on new pools, I will outline here the procedure for replacing a fixture when it rusts out or otherwise fails.

1. **Find J-Box** Turn off the power source at the breaker. Find the junction box. On older pools, this will be in the deck directly above the light niche, often under a 4-inch-diameter (10-centimeter) stainless steel (or bronze) cover plate that is held in place by three screws that probably have rusted or stripped heads, making them impossible to remove.

Later building codes required the junction box to be at least 5 feet (1.5 meters) from the edge of the water and 18 inches (45 centimeters) above the surface of the water, so on more recently constructed pools, look in the garden directly behind the light niche and you will often find it sticking up there.

2. **Disconnect** Figure 7-15 shows a typical junction box (simply called a J-box). Remove the four screws and take off the cover. Three wires come into the box from the breaker or switch and three go out of the box to the light fixture. The three to the light fixture will be individually insulated, colored white, black, and green, and are bound together with a single rubber sheath that waterproofs the package. This is the cord of the light fixture as described previously. Disconnect the three wires. Unscrew the cord clamp.
3. **Remove Fixture** Lean into the pool (yes, you have to get at least an arm wet) and remove the face rim lockscrew (retaining screw) from the faceplate that holds the fixture to the niche. The top of the fixture will float outward, the bottom is hooked into the niche and can be lifted out.
4. **Cords** Uncoil the excess cord to give you enough slack to raise the fixture out of the water onto the deck. Go back to your J-box and tug at the cord of the fixture. If you see the cord move at the fixture end, you know the replacement is an easy one. If it's tight, follow the steps outlined later (see tight cords).
5. **Cut** Cut the cord where it enters the old fixture. Strip off the rubber sheath, back about 6 inches (15 centimeters). Remove the string or paper threads that run alongside the wires (these were put in the cord to add strength for when you pull on it). Remove the insulation from each single wire back about 6 inches (15 centimeters).
6. **Pull New Cord** Remove the new fixture from the box and float it in the pool. I have laid the new fixture on the diving board or deck and it always gets tugged off or kicked, shattering the lens, so just get in the habit of laying it in the water and you won't face this problem. Take the wires of the new fixture and strip the cord back as described in step 5. Now bind the wires of the new cord to the ones of the old, folding each wire over as if to make a hook (Fig. 7-15),

then connecting the two (like two fishing hooks) and twisting the loose end around the base wire. Use electrical tape to cover the exposed wire, wrapping it tightly and thoroughly.

Don't tape so much that you make a connection thicker than the cord itself. It won't pass through the conduit easily. The idea is to make a union of the wires that will not separate when you pull.

Lay the cord out freely into the water. Pull the old cord at the J-box until you have pulled the connection and the new cord through. Some niches have the conduit connector in the top of the niche pointing upward at an angle, so when pulling the cord, it might snag at your connection.

To solve this problem, have someone lean into the pool and angle the cable down into the water, at the same angle as the conduit connector itself. That person can feed the line as you pull from the other end. Keep pulling until you have enough cord left with the fixture for it to easily reach up to the deck for future bulb-changing access. Now untape your connection and remove the old wire.

7. **Reassemble** Reach into the pool again and coil the excess line around the fixture, then reset the fixture in the niche. Replace the lockscrew on top.

   Sometimes, particularly if the fixture is way below the waterline, you might have to get into the pool to loosen or reset the fixture. If you're on the deck and having trouble lining up the face rim lockscrew with the hole of the niche, use a piece of coat hanger wire. Slide it through the hole on the faceplate and into the hole of the niche. Slide the fixture into place and the holes will line up for the screw. Remove the wire and insert the screw. Cut off any excess cord at the J-box, leaving enough to make a good connection with the supply wires. Reconnect the wiring and close the J-box. Test the fixture.

REPLACE THE BULB

RATING: EASY

The most important factor in bulb replacement is to maintain the waterproof integrity of the fixture. Follow each step carefully to avoid

TRICKS OF THE TRADE: TIGHT CORDS

- **If the cord won't budge from either direction, the cord might have swelled from age, heat (electricity creates heat), or moisture. Try pouring tile soap down the conduit from the J-box. You might have to let it sit overnight to slither all the way along the conduit and be effective.**
- **Another trick is to use leverage. Place a long 2-by-4-inch (5-by-10-centimeter) board down into the pool and run the cord up along its length. Using the edge of the pool coping as a fulcrum, try to pry the cord away from the fixture.**
- **Another trick is to cut the cord as low as possible at the J-box side of the conduit. Using your drain flush (blow bag) and the garden hose, try forcing water from the J-box to the pool through the conduit to loosen the cord.**
- **Sometimes the cable comes out, but the new one tears off as you are pulling it. Remove the new cord and use an electrician's snake to run through the conduit, starting at the J-box and working toward the pool. When it comes out the other end, hook the wires of the new fixture to it and try to pull them through again.**

leaks that will not only damage the new bulb and fixture, but might also lead to electrical shock of the next swimmer.

1. **Preparation** Turn off the power, then get the fixture up on the deck as described previously.
2. **Disassembly** Remove the lens (Fig. 7-16, item 2) that is held in place by a clamp or a set of screws. Gently pry the lens away from the fixture, taking care not to gouge the lens gasket (item 3).
3. **Replace the Bulb** Inside some fixtures you will find a bare coiled spring wire. This is nonelectrical but is designed to break a circuit. Notice that without a bulb in place, the spring lays to one side of the fixture. Hold it up against the opposite side and screw in the new bulb. The spring lays on the bulb itself.

   If the bulb bursts when in use, the spring sweeps across the filament, cutting the electricity in the circuit. In this way, if water has gotten into the fixture, a live electrical circuit won't stay in contact with the water, ultimately electrocuting someone in the pool.
4. **Test** Now place the fixture on the deck and turn on the light to make sure your new bulb works. Never operate a closed fixture out of the

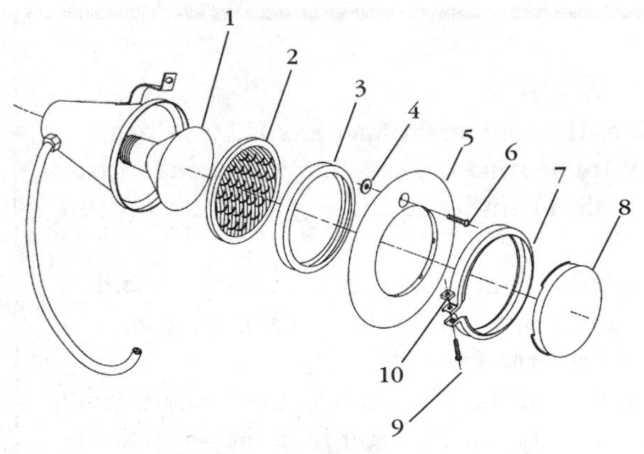

1 Bulb
2 Lens
3 Lens gasket
4 Washer
5 Stainless steel face plate
6 Assembly screw
7 Clamp assembly (including 9, 10)
8 Colored lens cover

FIGURE 7-16 Light fixture assembly. *PacFab, Inc.*

water, but with the lens off, the heat can escape without problem.

5. **Reassemble the Fixture** For the $2 extra, I always use a new gasket. After long use, heat, and harsh chemicals, the old one is probably compressed, and if it doesn't fail immediately, it might fail before the next bulb change. In reassembling the faceplate, if it has the series of screws, tighten them on opposite sides until you have done them all. This applies even pressure and prevents gaps in the gasket that will ultimately leak.

6. **Look for Leaks** Lay the fixture back in the water. Hold it underwater for several minutes to make sure it doesn't leak. A few bubbles might rise from air trapped under the lip of the faceplate, but a steady stream means the fixture is filling with water. Take it apart again and dry it thoroughly. Go back to step 4 and be more careful.

7. **Reset** If it passes the leak test, turn it on for a few seconds before putting the fixture back in the niche. During reassembly or testing you might have banged the unit around enough to break the bulb or its delicate filament, or you might have a bad bulb. I hate getting wet and wasting the time getting the fixture back in the niche, only to find out that the bulb was damaged during the process. Remember, the closed fixture depends on water for cooling, so conduct this test for literally no more than two seconds. Now reset the fixture in the niche as outlined previously.

TROUBLESHOOTING

RATING: ADVANCED

Replacing a bulb or fixture as outlined previously is about all you will need to do with lights and water. But how do you know if it's the bulb, cord, fixture, or breaker? If the light won't come on, try this procedure.

TRICKS OF THE TRADE: LIGHT SAFETY

A few words of wisdom about lights and pools. Electricity and water mix only too well, often with deadly results. Therefore, since you can be injured or your work might make you liable for someone else's injury, take note.

- **If a fixture, lens, or gasket looks suspicious, replace it. It's not worth the time, hazard, or money to "try that old one, one more time."**
- **Use the bulb, gasket, or replacement lens that fits the fixture. Try to get the same manufacturer's replacement parts or use the generic brand designed for that make and model. You might be able to force a bulb, gasket, or lens into place and it might stay watertight for a few days, but what happens when it ultimately overheats and leaks, bringing water, swimmer, and electricity into contact?**
- **Stubby bulbs are popular replacements. They are squat, short bulbs so they fit almost any fixture. The problem is that by being short, the heat is brought closer to the end of the fixture than the originally designed long-stem bulb. Also, bulbs with a reflector material painted on the underside of the bulb will reflect more light into the pool, so you get the appearance of a higher wattage bulb. The reflective surface is also designed to direct heat away from the rear of the fixture. This is significant because the resin at the rear of the fixture, that makes it waterproof, can melt.**
- **For the same reason, don't put higher wattage bulbs in a fixture than it was designed to take. If it isn't marked and you can't read the old bulb, don't use greater than 400 watts in your replacement.**
- **When changing a bulb, as with any repair, you are required to work to local building and health department codes. This might mean bringing an older installation up to code by adding a ground fault interrupter (GFI) as outlined in the chapter on basic electricity.**

1. Make a visual inspection. Look at the lens of the fixture (remove any color overlays first). If you can see water or black soot (the result of a bulb bursting in the fixture), you know it will have to come apart. If the water has filled the fixture, it might only look cloudy, so it pays to know what a dry, well-operating fixture looks like underwater.
2. Turn off the power supply at the circuit breaker. Open the J-box and expose the wiring connections. On old J-boxes mounted in the deck, the gasket often fails and allows water to fill the box. The water shorts out the wiring. Clean and dry out the box and replace the gasket and lid firmly.

3. If that is not the obvious problem, turn the breaker and light switch on and use your multimeter to verify that power is getting to the light fixture cord. If not, trace back the problem to the breaker or switch. Sometimes there is more than one switch, such as a remote control, and the problem might just be a second switch being off.
4. If the breaker trips off when you turn on the light, disconnect the light cord from the power wiring at the J-box and reset the breaker. If it still trips off, the problem is the breaker or GFI, or more rarely, the wires between the breaker and J-box have shorted together. If the breaker stays on, the problem is in the fixture or its wiring.
5. Set your multimeter to test circuit continuity. Touch one lead to the white wire of the fixture cord and one to the black. If you have continuity, the bulb and wiring are good.
6. Touch one lead to the green and the other to the white, then the black. If continuity exists between white and green or black and green, the wiring is bad. You can test bulbs that way too—touch one lead to the threaded part of the bulb base and one lead to the tip of the base. A good bulb will have a complete circuit (continuity), a bad one will not.

Low-Voltage Lights

Designed like their 120/240-volt cousins, low-voltage fixtures employ a transformer to drop the voltage to 24 volts. As a result, the highest wattage fixture does not deliver much light, so they are used in spas or fountains where lower wattage is sufficient to light a smaller area. Installation, repair, and replacement techniques are identical to those described previously.

Fiberoptics

The most significant recent development in water lighting is the use of fiberoptics. The hardware looks the same, but instead of running electricity to a fixture, the cord in the conduit contains thin plastic fibers that conduct light, not electricity. They terminate in a lens in the water to shed the light into the pool or spa. The light source and electricity are located safely away from the body of water.

It was not until 1970 that practical transmission of light via fiberoptics was achieved up to 250 meters and not until 1980 that practical transmission extended to significantly more. Therefore, homeowner applications are relatively new, and until they become more widespread and mass-produced, they will remain relatively expensive.

Materials used today are acrylic fibers bundled in multiple-strand cables instead of the old-style solid-core cables. The solid-core style actually transmits more light but is more sensitive to moisture and hardens over time, while the multistrand cable is more flexible and less prone to failure from intrusion by moisture. Of course, multiple sections of the solid-core cable can be spliced end to end as needed, while it would not be practical to do that with multistrand cables.

In considering whether to install a fiberoptic system instead of traditional pool lighting, there are many advantages and disadvantages.

TRICKS OF THE TRADE: FIBEROPTICS

- **Shorter runs (from the light generator) result in less loss of light and a less costly installation.**
- **Cuts are critical: You may be cutting 200 to 400 individual strands within a bundle and each one can be scratched if cut wrong, resulting in light loss. Never use knife blades or mechanical cutters.**
- **Do not bend the final 12 inches (30 centimeters) of cable before the lens.**
- **Install several smaller lights for better results instead of relying on a single larger one.**
- **Locate the generator above grade and ensure that it stays waterproof.**
- **Don't skimp. More strands per cable and a larger generator are more costly but will yield superior results.**

Pros	Cons
No electricity near water	More costly
Allows more special lighting effects	Not effective in dark bottom pools
Maintenance easier (no bulbs to replace in underwater fixtures)	Light loss up to 2% per foot, so not good on long runs from the generator
Easier to change colors	Space needed to locate generator

Installation and troubleshooting of fiberoptics is a specialty and requires manufacturer training. The "do's and don't's" in the sidebar on this page are provided to assist you in better understanding both the system and the installer.

Covers

Pool and spa covers have come into increasing use over the past ten years in an effort to save heating energy costs and water through evaporation. As water evaporates, the minerals in it are left behind in the remaining pool or spa water. Chlorine, which is made from salt, leaves minerals in the water as well. If you live in say, Malibu, where the water comes from wells or other very hard sources, this is a very real problem. Covers slow down the rate of evaporation and, therefore, reduce chemical use. Covers keep out leaves and dirt which also absorb chemicals. Keeping out the dirt means you vacuum less and clean your filter less.

Shielding the water from the sun further cuts down on chemical use. The result is the water in the pool or spa doesn't get so hard, so quickly. That saves money in water bills, chemical costs, expensive draining and acid washing, and, as if that weren't enough, the heater does not scale up inside from hard water.

Covers also help retain heat. Some estimates say a cover alone will heat the pool or spa 10° to 15°F (5° to 7°C), meaning money is saved on gas or electricity. When you do turn on the heater, the water starts out warmer and heats up faster—a real benefit to a spa when you have a hot date on a cold night.

Finally, covers can provide safety. Heavy-duty spa covers can prevent kids from falling into spas, especially when combined with straps or locks to ensure they can be removed only by adults.

Okay, I'm sold. What are the choices? Here are some covers that will protect your pool or spa.

Bubble Solar Covers

Have you ever unpacked something sent in the mail that is wrapped in that plastic bubble wrap? If you're like me, you squeeze the little bubbles between your thumb and finger to pop them, right? Well, someone realized that large sheets of this stuff would make good pool and spa covers.

Figure 7-17 shows a typical bubble cover (also called a *sealed air* or *solar* blanket). In profile, the cover has one flat side and one bumpy side. In fact, the cover is made from two sheets of blue plastic (usually available in 8 or 12 millimeter thickness), heat welded together with air bubbles in between.

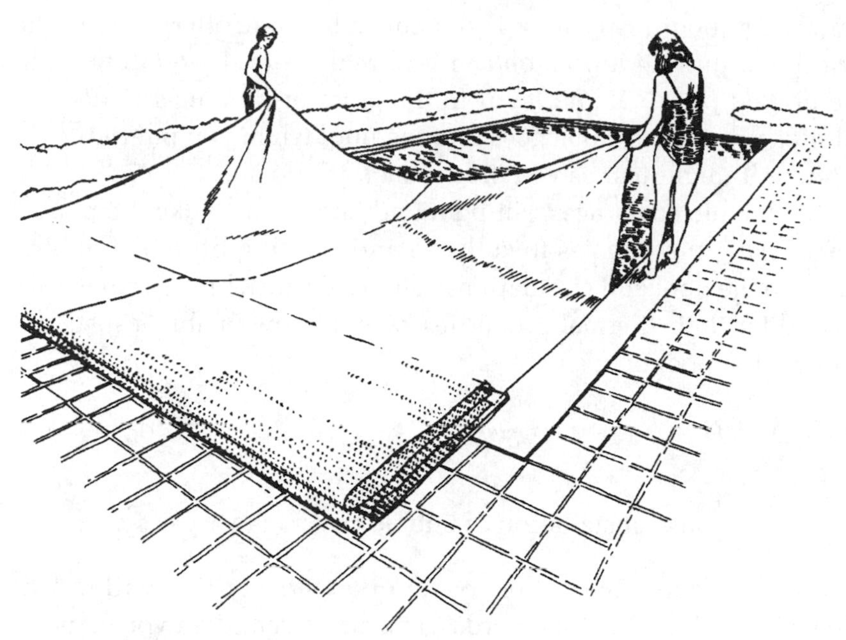

FIGURE 7-17 Solar blanket cover. *Cantar/Polyair Corp.*

The sun warms the air bubbles which transfer the heat to the water (thus the term solar cover). Similarly, the trapped air acts as an insulator for the heat coming up from the water. Always lay a bubble cover on the water with the bubble side down. In this way, the spaces between the bubbles also act as pockets for trapped air, further insulating the water.

Because they are thin, lightweight, and flexible, bubble covers can be cut to any size and are sold in large sheets from 5 feet by 5 feet (1.5 meters by 1.5 meters) for spas up to 30 feet by 50 feet (10 meters by 15 meters) for pools, with many intermediate sizes to fit any job. They are easily cut with scissors or a razor knife. Bubble covers are cheap, costing about 12 cents per square foot (930 square centimeters) wholesale. They last two to four years depending on water chemistry, weather conditions, and user wear and tear.

The disadvantage of a bubble cover is that in heavy winds they can blow off or away into the next county. Also, they don't really keep out dirt or debris, because as you remove the cover the dirt falls into the pool. Much of the dirt does stay on the cover, meaning you have to spread it out somewhere and clean the cover as well as the pool.

On a large pool (especially if the cover has no roller), taking the cover off and putting it back on can be a real chore. If you drag it off, where do you put it? If you lay it in the grass or on a nearby deck, it might pick up more dirt that goes into the pool when you put the cover back on, or it might tear as you drag it back into place.

As the bubble cover ages, sunlight and chemicals make the plastic brittle, causing the bubbles to collapse and sending little bits of blue plastic into the pool and circulation system. In short, bubble covers are only good for their thermal properties, which are valuable, especially if you heat the pool or spa a lot.

INSTALLATION

RATING: EASY

As you might guess, installation of bubble covers is very easy.

1. **Measure** Measure the pool or spa and buy a cover that will overfit the water surface. In other words, get a size larger than you actually need.
2. **Layout** Lay the cover out on the water surface and leave it for two or three days (some manufacturers recommend as many as 10 days, but unless it is very cold, I have not found any difference after two or three days). You can remove it to use the pool or spa during this period, but the idea is to give the material time to relax to its full size in the sun and to shrink to any degree that it might (I have found shrinkage is not more than 5 percent in any direction).
3. **Cut** Using shears, scissors, or a razor knife, cut the cover to the water surface size of the pool or spa. As you walk around the cover, cut slightly less than you think you should—you can always go around again and trim off a bit more, but it sure is hard to add any back on if you cut too much. There are no frayed edges or seams to worry about.

ROLLERS

As noted, handling large bubble covers can be inconvenient. You can try to fold the cover into large folds before taking it off, which is easier with two people of course. You can also cut the cover into two smaller pieces, which might be easier to handle one at a time. Or you can buy

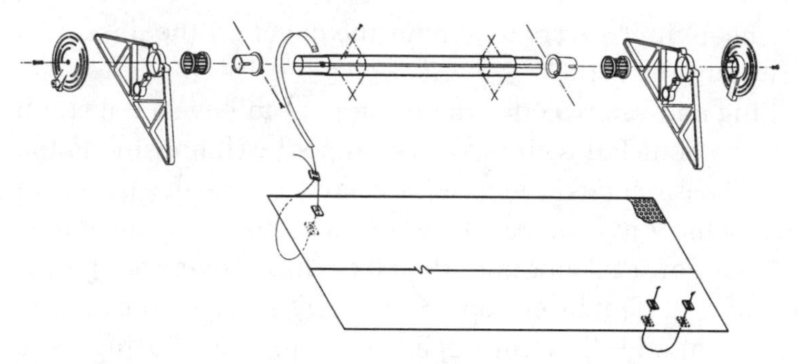

FIGURE 7-18 Cover roller (detail). *Odyssey Reel Systems Ltd., San Clemente, Calif.*

a roller (also called a reel system). Figure 7-18 shows a typical roller. The concept is to attach the cover to the barrel of the roller with straps so that when you rotate the barrel, the straps roll up first, pulling the cover along and rolling it on the barrel.

Even these can be cumbersome without two people, but they do make storage, cleaning, and handling the cover much easier. Replacement parts are available for rollers and wheels can be added to make it easier to move the entire assembly out of the way when using the pool. Plastic covers are also sold to cover the cover. If left on the roller for extended periods, this cheap plastic cover keeps sunlight and dirt off the bubble cover, extending its life. Rollers cost about $250 wholesale.

Foam

Much like bubble covers, foam covers are sheets of lightweight compressed foam [⅛ inch (3 millimeters) thick] that float on the surface of the pool or spa. Because foam is much more expensive (about 75 cents per square foot wholesale) than bubble plastic, these are primarily used for spas. Installation is the same as with bubble covers, described previously.

Sheet Vinyl

Particularly in cold climates where it is desirable to cover a pool in winter, sheet vinyl is sold like a bubble cover without bubbles. Sheet vinyl has little insulating value and doesn't even float very well, but it is very cheap. Sandbags or plastic ballast bags that you fill with water are sold with sheet vinyl covers to anchor the cover on the deck as it lays over the water (Fig. 7-19C).

Some building and safety codes require a pool to have a cover, on the theory that someone falling into a covered pool will not sink to the bottom as fast. These codes don't require you to use the cover, of course, so most of the vinyl covers I have sold were to satisfy the building inspector then never taken out of the box. These covers are generally \$10 to \$20 for a 20-foot by 40-foot (6-meter by 13-meter) cover and can be purchased at any discount department store as cheaply as at your supply house.

A higher grade variation of the vinyl cover is a heavy-duty material reinforced with wire to make them kid-proof. These are hooked onto fittings in the deck and are a real pain to take off and put back on. They are a beast to handle. The hooking system usually requires dozens of fastening points which means lots of stretching, tugging, and fighting.

Electric Covers

Somewhat a misnomer because you don't cover the pool with electricity, I call a sheet vinyl cover on a track, rolled up on a roller that is operated by a motor, an *electric cover* (Fig. 7-19A).

Electric cover systems, with the cover material stretched between tracks or rollers on the edge of the coping and a motor-driven pulley system to roll it up on a barrel roller (which is often concealed under the deck or in a bench-box at one end), are sold by several major companies.

Electric covers are large, complex pieces of machinery. The cover can easily tear if it gets caught or bound up, because the motor doesn't know when to stop it. The pulley system and ropes can break or get bound up and the cover won't open or close, often burning out the motor. Switches and controls fail. Manufacturers sell small submersible pumps with their cover systems, because the owner is responsible for pumping the water off the cover after rainstorms.

Electric covers are, however, just what I had in mind when writing about all of the positive features of covers at the beginning of this section. They protect, save, and insulate, and, best of all, they keep dirt and debris out of the water. Of course, they can be a pain if a lot of debris gets on top of them because you have to clean the cover before removing it. They also provide protection against someone falling into the pool. These systems are generally \$5000 or more installed.

A

B

FIGURE 7-19 (A) Reinforced vinyl cover (electric roller system). (B) Strength test for an anchored mesh safety cover. (C) Winter pool cover. *A: Anthony & Sylvan Pools, Doylestown, Pa. B: Loop-Loc Ltd. C: Cantar/Polyair Corp.*

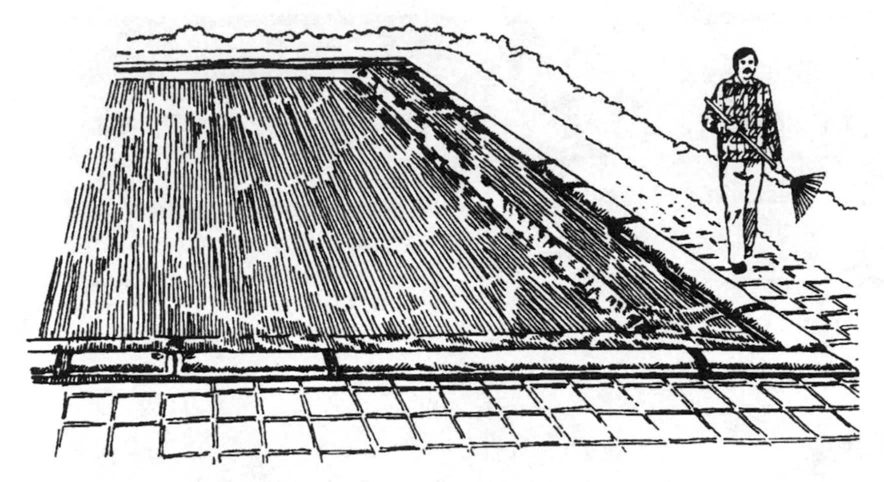

FIGURE 7-19 *(Continued)*

FAQs: ADDITIONAL POOL EQUIPMENT

Why Do I Need a Safety Barrier or Fence?

- Many localities require some kind of effective safety barrier to prevent children from gaining unsupervised access to the pool and possibly drowning. Whether required or not, a solid fence around your pool or a rigid cover on your spa is a wise investment.

Isn't It Dangerous to Have Electrical Equipment Near a Pool?

- Yes, it's very dangerous. That's why all jurisdictions establish building codes that require setbacks for electrical equipment and outlets (most say a minimum of 5 feet—1.5 meters—away from the water's edge). Always remove electric robot auto-cleaners from the pool before swimming. Low-voltage lights are another important safety investment, both for lighting your pool and any landscape lights adjacent to it.

Can I Install a Diving Board on My Above-Ground Pool?

- Diving boards are never permitted at any above-ground pool, because even those with deep ends are not safe for diving or jumping. Slides are made for above-ground pools, but only to be used on pools designed for them and only to be used as intended (sliding, not diving or jumping).

Do I Still Have to Vacuum if the Pool Has an Automatic Cleaner?

- **Yes. No automatic cleaner works 100%, because corners and other variations in shape cause the cleaner to miss some spots. Vacuums designed for above-ground pools have brushes built into them, so vacuuming also performs valuable brushing that keeps the liner clean (making it last longer) and free of algae.**

CHAPTER

8

Water Chemistry

Let's begin with the easy part, the part that all pool and spa people, health departments, manufacturers, and builders can generally agree upon. See the sidebar on the next page for the components of healthy water (each of which will be explained in detail, so don't fret if it looks Greek to you now).

Oh sure, you'll get slight variations from one expert to the next even on these basics, but no one would completely disagree with these recommendations and no health department would close you down within these ranges. In fact, if your pool or spa tested out exactly within these parameters, you'd probably get a medal and your picture in a pool service magazine.

So if everyone generally agrees on the guidelines listed, where does the disagreement enter into it? The heated debate begins when you discuss how to achieve and then maintain these parameters; how to correct a severe imbalance of one or more of the components without throwing other components out of alignment; how to kill an unexpected algae bloom; how to eliminate blue staining from the plaster, etc., etc., etc. Well here's my approach.

BALANCED WATER

- **Chlorine residual: 1.0–3.0 ppm**
- **Total alkalinity: 80–150 ppm**
- **pH: 7.4–7.6**
- **Hardness: 200–400 ppm**
- **Total dissolved solids: Less than 2000 ppm**
- **Cyanuric acid: 30–80 ppm**

Demand and Balance

If you get nothing else from this chapter, learn this section on demand and balance—you will be ahead of many water technicians who think they know a lot about water chemistry.

Water chemistry is a process of balance. Change one component, even to bring it into a correct range, and you might adversely affect another component, thus adversely affecting the entire pool or spa. Imagine that the water quality parameters are stones, each of equal size and weight, evenly distributed around the edge of a dinner plate. Now imagine balancing that plate on one finger. You can do it if you find the exact center of the plate, where each stone balances the others. But now imagine that one stone is doubled in weight or removed. It changes the balance of the plate—the other stones will slide into new positions or off the plate completely, eventually making the entire plate fall from your finger and crash to the ground.

This introduces the concept of *balance*, but what is the *demand* part? Water is a solvent. It will dissolve and absorb animal, vegetable, and mineral until it can no longer hold what it dissolves (the *saturation point*). After this, it will dump the excess of what it has dissolved (the *precipitate*). With this process in mind, understand that water makes demands on anything it comes in contact with until those demands are satisfied.

As you will see in the discussion of pH, for example, if the water is very acidic, it will demand to be balanced with something alkaline. If such water is in your plastered pool, the alkaline lime in the plaster will be dissolved into the water until that balance is achieved. When the water is no longer acidic, it will start dumping excess alkaline material, depositing it on tiles and inside pool equipment, as well as back on the plaster as rough, uneven calcium deposits. So demand and balance are inextricably related as will be increasingly apparent with each section of this chapter.

Components of Water Chemistry

Our original ideal parameters are worth repeating before I discuss each one individually.

- Chlorine residual: 1.0–3.0 ppm
- Total alkalinity: 80–150 ppm
- pH: 7.4–7.6
- Hardness: 200–400 ppm
- Total dissolved solids: Less than 2000 ppm
- Cyanuric acid: 30–80 ppm
- Temperature, wind, evaporation, and bather load

I've added one more component to the list. Obviously not a chemical component, temperature has an impact on some of the other components, either directly or indirectly, as do evaporation, wind, and dirt. More on that later, but for now just add it to your list.

Sanitizers

To discuss the first item on the list, chlorine residual, let's review the concept of sanitizers in general. There are many alternate sanitizing methods in common use, and there are a great number of "wonder" products for killing off special algae or solving stain problems. Chlorine is used as a benchmark because it is the most common sanitizer.

The purpose of any sanitizer is to kill bacteria in the water. Bacteria carry disease and stimulate algae growth. Sanitizers accomplish this by *oxidizing* the bacteria and other waste in the water. Rust is oxidization in progress. Oxygen is, in essence, dissolving the material with which it comes in contact. Sanitizers, oxidizing in your pool or spa, are essentially "rusting away" the bacteria and other waste material in the water.

The most simple form of chlorine, which is found in nature as chloride mineral salts, is as a gas. It is made by passing electricity through a saline (salt) solution, one by-product of which is sodium hydroxide (caustic soda). Liquid chlorine (sodium hypochlorite) is manufactured by passing chlorine gas through this solution of caustic soda.

Dry chlorine is subsequently made by removing the water from such a solution. Before I examine the properties of each, let's look at some properties common to all, then some evaluation criteria for each form. The chemistry that follows is about all you need to know for this entire subject to make sense, so don't be confused or discouraged by the notations and formulas—it's not really much to learn and I promise there is no more.

THE CHEMICAL PROCESSES OF CHLORINE

Chlorine, in any form, mixed with water forms hypochlorous acid (HOCl), the bacteria-killing form of chlorine, and hypochlorite ions (OCl^-), a weak form of chlorine. When chlorine gas is added to water, it also creates muriatic acid (HCl). In scientific notation it looks like this:

$$CL_2 + H_2O = HCl + HOCl$$

The HOCl part of the equation (the hypochlorous acid or killing part of the chlorine) further breaks down when it goes in your pool or spa. It breaks down into a positive ion of hydrogen and a negative ion of hypochlorite, which is a noneffective oxidizer of bacteria. The process looks like this:

$$HOCl = H^+ + OCl^-$$

If these resulting products are ineffective, then obviously the goal is to stop the chemical reaction at HOCl without further breakdown. How is that done? By maintaining a slightly lower than neutral pH in the water. As pH increases, chlorine becomes less effective because the process described previously is allowed to continue. Simply stated, available chlorine in the water does this:

- at pH 7.2—80 percent killing HOCl + 20 percent weak OCl^-
- at pH 8.0—20 percent killing HOCl + 80 percent weak OCl^-

See what I meant about balance, how one aspect of water chemistry affects another?

CHLORINE DEMAND

As I mentioned in introducing the ideas of water chemistry, water demands a lot. *Chlorine demand* can be defined as the amount of any chlorine product (in any form) needed to kill all the bacteria present in a body of water.

CHLORINE RESIDUAL

Chlorine residual is the amount of chlorine (expressed in parts per million) that is left over after demand is satisfied.

CHLORINE EVALUATION CRITERIA

Here are a few criteria for comparing various forms of chlorine and, for that matter, each of the other sanitizers I'll deal with. In each case, I will note any special properties of the sanitizer (for example, if the product is reasonably stable I will make no comment, but if it is very unstable I will say so).

Chlorine Availability: What does *available* mean—that you can ask it out for a date at any time and it will say yes? Well, sort of. Chlorine in its various forms combines with other elements present in the water, meaning some of the chlorine is locked up or unavailable for oxidizing bacteria, while some is available for that sanitizing purpose.

In considering the effectiveness of any sanitizing program in a body of water, this availability question must be considered to know what is really going on. In other words, you might add 5 gallons (19 liters) of liquid chlorine to your pool, but have less effective, active, sanitizing *available* chlorine than if you had added 2 pounds (0.9 kilogram) of chlorine gas. I will explain in a moment.

Stability: Chlorine is by nature very unstable, which is to say that it is easily destroyed by contact with ultraviolet (UV) light (sun). Efforts are made, as you will see, to make it more stable using various methods.

pH: Since you add some form of sanitizer to your water with such regularity, you must consider the pH of the product, because it will alter the pH of the water if enough is added.

Convenience: How easy is it to administer the product?

Cost: This is self-explanatory.

By-products: Does the form of chlorine you are using add anything to the water besides chlorine?

Before continuing, if this is your first time through this chapter, you might want to skip ahead to the section on pH and get into the chemistry detail the next time through when such concepts are not all new to you. If you already have a general knowledge of what's ahead, press on.

CHLORINE GAS

Chlorine gas is about two and one-half times heavier than air and light green in color. Chlorine gas lowers the pH of water when introduced into it. The advantages of chlorine gas are that it is cheap (about one-fifth the cost of liquid chlorine, the next cheapest form) and that it is 100 percent available. Because it is pure chlorine, there are no by-products added to the water when using it. So if chlorine gas is 100 percent available and cheap, why would there be any other form, why doesn't everyone use it? Because it is cumbersome to handle and administer and it is potentially lethal.

Chlorine gas can cause severe eye, nose, and throat irritation (see the sidebar on chemical safety for an anecdote about what happened to me when I inadvertently created chlorine gas in my shop one day), and in extreme cases it can be lethal (chlorine gas was used in World War I to kill soldiers in the trenches of Europe). Of course, it takes an extremely concentrated dose of gas to be deadly, but the point of describing this is to make you take seriously the handling of chlorine gas in any strength. A maintenance man in Arizona working around a leaking tank of swimming pool chlorine gas died from inhaling the fumes. It can happen.

That's why chlorine gas is not universally loved in the water maintenance business. It's cumbersome (pressurized tanks of gas, valves, wands, or expensive injection devices attached to the circulation equipment), and it's deadly if mishandled.

The National Association of Gas Chlorinators provides the following facts:

Effect	Chlorine concentration in air by volume
Slight symptoms produced at	1 ppm
Detectable odor	3.5 ppm
Maximum exposure level over 1 h without symptoms	4 ppm
Danger level in 30–60 min	40–60 ppm
Kills rapidly	1000 ppm

As can be seen, symptoms (difficulty breathing, coughing, eye and throat irritation) can occur at concentrations as low as 1 ppm. The good news is that since the odor should be detectable at 3.5 ppm, you are likely to be aware of the presence of chlorine gas well before it reaches lethal levels. However, pay attention to those odors—ventilate the room or area immediately when you detect strong chlorine odors.

LIQUID CHLORINE

Liquid chlorine, sodium hypochlorite, has a high pH because of the caustic soda used in its manufacture (see above). Although the caustic soda helps to keep the chlorine from escaping the solution, it gives liquid chlorine its high pH, between 13 and 14. Chemically, liquid chlorine is $NaOCl$.

The fact that chlorine is the product of salts makes liquid chlorine very salty. In fact, you are adding 1.5 pounds (0.7 kilogram) of salt to the water for every gallon (4 liters) of liquid chlorine you add. This adds to water hardness.

Liquid chlorine is produced at about 16 percent strength and by the time you use it, it is about 12.5 percent available. When you add sugar to your coffee, it tastes sweet. If you add twice as much, it tastes twice as sweet. If the object were to put as much sugar in solution in the coffee as possible, you would keep adding sugar until the coffee could simply not contain anymore.

Why not do the same thing with liquid chlorine? Why not have, say, 65 percent chlorine? Then you would only need to carry 1 gallon of the product for every 4 gallons you now carry—it would be a more concentrated, convenient product. Unfortunately, because chlorine is

TRICKS OF THE TRADE: CHEMICAL SAFETY

One day in my pool supply store I was in one of those cleanup moods and thought I would combine two half-filled bottles of chlorine into one. What I didn't know was that one bottle contained muriatic acid and one contained chlorine. When I poured one into the other, holding them close in front of my face, the resulting explosion of chlorine gas (released by contact with the muriatic acid) choked me, burned my eyes and lungs, and I thought I was going to die. Fortunately I didn't, but the lesson was one I hope I can learn for you—don't make the same mistake.

By the way, liquid pool and spa chemicals come in color-coded plastic bottles so, in theory, you can't make that kind of mistake. The color varies by manufacturer, but acid usually comes in red or green bottles while chlorine comes in yellow, light blue, or white bottles. Whatever the color scheme, don't ever do as someone in my shop did that nearly fatal day—don't ever put a different chemical in a bottle meant for something else.

so unstable, producing it at such strength is pointless. Within hours, that 65 percent solution would deteriorate to less than 15 percent anyway. So that's what you get from the factory to start with, ending up with about a 12 percent solution after some normal deterioration.

This brings me to another important point about liquid chlorine. Air, sunlight, and age accelerate the deterioration process, so keep your supplies fresh and covered. By the way, household laundry bleach, which is nothing more than liquid chlorine, is produced to around 3 percent strength. You realize how profitable the household bleach business must be when you note that a gallon costs about the same as a gallon of pool or spa liquid chlorine, yet the pool chlorine is four times stronger.

The advantages of liquid chlorine are that it is easy to use (just pour it into the body of water being treated) and it goes into solution immediately, because it's already a liquid. When pouring liquid chlorine, pour it close to the surface of the water to prevent splashing (bleaching your shoes) and to prevent unnecessary dissipation of strength (by contact with air).

DRY (GRANULAR OR TABLET) CHLORINE

There are essentially two types of dry chlorine sanitizers. The more popular of these two are called *cyanurates*—they contain stabilizer (cyanuric acid) to help prevent breakdown, generally about 1 pound (450 grams) of cyanurate to each 4 pounds (1.8 kilograms) of chlorine product. But more on that later. First, let's review the other type of dry chlorine.

Calcium Hypochlorite: Commonly sold under brand names including HTH, this product is popular for its relatively low cost and convenience of use. It is available in granular or tablet form.

Calcium hypochlorite is unstable, slow dissolving, and leaves substantial sediment after the chlorine portion of the product enters into solution in the water. It is 65 percent available chlorine, so it is rather potent (remember, liquid chlorine is only 12 percent available). Its pH is 11.5, so it tends to raise the pH of the water.

Calcium hypochlorite is often used as a shocking agent (see the discussion of shocking later). When shocking a pool, your main concern is available chlorine, delivered for immediate sanitizing use. Thus, a product that contains no stabilizer is preferred, because there is no point in paying the extra cost for chemicals that you don't need for this particular application. For regular sanitizing and for maintaining the chlorine residual, one of the cyanurates might be better because they are longer lasting with fewer by-products.

Cyanurates: There are two types of cyanurates, which are chlorine sanitizers containing stabilizer (cyanuric acid). The first form is *dichlor* (sodium dichloro-s-triazinetrione). Okay, I know I promised not to muck up this chapter with unintelligible chemical jargon and this product sure sounds like just that. But this is really simple. Most technicians refer to this product as dichlor. Triazinetrione just means stabilized—the product contains a stabilizing agent.

So what are the properties of dichlor?

- 56–63 percent available chlorine
- No sediment or by-products as with HTH-type products
- A pH of 6.8, so it is slightly acidic

When you buy dichlor products they won't be called dichlor. They will be called by various brand names, but read the label.

The second form of dry chlorine is *trichlor* (trichloro-triazinetrione). Trichlor is the most concentrated (and therefore the most expensive) form of chlorine produced—90 percent available chlorine. To achieve both stability and strength, trichlor is produced mostly as tablets that slowly dissolve in a floater, although dry granular trichlor is produced as a super-killing agent such as Algae-Out for problem algae blooms.

Trichlor's pH is 2.8 to 3.2, very acidic. So make sure the floater (floating duck, dispenser, etc.) that you use in the pool is well away from any metal and/or skimmer or intake. Never put trichlor tablets in the skimmer as I have seen many pool technicians do. The acidic properties will have the same results as dumping acid directly into the system—the metal will dissolve. When tablets are left in a skimmer when the pump is off, the tablet dissolves creating a few very acidic gallons of water. When the pump comes back on, this acid water is sucked directly into the metal plumbing and components of the circulation system. This daily acid bath will quickly dissolve metals, destroying plumbing and equipment and depositing metal on the plaster walls of the pool or spa.

One major advantage of trichlor is that it dissolves slowly, so if a pool is not serviced frequently, chlorine will still be released into the water daily. A disadvantage of both dichlor and trichlor is that they are 50 to 58 percent cyanurate, so continued heavy use leads to a buildup of stabilizer, requiring at least a partial draining of the pool.

There is one more form of dry chlorine, available in powdered form, called lithium (lithium hypochlorite). Lithium is very soluble; in fact, when broadcast over the surface of the pool it will dissolve before it reaches the bottom. This makes it good for vinyl-lined or dark plaster pools where you might otherwise be concerned about discoloration.

Lithium is not in common use, however, because it costs more than double to maintain your pool with lithium as opposed to liquid chlorine.

CHLORAMINES AND AMMONIA

As you might have guessed from the name, *chloramine* is a combination of chlorine and ammonia. Remember I said that chlorine likes to combine with other elements in the water. It is this fact that makes it an effective oxidizer of bacteria, as it tries to combine with the bacteria, thus killing it. But when ammonia is present in the water, the two will combine together to form chloramines.

You don't need any chemical formulas here. Just know these facts about chloramines:

- Chloramines are very weak cleaners (weak oxidizers of bacteria) because the chlorine is locked up with the ammonia and is not available to kill bacteria. The colloquial term is *chlorine lock*.

- Chloramines are formed when ammonia is present in the water from human sweat or urine (chemically very similar) and when there is insufficient chlorine present to combine with all the available ammonia and to oxidize bacteria. One active swimmer produces one quart of sweat per hour.
- Chlorine odors in a pool (and irritated eyes) are caused by chloramines, not too much chlorine.
- *Breakpoint chlorination* is the point at which you have added enough chlorine to neutralize all chloramines, after which the available chlorine goes back to oxidizing bacteria and algae instead of combining with ammonia in the water.

SUPERCHLORINATION

As the name suggests, adding a lot of chlorine to a pool is superchlorination. Why would you do it?

Even if you maintain a sufficient residual chlorine in your pool at all times, the ammonia and other foreign matter in the pool might keep your chlorine from being 100 percent available. That's why algae grows in a pool even though it has a high chlorine residual when you test the water. Superchlorination is, therefore, adding lots and lots of whatever chlorine product you use so that there is plenty in your water for the foreign matter to absorb, leaving enough to oxidize bacteria (kill algae).

Superchlorination is recommended by various experts as something to do monthly, three times a year, etc.—in other words, every pool person has a different opinion based on experience. The truth is that there is no one answer for every pool. You only want to superchlorinate when you have a substantial volume of ammonia or other foreign matter in your pool to soak up. Since you can't really know how much ammonia is in the water that might be locking up your chlorine, you can guess based on how dirty the pool gets every week and how much it is used. Of course, the more swimmers, the more sweat and urine are likely to be present. Therefore, it's a judgment call. Each chlorine product requires different amounts to achieve superchlorination. Remember, you have to close the pool to swimmers until the residual comes back into normal levels. So it pays to use enough, but no more. Follow the directions on the package for superchlorination. With liquid chlorine, I use 8 gallons (30 liters) of chlorine per 20,000 gallons (75,700 liters) of water. Keep the water circulating 24 hours a day until the residual reads normal.

Another time to superchlorinate is when the pool has been allowed to get exceptionally dirty or full of algae. A quick way to kill off a bad algae bloom (growth) is the same method I described to eliminate chloramines from your pool. Add 1 gallon (3.8 liters) of anhydrous ammonia and 8 gallons (30 liters) of liquid chlorine for every 20,000 gallons (75,700 liters) of pool water. In the chloramine discussion, I said that chloramines are not an effective cleaner. That was only half the story. Chloramines will, in fact, kill and dissipate quickly. They have a reputation as a poor cleaner because chloramines are totally unstable—they leave no residual, they kill and are gone.

So for especially bad algae problems, I use that method with the pump running 24 hours a day and continue to add chlorine daily [usually 8 gallons (30 liters) per day per 20,000 gallons (75,700 liters)] until the chloramines are absorbed and a normal residual is restored. The white powder you see in the pool the next day is the bleached, dead algae.

TRICKS OF THE TRADE: BREAKPOINT CHLORINATION

My method of achieving breakpoint chlorination is to add 1 gallon (3.8 liters) of anhydrous ammonia and 8 gallons (30 liters) of liquid chlorine to every 20,000 gallons (75,700 liters) of pool water and turn off the equipment for 24 hours. After that time, add another 8 gallons of liquid chlorine, turn on the equipment, and allow swimmers back into the pool when the chlorine residual reads normal.

Now this might seem odd—you're trying to use chlorine to eliminate ammonia, so why add more ammonia to the pool? Ammonia and chlorine combine to form chloramines. When ammonia decomposes, it forms nitrogen compounds, which are nutrients for algae. When the algae ingest the nitrogen, they also ingest chloramines, which kills them from the inside out.

SUMMARY SCORECARD OF CHLORINE

To give you an overview of the chlorine products just discussed, here's a simple summary of their properties. They are arranged in order of relative cost, with gas being the cheapest. Note the relationship between cost and stability—the more stable you try to make this inherently unstable product, the more you must add to it and therefore the more expensive it becomes.

Product	pH	Available chlorine	Common form	Stability in water
Cl_2 gas	Low	100%	Gas	Very unstable
Sodium hypochlorite	13+	12.5%	Liquid	Unstable
Calcium hypochlorite	11.5	65%	Dry granular	Stable
Dichlor	6.8	60%	Dry granular	Very stable
Trichlor	3.0	90%	Tablet (or granular)	Very, very stable

In terms of relative effectiveness, this comparsion is useful: 1 lb chlorine gas = 1 gal sodium hypo = 1 lb, 8.5 oz calcium hypo = 2 lb, 13.5 oz lithium hypo = 1 lb, 12.5 oz dichlor = 1 lb, 2 oz trichlor, or 0.45 kg chlorine gas = 3.785 L sodium hypo = 680 g calcium hypo = 1.3 kg lithium hypo = 795 g dichlor = 510 g trichlor.

FORMS OF CHLORINE: COMPARATIVE COST

Another important evaluation factor of which type of sanitizer to use is cost. The delivery system used, labor involved, transportation costs, and pH neutralizers needed (or other compensating chemicals required if the sanitizer chosen has "side effects" on the water as previously discussed) all must be factored in the real cost of the sanitizer.

Gas chlorine is the cheapest and most effective sanitizer, but it is also the most dangerous and difficult to use. To compare other sanitizer forms, assume we are comparing equal sanitizing action (not equal actual volumes or weights). Therefore, as a rule of thumb, here is a general guideline:

Liquid chlorine	costs 2 times the equivalent amount of gas
Calcium hypochlorite	3
Trichlor	3
Lithium hypochlorite	5
Dichlor	5

CHLORINE ALTERNATIVES

There are many alternatives to chlorine, although none are as widely used or as easy to use.

Bromine: Bromine is a member of the chemical family known as *halogens*. In fact, so is chlorine, iodine, and fluorine. All are oxidizers. So if bromine is a cousin of chlorine, why isn't it used more? First of all, it cannot be stabilized and is therefore expensive to use. Having said that, it must be noted that bromine is more stable than unstabilized chlorine *at high temperatures*, and that's why many people use it instead of chlorine in their spas and hot tubs.

Nevertheless, a stabilized chlorine product is more effective and cheaper than bromine, so the only contribution made by bromine to the world of pools and spas is for people who use their spas a lot and object to the chlorine odor. As noted in the section on chloramines, there won't be a chlorine odor if the correct amount is maintained and if the user doesn't urinate or sweat in the water.

The ultimate irony about bromine, however, is the fact that for it to work it must have a catalyst, which is usually chlorine. The small amount of chlorine added to bromine products is not detectable, so most technicians don't know that. You might have guessed, I'm not a fan of bromine.

Ultraviolet (UV) Light: You know what ultraviolet light is: light waves produced at the highest end of the spectrum—the rays that when they come from the sun turn your skin to fried chicken. While very effective at producing skin cancer, the problem with ultraviolet light as a sanitizer is that it is reactive, not preventive.

After the water has passed through your pump, filter, heater, and other equipment, it passes through a chamber where it is exposed to a beam of ultraviolet light. Any bacteria in the water passing through this light is killed. Do you begin to see the problem? The bacteria must be present in the water to be killed, whereas with a residual level of any chemical sanitizer already in the pool water, the bacteria is prevented from growing in the first place. Also, algae growing in the pool, which is not suspended in the water passing through the circulation system, are totally unaffected because the ultraviolet light has no way to kill the algae inside the pool itself. So ultraviolet light is not effective by itself. You still need to use chlorine or some other chemical agent to establish a residual in the water. For this reason, in many health department jurisdictions, ultraviolet by itself is not an approved sanitizer.

So if you have to use chemicals anyway, why buy the ultraviolet gadget? Some will say that it allows you to use less chemicals and therefore helps eliminate chlorine odors, etc. Well, remember, it's not the chlorine that causes the odor, it's the chloramines that are caused by ammonia in the water. In short, ultraviolet is just a gimmick, not a truly effective sanitizer.

Another deterrent to UV systems is cost (typically over $700). One more thing. Don't confuse UV sanitizing systems with UV ozonators, which use ultraviolet light to create ozone which is then injected into the plumbing system. Ozonators of several varieties are discussed next.

Ozone: Ozone consists of three atoms of oxygen which is created naturally in the earth's atmosphere by, among other things, lightning. It has been used for decades to purify municipal drinking and sewage waters and more recently for pool and spa sanitizing applications.

Ozone oxidizes contaminants in the water, but it is very unstable. Thus, it will react with contaminants immediately and break down, leaving little to attack algae on the walls of the pool. Therefore, a residual of chlorine, bromine, or another sanitizer is always required for a total sanitizing package. Of course, the purpose of using ozone is to reduce the amount of chemicals needed to achieve the minimum required residual.

For pools and spas, ozone is created in one of two ways. One uses ultraviolet light (not to be confused with the ultraviolet systems described above) in a generator which contains UV bulbs that resemble tubular fluorescent bulbs. As air passes the bulb a photochemical reaction takes place that produces ozone. The benefit of the UV ozone generator is that it is cheaper than its counterpart (described next); the drawback is that it produces far less ozone.

The second method of ozone production is called corona discharge (CD). The CD ozone generator (Fig. 8-1) forms an electrical field (or "corona") which converts oxygen (O_2) to ozone (O_3), much like the creation of ozone by lightning. Generally, this produces ten times more ozone for the same amount of electricity as a UV ozone generator, but

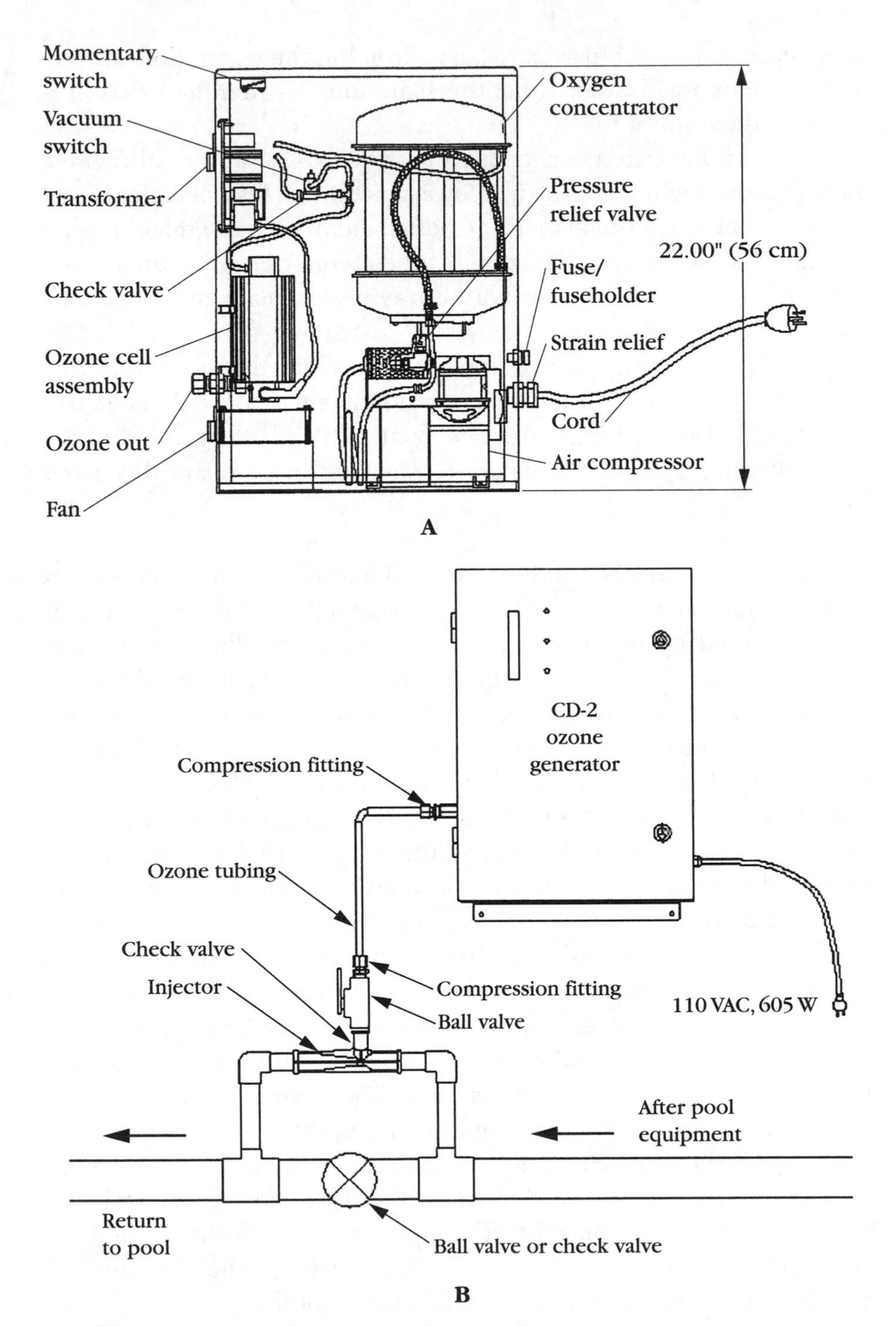

FIGURE 8-1 **(A) Corona discharge ozone unit. (B) Plumbing of an ozone unit.**
Del Industries.

the CD units are more costly because of the equipment needed to dry the air which passes through them. CD units require dry air because moisture combined with the electrical discharge can create corrosive nitric acid inside the appliance. Smaller CD units for spas do not use an air drier component because of the smaller volume of ozone needed to sanitize the smaller body of water and are thus much less costly.

Ozone has a neutral pH and generally has little impact on other parameters of water balancing. It is still necessary to shock the pool periodically as previously described. Other maintenance procedures are also unaffected, such as the need to brush and vacuum the pool, although ozone facilitates the flocculation of particles in the water, making them easier for the filter to remove.

Purchasing and installing an ozone system is easy (Fig. 8-1). There are several manufacturers who provide models based on the size of the pool or spa and/or bather load. Each comes with plumbing and wiring directions which are easy to follow if you comprehend the plumbing and electrical chapters presented elsewhere in this book.

Commercial pools using ozone as a sanitizer require a bit more equipment to finish the job including a contact tank, mixing tower, injector and booster pump, ozone monitors, and other components too complicated to describe in this book. Suffice it to say that ozone is great for commercial pools but the system should be installed and maintained by a professionally trained technician.

Ozone generators for residential pools and spas have come a long way in the past few years and are a genuine and reliable product to assist with reductions in the use of harsh chemicals which many customers find objectionable. But remember, ozone is unstable and can't provide a sanitizer residual in the water, so it can only be used in tandem with chemical sanitizers.

Ionizers: An ion is any atom with an electrical charge. An ion with a positive charge is called a cation; one with a negative charge is called an anion. To ionize something is to transform it into one of these two forms. Ionization as it applies to pool and spa water is an electrolytic reaction caused by applying electrical current to metals (copper or silver) which in turn act as sanitizers. Ionizers consist of a chamber which houses electrodes of these metals. Low-voltage current is passed between the cathode (positive) and the anode (negative) electrodes. As the current moves, it produces metal ions that are flushed into the pool. The electrical current is so minute that it does not cause shocks.

As with ozone generators, ionizers cannot be used as the only sanitizer, but they will reduce the need for chemicals significantly. There are two major concerns with ionizers, however. First, the metals have the potential to stain plaster. Second, many users say they simply don't work. These two problems are directly related. By keeping the copper or silver level low enough to avoid precipitation and staining, it may be too low to make a noticeable difference. By raising it to levels that will certainly do the sanitizing job, staining is almost certain.

Therefore, the sophistication of the ionizer will directly determine the user's satisfaction. Some units float in the pool like chlorine tablet "floaters," powered by solar cells. These are not adjustable and their output is minimal. More sophisticated units that are plumbed into the circulation system and hardwired for electricity can be more precisely controlled but are more costly as a result.

Potassium Monopersulfate: Not very common, yet commonly available, PM is a very efficient oxidizer used most often in products sold for superchlorination. It is also an effective catalyst for bromine and is sometimes found in small quantities in bromine tablets (or sticks). But you can't go into your supply house and ask for PM in the way you might ask for chlorine. You'll simply find it when you read the label of various sanitizing products.

Biguanicides: A relatively new sanitizer most familiar by the trade name Baquacil, biguanicides are a general term referring to a disinfectant polymer that more accurately goes by the name polyhexamethylene biguanicide (PHMB). It is an effective sanitizer but not an oxidizer, so it cannot be used alone. A hydrogen peroxide product is applied as a monthly shock and a quaternary ammonium-based supplement is needed weekly. PHMB concentrations need to be kept between 30 and 50 ppm and can be tested with a special test kit using reagent drops. Other water balance parameters will be the same as for any other sanitizer.

PHMB cannot be mixed with chlorine products or, for that matter, any other chemicals except those designed as part of the package. The result if you do? Chocolate-brown-colored water and stains on the plaster. This means if you change a chlorine-treated pool to PHMB, you must follow package instructions very carefully to first neutralize other chemicals and then remove any metals that may be present in the water. PHMB also reacts with household detergents and trisodium phosphate (TSP), so be careful with tile cleaners or any other cleaning products in a pool that uses PHMB products.

The good thing about PHMBs is that they have no equivalent to chloramines, so there is no objectionable odor or "combined" chemical problem to deal with. PHMBs are good products and viable alternatives which customers will appreciate, but more than with any other water treatment product, *read* and *follow* the labels!

pH

pH is a way to assess the relative acidity or alkalinity of water (or anything else for that matter). It is a comparative logarithmic scale (meaning for every point up the scale, the value increases tenfold) of 1 to 14, where 1 is extremely acidic and 14 is extremely alkaline. In the middle, 7.0, is neutral, but for our purposes, 7.4 is considered neutral, neither acidic nor alkaline. In fact, the Los Angeles health department requires a pool's pH to be between 7.2 and 7.8.

pH has an amazing effect on water. If it is allowed to become very acidic (from adding too much acid, from too many swimmers sweating or urinating, from too much acidic dirt and leaves in the water) the water becomes corrosive, dissolving metal it comes in contact with. If it is allowed to become too alkaline, calcium deposits (scale) can form in plumbing, equipment, and on pool walls or tiles.

More than that, pH determines the effectiveness of the sanitizer. For example, you might have a chlorine residual of 3 ppm in a pool with a pH of 7.4. But increase the pH just four-tenths to 7.8 and you will still have a chlorine residual of 3 ppm, but the ability of the chlorine or bromine to oxidize bacteria and algae will decrease by as much as one-third. Therefore, maintaining a proper pH is not just a factor of interesting chemistry or saving your equipment, it is a matter of effective maintenance.

pH is adjusted with various forms of acid (to bring down a pH that is too high) or alkaline substances (to raise a pH that is too low). With plaster pools and spas, the calcium in the plaster creates a situation where the water is always in contact with alkaline material, slowly dissolving it (remember, water is the universal solvent), causing the water to be slightly alkaline. Therefore, in plaster pools and spas you will always be adding small amounts of acid to keep the pH down.

Acid is most readily available in liquid (a 31-percent solution of muriatic acid) or dry granular forms. As noted elsewhere in the book, never mix acid and chlorine—the result is potentially **deadly** chlorine gas.

Because each product is different, I won't try to guide your use of acid; just follow the instructions on the package or in your test kit for the amount to add. Keep this general guideline in mind, however. If you add acid in small amounts, you can always add a bit more to get the desired

TRICKS OF THE TRADE: pH

- **When you add acid, particularly liquid acid, pour close to the surface of the water so the acid doesn't splash on the deck or your shoes where it will do damage. Keep it away from your face and hands as well. If you get some on you, it might not burn right away, but don't be deceived—it will start eating your skin. Wash the area immediately with fresh water and seek emergency first aid if you are exposed to a lot or if you get it in your eyes.**
- **Also, when you add acid to the water, distribute it as much as possible, walking around the pool or adding it near a strong return outlet while the pump is running. The idea is to dilute it as soon as possible. Never add acid near a main drain, skimmer, or other suction inlet. The idea is not to have full-strength acid sucked directly into your pump and other equipment which could be damaged.**
- **When adding soda ash or other alkaline material, dissolve it in a bucket of water first and add it slowly. Adding too much or adding it too quickly might result in the water turning milky blue. Follow the package directions. By the way, soda ash has a pH of 12 and bicarb has a pH of 8. These products are described later, but for now, use soda ash to raise pH and total alkalinity. Use bicarb to raise total alkalinity alone.**
- **In both cases, when adding acid or alkaline, allow several hours for the pH to stabilize before testing again—the next day is even better. If you test right after adding chemicals, you will get a false reading. Similarly, don't test pH after adding chlorine or other sanitizers. Remember, each has a significant pH itself, so if you test right after adding sanitizer, you might be testing water saturated with that product and not truly reflective of the water's pH.**

results. But if you add too much, the water becomes corrosive, which means you must now add alkaline material to bring the pH back up.

pH is tested using a chemical called phenol red. For easy reference, all discussion of testing techniques is in the section on chemical testing.

Total Alkalinity

Total alkalinity is one of those concepts that frequently gets confused with pH or hardness. While the total alkalinity of a pool or spa has an impact on those and other factors, it is not the same thing.

Imagine a waiter brings you a cup of coffee already sweetened with sugar. You sip the coffee and rate the flavor as slightly sweet, moderately sweet, very sweet, very, very sweet, and so on. But does that rating tell you how much sugar was actually put into the cup? To know that, you would ask the waiter and he might tell you, for example, it was one tablespoon (15 milliliters).

Total alkalinity is to water chemistry what that tablespoon of sugar is to the cup of coffee; while pH is to water chemistry what the flavor rating is to that coffee. A pH test will tell you the relative acidity or alkalinity of the water, while the total alkalinity will tell you the quantity of alkaline material in the water. It is a measurement of the soluble minerals present in the water (like the measurement of sugar in the coffee).

To use another analogy, you might say that a pH reading is similar to a temperature reading on a thermometer, which simply states the present temperature; while a total alkalinity measurement represents the volume of heat that brought you to that temperature. The usefulness of such information is that you can then predict the added quantity of heat required to reach a desired temperature. Therefore, total alkalinity readings will tell you the amount of acidic material required to reach a desired pH.

In other words, *total alkalinity* is a measurement of the alkaline nature of the water itself and, therefore, the ability of that water to resist abrupt changes in pH. In fact, adjusting water chemistry to a proper total alkalinity acts as a buffer against fast and extreme changes in pH (called *spiking*).

The actual test method is discussed later, but it is important to know at this point that an appropriate reading for your water's total alkalinity is 80 to 150 ppm. Less than 80 ppm means that too much acid has been added even if the pH reading is high. Therefore, you always adjust the total alkalinity level first, then the pH.

Proper maintenance of the total alkalinity of a pool will pay great dividends. You will use less chlorine, less acid, and see fewer algae problems.

Hardness

Hardness (or calcium hardness) is a component of total alkalinity. It is a measurement of the amount of one alkaline, soluble mineral (calcium) out of the many that might be present.

So why single out this one alkaline mineral for special measurement, especially if you have measured and balanced total alkalinity? Because, when in sufficient quantity, calcium readily precipitates out of solution and forms salty deposits on the pool, tile, and in equipment. This deposit is called *scale* and is the white discoloration you see on rocks or tile at the waterline of a pool or spa (Fig. 1-20). If you live in an area with hard water (water that contains large amounts of minerals) you might also have seen it in your home—in drinking glasses or flower pots that are left with standing water for long periods.

The acceptable range when testing for hardness is 100 to 600 ppm. Over 600 ppm and you will see scale. The only cure for water that hard is to partially or totally drain the pool and add fresh water.

Total Dissolved Solids (TDS)

Now that I've discussed two measurements of minerals in the water, there is one overall category that helps you keep track of the big picture. *Total dissolved solids* is, as the name implies, a measurement of everything that has gone into the water and remained (not been filtered out), intentionally or not.

The total of minerals (including calcium), cyanurates, chlorides, suntan lotion, dirt, etc., etc., equals TDS. The main contributor to TDS increases in a pool or spa is evaporation. When the water evaporates, it leaves its contents behind. You add more water, and the solids keep building up as it evaporates. As noted earlier, the liquid chlorine you add to the pool is made from salt, so when the liquid evaporates the mineral remains behind, adding to the total amount of dissolved solid material in the water.

In most places, water from the tap already has a TDS reading of 400 ppm. You add about 500 ppm more each year with chemicals. Depending on where you live, the evaporation rate might add another 500 ppm per year. In California, for example, between wind and temperature you can expect to lose the equivalent of the entire volume of your pool each year to evaporation.

When a body of water reaches a reading of 2000 to 2500 ppm, it's time to empty it. There is no other way to effectively remove all those solids floating around in your water.

If you can't see, taste, or smell TDS, why make such a big deal about it? Because TDS acts like a sponge, absorbing chlorine and other chemicals you put in the water, rendering them ineffective or requiring you to use far more of each chemical to attain the same results. That becomes expensive and creates further problems (remember, balance).

Cyanuric Acid

Cyanuric acid (also called "conditioner") simply extends the life of chlorine in water by shielding it from the ultraviolet rays of the sun, which would otherwise make the chlorine decompose. Cyanuric acid is a powder that is added once or twice a year in most pools, maintaining a level of 30 to 80 ppm. Above 100 ppm serves no function and creates other chemistry problems, so this is prohibited by most health department codes.

Because cyanuric acid helps prevent chlorine decomposition, it is also frequently called *stabilizer*. Because it extends the life of the chlorine, the result is that you use less chlorine and save money. You will thereby be adding less mineral to the water, slowing the rate of increase of TDS.

Cyanuric acid itself does not decompose or burn out in pool or spa water as is commonly believed. The only way to dissipate it is to remove it through water replacement. Much is also lost when bathers splash and leave the pool and when removing leaves, dirt, or other debris during cleaning.

As with the other components of water chemistry, the testing methods are detailed later. The method of administering cyanuric acid to water is detailed in the chapter on cleaning and servicing.

Weather

Perhaps the least discussed and yet most important aspect of water chemistry is the impact of varying weather conditions. Although you can't do much about these factors, understanding the role they play in your pool or spa will help you make decisions about adding water and chemicals and whether or not to use a cover.

SUN

As mentioned, ultraviolet rays in sunlight speed the decomposition of chlorine. Without stabilizer, 95 percent of chlorine can be lost in two hours on a sunny day. In addition to protecting chlorine with stabilizer, a pool cover might pay for itself in the chemical savings over many years. Similarly, wind, low relative humidity, and high temperatures will speed evaporation and, as I have just discussed, that increases the frequency of draining your pool.

TEMPERATURE

In Malibu, pool water will stay in the 70s (near 21°C) during the summer and below 60°F (15°C) in winter. Although unscientific, I have found that below 65°F (18°C) you will have little or no problem with algae growth. Above that temperature you are always fighting it.

WIND AND EVAPORATION

Wind not only speeds evaporation, but it also carries dirt into the pool, adding to the TDS (not to mention adding to your cleaning problems). Evaporation becomes a real problem when customers see their water bills going up and begin to question if there is a leak in the pool.

DIRT

The dirt, leaves, and other debris carried into a pool by wind also impact on chlorine use. Such debris will absorb chlorine, effectively removing it from the water. You will need to add two to three times as much chlorine to a dirty pool than you will to a clean one to reach the same results. One more reason to keep the pool clean, as well as the skimmer basket, strainer basket, and filter.

RAIN

Rain is another weather problem. In some areas, such as Southern California, the rain comes in enough volume during certain months that you have to pump water out of the pool to prevent overflowing and flooding. In doing so, your carefully balanced chemistry is quickly unbalanced. The extra water saves the customer's water bill, but plays havoc with maintaining chlorine residual and other levels, requiring far more tweaking of water chemistry during rainy months than dry.

In summary, it is important to realize that varying weather conditions and bather loads have varying impacts on your water chemistry. [Two people in a 400-gallon (1500-liter) spa can destroy up to 40 ppm per hour of chlorine residual.] Simply being aware of that probability will keep you testing and adjusting your water frequently enough to stay ahead of rapid changes that can be caused by weather influences.

Algae

Algae are one-celled plants, of which there are over 20,000 known varieties. Algae include microscopic ocean plankton, giant kelp that grow 2 feet (60 centimeters) a day, and virtually everything in between. The word algae is derived from the Latin word for seaweed and is already plural (the singular being alga).

Sunlight speeds algae growth, appearing in the pool as a green, brown, yellow, or black slime often resembling fur. It thrives in corners and on steps where circulation might not be as thorough as elsewhere in the pool.

When you brush algae from pool surfaces, some will immediately flake off while much will remain tenaciously attached to the plaster, appearing as a stain. It is fairly easy to brush algae off of smooth surfaces, such as fiberglass, and very difficult to brush it off of rough plaster where it can burrow in the cracks and crevices. Black algae (actually a blue-green), for example, is completely impervious to brushing, forming a hard protective shell over itself.

Algae can have an impact on the water chemistry components themselves. For example, algae can break down bicarbonates in water, raising the pH significantly and adding to the alkalinity at the same time.

Forms of Algae

There are three forms of algae that you will encounter in pools and spas.

GREEN ALGAE (CHLOROPHYTA)

The most common algae is green in color and grows as a broad slime on pool and spa surfaces. The slime can be removed by brushing, but that doesn't kill the plant. Superchlorination, combined with other procedures, will solve most green algae problems.

Green algae will first be seen on steps or in pool corners in very small patches. That is the time to take action, because any algae is a sign of a pool or spa that is not being properly maintained. I have seen pools, especially if they are heated to about 80°F (27°C), that have gone from those first few patchy signs of algae to entire green coverage in 24 hours. In other words, it can grow and get out of control very quickly.

YELLOW ALGAE (PHAEOPHYTA)

Yellow algae, which at times appears brown or muddy in color, is also called mustard algae. It does not grow as rapidly as green algae, but it is more difficult to kill. It grows with the same broad, fur or mold-like pattern as green algae. Brushing has little visible effect, although it will remove the outer slimy layer, exposing the algae underneath to chlorine. As with green algae, the best killer is superchlorination combined with remediation techniques and general good maintenance.

BLACK ALGAE (CYANOPHYTA)

Black algae is actually dark blue-green in color and is the pool technician's worst nightmare. I don't know any pool person, no matter how fastidious about their maintenance procedures, who has not encountered and dealt successfully with green or yellow algae. But at the first sign of black algae, you need to consider the pool or spa as a patient that is in critical condition.

Black algae grows first in small dots, appearing to be specks of dirt on the bottom. It seems to be less influenced by depth than green or yellow algae, growing sometimes from the deepest part of the pool first. As time passes, these specs begin to enlarge and then appear all over the pool. The growth rate is slow at first, but then exponential. In other words, it starts slowly, but once black algae takes hold it quickly makes up for lost time. I was called to a pool in a rural canyon area because the homeowner thought she had a problem with green algae. Sure enough, when I arrived, I found a nice black plaster pool with early green, slimy algae growth on the steps and shallow corners. When I began brushing, I noticed patches of discoloration in the black plaster which led me to try a few other tests. I reached into the water with a screwdriver and scraped a small area of the plaster, discovering that I was actually dealing with a white plaster pool covered entirely by black algae, over which green algae had started to form.

The homeowner failed to mention that the pool was white plaster and that about a year ago it slowly started turning to black. In other words, it took awhile, but the black algae had completely taken control of the pool. Of course, the homeowner had failed to apply virtually any basic maintenance during that period and the only remedy was draining the pool and replastering. That's right, we were unable to acid wash or otherwise remove the algae because black algae grows so deeply into the plaster and concrete below that the only solution is to remove the plaster and start over.

Having seen that kind of plant power, I have the greatest respect for black algae and never want to have to explain to my customer why I allowed it to grow in his pool while he was paying me for weekly maintenance. Believe me, you don't either.

Part of the reason for black algae's virulence is that instead of creating a slimy substance as a protective barrier like green or yellow algae, black algae covers itself with a hard substance that resists even vigorous brushing. Only a stainless steel brush will break open the shell, allowing sanitizers or algicides to penetrate the plant. A more detailed method of black algae removal is discussed in the next section.

PINK ALGAE

Actually not an algae, but you will hear this fungus referred to as such, it appears as a reddish slime at the water line. It is easily removed with brushing and normal sanitizing.

Algae Elimination Techniques

RATING: EASY

The best cure for algae is keeping a clean pool, filter, and baskets. Keep your chemical components balanced and brush the pool frequently, even if there is no apparent algae or dirt. The brushing exposes microscopic algae growth to sanitizer before it has a chance to really blossom, preventing a major problem.

If you do encounter a situation of advanced algae growth, there are two approaches to deal with it. One is a general elimination program that will work with most algae blooms. The other is to combine that program with a special algicide. Please note, however, that there are no miracle cures. Despite the claims of some manufacturers, there are no chemicals you can simply add to the water and walk away, expecting the algae to disappear by the next service call. Every elimination technique requires hard, repetitive work—another good reason to take every preventive measure in the first place.

GENERAL ALGAE ELIMINATION PROCEDURE

RATING: EASY

The following steps will help with any type or spread of algae growth and should be undertaken before adding algicides.

1. **Clean the Pool** Dirt and leaves will absorb sanitizers, defeating the actions you are about to take. Be sure you have a clean skimmer and strainer basket and break down the filter as well, giving it a thorough cleaning. As the algae dies or is brushed from the pool surfaces, it will clog the filter, so it is important to start clean. You might need to repeat this process several times during a major algae treatment.
2. **Check the pH** Adjust it if necessary (see water treatment section).
3. **Sanitize** Add 1 pound of trichlor for every 3000 gallons of water (1 gram per 25 liters). Brush the entire pool thoroughly, stirring up the trichlor for even dissolution and distribution. Never use trichlor or other granular sanitizers at such strength on dark-colored plaster, painted, or vinyl-lined pools. These surfaces will discolor. Instead, use liquid chlorine as outlined in the section on chloramines.
4. **Circulate** Run the circulation for 72 hours, allowing the trichlor to attack the algae at full strength, brushing the pool at least once each day. Adjust the pH as needed and continue to add liquid chlorine to maintain a chlorine residual of at least 6 ppm.
5. **Filter** As noted earlier, the dead algae might clog the filter, requiring teardown and cleaning several times. Keep an eye on the pres-

sure. When the chlorine residual returns to 3 ppm, resume normal maintenance. You will need to vacuum the pool frequently during this period to remove the dead algae and the inert ingredients of the trichlor, both of which will appear as white dust when you brush or otherwise disturb the bottom.

You will also need to brush frequently, not only at first, but for at least one week after you can no longer see any trace of algae. Believe me, it's still there and will rebloom if you let up.

TRICKS OF THE TRADE: SPOT ALGAE REMOVAL

- **Use a length of 2-inch (50-millimeter) PVC pipe. Insert the pipe into the water, with one end over the algae. Pour a cup of trichlor into the other end of the pipe and allow it to settle over the spot. You can also maneuver a chlorine tablet to lay over the algae.**
- **On deeper locations or where the algae is on a vertical location, fill a stocking with trichlor or tablets and hang it on the pool wall so that the stocking makes contact with the spots. It isn't necessary to cover every inch of every algae location. The act of locating such a concentrated dose of chemical within a few inches or millimeters will have the desired effect.**

SPECIAL ALGAE FIGHTING PROCEDURE

The second general procedure to understand is for particularly stubborn algae, such as yellow (occasionally) or black (always). This method applies even when you can only see a few small patches, because there are other contaminated areas that you can't see yet.

1. **Prep** Follow the routine outlined in the general elimination guideline.
2. **Treat** Use a stainless steel wire brush and vigorously brush the surface of the patches. Brush the remainder of the pool normally. In addition to the chemical application already recommended, pour some trichlor directly over the top of the algae spots. You might need to turn off the pump to accomplish this, so that the currents in the pool don't redistribute the trichlor you are trying to get on top of the algae.
3. **Brush** Continue daily brushing of the spot with the wire brush and reapply chemicals until the algae is gone. On areas of especially difficult access, I have often put on a mask and fins and gone into the pool to make a more effective treatment of chemical or brushing.

If these measures are not effective on your particular situation, or for more certain, rapid results, you can also use an algicide.

ALGICIDES

RATING: EASY

Algicides kill algae, while algistats inhibit their growth. In fact, most products can be used in varying strengths for each purpose. There are countless companies producing countless products designed to prevent and/or kill algae in pools, spas, or water features. Brand names vary and because many of them use a chemical as part of their name, they might be confusing.

Algicides fall into two basic categories, depending on their intended use. There are those made for pools and those made for spas. Since all algicides are regulated by the U.S. Department of Agriculture as pesticides (because they are designed to kill living organisms), they are required to be registered for pool or spa use, and it is a felony to use the product for purposes other than intended. In short, a great deal of testing and engineering has gone into each of these products, so follow the product instructions. The various types of algicides, based on their active ingredient(s), follow.

Copper Sulfate (Liquid Copper): Effective on all types of algae, but recommended for use in decorative ponds or dark-colored pools because of staining.

Copper sheets were nailed to the bottom of sailing vessels for centuries to eliminate marine organism growth, such as barnacles. It remains a leading ingredient in paints applied to boat hulls. In the last hundred years, copper sulfate (also called *bluestone*) has been used as an algicide in lakes and decorative ponds.

The copper actually destroys the algae's ability to breath and consume food, killing it in the process. It also tends to coat the pool surfaces, acting as a preventive long after the original problem it was applied for has disappeared.

The only drawback to this product is that it stains the porous surfaces of plaster pools, along with swimmers' hair, that distinctive blue-green copper color you see on buildings that have copper roofs or trim. Modern products containing copper sulfate also contain other chemicals (called chelating agents) to prevent the staining, but in the simultaneous presence of strong oxidizers, like chlorine, the staining returns. Products containing copper sulfate are extremely effective, but I only use them on decorative ponds or dark-colored pools where the potential stains will go unnoticed.

Colloidal (Suspended) Silver: Effective against all types of algae in all situations.

Silver works the same way as copper, without the side effects. I have heard of extreme cases where silver, which is sensitive to light, has turned black in a pool if administered in sufficient quantity and catalyzed by the presence of excessive amounts of acid or stabilizer. In practice, I have never seen this occur and find products such as Silver Algaedyne to be remarkably effective.

Polymers: Effective on all types of algae to some degree, but works best on green and yellow algae.

Polymeric algicides might be any of a number of chemical compounds, preceded by the prefix *poly*, meaning many parts which duplicate each other. In short, they work by repeating the job over and over.

The way polymers accomplish the job is to invade the outer membrane of the algae and effectively smother it. Polymeric compounds contain a strong, positive electrical charge and are attracted to algae, which is naturally negatively charged. Herein lies the drawback. They are also attracted to dirt in the water or filter, so their strength might be diluted. Also, because they work by attacking the exterior of the plant, they're not very effective against black algae, which has a very strong outer shell, or yellow algae, which has a stronger exterior than green, unless you have done a complete job of wire brushing (which might be impossible in some pools). Also, polymers don't coat pool surfaces and build much preventive effect. In other words, use them on green algae problems or where you're sure you have brushed thoroughly, but not with any hope of preventive maintenance. When you do use them, apply 20 percent more than the label suggests to allow for dirt in the pool.

Quaternary Ammonium Compounds (Quats): Good on green algae in early stages.

Quats work like polymers, but the positive electrical charge is not as strong. Generally I have been unimpressed with the results of these products. Further, many of them will foam up if agitated, so don't use them in spas because they will create foam when you turn the jets and blower back on. Filtration will ultimately remove the foam.

Chlorine Boosters: Great against stubborn yellow or large green algae blooms.

The most effective chlorine booster is anhydrous (or aqua) ammonia. As described previously, it turbocharges the chlorine so it is more effective and kills algae from literally the inside out. It is good and inexpensive, especially for large pools.

TRICKS OF THE TRADE: COMMON WATER PROBLEMS AND CURES

Rating: Easy

Cloudy Water

- **Inadequate filtration.**
  - **Increase daily filtration time.**
  - **Clean the filter.**
- **Too many total dissolved solids.**
  - **Check the total alkalinity and pH; adjust as needed.**
  - **Drain and refill with clean water.**

Algae in Pool or Spa

- **Follow "Algae Elimination Techniques" as described.**

Blue-Green Water

- **Copper from plumbing or heater components, stripped away by acidic water.**
  - **Immediately shut down the circulation equipment until you can ascertain the problem and solution.**
  - **Test and balance pH. Apply alkaline (pH raisers) and brush the spa. After several hours, take another pH reading. If it is now normal, turn the equipment back on; otherwise, keep balancing.**
  - **Apply a metal chelation agent to the water. These products, available in various brand names, will attract and combine metals so the metals can be filtered out. Follow label instructions of the product you choose, and in all cases, run the filter for 72 hours once you have balanced the chemistry.**

Brown-Red Water

- Iron in the water from metal fixtures corroded by acidic water.
  - Check and adjust first the total alkalinity, then the hardness, and finally the pH.
  - You may need to drain part of or all the water and add fresh.
  - Apply a metal chelation agent to the water (see above).

Corrosion of Metal Light Fixtures, Rails, etc.

- Water too acidic: Take steps outlined above.
- If metal turns black, electrolysis may be the problem.
  - Look for electric current (perhaps a slight water leak in a light fixture or J-box); and water with enough minerals (salts) to conduct the weak current.
  - Correct as needed.

Scale Forming on Pool or Spa Walls or at Waterline

- Buildup of calcium carbonate precipitated out of water from evaporation or heat.
  - Check the hardness. If it is near or exceeds the standard of 2000 ppm, drain and replace some of or all the water.
  - If the hardness is within acceptable limits, the problem may be high pH or total alkalinity. Check and adjust both.

Eye/Skin Irritation; Colored Hair or Skin

- Low pH and/or too many chloramines: Adjust the pH and shock-treat the spa.

Odors

- Chlorine odor from too many chloramines: Adjust the pH and shock-treat the spa.
- Musty odor from algae growth or high bacteria: Shock-treat the water to kill any bacteria or drain and refill spa.
- Mildew odor from accumulation of mildew on spa covers or in deck crevices where water has been standing: Follow your nose to the source and clean contaminated areas with sanitizers.

Foamy Water (Typically in Spas)

- Too much soap, body oil, or lotion from bathers; too much spa cleanser.
  - Extend filtration time until foam is gone; clean filter.
  - Drain and refill spa.

Water Testing

RATING: EASY

Having reviewed various potential problems of pool and spa water and learned that diagnosis is based on testing the water, we will now examine the methods of doing so. Each component of water chemistry has unique test requirements, but the most popular basic methodology is similar.

Test Methods

There are three basic approaches to water testing: using chemical reactions and comparing the resulting colors, estimating values with electronic devices, and making observations of the relative cloudiness (turbidity) of a water sample.

COLORIMETRIC

The most common approach to testing water is to collect a small sample in a clean tube and add some chemicals to the sample. These chemicals, called *reagents* (Fig. 8-2A), let you evaluate the sample by its changing color. By comparing the intensity of the color with a color chart of known values, you can determine the relative degree of each water chemistry parameter. To be legal for use in some health department jurisdictions (and to be practical for accurate reading), a color chart must have at least four different shades of the color in question. For example, in one such test, a sample turning yellow denotes the presence of chlorine. The stronger the intensity of the color, the more chlorine is in the sample.

TITRATION

Also a color-based test, *titration* is the process of adding an indicator reagent (a dye) followed by a second reagent (called the *titrant*) in measured amounts, usually a drop at a time, until a color change occurs. By counting the number of drops of reagent required to effect the color change, you can estimate the value in question. Some manufacturers supply tablets in place of liquids for titration testing. The amount of acid demanded by a body of water can be calculated with this method.

TEST STRIPS

Paper strips impregnated with certain test chemicals are also used as colorimetric test media (Fig. 8-2B). With this method, you dip a test strip into the pool or spa and move it around in the water for 30 sec-

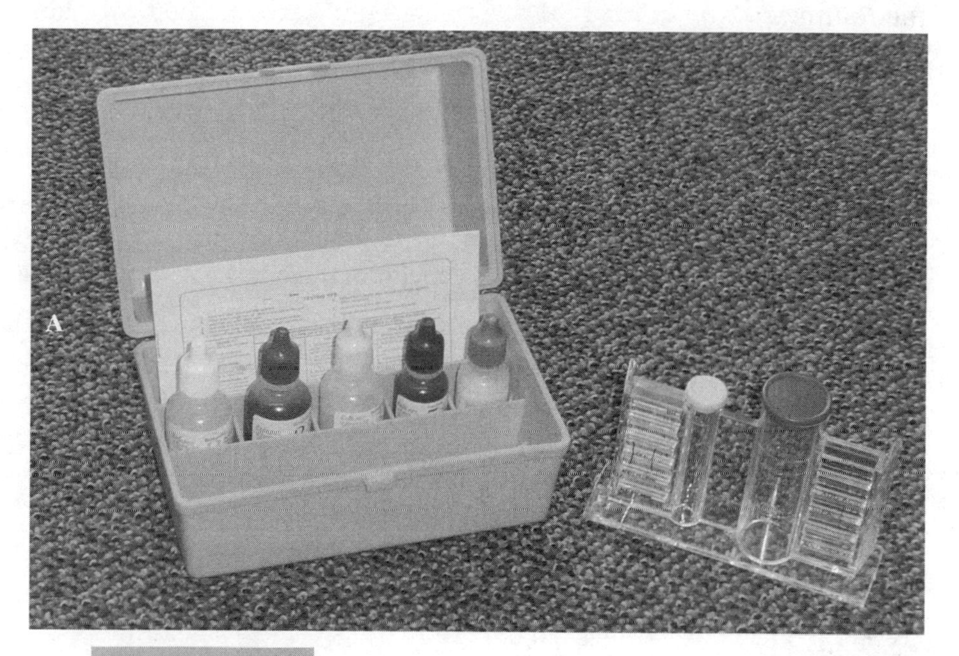

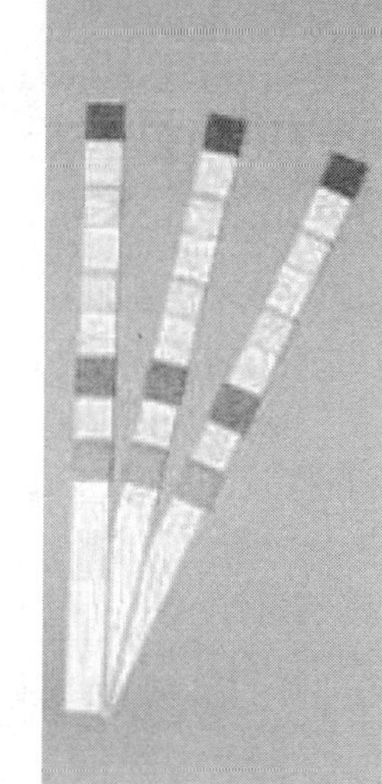

FIGURE 8-2 (A) Typical pool water test kit with reagents and color chart; (B) test strips.

onds. When you remove the strip, you compare its color to the color chart of known values to determine the test results. Test strips are produced for individual tests or have several on one strip. Because they are more costly to use, test strips are not yet widely used, but because of their simplicity and accuracy they are likely to be more popular in the future.

ELECTROMETRIC

Testing using electronic probes attached to calibrated digital or analog readouts is the most accurate method of chemical testing (Fig. 8-3). Because of their high price, these methods are not yet widely used, although they are becoming more popular every year. Especially "orp" meters, which measure the potential of the sanitizer in water to oxidize.

TURBIDITY

A test that uses the cloudiness of a water sample to detect a substance is also commonly used. Called a *turbidity* (cloudiness) test, this is used, for example, to measure the amount of stabilizer in a pool. In one

FIGURE 8-3 Electronic testing meter.

type of turbidity test, a vial is provided with a dot on the bottom. A water sample is taken and combined with a reagent called melamine, which clouds the sample. When viewing the sample from above, you compare the appearance of the dot with a chart of known values to determine the results. In a similar test, a vial is provided with a plastic ladle. The ladle has a dot on it. You raise the ladle up through the sample until you can see the dot, then note the value on the scale that is printed on the vial to obtain the test result.

Now let's examine each parameter of water quality and identify which test is used for each.

Chlorine

Several reagents are used for testing not only the presence of chlorine, but also the free, available chlorine within the total.

OTO

The first bottle in most reagent test kits is OTO, or more exactly, orthotolidine. When combined with a sample that contains chlorine, OTO (which contains muriatic acid) turns color, from light pink to yellow to deep red, depending on the strength of the chlorine. You compare the color of the sample with the color chart to determine the exact parts per million of chlorine in the water.

The only drawback to this simple procedure is that OTO only measures the total chlorine content. It does not distinguish between free, available chlorine and that which is present but combined to form chloramines. It is, therefore, a good indicator test, but not a complete result. If you are quick, the reading in the first 15 seconds is available chlorine. Final color change for total chlorine actually takes up to 3 minutes, although most of the color develops within 30 seconds.

The accuracy of OTO testing also makes this test more valuable as an indicator than as a precise, final answer. The pH, alkalinity, turbidity, temperature [OTO works best on samples at 34°F (1°C)], and overall amount of chlorine in the sample can affect the accuracy of the result. For example, chlorine content in excess of 6 ppm cannot be measured and appears as a deep red color. Also, OTO measures total oxidizer in the sample, so if any other oxidizing agent is present the reading will be false.

Having mentioned these limitations of OTO testing, it is the most widely used test in the pool and spa industry for chlorine monitoring because of its simplicity and low cost. As a regular, weekly test it is quite adequate to indicate the presence and approximate residual of total chlorine in the water. When unexplained water problems appear and as a periodic evaluation, it is wise to supplement regular OTO testing with DPD or other more precise methods. Paper test strips are rapidly replacing the use of liquid reagents as the cost comes down. Both methods are fine if you follow product directons.

OTO can also measure bromine in the same way, using a slightly different color chart and with similar limitations. (Multiply the chlorine reading by 2.25 to get the bromine ppm.) Bromine results will turn a darker yellow.

DPD

A second colorimetric test, which determines the actual free chlorine in a sample, is conducted with a reagent called DPD (diethyl phenylene diamene). By subtracting the results of the DPD test from the OTO test, you will know the amount of combined chlorine—chloramines—in the sample.

DPD produces colors between pink and red for both chlorine and bromine. DPD reagents are available in liquid or tablet forms, with several manufacturers providing additional reagents for determining results of total, combined chlorine and then singling out free, available chlorine (or bromine). Each manufacturer provides detailed instructions for their particular product.

DPD has another advantage over OTO. DPD reagents are produced with pH buffers that make the test less sensitive to false readings at varying pH levels. Temperature and the presence of heavy metals or other oxidizers can also affect the accuracy of DPD tests. As with all sampling, however, it is wise to test samples between 60° and 85°F (15° and 29°C). Also, if the sample has a particularly high hardness value, the DPD reagent might precipitate mineral out of solution. If this occurs, it is easily corrected by placing the reagent in the sample vial first, then adding the water. More sampling parameters are discussed later.

One final note about DPD reagents. They contain the chemical aniline, which is extremely toxic and can be absorbed through the skin. Although you won't use much reagent with each test, if you use DPD on many samples daily, repeated contact can be harmful. Commonsense handling will solve this potential problem—avoiding direct skin contact and washing any spills from your skin immediately. Also, avoid touching your eyes if DPD is spilled on your hands.

pH

pH is also tested with reagents and a color comparison chart. The reagent for this test is phenol (phenosulfonephthalein) red. Before adding phenol red to a sample you must neutralize any chlorine in the sample to avoid a false reading. Various forms of chlorine exist at various pH levels and their presence in the sample can give a false reading. When adding phenol red to a sample with extremely high levels of chlorine or bromine, a chemical compound is created (chlorophenol or bromophenol) that turns purple. This color tells you that the sample is unreadable and no inference should be taken from such a test result. Before adding phenol red to a sample, you add a predetermined amount of the neutralizer sodium thiosulfate.

Phenol red is accurate in the pH range of 6.9 to 8.2 and is the most accurate around 7.4, where your water sample should be anyway. Therefore, at the extremes, use the phenol red test as an indicator only, trusting the test results only when they indicate your water is returning to the middle range.

All test reagents should be fresh to ensure accurate results, and phenol red is the one that will quickly go bad with age. As a water technician, you will be using test chemicals in sufficient quantity that they will always be fresh, but you need to remember this when a customer tells you his readings are different from yours and the test kit he is using has been sitting in the garage for the past several years.

Perhaps the best use of electronic testing devices is in determining pH. They are accurate, compensate for extreme temperatures (useful when testing spa water), and are not subject to problems from high chlorine or bromine levels. Some technicians say electronic testers are influenced by high sanitizer levels, but my experience is to the contrary. If you choose to invest in an electronic pH meter, try it on a sample of water known to have a high chlorine content. Then try it again after neutralizing the chlorine. I think you will find the results identical and your confidence in your pH meter complete. Follow the pH test with a base-demand or acid-demand test.

Total Alkalinity

The most common chemical test for total alkalinity of a water sample is the titration method described earlier. There are no significant limitations in total alkalinity titration tests such as those described for other reagent tests.

Test strips are available for total alkalinity testing. With both pH and total alkalinity testing, there is one test to determine the relative value (what is the pH and what is the total alkalinity value?), followed by another test to determine the acid or base demand. In short, a second titration test determines the amount of acid or alkaline (base) material required to bring the water back to desired levels.

TRICKS OF THE TRADE: WATER TESTING

Rating: Easy

- **Take your time. Hurried testing leads to inaccurate results, which leads to improper application of chemicals. Follow the directions in the kit. Some might seem redundant or unnecessary, but believe me, they are stated for a reason, so follow them. Also, allow time for hot samples to cool for more accurate results.**
- **Never flash test. Old-time pool and spa technicians often carry a bottle of OTO and one of phenol red in a leather case strapped to their belt. They arrive at a pool and dash a couple of drops of each into the pool, saying they can analyze enough of the chemistry from that. If you have followed the preceding discussion of the many factors influencing the accuracy of various test methods, you will know that this is simply false. On top of the many reasons for more careful analysis techniques, there is a very different color appearance when viewing a dispersing reagent in the pool as compared with a captured result in a vial held at eye level.**
- **Conduct all tests before adding any chemicals to modify the water chemistry. If you test for chlorine residual and add significant amounts of chlorine in any form, then test the pH, your pH reading will reflect the value of the chemicals just added, not the water. After adding chemicals to the water, allow adequate time for distribution, then test to see if your actions were adequate. Allow at least 15 minutes for liquid chlorine to circulate and at least 12 hours for the pH to adjust before testing again.**
- **Because test kit reagents can be replaced, you will probably use your kit for many years. The color chart might, however, fade over time. Compare your chart against a new one from time to time, or simply buy a new test kit annually (they're not that expensive).**

Hardness

Hardness is tested using a colorimetric and titration test combined or test strips as described previously. With combined testing, a buffer solution is added to the sample to increase the pH to about 10.0 to facilitate the accuracy of the subsequently added dye solution. The dye reacts with calcium and magnesium in the sample to produce a red color. A reagent called EDTA (ethylenediamine tetra-acetic acid) is then added a drop at a time until the solution turns blue. The number of drops is compared to a chart to evaluate the hardness of the sample.

Older kits using this method express the result in grains per gallon (gpg) rather than the more common parts per million (ppm). To convert grains per gallon to parts per million, multiply the gpg by 17.1. For example, if your older kit gives you a reading of 10 gpg, that is equal to 171 ppm.

Total Dissolved Solids (TDS)

There is no simple, inexpensive way to test for total dissolved solids in a water sample, so many technicians take a sample to a lab or local pool retailer that is equipped with electronic equipment to analyze for TDS. With the recent advent of less costly electronics, however, TDS meters have dropped below the $100 mark and are an excellent investment for the professional water technician. TDS is evaluated by the conductivity of the water when two test leads (electrodes) are placed in a sample and a small current is passed between them.

Heavy Metals

As noted previously, there are times when you might suspect that there is iron or copper in your pool or spa, particularly when stains appear on the plaster or when fixtures corrode.

I have rarely tested for heavy metals, however, because if you keep the other chemistry components in line it is not likely that you will have problems with metals. The metals enter the pool or spa from improper chemistry that creates etching conditions, corroding equipment and fixtures. If you have this problem, the causes are not difficult to determine and correct, so knowing the volume of metals in the water is not meaningful. If you are interested in pursuing testing for metals, it is conducted by standard colorimetric testing.

Cyanuric Acid

Cyanuric acid is evaluated with the turbidity test described earlier. There is an electronic device available, but it is expensive. Also, you can use test strips that have recently been added to the water testing marketplace.

Test Procedures

Regardless of the type of testing you are conducting, some basic precautions and procedures are necessary.

Use only fresh reagents or test strips. Keep them out of prolonged direct sunlight and store them at moderate temperatures. Look at reagents before use. If they have changed color, appear cloudy, or have precipitate on the bottom of the bottle, discard them.

Thoroughly rinse the sample vials and any other testing equipment using the water you are about to test. Never clean equipment with detergents because chemical residue from these products can deliver false readings.

Consider where and when you are sampling to ensure truly representative results. Samples should be collected several hours after any chemicals have been added and thoroughly circulated through the body of water to ensure that their effect has been completely registered. The water should be circulated for at least 15 minutes before sampling. Sample away from return outlets and away from dead zones in the pool to make sure you are collecting a representative sample. There are times, however, when you might want to know more about various zones in the pool, to evaluate circulation for example, in which case you might specifically test these areas. Finally, collect samples from at least 18 inches below the surface to avoid inaccuracies caused by evaporation, direct sunlight contact, dirt, etc.

Make color comparisons in bright white light against simple white backgrounds to ensure accurate colorimetric comparisons. Most test kits come with a white card to place against the back of the test vial.

Observe and record the results immediately; never take samples for later evaluation. The chemistry of the water can change when a sample sits around.

Handle samples and reagents carefully. As noted earlier, some reagents can impact health, so avoid direct contact and never pour them into the pool or spa, even after the test is complete. Avoid contact with the sample because acids and oils from your hands can contaminate a sample, leading to false readings.

This brings me to an important related point. As a water professional, your most valuable tool is your test kit. All of your water quality actions as a technician will be determined by this kit and, as a result, the accuracy of your evaluations is the basis for your recommendations. Inaccurate results can be more than embarrassing or inconvenient. Dealing improperly with water chemistry problems can result in very expensive repairs, health considerations, and even legal liability. Your reputation and livelihood depend on your test kit. The point is that because so much is dependent on that test kit, buy the best equipment you can afford, make necessary replacements regularly, and don't skimp.

Langlier Index

In the 1930s, a researcher developed a water quality index that now bears his name and which reflects the balance of water chemistry components. Although not literally useful in the pool and spa industry (the index was designed to evaluate closed water systems, not open pools and spas), many certification courses require you to know the Langlier index. The basic principle is to assign a value to each component of water quality, resulting in a positive or negative numerical result. A positive value indicates water that is saturated by alkaline material and likely to precipitate scale; a negative value indicates corrosive water that is likely to etch.

As noted earlier, water is the universal solvent. Over time it will dissolve anything. At some point, however, a body of water becomes saturated with the materials being dissolved. After that point, these materials precipitate out of the solution, back to their natural state. For example, you can continue to add sugar to your coffee, dissolving each spoonful. When the coffee can no longer hold more sugar, any additional sugar will not dissolve. Further, if you change other parameters, such as the temperature or pH, some of the previously dissolved sugar will come out of solution and precipitate back into granular form, depositing on the bottom of the cup.

When the coffee (or water) can no longer dissolve materials, it is said to have reached the *saturation point.* In the pool, the water takes the place of coffee and calcium takes the place of sugar. Water might be undersaturated (not yet full of dissolved calcium), saturated (dissolved as much as it can, called *equilibrium*), or oversaturated (returning calcium to the mineral state in the form of scale precipitating back into the pool).

Langlier's index is based on the water's state of saturation. He noted that extremely undersaturated water will be hungry, trying to dissolve minerals that it contacts (corrosive); balanced, neither dissolving mineral nor depositing any back into the pool; or extremely oversaturated, precipitating scale on pool surfaces and in the equipment.

Langlier's formula to determine where the water stands on this scale is simple (Fig. 8-4). After sampling and testing each water quality component, you assign a numerical value called a *factor*. The saturation index shows the formula for application of these factors.

I have seen several variations on the original Langlier index. Some forms include TDS in the equation and combine hardness and total alkalinity into one factor. Whichever you use, the result is still a simple equation such as the one in Fig. 8-4, with the result being positive, neutral, or negative.

TEMPERATURE			CALCIUM HARDNESS		ALKALINITY	
°F	°C	Factor	ppm	Factor	ppm	Factor
40	4.4	2.7	10	4.0	10	3.7
50	10	2.6	30	3.5	30	3.2
60	15	2.5	50	3.3	50	3.0
65	18	2.4	75	3.1	75	2.8
70	21	2.3	100	3.0	100	2.7
80	27	2.2	150	2.8	150	2.5
90	32	2.1	200	2.7	200	2.4
100	38	2.0	300	2.5	300	2.2
110	43	1.9	500	2.3	500	2.0
120	49	1.8	700	2.2	700	1.9
130	54	1.7	1,000	2.0	1,000	1.7

Formula:

pH - temperature factor - calcium hardness factor - alkalinity factor = saturation index value

Interpretation:

Positive value (+) = scaling condition in water
Negative value (-) = etching condition in water
Zero value = perfectly balanced water (theoretical)

FIGURE 8-4 Langlier index. *Los Angeles County Health Department.*

Obviously, you can correct an extremely high positive or negative result by changing any of the factors. Herein lies the impracticality of the index as a water quality tool. For example, raising the temperature of the water by 90°F (50°C) results in a change factor of only 1.0, and of course, you are unlikely to balance your water components by raising the temperature. Similarly, significant lowering of hardness (or in the alternate index changing the TDS) requires replacing the pool water. Therefore, the corrective measures available to you are pH control (which you want about 7.4) and alkalinity (which should be 80 to 150 ppm). Again, to bring the index into line you would not take either pH or alkalinity to great extremes.

In short, the Langlier (or similar) index is a pleasant parlor game, to add up the values of your water testing and see how balanced it is. If you have maintained proper chemistry and/or taken appropriate corrective measures, the result should always be neutral, so this is really not a useful analysis tool in the real world.

Water Treatment

We have reviewed the various testing methods to determine the chemical needs of a body of water, but what about administering the actual chemicals that will deliver the desired results? There are as many products for sanitizing and balancing the chemistry of water as there are for testing it. It pays to follow the directions on product labels, but some general guidelines apply.

Liquids

RATING: EASY

Liquid chlorine and muriatic acid are generally sold in 1-gallon (4-liter) plastic bottles, four to a case. When adding liquid chlorine or acid to a pool, pour as close as possible to the water's surface to avoid splashing it on your shoes or the deck, discoloring both. Air and sun contact will diminish the concentration of chlorine, so keeping it close to the surface of the water also minimizes those impacts. Add the chlorine slowly near a return line while the circulation is running to maximize even distribution. Pour some near steps or in any known dead spots in the pool, or directly over any appearance of algae. Never pour liquid chlorine into the skimmer. Avoid skin or eye contact with chemicals.

Amount of product required to obtain a change in 1000 gallons (3785 L) of water:

Item	Raise or lower 10 ppm	Product	Amount required
Calcium hardness	Raise	Calcium chloride	2 oz. (57 g) dry weight
Total alkalinity	Raise	Bicarb of soda	2.5 oz. (71 g) dry weight
Total alkalinity	Lower	Sodium bisulfate (dry acid)	2.5 oz. (71 g) dry weight
Total alkalinity	Lower	Muriatic acid	1/4 cup (62 mL) liquid
Conditioner	Raise	Cyanuric acid	1.5 oz. (43 g) dry weight

Example: 20,000 gallon pool needs 30 ppm increase in total alkalinity using bicarb.
20 (thousands of gallons) × 3 (30 ppm divided by 10 ppm) × 2.5 oz. = 150 oz. required (9 pounds, 6 oz. or 4.3 kg)

Amount of chlorine needed to raise residual in 1000 gallons (3785 L) of water by 1 ppm (multiply results by 30 for superchlorination procedures):

Available chlorine of product used	Product required to raise 1 ppm
12%	1/8 cup (31 mL) liquid
50%	1/4 oz. (7 g) dry weight
80%	1/6 oz. (4.5 g) dry weight

Example: 20,000 gallon pool needs 2 ppm residual increase using liquid chlorine.
20 (thousands of gallons) × 2 (2 ppm) × 1/8 cup (per ppm) = 5 cups (1250 mL)

Note:
— 16 oz. dry weight = 1 pound (454 g) dry weight
— 64 oz. liquid = 1 gallon (3.8 L)
— 8 oz. liquid = 1 cup (250 mL) liquid
— Use accurate measurements, such as a measuring cup or the cap of a product container after measuring the volume of the cap.

FIGURE 8-5 **Chlorination, conditioner, alkalinity, and calcium hardness chart.**

The amounts of various products needed to raise or lower chlorine, alkalinity, hardness, and stabilizer levels are detailed in Fig. 8-5. Because of variations in products, however, it is always important to read the product label to ensure the desired results.

Granulars

RATING: EASY

Sanitizers, acids, and alkaline materials are available in granular form. As with liquids, study of the product label will give you exact application methods and amounts.

Granular products tend to settle to the bottom of the pool or spa before dissolving, so if the vessel is vinyl or dark plaster, it is wise to brush immediately after application. Another method is to create a solution of the product by dissolving it in a bucket of water before pouring it into the pool.

None of these products should be poured directly into the skimmer and all should be applied with the circulation running. Since granular products tend to be extremely concentrated, they can cause skin irritation or breathing problems if direct contact is made. Handle them with care.

Granular products tend to have a longer shelf life than liquids, but time or prolonged exposure to sunlight will diminish their efficacy as well. Store and treat granular products as you would a liquid.

Tabs and Floaters

RATING: EASY

Chlorine tablets are sold to place in floating devices (sometimes shaped like ducks) which allow a slow dissolving process. Tablets are valuable when you can't service the pool or spa for some time and need a constant source of sanitizer. They are expensive and unregulated, which is to say you can't control the rate of dissolution. Some floaters have a valve that allows more or less water to flow in them, theoretically controlling the amount of chemical entering the water, but the results are trial and error. Like granulars, tablets left on the bottom of a pool or spa will bleach out any color, so use a floater unless the surface is already white.

TRICKS OF THE TRADE: FLOATERS

- **Use a floater for your chlorine tablets; never leave them in the skimmer. The low pH of tabs means you are assaulting your circulation system and any related metal plumbing with acid!**
- **To keep the floater from drifting to the skimmer opening, tie it to a return line nozzle away from the skimmer. This also makes collection easier when you need to replace the tablet.**
- **Don't use floaters if the pool is used by curious kids. You don't want them playing with a chemical delivery device!**

Mechanical Delivery Devices

Chlorine is the only pool and spa chemical generally added by a mechanical device, although large commercial pools might use them for other chemicals as well. Mechanical delivery systems generally fall into three categories: erosion systems, pumps, and salt chlorine generators.

EROSION SYSTEMS

RATING: EASY

As the name implies, these systems use the water passing over the dry chemical to erode or dissolve it into solution in the water. Like the floater, the erosion system is usually controlled by restricting the volume of water allowed to pass through the system and, therefore, the amount of erosion that can take place of the tablet or granular chlorine (or bromine) inside. Figure 8-6 shows a typical erosion chlorinator system.

Erosion systems are typically made of PVC plastic and are located in the equipment area, plumbed directly into the circulation lines. These chlorinators are easily installed following the directions supplied, and require only basic plumbing techniques. They must always be located after any other equipment in the system. If you were to place it between the filter and heater, the concentrated chemical would corrode the internal parts of the heater.

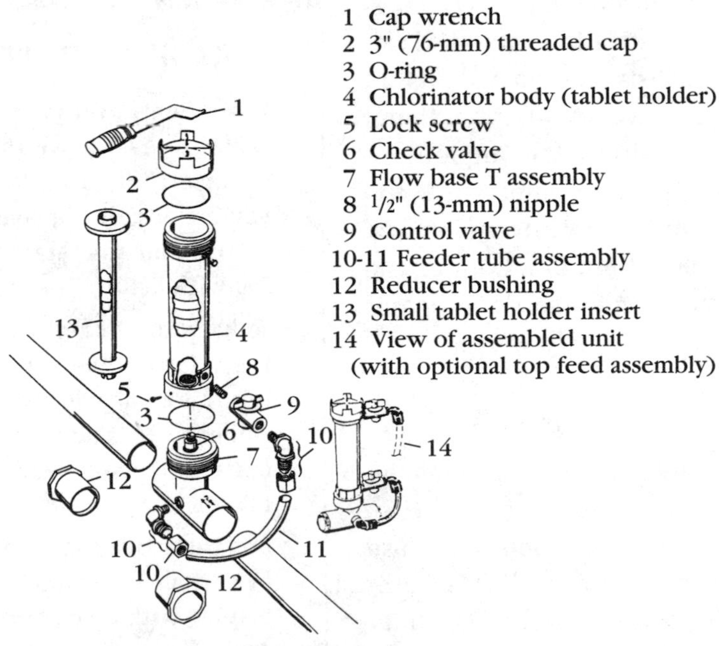

FIGURE 8-6 Typical inline erosion type chlorinator. *Rainbow Lifegard Products, a division of PacFab, Inc.*

PUMP SYSTEMS

RATING: ADVANCED

There are basically two types of pump chlorinators: peristaltic and standard pumps. Peristaltic pumps are generally mounted in the equipment area and pump a specific volume of liquid sanitizer into the plumbing lines during normal circulation. Figure 8-7 shows the working of a typical peristaltic pump. The rollers press against the flexible tube to create suction from a nearby bottle of liquid sanitizer (Fig. 8-7B). As with erosion systems, these pumps should be plumbed after any other equipment in the system (Fig. 8-7A).

Standard piston pumps are also used for chemical delivery, but they tend to corrode and require frequent repair or replacement. In the peristaltic pump, the liquid chlorine only touches the plastic tube (Fig. 8-7C), while the standard pump draws chemical into it and pumps it out into the circulation plumbing. The corrosive action of the concentrated chemical will, over time, corrode the pump, especially if even a tiny leak develops (and these systems are prone to leaks). A better pump uses a rubber diaphragm instead of a piston to create pressure and suction.

Salt Chlorine Generators

Salt chlorine generators are becoming more common each year for both a good reason and a bad one. The bad reason is this: some people mistakenly believe that adding salt to your pool eliminates chlorine and the salt acts as a sanitizer. In truth, salt is added to your pool and the generator converts it into chlorine, so the pool is still very much a chlorinated body of water. Recall that chlorine is made by passing electricity through salt water.

The better reason to add a salt chlorine generator to your pool equipment is because it makes the chlorine for you and all you do is periodically add salt. Users of these systems also report that the increased salinity of the water gives it a silky texture and reduces objectionable chlorine odors. As you know from this chapter, however, chlorine odors are not caused by the chlorine alone, so the key to eliminating odors is maintaining a good water balance and a clean pool!

So just how does a salt chlorine generator work? First, ordinary table salt is added to the pool water at a concentration of about 3000

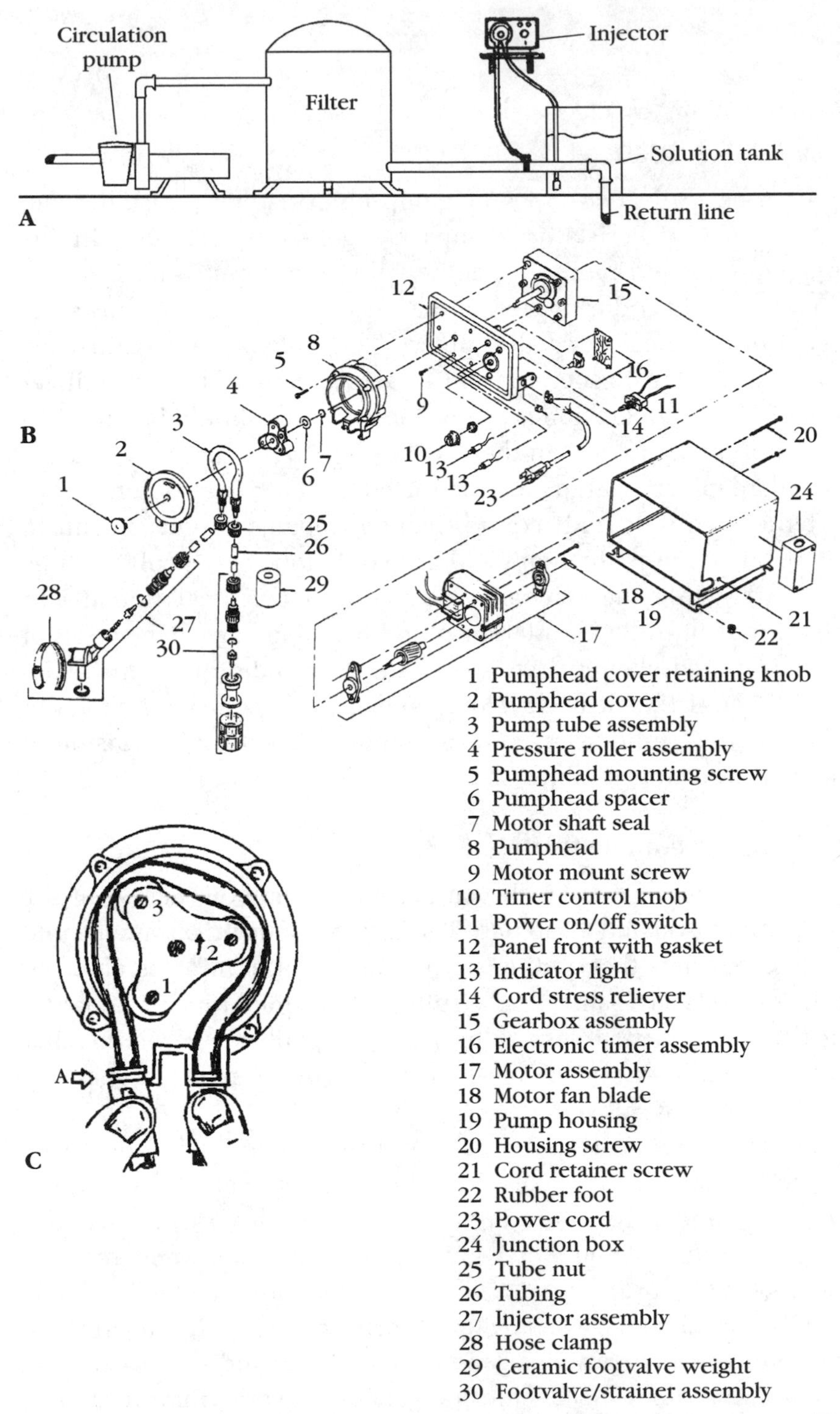

FIGURE 8-7 Typical peristaltic chlorinator. *Blue White Industries, Westminster, Calif.*

parts per million (by comparison, ocean water is around 38,000 parts per million). Your generator will come with a salt concentration test kit or you can buy one at the pool supply store along with your other water-testing supplies.

Next, a computerized control unit converts household AC current (120 or 240 volts) to low-voltage DC current and sends it into an electrolytic cell that is mounted in a plumbing fitting (added to pool plumbing after the heater). This current is passed from one electrode to the other (anode to cathode) and chlorine is generated as the salted pool water passes between the electrodes. Chlorine production can be varied by either adjusting the production level on the controls or by varying the number of hours the generator operates each day.

Salt is very corrosive and has a tendency to build up on the surfaces it touches, so salt chlorine generators typically self-clean after every few hours of use. Most units have sophisticated sensors to gauge the flow rate through the system and shut the unit down if other problems are detected.

TRICKS OF THE TRADE: SALT CHLORINE GENERATORS

Rating: Advanced

- **To avoid corrosion in plumbing and equipment and to prevent clogs, always add salt to the pool by broadcasting it over the pool (as you would do with dry chlorine) when the circulation is running. Never add salt through the skimmer.**
- **Use granulated, table-quality salt.**
- **Brush the pool and circulate the water for 24 hours to dissolve the salt completely BEFORE starting the generator.**
- **When you start your generator, set the production level to 50% and check the chlorine level the next day. As with other forms of chlorine, the residual should have reached between 1 and 3 ppm. Adjust the production level of the generator as needed for more or less chlorine.**
- **Every two months, the electrolytic cell should be removed and scale cleaned off the electrodes. Clean the unit carefully according to manufacturer's guidelines to prevent damaging the delicate components.**
- **Turn the generator off when adding acid or soda ash to avoid reaction of these substances with the electrodes. After the acid or soda ash has been thoroughly distributed throughout the pool water, resume normal generator operations.**

TRICKS OF THE TRADE: CHEMICAL SAFETY

Throughout this chapter, I have recommended that you take care when handling or dispensing chemicals or even small amounts of reagents. Here are a few additional safety tips:

1. **Study the material safety data sheet (MSDS) found with each hazardous chemical sold. Familiarize yourself with the potential hazards and the remedies for exposure.**
2. **Sodium bicarb can be broadcast over the water, but soda ash should be dissolved in a bucket of water first. Dry acid (sodium sulfate) should also be dissolved in water. Cyanuric acid should be dissolved in a bucket of hot water, with ⅓ pound (151 grams) of soda ash for every pound (454 grams) of cyanuric.**
3. **When dealing with new chemical treatments, read all labels carefully. Apply in small amounts first, check for any unexpected results, and then continue.**
4. **Always take your time. Caution is the way, especially when mixing chemicals (see Fig. 8-8).**

	Other Chlorine Products	Bromine	Biguanicide
Chlorine	OK	OK	NO
Bromine	OK	OK	NO
Biguanicide	NO	NO	OK
Muriatic acid	OK	OK	OK
Soda ash	OK	OK	OK
Clarifiers	OK	OK	OK
Cyanuric acid	OK	NO	NO
Phosphate-based vinyl cleaners	OK	OK	NO
Monopersulfate shock treatments	OK	OK	NO
Algaecides	OK	OK	CAUTION
Alkalinity/Calcium hardness adjusters	OK	OK	OK
Ozone	OK	OK	OK
Ultraviolet	OK	OK	OK
Chelating agents	OK	OK	CAUTION

FIGURE 8-8 Pool chemical mixing guidelines.

FAQs: WATER CHEMISTRY

How Often Must I Add Chemicals to the Pool?

- In summer, when ultraviolet rays from the sun and heavy bather loads are depleting your sanitizer, you may need to add more (and keep pH balanced) every other day. In cold weather, chemical treatments may only be needed every other week.

Are There Alternatives to Chlorine?

- Yes. Bromine, biguanicides, and ozone are the most popular alternatives to chlorine. Some still need low concentrations (residuals) of chlorine in the water to ensure algae doesn't get started.

What Happens if I Spill Chlorine or Acid in My Pool or Yard?

- Chlorine and acid together form potentially lethal chlorine gas. Be very careful when handling all pool chemicals, but especially those two. Dilute any spilled chemicals with plenty of fresh water as soon as possible to prevent killing grass or severely staining patios and decks.

Why Does the Algae Keep Coming Back?

- Algae won't grow in a pool that is properly cleaned and sanitized. If you are doing both of those tasks diligently, your water may have become too hard or the pH may be too high. Keep an eye on all chemistry parameters for best results, and brush the bottom and sides of your pool often. Finally, recurring algae is often a sign of insufficient daily circulation. Increase the time of your filter run and look for improvements.

CHAPTER

9

Cleaning and Servicing

Only experience and your personal style will determine how you approach a routine cleaning, a major cleanup, or other special service work. What follows is a description of what jobs you will face and what tools you will need to do the job.

Tools

Figures 9-1 through 9-5 show the basic service equipment carried by a professional water technician.

Telepoles

The heart of the cleaning system is the telepole (telescoping pole). The one you will use most on pools is 8 feet long, telescoping to 16 feet by pulling the inner pole out of the outer one (Fig. 9-1). The end of the pole has a handgrip or a rounded tip to prevent your hand from slipping off the pole. The tip might also include a magnet for picking up hairpins or nails from the pool bottom. There are several sizes, from a 4-foot (1.2-meter) pole that telescopes to 8 feet (2.4 meters), all the way up to a 12-foot (3.6-meter) pole that telescopes to 24 feet (7 meters). To lock the two poles together, there is a cam lock or compression nut ring.

The cam lock (Fig. 9-2) is a simple device. The hub is mounted to the end of the inside pole by drilling a hole in the center of the hub, passing a screw through the hole, and screwing this assembly into the tip of the pole. The hub is serrated, like a cog wheel, to grip the interior surfaces of the outer pole. The screw is set just off center so that it can create a circumference larger than the interior diameter of the pole. When you twist the two poles of the telepole in opposite directions, the cam swings to one side or the other in this large circumference, locking the two halves together. By twisting in the opposite direction, you unlock the poles.

When you purchase your first telepole, take it apart and observe how this cam system works. Sooner or later, scale, corrosion, or wear and tear will clog or jam the cam. Rather than buy an entirely new telepole, you can take it apart, clean it up, replace the cam if necessary, and get on with the job.

The other locking device for telepoles is a compression nut ring, like the plumbing compression rings described previously. By twisting the ring at the joint of the two poles, pressure is applied to the inner pole, locking the two together.

TOOLS OF THE TRADE: CLEANING

Copy and laminate this handy checklist of the basic tools you'll need to service a pool or spa. By referring to it before leaving your truck, you'll take just the right tools each time and not make numerous trips back out for things you forgot!

- Telepole
- Vacuum head
- Vacuum hose and barbed hose connector
- Leaf rake
- Leaf vacuum
- Tile soap
- Tile brush
- Wall brush (nylon)
- Wall brush (stainless steel)
- Test kit
- Submersible pump and extension cord
- Drain flush bag
- Garden hose
- Multitool (Leatherman type)
- Thermometer
- Tennis ball
- Skimmer suction diverter
- Pumice stone
- Backwash hose with 2-inch (5-centimeter) hose clamp
- Spa vacuum
- Notepad and pencil
- Waterproof marker
- Foam knee pads
- Sanitizer and pH adjusters

FIGURE 9-1 Telepole.

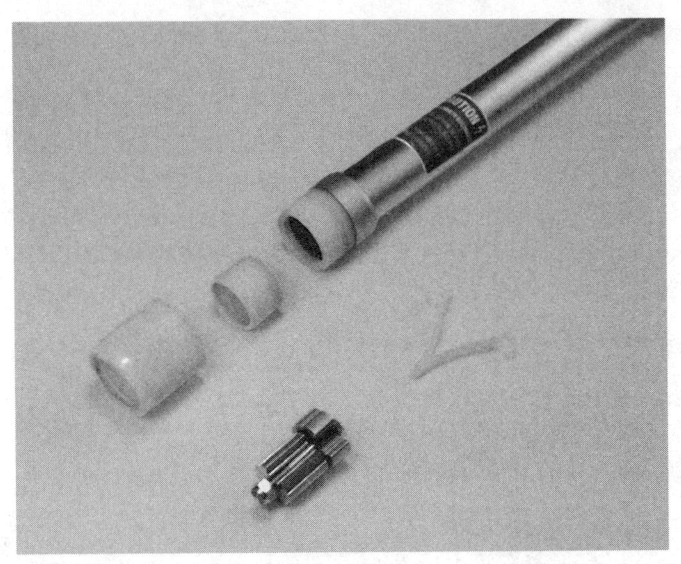

FIGURE 9-2 Twist-grip compression nut lock, cam lock, and tool clip.

I have used both types of locking devices and sooner or later they both fail because of normal wear and tear. My telepole is a hybrid. I use a good-quality compression ring pole, but I have disassembled it and added a cam to the inner pole. This belt and suspenders technique gives me a pole that lasts forever (seemingly) and holds up no matter what pressures I apply, particularly important when doing a cleanup of an extremely dirty, debris- and leaf-filled pool.

At the end of the outer pole you will notice two small holes drilled through each side, about 2 inches (50 millimeters) from the end and again about 6 inches (15 centimeters) higher. The various tools you will use are designed to fit the diameter of the pole. You attach them to the pole by sliding the end of the tool into the end of the pole. Small clips inside the tool have nipples that snap into place in one of these sets of holes, locking the tool in place. Other tools are designed to slip over the circumference of the pole, but they also use a clip device to secure the tool to the holes at the end of the telepole.

Telepoles are made of aluminum or fiberglass. The latter is more expensive but well worth the money. They are not only impervious to corrosion and virtually unbreakable, but they won't kill you if you inadvertently touch them to an exposed overhead wire (not that rare) or insert them in a pool with an electrical short (rare). I carry the 8-by-

16-foot (2.4-by-5-meter) and 4-by-8-foot (1.2-by-2.4-meter) versions, because the long one is necessary for pools and the small one is more convenient for spas.

Leaf Rake

Figure 9-3 shows a professional, deep-net leaf rake. The net itself is made from stainless steel mesh and the frame is aluminum with a generous 16-inch-wide (40-centimeter) opening. There are numerous leaf rakes (deep net) and skimmer nets (shallow net) you can buy, but only the one pictured will last. The cheap ones are made from plastic net material and frames. Although the original price is about twice that of the cheap ones, metal ones last a long time and resist tearing when you are scooping out huge volumes of wet leaves after a windy autumn day. They also stand up to rubbing them along rough plaster surfaces, thanks to a rubber-plastic gasket that fits around the edge, unlike the plastic rakes that break or wear down when you apply such pressures.

The leaf rake shank fits into the telepole and clips in place as described previously. Be careful not to spill acid or other caustic chemicals on your leaf rake; both the metal and plastic mesh will deteriorate and holes will develop. Some leaf rakes are designed so you can disassemble them and replace the netting, which is fine if you have the time and patience to do it.

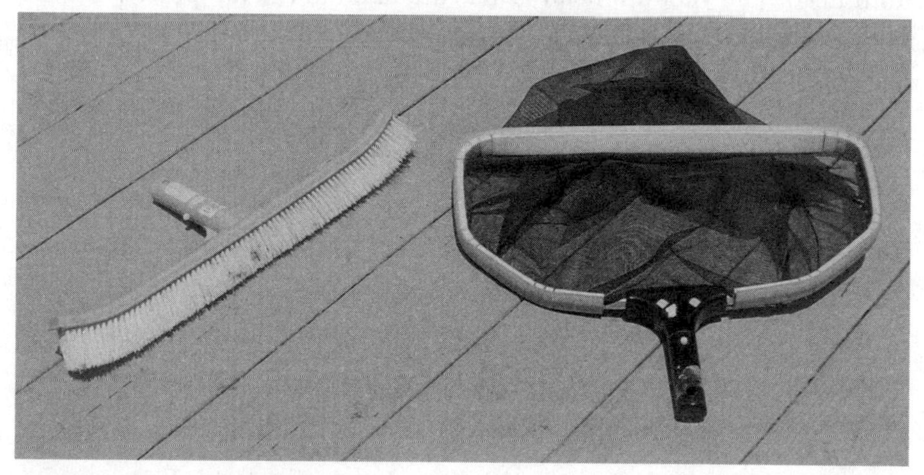

FIGURE 9-3 Brush and leaf rake.

Wall Brush

The wall brush (Fig. 9-3) is designed to brush pool and spa interior surfaces. Made of an aluminum frame with a shank that fits the telepole, the nylon bristles are built on the brush either straight across or curved slightly at each end. The curved unit is useful for getting into pool corners and tight step areas. Wall brushes come in various sizes, the most common for pool use being 18 inches wide. I carry an 18-inch (45-centimeter) brush and a 6-inch (15-centimeter) brush that I use for spas and tight places in the pool. I also carry a 6-inch (15-centimeter) wall brush with stainless steel bristles for heavy stains or algae problems. Don't ever use a wire brush that is not stainless steel in a pool or spa. Steel bristles can snap off during brushing and leave stains on the plaster when they rust. Also, if they are a bit rusty already, when you brush the plaster you will transfer the rust to the plaster, causing a stain.

Vacuum Head and Hose

There are two ways to vacuum the bottom of a pool or spa. One actually sucks dirt from the water and sends it to the filter. The other uses water pressure from a garden hose to force debris into a bag that you then remove and clean (leaf vacuum).

The vacuum head and hose (Fig. 9-4) are designed to operate with the pool or spa circulation equipment. The hose is attached at one end to the bottom of the skimmer opening and at the other end to the vacuum head. The vacuum head is also attached to the telepole. With the pump running, you glide the vacuum head over the underwater surfaces, vacuuming up the dirt directly to the filter.

Vacuum heads are made of flexible plastic, with plastic wheels that keep the head just above the pool surface. The flexibility of the head allows it to contour to the curvature of pool corners and bottoms. Adjustable-height wheels allow you to set the vacuum head to the best clearance for each pool's conditions. The closer to the surface, the better the removal of dirt. But if the suction is too great, it might suck the vacuum head right onto the surface, rendering it immobile. In this case, adjust the head height upward.

Wheels for vacuum heads are made of plastic or high-tech composite resins. Their bearing systems can be as simple as a hole in the wheel through which the axle is inserted or wheels with ball bearings

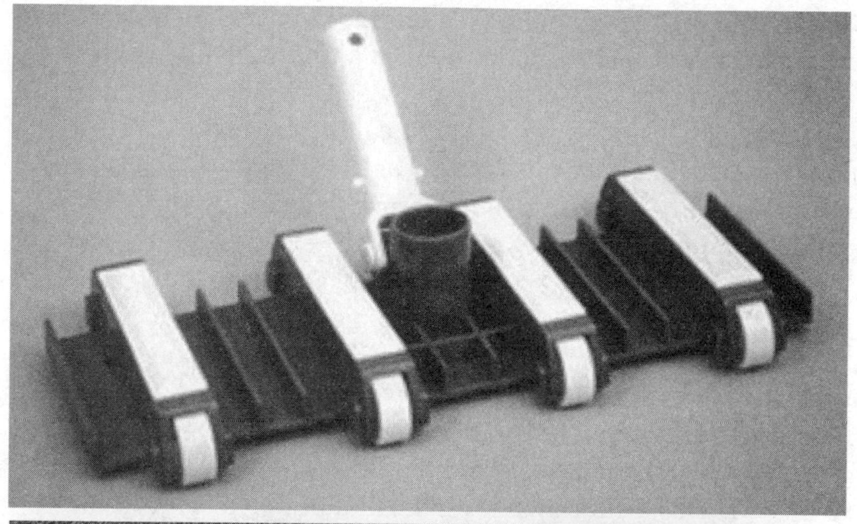

FIGURE 9-4 Vacuum head and hose.

to distribute the load and help the vacuum glide smoothly. I highly recommend you invest the few extra dollars for the resin, ball-bearing wheels for both your vacuum head and leaf vacuum. When you have tried each for a week, you'll agree.

Some commercial vacuum heads are made several feet wide and are built of stainless steel. Another type is a plastic helmet style, with a ridge of bristles instead of wheels. This vacuum head is used for vinyl pools, fiberglass spas, and when breaking in new plaster. In each of these cases, standard wheels can tear or score the surface. The brush vacuum is not only less harsh, but it brushes dirt loose from the surface being vacuumed for easier removal.

Hoses are available in economy models (thin plastic material) through "Cadillacs" (heavy rubber-plastic material with ribs to protect against wear), and in various lengths [10 to 50 feet (3 to 30 meters)]. The hose cuff is made 1¼- or 1½-inch (38- or 40-centimeter) diameter to be used with similar vacuum head dimensions. Cuffs are female threaded at the end that attaches to the hose so you can screw replacement cuffs onto a hose. The best cuffs swivel on the end of the hose, so when you are vacuuming there is less tendency for the hose to coil and kink. Another valuable hose fitting is the connector. It is designed with female threads on both ends to allow joining of two hose lengths—a useful feature when you encounter a large or extremely deep pool.

Leaf Vacuum and Garden Hose

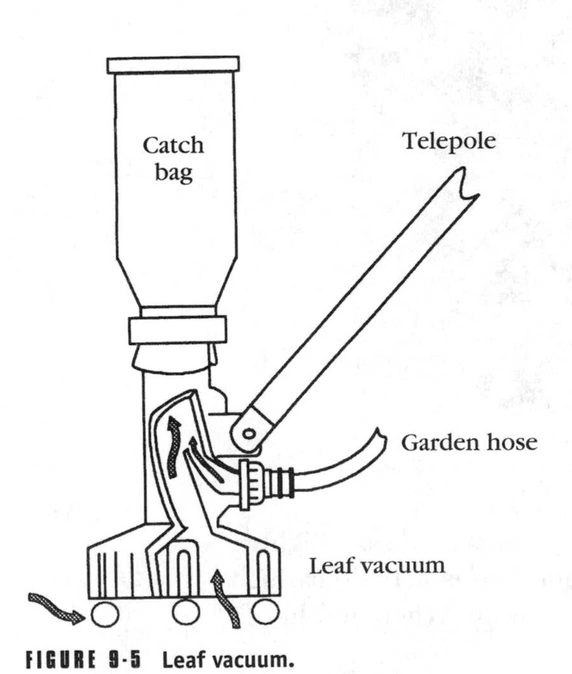

FIGURE 9-5 Leaf vacuum.

Some products become so successful in setting the standard for the industry that the brand name becomes interchangeable with the product's descriptive name. Kleenex and Jell-O are examples of facial tissue and gelatin, but we all use the brand name regardless of the actual maker of the item. So it is with Leaf Master, the original water-powered leaf vacuum (also called a leaf bagger). The leafmaster (Fig. 9-5), which is attached to the telepole and a garden hose, operates by forcing water from the hose into the unit where it is diverted into dozens of tiny jets that are directed upward toward a fabric bag on top of the unit. The upwelling water creates a vacuum at the base of the plastic helmet, sucking leaves and debris into the unit and up into the bag. Water passes through the mesh of the bag but the debris is trapped.

Fine dirt passes through the filter bag, but a fine-mesh bag is sold for these units that will capture more dirt. When the bag has a few leaves in it, they will also trap much of the sand and other fine particulate matter that would otherwise pass through.

The only other drawback to the leafmaster is if you are in a location where water pressure from the garden hose is weak. The result is weak jet action and weak suction. The other result is that as debris fills the bag, the weight of it (especially wet leaves) tips the bag over, scraping the pool floor, stirring up debris, or tangling with the hose. The latter problem is easily solved by putting a tennis ball in the bag before placing it in the pool. The tennis ball floats, keeping the bag upright.

The water pressure problem can be solved if the pool is equipped with an automatic cleaning system that uses booster pump technology. The fitting for the automatic cleaner at the pool wall is usually a threaded two-part adapter. By removing the automatic cleaner's hose and unscrewing the adapter fitting, you are left with a ¾-inch (19-millimeter) female threaded opening. I carry a short plumbing nipple, available at the supply house, that is ¾-inch (19-millimeter) NPT threads on one end and ¾-inch (19-millimeter) garden hose threads on the other. By screwing this adapter into the automatic cleaner opening, I can attach the garden hose directly to the automatic cleaner booster pump line. I then turn on the booster pump and my leafmaster is turbocharged. Not all automatic cleaner fittings are the same, but a little disassembly and inspection will tell you what fittings you need to create an adapter that will work for that particular customer's pool.

The alternative is to create an adapter that fits the garden hose and terminates in a quick-disconnect, similar to the one on the automatic cleaner hose, for attachment to the automatic cleaner wall fitting without disassembly of that fitting. Again, a little experimentation will create the fitting you need for each style of automatic cleaner, but it will be well worth the trouble if the home has weak water pressure.

Leafmasters are made in rigid plastic or aluminum and, for my money, the original still works the best. One word of caution, however. You will frequently see leafmasters by the side of the road or freeway, blown out of the back of open pickup trucks. The design is a natural flying saucer, so make sure yours is secure when underway.

Your leafmaster will serve you better if you customize it slightly. Add resin, ball-bearing wheels, a shutoff valve at the garden hose fitting, and a double bag for catching all the dirt.

You would think that most homeowners would own and keep handy a simple garden hose. Well most don't, or it is too short, too kinked, or too leaky. Bring your own. Pool supply houses sell garden hoses that are resistant to kinks and float. A hose that scrapes the pool bottom will stir up dirt and debris before you can vacuum it.

Tile Brush and Tile Soap

Tile brushes are made to snap into your telepole so you can scrub the tile without too much bending. Mounted to a simple L-shaped, two-part aluminum tube, the brush itself is about 3 by 5 inches (76 by 127 millimeters) with a fairly abrasive foam pad for effective scrubbing. I have found that these brushes are valuable for wiping algae off of ladders or other tricky spots in spas, but the elbow grease required to remove body oil, suntan oil, and scale from tiles is much more than you can get at the end of the telepole. Therefore, I also carry a barbecue grill cleaning pad. It has a convenient grip handle and an abrasive Brillo-type pad, which is much more effective at cleaning tiles. Because this also requires that you get on your hands and knees around the entire circumference of the pool or spa, carry a foam knee pad as well.

Tile soap is sold in standard preparation at the supply house, but I recommend mixing it into another container with one part muriatic acid for every five parts soap. This will help cut the stubborn stains and oils, but it will also eat into the plastic on the tile brush pads and plastic barbecue grill brush handle, so keep rinsing them in pool water after each application and scrubbing. Don't use other types of soap in place of tile formulations, because they might foam and suds up when they enter the circulation system (especially in spas).

Keep your tile soap in a squirter bottle, such as a kitchen dishwashing detergent bottle. You can control the amount of soap that flows onto the brush pad and this bottle serves another valuable function. When the wind is blowing ripples on the surface of the water, you won't be able to see the bottom to know what you are vacuuming. The squirter bottle allows you to send a stream of tile soap across the pool and, as the soap spreads out over the water, it calms the ripples. You don't need much, though you might have to repeat the process two or three times in the course of a service call to finish your work.

Test Kit and Thermometer

As discussed in the previous chapter, buy the best test kit you can afford and keep it in good working order. Your chemical testing is by far the most important aspect of your work. No one I know was ever injured or killed by a dirty pool, but many have been from improper handling and application of pool chemicals.

Because test kits and methods were discussed previously, I will not repeat the information here.

A thermometer is also an important item in your test kit. I keep a floating and a standard thermometer (with string). As noted in the chemistry chapter, certain tests are best performed at certain temperatures. You will also need a thermometer to check heater performance, spa temperatures, and to answer customers' questions about their water (especially useful at apartment building pools and spas when one resident wants the water warmer and another cooler, but both think 72°F (22°C) is what they mean).

Spa Vacuum

The spa vacuum is a miniature version of a leafmaster. It works on the same principle using a garden hose for water pressure to create suction. The dirt and debris are forced into a small sock and, like the leafmaster bag, fine dirt passes through the bag. I use a fine-mesh sock or ladies hosiery inside the bag provided when I want to pick up fine dirt.

The spa vacuum attaches to the telepole and is provided with various attachments, much like a household vacuum cleaner, for getting into crevices or brushing while you vacuum. I don't find any of these too useful and generally use the vacuum by itself. The spa vacuum is also a useful tool for sucking up small hairpins, nails, coins, or other hard to grab items from the bottom of pools.

There are two other types of spa vacuum that are not widely used, the hand-pump model that creates its own suction (cumbersome) and a battery-powered unit that has a built-in pump (weak and the batteries run down fast).

Pumice Stones

The soft pumice stone, made from volcanic ash, is abrasive enough to remove scale from tiles and other deposits or stains from plaster surfaces without scratching them excessively. Pumice stones are sold as blocks, about the size of a brick, and as small bladed stones that attach to your telepole for reaching tight spaces and underwater depths. Carry both. Since pumice stones disintegrate easily, it is wise to use them before you vacuum a pool or spa. Alternately, you can brush the residue to the main drain where it is carried to the filter. A good alternative to pumice, especially on fiberglass, which scratches easily, is a block of styrofoam or similar plastic foam.

Pool Cleaning Procedures

Every technician, and every homeowner for that matter, approaches pool cleaning differently. After many years and hundreds of thousands of service calls I have discovered that there are a few basic procedures that are efficient and save time which any technician would do well to follow. Always start by determining the surface composition. You will work very differently on plaster than on a painted surface, for example.

Deck and Cover Cleaning

RATING: EASY

Most technicians overlook the fact that if they spend 30 or more minutes cleaning a pool, it will quickly appear as if they were never there when the leaves and dirt on the adjoining deck blow in on the first breeze. A quick sweep or hosing of any debris near the pool, at least 10 feet (3 meters) back from the edge, will keep your service work looking good after you have left. Similarly, remove as much debris as possible from the pool or spa cover before removing it. If the cover is a floating type without a roller system, be sure to fold or place it on a clean surface. Otherwise, when you put it back in place it will drag leaves, grass, or dirt into the pool. If it is a mechanized cover system, any small amount of standing water on top of the cover will slide off as you roll it up. If the motor is laboring you will need to use the water removal pump, the sump-type pump provided with these systems.

QUICK START GUIDE: POOL CLEANING

Rating: Easy

1. Prep

- Set all service tools and chemicals near the pool.
  - Don't forget the garden hose for adding water and/or using your leaf vacuum.
- Remove pool cover.
- Make sure pump is on and water circulating normally.
  - Check/adjust water level.
  - Check equipment area for leaks.
  - Empty strainer pot as needed.

2. Topside cleaning

- Clean the skimmer basket.
- Skim pool water surface.

3. Interior cleaning

- Vacuum the bottom with suction vacuum or leaf vacuum.
- Clean scum from vinyl at waterline.

4. Chemistry

- Check sanitizer and pH levels.
  - Check other water parameters if scheduled.
- Adjust levels as needed with preferred chemicals.
- Brush bottom and sides to remove algae and distribute chemicals.

5. Clean up

- Pack away service tools and chemicals.
  - Be especially cautious with chemicals—keep them away from kids.
- Replace cover as needed.

Also be careful to avoid abrasive or sharp surfaces as you drag the cover off of the pool or you might be responsible for an expensive replacement. Finally, hose off the cover before returning it to the pool.

Water Level

RATING: EASY

If you add an inch or so of water to the pool each time you service it, you will probably keep up with normal evaporation. If you wait a few weeks until the level is several inches (over 10 centimeters) low, it will take hours to fill, requiring you to turn on the water supply and go back later in the day to shut it off. You can leave a note for your customer to fill the pool, but most homeowners will forget or resent doing part of your job. Never leave the water on and leave a note for the customer to turn it off when they get home. The note might get lost or their plans might change and they could arrive home to a flooded backyard.

If the pool has a dedicated fill line, turn it on and leave your truck keys on the valve. This is one way to make sure you don't forget to shut it off. If the fill line is above the water surface, it will splash into the pool, obscuring the bottom and making it difficult to see your vacuuming work. In this case, use a garden hose instead of the fill line and put the tip of the hose below the water surface. You can also put it in the skimmer to diffuse the flow.

After rains you might need to lower the pool level. In this case, use your submersible pump and a backwash hose or spare vacuum hose for the discharge. Alternatively, you can run the pool circulation system and turn the valves to waste. If you use this method, don't forget to return the valves to normal circulation before you leave or you might get a late night call from an irate customer with an empty pool and burned-out pump.

Surface Skimming

RATING: EASY

It is much easier to remove dirt floating on the surface of the water than to remove it by any means from the bottom. Using your leaf rake and telepole, work your way around the pool raking any floating debris off the surface (Fig. 9-6). As the net fills, empty it into a trash can or plastic garbage bag (which you will notice is on your equipment list). Never empty your skimming debris into the garden or on the lawn. Even if dumping a very small amount, a customer who sees you shaking out your leaf rake over his prize roses will quickly terminate

FIGURE 9-6 Using a leaf rake.

your services. Also, the debris is likely to blow right back into the pool as soon as it dries out.

There is no right or wrong way to skim, but as you do, scrape the tile line, which acts as a magnet for small bits of leaves and dirt. The rubber-plastic edge gasket on the professional leaf rake will prevent scratching the tile. This action is often overlooked and is a rich source of small debris that will soon end up on the bottom.

One last tip. If there is scum or general dirt on the water surface, squirt a quick shot of tile soap over the length of the pool. The soap will spread the scum toward the edges of the pool, making it more concentrated and easier to skim off.

Tiles

RATING: EASY

Many technicians leave the tiles for last, but if they're fairly dirty you will remove the material from the tile and it will settle on the bottom you have already cleaned. Also, if you need to remove stubborn stains with a pumice stone, the pumice itself breaks down as you scrub, depositing debris on the bottom. Therefore, do the tiles first.

Tile cleaning is a real pain in the knees. Invest in a set of knee pads or a gardner's kneeling pad. Although body and suntan oils and other light dirt will come loose with a standard tile brush, most of the time you will have to get down on your hands and knees and do some real scrubbing. Generally even if the tiles appear clean, I scrub them once a month to knock loose any scale or other deposits before they become noticeable.

Whichever brush you use, apply a squirt of tile soap directly to the brush and start scrubbing (Fig. 9-7). Remember, don't use soaps not designed for tile work because they will foam up. I carry a foam knee pad, available at any gardening store, and work my way around the

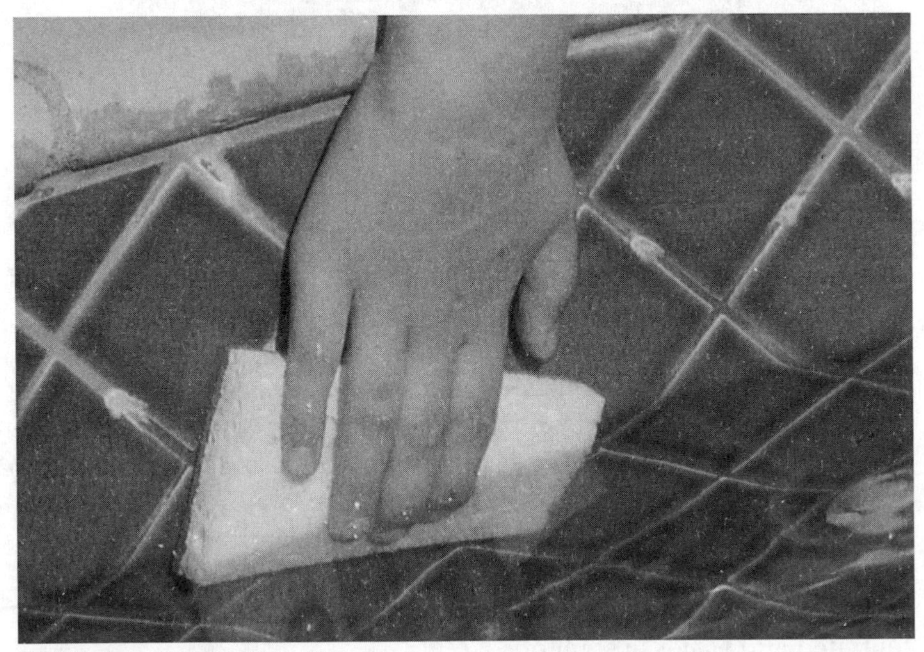

FIGURE 9-7 Scrubbing the tiles.

pool. To break up this tedious job a bit, you might do half the pool, then do your chemical testing or equipment check before finishing. It gives your knees a welcome pause.

When cleaning tile, scrub below the waterline as well as above. Evaporation and refilling means the waterline is rarely at the same level, so clean the entire tile line. It takes very little extra effort while you are there. Refresh the soap on your brush as needed.

Finally, you need to scrub to remove oil, dirt, and scale, but you want to avoid scratching the tile. Small scratches might not be visible but they do remove the glazing that makes the tile impervious to staining. Once the tiles are scratched, oils and scale are absorbed by the porous pottery material below and the tile will never come clean. Some minor scratching can't be avoided and won't penetrate the entire glazing layer anyway, but use only as much force as needed to clean the tile. Never use really abrasive brushes or scouring pads to clean tiles.

Equipment Check

RATING: EASY

At this point, many service technicians will start vacuuming the pool. Unfortunately they often get started only to discover that the suction is inadequate. It pays to check your equipment before vacuuming.

My technique is to review the circulation system by following the path of the water. Since you have already skimmed and cleaned the tile, you can now clean out the pool's skimmer basket without concern that it will fill up again while you work. Empty the contents of the skimmer basket into your trash can or garbage bag.

Next, open the pump strainer basket and clean it. If the pump has a clear lid, you might be able to observe if this step is necessary. Reprime the pump. Check the pressure of the filter. There is no point in checking it before cleaning out the skimmer and strainer baskets, because if they are full the filter pressure will be low and will come back up after cleaning the baskets. If the pressure is high, the filter might need cleaning. As mentioned in the chapter on filters, I don't believe in backwashing except as a very temporary measure. For example, if the filter is dirty and you need to vacuum the pool, you would not want to tear down and clean the filter, then vacuum dirt to it. In such cases you might want to backwash, if the filter uses that technology, vacuum, and then do the teardown.

Check the heater for leaves or debris. Look inside to make sure rodents haven't nested and to verify that the pilot light is still operating (if the unit is a standing pilot type). Turn the heater on and off a few times to make sure it is operating properly. While the heater is running, turn the pump off. The heater should shut off by itself when the pressure from the pump drops. This is an important safety check. Of course, some customers shut down the heater in winter and don't want the pilot running, so be familiar with your customer's needs before taking action.

Check the time clock. Is the time of day correct? Is the setting for the daily filter run long enough for prevailing conditions? If the pool includes an automatic cleaning system with a booster pump, is the cleaner's time clock set to come on at least one hour after the circulation pump comes on and set to go off at least one hour before the circulation pump goes off? Always check the clocks because trippers come loose and power fluctuations or outages can play havoc with them. Service work on household items unrelated to the pool can also affect the clocks, when other technicians turn off the breakers for a period of time to do some work. Also, electromechanical time clocks are not exactly precision instruments. One might run slightly faster than another, so over a few weeks one might show a difference of an hour or more, upsetting your planned timing schedule.

At each step of the equipment check, look for leaks or other early signs of equipment failure. Clean up the equipment area itself. Remove leaves from around the motor vents and heater to prevent fires, and clear deck drains of debris that could prevent water from draining away from the equipment during rain. This is your area, because when the customer looks at the equipment, he will judge you by the appearance of it.

Vacuuming

RATING: EASY

If the pool is not dirty or has only a light dusting of dirt, you might be able to brush the walls and bottom, skipping the vacuuming completely. If the pool or spa is dirty, however, you have two ways to clean it: vacuuming to the filter or vacuuming with the leafmaster.

VACUUM TO FILTER

RATING: EASY

Vacuuming to the filter means the dirt is collected from the pool or spa and sent to the circulation system's filter.

1. **Maximize Suction** Make sure the circulation system is running correctly and that all suction is concentrated at the skimmer port. Use your bronze skimmer diverter for this process (as described in the chapter on basic plumbing) if dealing with a single port skimmer. If the system includes valves for diversion of suction between the main drain and the skimmer, close the main drain valve completely and turn the skimmer valve completely open. If there are two skimmers in the pool, close off one by covering the skimmer suction port with a tennis ball, concentrating the suction in the other one. On large pools, you might have to vacuum each half separately.

2. **Set Up** Attach your vacuum head to the telepole and attach the vacuum hose to the vacuum head. Working near the skimmer, feed the head straight down into the pool with the hose following (Fig. 9-8A). By slowly feeding the hose straight down, water will fill the hose and displace the air. When you have fed all the hose into the pool, you should see water at the other end (which should now be in your hand).

3. **Start Up** Keeping the hose at or near water level to avoid draining the water from it, slide the hose through the skimmer opening (Fig. 9-8B) and into the skimmer. Attach the hose to the diverter (or with two-port skimmers, insert the hose cuff into the skimmer's suction port). The hose and vacuum head now have suction.

   On older pools, the suction port might be in the side of the pool below the skimmer. In this case you might need to put a tennis ball over the skimmer suction port to concentrate the suction at this wall port.

   In any case, make sure the hose does not contain a significant amount of air. When the air reaches the pump, you will lose prime. If this occurs, remove the vacuum hose, reprime the pump, then try again.

FIGURE 9-8 (A) Preparing the vacuum; (B) preparing the hose; and (C) vacuuming the pool.

4. **Vacuum** Vacuuming a pool or spa is no different than vacuuming your carpet. Work your way around the bottom and sides of the pool (Fig. 9-8C). If the pool is dirty, it will be very obvious where you have vacuumed and where you haven't. The only word of caution is to avoid moving the vacuum head too quickly. You will stir up the dirt

rather than suck it into the vacuum. If the suction is so strong that it sucks the vacuum head to the pool surfaces, adjust the skimmer diverter or valves to reduce the flow. You might also need to lower the wheels on the vacuum head, raising the vacuum head itself.

5. **Adjust** Conversely, if the suction is weak, you might want to lower the vacuum head or you might need to move the head more slowly around the pool to vacuum it thoroughly.

   If the pool is very dirty, the sediment might fill the strainer basket or exceed the capacity of the filter. You need to stop when suction becomes weak to empty the strainer basket or, in extreme cases, clean the filter.

   If the pool contains both fine dirt and leaves, the leaves will quickly clog the strainer basket. You can now purchase a leaf canister, which is an in-line strainer that collects the leaves and allows fine dirt to pass on to the filter. You can empty the canister without losing prime in the pump, making your cleaning job faster.

   If the pool has an adjacent spa or wading pool operating from the same circulation system, you can vacuum this at the same time by quickly removing the vacuum head from the pool and putting it immediately into the spa or wading pool. If you are quick enough, only a small amount of air will enter the hose and you might lose prime for a minute, but there will be enough water in the line for it to reprime itself.

   If you do vacuum both units at once, make sure the water you have vacuumed from the spa or wading pool is replaced. Normally, adjacent spas or wading pools have pool return lines in them so that water falls from one to the other. This will refill the spa or wading pool. If the systems operate from the same circulation system by changing valve positions, but are not circulating together, you will need to vacuum each separately so that water removed during vacuuming returns to the place of origin.

6. **Purge** When you are finished, position yourself near the skimmer and remove the vacuum head from the water. The suction will rapidly pull the water from the hose. Be prepared to pull the hose off of the skimmer suction port before the air reaches there, otherwise you will lose prime. You can also let the vacuum run for a minute after you finish to make sure the hose contains clean water, then remove the hose from the skimmer suction port and drain the

TRICKS OF THE TRADE: HEAVY-DUTY CLEANING

- **When a pool is exceptionally dirty you have another alternative to vacuuming and cleaning the filter several times. I carry a portable pump and motor, which is the same 1½-hp unit used in the circulation system, that simply plugs into a household electrical socket. Using this pump, I vacuum the pool and run the discharge through a second vacuum hose to a waste drain or the street. This method wastes a lot of water, but for extreme cases it might be the only alternative to draining the pool, which wastes even more. This method may also be illegal in some jurisdictions unless the chlorine is neutralized first.**
- **Twice in recent years, Malibu endured devastating brush fires and all my pools were blanketed in soot and ash, as much as a foot thick on the bottom of some pools. I devised this portable pump after the first pool took six hours to vacuum to the system's filter. Since then, PoolTool and other companies have developed portable systems, but you can put one together yourself with a standard 120-volt pump and motor, a length of extension cord, and some hose cuff fittings on the suction and discharge sides of the pump. If the circulation system is equipped with a waste line, you can work this method using the pool's own pump, diverting the water directly to waste instead of the filter. Remember to refill the lost water so the circulation system does not run dry.**

water from the hose back into the pool. Unfortunately, if any sand or small stones are hung up in the coils of the hose, they will wash back into the pool when you empty it that way, so I suggest the pump evacuation method just described. You can also pull the vacuum head from the pool and the suction end of the hose from the skimmer simultaneously, remove the hose from the water, and drain it on the deck.

7. **Clean Up** Remove your equipment from the pool. Check the pump strainer basket and filter to see if they have become clogged with debris. Clean as needed. Replace the skimmer basket.

VACUUM TO LEAFMASTER

RATING: EASY

If the pool is littered with leaves or other heavy debris, you might need to use the leafmaster instead of the vacuum, or use the leafmaster first, then allow the fine dirt to settle and vacuum it to the filter. Most technicians would rather leafmaster than vacuum to the filter, because the leafmaster is faster and easier to use.

1. **Set Up** Attach a garden hose to a water supply, then to the leafmaster. Clip the leafmaster onto the telepole.
2. **Vacuum** Place the leafmaster in the pool (Fig. 9-9A). Turn on the water supply and vacuum. As with vacuuming to the filter, leafmaster vacuuming is just a matter of covering the pool floor and walls. Because the leafmaster is large, you can move it quickly and vacuum the pool in about half the time required to filter vacuum (Fig. 9-9B). Be careful not to move the leafmaster so fast that you stir up the debris. If your garden hose is not the floating variety, work in a pattern that keeps the hose behind your work so you will not stir up debris before you can vacuum it.

   Empty the collection bag periodically if it becomes too full (see step 3 for directions), but remember, some leaves in the bag will help trap fine dirt or sand in the bag, so leaving some debris in the bag is helpful.
3. **Removal** When you are finished, remove the leafmaster by turning it slightly to one side and slowly lifting it through the water to the surface (Fig 9-9C). If you pull it straight up, debris will be forced out of the bag and back into the pool. Never turn the water supply off before removing the leafmaster from the pool, for the same reason. The loss of vacuum action will dump the collected debris right back into the pool. When the leafmaster is on the deck, turn off the water supply and clean out the collection bag.

   Always be sure your collection bag is securely tied to the leafmaster. I can't tell you how frustrating it is to have the bag come loose as you lift the leafmaster from the pool, spilling the debris back into the pool. It can also come loose during vacuuming, especially if there is a lot of debris and strong water pressure.

   Remember, if the water pressure is poor, use the automatic cleaner booster technique described earlier or keep a tennis ball in the collection bag to keep it upright.

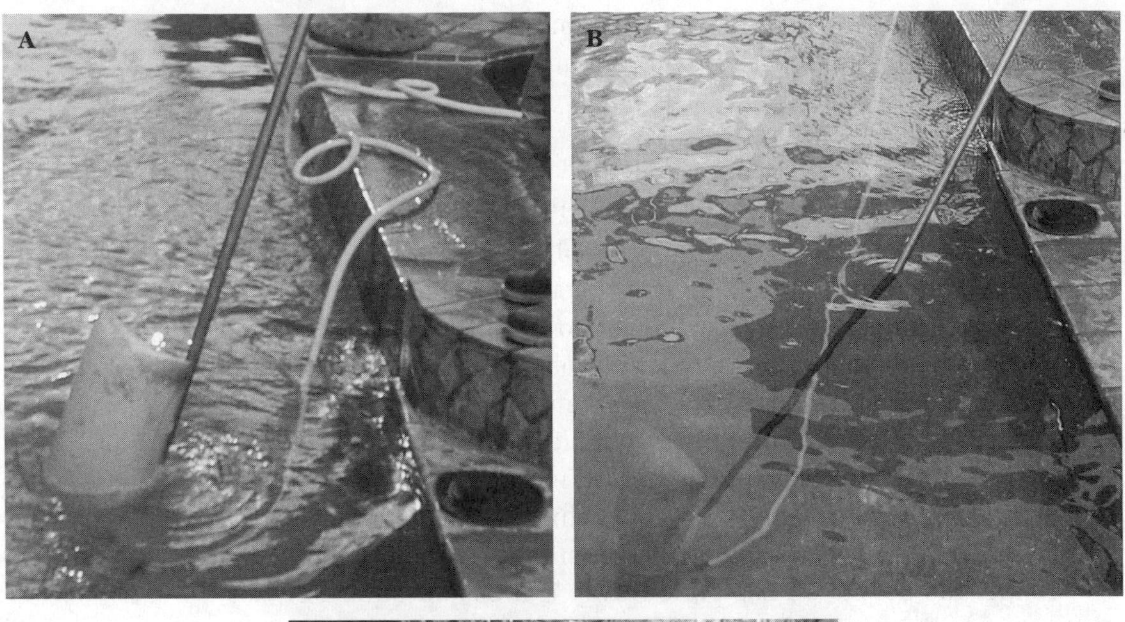

FIGURE 9-9 (A) Preparing the leafmaster; (B) vacuuming with the leafmaster; and (C) removing the leafmaster so debris doesn't fall back into the pool.

PERIODIC SERVICE AND MAINTENANCE SCHEDULE

Adjust as needed during months of heaviest use or extreme weather. ("X" = times per)

	Day	Week	Month	Year
Check water level	1X			
Check pH & sanitizer		2X		
Check Hardness, TDS, Total Alkalinity			1X	
Test for heavy metals				2X
Check conditioner				2X
Check skimmer basket		2X		
Check pump strainer basket		1X		
Check filter pressure		1X		
Look for leaks in plumbing and equipment; check time clock settings		1X		
Surface skim the pool	1X			
Vacuum the pool (pool has no autocleaner)		2X		
Vacuum the pool (pool has autocleaner)		1X		
Operate autocleaner (3 hours)		2X		
Brush pool walls and bottom		1X		
Clean waterline		1X		
Empty autocleaner catch bag		2X		
Clean solar panels			1X	
Winterize				1X
Reopening and/or equipment tuneup				1X
Teardown and clean filter				3X

Chemical Testing and Application

RATING: EASY

1. **Test** Follow the general testing guidelines described previously, testing for chlorine residual, pH, total alkalinity, and acid (or base) demand. Unless you detect a problem, these are the basic tests you will conduct at each service call. Once a month, or if you find unusual readings in the basic four tests, you will test for calcium hardness and stabilizer.
2. **Treat** Apply the chemicals as described previously. I use liquid chemicals for ease of distribution and price, but you will make your own judgments as you proceed. In any case, it is purely a personal preference. Apply sanitizer, following the guidelines described previously, then move ahead to the brushing step to give it time to circulate and distribute. Brushing accelerates that distribution. Apply acid or alkaline materials next. You don't want to apply them together because combinations of pool chemicals can be deadly.
3. **Caution** Be careful with chemical bottles on pool decks. Like the ubiquitous coffee cup ring, pool chemical bottles will leave stains on decks that are difficult to remove and even more difficult to explain. If you must set them on the deck, immerse the bottom of the bottle in the pool for a moment to rinse off any chemical. Splash some water on the deck where you set the bottle down so any residue will be neutralized and not leave stains.

Brushing

RATING: EASY

Having now spent 30 to 45 minutes on the service call, many technicians skip the brushing of the pool or spa. This is a big mistake. Dirt and minerals will build up over time, making the pool appear dingy and dirty even when you have cleaned it thoroughly. Also, algae starts out invisible to the naked eye, and brushing removes it from surfaces, suspending it in the water where chemicals will kill it before it can really take hold.

On pools or spas that are not very dirty, you can skip vacuuming and brush from the walls to the bottom and from the shallow to the deep end, directing the dirt toward the main drain where it is sucked

TRICKS OF THE TRADE: SPA AND WATER FEATURE CLEANING

The steps outlined for pool cleaning will work just as well for spas and water features. Here are a few special tips about what you might encounter that is unique to these bodies of water.

- **Many spas are made of fiberglass, so take care when vacuuming to avoid scratching the surfaces.**
- **Vacuum the corners of water features and small spas with the spa vacuum described earlier.**
- **If you drain a spa or water feature, be sure the equipment is turned off at the breaker so the time clock won't turn it on before you are ready.**
- **Before you clean the spa or water feature, clean the filter and run some fresh water (from the garden hose) through the circulation system to purge any dirty water from the lines. Nothing is worse than draining, cleaning, and refilling a spa only to turn the circulation back on and watch dirty water contaminate your work.**
- **Be extra careful with chemical testing and application. Most spas and water features contain a tiny fraction of the volume of water in a pool, so they can't absorb a mistake the way a pool might. It is better to add chemicals more slowly and in less quantity than you think necessary. You can always add more, but it is a real problem to remove any excess.**

to the filter. If you plan to use this technique instead of vacuuming, divert all suction to the main drain as described previously.

Winterizing

If you work in a cold climate, I probably don't need to tell you that a body of water needs to be prepared for the winter months. But even in warmer climates, you might discover that some of the following information is useful.

The most important concern from cold temperatures is freezing. Water expands when it freezes, meaning that if it is trapped inside pipes and equipment it will expand and crack them. PVC pipes, fittings, and ABS plastic pump parts are most susceptible, but soft copper heat exchangers and galvanized fill-line pipes are not exempt either. Expanding frozen water will also crack tiles and plastic skimmers.

The second problem, which relates to the seasonal winter closure of a pool or spa, is potential damage from algae and debris. Although algae does not grow well in temperatures below 55°F (13°C), especially if the water is shielded from sunlight by a cover, prolonged periods of stagnation will permit and promote algae development.

Temperate Climates

RATING: EASY

For normally temperate climates, the best protection during winter is to keep an eye on weather forecasts. When temperatures are predicted to drop near freezing, set the circulation to run 24 hours per day, preferably with the heater on at the lowest thermostat setting. This is especially important for pools and spas in valley or mountain regions where temperatures might be slightly colder than those predicted in metropolitan areas.

Continue normal service during the winter. Wind and rain means that pools will flood, and stain-causing leaves and debris can harm the surfaces or clog the system. In short, the work in winter doesn't lessen, it simply changes slightly. You will use less chlorine to maintain the same residual and you can reduce filtration times to a total of four hours per day (staggered as described earlier). Some specialty control systems, especially for spas, include freeze protection. Temperature sensors detect freezing conditions and operate the circulation system and/or heater automatically. Of course, if the pool or spa is kept heated, it's business as usual.

Colder Climates

RATING: ADVANCED

In climates where winter means actually wearing long pants and jackets, most pools and spas are closed for the season. The shutdown process requires some planning to avoid damage to the installation and to make reopening in the spring a simple task. The following steps will help you make these preparations for your customers and provide some options to achieve the goals regardless of your skills or the availability of specialized equipment. Above-ground plastic spas can be drained completely, although in-ground gunite and plastic spas should be treated just like pools. In either case, the plumbing and equipment must be prepared for winter.

1. **Balance the Chemistry** Etching or scaling conditions of water will harm the pool or spa even when the circulation is off, so before

closing make sure the chemistry of the water is balanced.

2. **Cleaning** Dirt and debris left in the water during long periods of stagnation will leave stains and/or be much harder to remove several months later. Therefore, thoroughly service and clean the pool before closure.

3. **Algae and Stain Prevention** To prevent algae growth and staining during the closure, superchlorinate the water. A simple formula is to triple the normal superchlorination that you use for the particular installation. A more precise approach is to raise the residual to 30 ppm for plaster surfaces, 10 ppm for vinyl or plastic. Also, add a chelating agent to prevent metals from dropping out of solution and staining the surfaces. Finally, add an algicide that will inhibit black algae growth (I prefer silver-based products).

   Do not leave tablets or floaters in a pool or spa during closure. Since the water isn't circulating, the dissolving chemical isn't either, and the extreme concentrations in one area can do structural or cosmetic damage, particularly to vinyl or plastic surfaces.

   Make sure that you have adequately circulated the chemicals before shutting off the system for the winter.

4. **Shut Down the Equipment** Turn off the circulation equipment at the breakers and tape over them to prevent someone from turning them back on. Turn off all manual switches and time clocks and remove the trippers as an added precaution in case someone does turn the breakers back on. Pay special attention to lights, which might be on household circuits rather than wired through the equipment breakers. Some technicians don't disconnect underwater lights if the water level will remain above the fixture all winter. The light provides a little warmth to the water

TOOLS OF THE TRADE: WINTERIZING

- **Submersible pump**
- **Pool-safe liquid antifreeze**
- **Tapered rubber winterizing plugs (one for each return outlet in the pool)**
- **Air pillows**
- **Plastic jug with sand or pebbles (that fits inside skimmer)**
- **Chemical test kit**
- **Sanitizer**
- **Tape (to obstruct electrical switches and breakers when shut off)**
- **Compressed air, hose, blow bag**
- **Pliers to remove drain plugs from pump, heater, solar panels**
- **Any hand tools needed to remove your equipment for storage**
- **Winter cover (and clips, cords, or weights to secure it)**
- **Yellow caution tape**

QUICK START GUIDE: WINTERIZING

Rating: Advanced

1. Prep

- **Bring all tools and supplies to the pool area.**
- **Clean and service the pool normally before lowering water level.**
  - **Increase sanitizer level per package directions.**
- **Shut OFF:**
  - **Pump (tape over switch or breaker, so no one can turn it on before spring).**
  - **Electrical source to heater electronic ignition.**
  - **Gas supply to heater.**

2. Drain

- **Lower the water level below skimmer and return outlet.**
- **Clear lines of water and plug them.**
- **Fully drain all equipment, even if stored indoors (unless stored in heated location).**
  - **Remove and store equipment (if possible).**

3. Secure the Pool

- **Install airbags (including small airbag or plastic jug in skimmer).**
- **Install winter cover.**
- **Block off pool entry with caution tape or other barriers.**
- **Pack up and store all service tools, especially chemicals.**

4. Reopening in Spring

- **Perform the same tasks in reverse order.**

periodically and draws the owner's attention to the pool or spa for regular inspection. I have seen light lenses crack, however, as the extremely cold water and extremely hot light fixtures contact, so I don't recommend this practice.

5. **Lower the Water Level** Pump out 24 to 36 inches (60 to 90 centimeters) of water, or at least enough to drop the level 18 inches (45 centimeters) below the tile and skimmer line. When water freezes and expands, it can crack the tile and plastic skimmer components, so the goal is to lower the level to a point where

winter rain or snow will not raise it back up to those delicate areas. If the equipment at the installation permits, lower the level by vacuuming to the waste line on your final cleanup. This will save you time by accomplishing two goals at once.

The vessel should not be drained completely for the season, because hydrostatic pressures might cause cracks or make the entire vessel pop out of the ground.

6. **Clear the Lines** Perhaps the most important objective of winterizing is to protect the plumbing from freeze damage. There are several methods, some easier than others. The method you choose might depend on your equipment, skills, and the availability of water in your area.

The most effective method of protection is to drain the pool, completely evacuating the lines. Of course, you then refill it to the level discussed in step 5. If the water needs draining anyway for chemistry reasons, as discussed in the chemistry chapter, this is an opportunity to accomplish two tasks at once. The cost or availability of water in your area might prohibit annual draining, but if you can, it is the most effective way to be sure all water has been removed from the system. Even small amounts of frozen water can crack pipes and fittings.

With the pool empty, remove the collar/nozzle fittings from the return lines and plug them with tapered expandable rubber winterizing plugs (Fig. 9-10). Be sure to use tapered plugs, because they will pop out if freezing water in the pipe expands. You're better off with a popped plug than a cracked pipe. The main drain might still have water in the bottom of the plumbing, so try to suck it out with a wet/dry shop vacuum or mop it out with a sponge. Just to be sure, pour a cup of antifreeze into the main drain before plugging it. For antifreeze, use a mixture of one part propylene glycol to two parts water. Your supply house will have propylene glycol or premixed products. Never use automotive

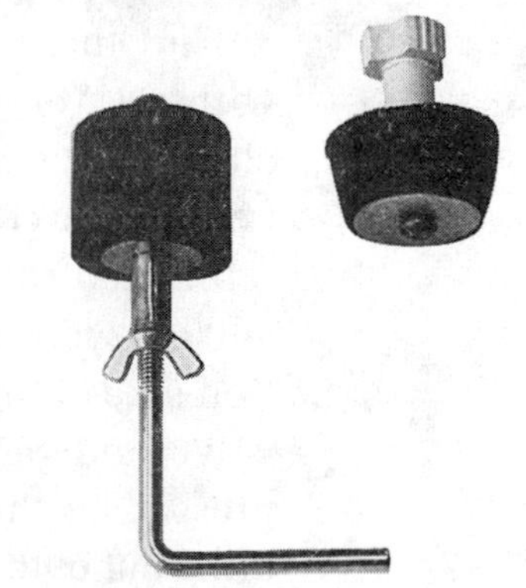

FIGURE 9-10 Test plug and tapered winterizing plug.

antifreeze, which is corrosive. With all of the lines plugged, refill the pool to the level described previously.

If you are unable to drain the pool, you can blow the lines out with air (described later) and plug them underwater. Some references suggest you can plug the lines first and then use a vacuum to suck water from the lines, but it is unlikely that all of the water (or even most of it) will be removed. If you have no way to blow the lines, you might try using a wet/dry vacuum, preferably a high-powered model available at tool rental shops. After you have vacuumed as much water out of the lines as possible, pour the antifreeze solution into the pipes by pouring it into the strainer pot and allowing it to flow back toward the pool. If there is a check valve in the line, the mixture will not get any closer to the pool than the valve. If that part of the plumbing is exposed or you can access the suction lines by disassembly of the three-port valves or gate valves, pour the mixture in at those points.

The return lines can also be filled with antifreeze by starting the pump with the strainer pot lid open. With the pump running, pour the mixture into the pump. It will be distributed throughout the equipment, plumbing, and return lines. When you see antifreeze discharging from all of the return lines, turn the system off and plug the return outlets. You might want to add some food dye to the mix to make it easier to see when it is discharging to the pool. Some water will remain in the lines, so as you can see, without draining the pool, clearing the lines is guesswork at best.

The other method of clearing lines underwater is to blow them with air. You can buy or rent a tank of compressed air (useful for clearing obstructions from lines also) from a plumbing supply house or a company that provides compressed gasses. Attach a reinforced rubber line with a garden hose attachment on the end and add your blow bag to the end. Insert the blow bag into the suction side of the pump (be sure to get on the pool side of any check valve and remember to open any gate valves) and force compressed air through the lines. Go to the pool while the air is bubbling out of the lines and plug them. The lines will be filled mostly with air, but some water remaining in the lines is inevitable with this method. Do the same on the return lines, inserting the blow bag at the heater outlet pipe or disassembling a valve on the return side of the equipment system.

The blow-out method requires diving down to the main drain to plug it. If you are unable to do that, you can try to flood the area with antifreeze by using the acid spotter described previously. You can also insert a long 2-inch PVC pipe into the pool, placing one end on top of the main drain (Fig. 9-11). Pour the antifreeze mixture into the pipe. Since the mixture is heavier than water, it will sink to the bottom and flow into the main drain plumbing.

7. **Remove Equipment** Any equipment in the pool or circulation system that might be damaged by exposure to the elements (or which might be stolen) should be removed and stored. Especially at commercial pools and spas, remove ladders, rails, safety equipment, deck furniture, and anything else that might be subject to weathering over the winter or which might be stolen.

The circulation equipment should be disassembled and important components stored. If the pump and motor is plumbed with

FIGURE 9-11 **Adding glycol to main drain.**

unions, you can easily remove the entire unit and disconnect the electrical connections. If there are no unions, you might want to cut the plumbing and add unions when you reinstall the unit in the spring to make the process easier next season. If the plumbing makes removal of the pump difficult, unbolt the wet end and motor from the volute and remove those.

The filter should be disassembled, cleaned, and drained thoroughly and the grids or cartridges put in storage. Freezing water can cause fabric deterioration. Sand filters should be cleaned and drained. Since rain might refill an open filter tank, close the tank and leave the drain plug out.

Close the gas valve to the heater and any supply valve (if it is a dedicated line for the pool heater) at the meter. The heater has drain plugs on both sides. I recommend removing the heat exchanger and burner tray and storing them after draining. Drain the water from the pressure switch tube as well.

Drain any solar panels and leave the plumbing fittings or gate valves open to the atmosphere. Even in winter the heat in a solar panel can be intense and expanding air can be as hazardous to the panels as freezing water.

Be sure the lines to the equipment are empty or are filled with antifreeze. Remove any spa jet or automatic cleaner booster pumps and store them. Be sure you have drained the automatic cleaner plumbing or have filled it with antifreeze. Usually the automatic cleaner plumbing outlet is at or near the tile line, so the water level will probably be lower than this, making draining easy and plugging unnecessary. Leave all valves open and disassemble any three-port valves. Finally, if the pool has a fill line, turn off the water supply and drain the line.

8. **Extra Precautions** Put a little dirt or gravel in an empty plastic bottle and leave it in the skimmer. The weight will help it stay upright in the skimmer if rain or snow refill it with water. Should the water subsequently freeze, the ice will compress the bottle, not crack the skimmer. Similarly, leave some plastic milk or chlorine jugs floating on the surface of the pool. Fill them about one-quarter with water so they will grip if the water freezes and not just pop out onto the surface of the ice. Again, the goal is to create an expansion joint so that the ice will crush the jugs, not the pool or spa walls.

If there are exposed pipes that you suspect might still contain water, such as in the equipment area, under decks, or in self-contained spa cabinets, wrap them with insulation tape (a good idea anyway to reduce heat loss when operating the system). In extremely cold areas, wrap them with electric insulation tape, available at most hardware stores on a seasonal basis. These tapes actually warm the pipe with a low-level electric current.

If metal ladders, rails, and light fixtures cannot be removed, protect them against corrosion with a coat of petroleum jelly. Cover the diving board.

9. **Cover the Pool** Some cover is better than none, because it will inhibit algae growth and keep heavy debris out of the pool. Sheet vinyl covers sold for pools are very inexpensive and can be held in place with sand- or waterbags around the edge of the pool. If the pool has a mesh cover, put a sheet vinyl cover over that, because the mesh cover will not keep out sunlight and dirt.

10. **Close Off Access to the Pool** Whether it is a commercial or residential pool or spa, blocking access to a closed installation is an important safety precaution. Yellow caution tape strung around the pool, locked gates and fences, and extra signage will help keep the pool or spa from becoming an inadvertent hazard and will limit your liability. At the same time, remove any deck furniture or other loose items that might be stolen, thrown, or blown into the pool. Store the automatic cleaner and water supply hose indoors. The hose should be laid out as straight as possible or coiled in large loops to avoid permanent bends.

11. **Pack Up** Take this opportunity to properly dispose of any extra chemicals or test kit reagents that won't be used during the winter and will not be potent in the spring. Soda ash and acids are about the only water chemicals that will still be good after prolonged storage, but make sure they are packed in watertight containers and are stored in well-ventilated areas away from water or heat sources.

During the winter you still need to check on the pool. Snow or rain might raise the water level or sink the cover. Animals or heavy debris might fall in the pool and are better removed now than in the spring.

Reopening the pool or spa is essentially the reverse of the shutdown procedure, with emphasis on balancing the water before restarting the circulation system.

FAQs: CLEANING AND SERVICING

Should I Hire a Service or Clean the Pool Myself?

- **Most routine pool maintenance tasks are easily performed by family members, but weekly cleaning service is not expensive and can save money in the long run. Equipment repairs, annual tuneups, and replacements or additions are probably best left to the pros. Many people enjoy cleaning their own pool (it is cheaper that therapy!), so try it before deciding to hire a service.**

How Often Should I Vacuum the Pool?

- **At least once a week, unless you have an automatic cleaner that is working extremely well. Remember that vacuuming will simultaneously brush the pool, so it will not only keep the pool clean, but also prevent algae from getting started.**

How Do I Get Rid of the "Bathtub Ring"?

- **Discoloration around the waterline of the pool is normal, caused by the minerals in water that have evaporated. Suntan lotion and body oil will cling to the waterline, too. Use a tile cleaner and keep the waterline clean, which will prevent that buildup of scum in the first place.**

CHAPTER 10

Special Procedures

If you can perform the tasks and understand the water technologies presented thus far, you will have many years of enjoyment from your pool or spa. There are some special procedures that are used far less often, but they are important to know when the need arises.

Draining a Pool

Many of the procedures outlined in this chapter require draining the pool or spa. It might seem obvious that you use your submersible pump to do the job, but this simple task can create many problems if you don't take all factors into consideration.

1. **Shut Down** Turn off all circulation equipment at the circuit breakers, so there is no chance it will start up from a time clock. Be sure that underwater lights are also switched off and not connected to a time clock. Often there are switches inside the home: tape over these with a note to make sure all family members know there is a reason to keep that switch off.

2. **Safety** At a commercial pool, run yellow caution tape around the pool deck to keep unwary visitors from falling into an empty pool. Turn deck furniture on its side and use it as a physical barrier as well. Post signs around the pool deck about what is going on and

letting people know the pool will be closed for several days. With today's busy families, it is a good idea to do the same for residential pools.

3. **Drain the Pool** When draining a pool with your submersible pump (Fig. 10-1) and hoses, direct the flow to a deck drain if possible. This will send the waste water through an intended channel, rather than over a backyard garden or down a hill where erosion damage can occur from such a large volume of fast-moving water. Once you have started pumping, watch the flow into the drain for several minutes. Sometimes debris will back up in the line and it will overflow, but not until it has filled several hundred yards of pipe. It might take time, but the clogged drain can flood backyards, living rooms, and might actually flush the water back into the pool.

   If deck drains cannot accommodate the flow, connect several vacuum or backwash hoses together and run the waste water into the street where it is carried to storm drains. Again, watch the flow

FIGURE 10-1 Submersible pump.

FIGURE 10-2 Gravity drain plumbing.

and make sure that it does not back up the storm drain and flood the street.

If your equipment is below the pool, you can also drain the water by gravity. Figure 10-2 shows a three-port valve plumbed before the pump, which can be used to drain the pool by simply rotating the valve, so the flow is directed to the open pipe.

In areas where water conservation is a concern, you should wait to drain the pool until the chlorine residual has gone below 1 ppm, then let the flow irrigate lawns and gardens. Obviously a 20,000-gallon (75,700-liter) pool will flood the typical backyard garden if it is pumped all at once, but use as much as you can before switching the discharge to a regular drain where the water is wasted. Don't spend days draining the pool just to use this conservation method. You will conserve water, but you might damage the pool plaster by allowing excessive drying (which can cause cracks and shrinking). Of course, if

TRICKS OF THE TRADE: SUBMERSIBLE SAFETY

- **When lowering a submersible pump into a pool, never do so by the cord. Attach a nylon rope to the bracket or handle and lower it that way to avoid pulling electrical wires loose. Not only will the pump fail if the wires come loose, but when you plug in the cord it might electrify the water.**
- **Buy a ground fault interrupter (GFI) that can be plugged into the wall socket before plugging in the cord of the pump. You can also rig a remote on/off switch that plugs into the socket as well, allowing you to operate the pump while you are working in the pool. Check your local electronics store for a combination remote receiver/GFI or simply buy one of each and plug them in together before plugging your pump into the outlet. It will be the best $30 to $50 you ever spent. I have been in water with a pump that had a slight short and felt the tingling of the electricity conducted through water and have known technicians who have been killed by not taking this aspect of operational safety seriously.**

the pool has a surface other than plaster, take your time. Another technique along these lines is to punch holes along the length of a backwash hose (or old vacuum hose) and seal up the discharge end (tie a knot in a backwash hose or use plumbing fittings to close a vacuum hose). This then acts as a huge sprinkler, evenly distributing the water all along the length in various gardens, perhaps even across the lawns of several neighbors (with their permission, of course). In other words, be creative about your method of pumping and the potential uses of the waste water.

One other word of warning about deck drains. They are usually made of PVC, but because they don't carry water under pressure they are not usually pressure tested. If ground movement or other erosion has destabilized them, the pipe might have separated and much of your waste water will end up eroding the backfill around the pool. If you have any doubts about the integrity of the deck drain or its ability to handle the water, run the hoses into the street.

In some jurisdictions there might be restrictions on pumping out pools relative to the permissible volume and even the permissible chemical makeup. Extremely low-pH water might have to be neutralized before pumping it into municipal stormwater or sewer lines. Check your local codes before turning on the pump.

Breaking-in New Plaster

There are two basic reasons to break in plaster rather than just turn on the pump and start swimming. First, you need to remove the plaster dust from the water, which will otherwise settle and build up as hard, rough scale. The pool is filled before the plaster dries, so the actual drying (curing) takes place underwater. Plaster shrinks somewhat as it dries, so if you allow the plaster to dry before filling the pool, the weight of the water against the brittle, dry plaster will cause it to crack. Adding the water before the plaster dries allows it to be pushed into place against the pool shell, pressed evenly by the weight of the water. It doesn't shrink, become brittle, or crack.

By allowing this curing process to occur underwater, you are left with the by-products of the process to remove from the pool or spa. Plaster dust—calcium carbonate—will adhere to the surfaces of the pool as scale, especially noticeable in dark-colored pools. The curing process, called *hydration*, can take up to four weeks, but most of the excess plaster dust is removed in the first week if the break-in process is properly executed.

The second reason to break in the plaster is to balance the water chemistry so that the water itself does not destroy a good plaster job. Tap water is never exactly balanced to reflect the ideal chemistry components for pools and spas (as detailed in the chapter on chemistry). Therefore, the break-in process is designed to create plaster-friendly water that is neither etching nor scaling.

Break-in Step by Step

RATING: ADVANCED

There are many variations to the steps needed for a successful break-in of new plaster, but there are not many disagreements about what those steps should be. What follows are my recommendations after breaking in literally hundreds of plaster jobs over the years.

1. **Plan** My first step always is to review the process with the builder and plasterer, making my recommendations but following any reasonable variations they might suggest.
2. **Inspect** As a part of setting up the job, inspect the plaster work. Look for trowel marks or footprints, and look for signs of sloppy work,

such as plaster on the tiles or deck. Examine the steps and anywhere that ladders, rails, or other hardware meet plaster surfaces. They should be seamless and smooth, without trowel marks or high spots. Bring any imperfections in the plaster job to the attention of the builder or plasterer for correction before filling.

3. **Test** Test the source of the fill water. In the chemistry chapter, I discussed the tests for heavy metals such as copper, iron, and manganese. In normal water maintenance you won't be testing for such metals, but now is the time for these tests. If you detect high levels of metals in the source water, you should be advised that additional chelating agents will be required and you can adjust your chemical balancing in general to avoid precipitating these metals onto the new plaster.

4. **Fill** Fill the pool or spa. This step seems the easiest part of the break-in process but it can be tricky.

   First, never just turn on the fill line if the pool has one. The water cascading several feet down will damage the soft plaster below. You must provide a cushion for the water, which can be a few feet of water provided by a hose before you fill with the fill line. You can also place a bucket under the flow of water to provide a cushion, but be sure to place rags under the bucket so that the weight of the water in the bucket doesn't leave circular scratches in the plaster.

   When filling with a hose, wrap the end with a rag so the hard fitting on the hose end does not scratch the plaster. Better yet, remove the fitting if you can, or cut it off and replace it later. Don't lay lengths of hose in the bottom of the pool either. The coils will leave marks, and if the pressure is enough to make the hose fishtail, the sweeping action will leave marks as well. Never walk on the fresh plaster; this will leave footprints that stand out like the proverbial sore thumb when the pool is full, and they are very difficult to remove.

   The objective is to fill the pool quickly and without stopping. If you stop at any point and then continue, you can expect to see a ring at the level you stopped. If you fill slowly, the plaster that is allowed to air dry will be a very different shade than the plaster that cured underwater. In extreme cases, the air-dried plaster might also crack.

For the same reason, don't add chemicals when filling the pool. The concentrations of chemicals while the pool was at one level versus another will tend to leave stains on the wall at the places where the chemistry changes took place. The exceptions to this are the addition of chemicals by the barrel method (described later), because you are evenly changing the chemistry of all of the fill water before it enters the pool, and chelating or antistaining agents.

To accelerate the filling process, add water from as many sources at once as you can. Use more than one hose, the fill line, and even a hose from a neighbor's yard (with permission, of course).

If the water tests out especially corrosive and low in calcium hardness, check to see if a water softener is a part of the household plumbing. If so, there are bypass valves on top of the softener to close it off from the system while you are filling the pool.

5. **Check the Equipment** It might take a day for the pool to fill, so this is a good time to make sure that you are prepared to circulate the water when that happens. Make sure that you clean the filter and that the pump works. Be careful not to force any dirty water into the pool that might have been standing in the filter or plumbing. At this point, turn the heater off and leave a note on it that it must remain off for at least three weeks after the pool is full.

6. **Precautions** Add a chelating agent to keep any metals in the water from precipitating out of solution and depositing on the new plaster surfaces. There are many products for this purpose with names like Metalgone or Stainout. Consult your supply house and follow the instructions on the label regarding application techniques and volume. Generally you will add a quart of liquid product as soon as you start the filling process. Repeat as directed by the product label.

7. **Circulate** When the pool or spa is full, turn on the circulation equipment. If the filter is a DE type, add the DE immediately so that the filter grids do not clog with plaster dust. Note the pressure on the filter so that when the pressure is 10 psi (689 millibars) higher you know to clean the filter. You might need to break down and clean the filter several times during the break-in process. Remove the trippers from the time clock and leave any necessary notes on it; the circulation equipment runs 24 hours a day for the first week.

8. **Filter** Purge all air from the system, especially from the filter. Filtration at this point is critical. You will recall from the filter chapter that air in the filter represents media area that is not filtering because it is not in contact with circulating water. Bleed the air from the filter regularly.

9. **Test and Balance** Test the water for pH, total alkalinity, and calcium hardness (at this point, there should be no chlorine residual to test). Plaster is made from calcium compounds, and if the water is low in calcium hardness specifically and total alkalinity in general, it will leach calcium out of the plaster, effectively dissolving it. When it is fresh and soft, plaster is especially unable to resist such corrosive water.

   The first component to balance is total alkalinity. Since this will also raise the pH, test and adjust that after you have adjusted the total alkalinity. Use sodium bicarbonate as directed on the product label to raise total alkalinity to about 120 ppm. *Bicarb*, as it is called, is extremely soluble and there is little risk of the product saturating the water and precipitating out of solution.

   Bring the calcium level up next (unless it is already in the 200- to 400-ppm range) by adding calcium chloride to the water (follow the instructions on the product). Make sure it is completely dissolved by adding it to a bucket of water, dissolving it, then adding it to the pool water. Add chemicals near return lines for even distribution, then brush the pool for more thorough circulation.

   Finally, adjust the pH. Generally you will need to lower the pH, but this must be done gradually. As noted in the chemistry chapter, pH readings take awhile to stabilize, particularly during the break-in process when no conditioner has yet been added to the water. Test the pH every four hours and, using the acid demand test, determine the amount of acid needed. Regardless of the result of that test, however, add no more than 1 pint (473 milliliters) of acid per 10,000 gallons (37,850 liters) of water at any one time. In this way you will avoid spiking the pH. Remember, you can easily correct water that needs more acid, but your plaster will immediately suffer if you add too much.

   The total alkalinity of the pool or spa represents the amount of alkaline material dissolved in the water. Because you are not finished dissolving alkaline material in the water, there is not much point in further adjusting this until the break-in process is complete. The calcium carbonate of the plaster dust, the calcium chloride you might have added, and other factors will affect the final total alkalinity measurement. Therefore, don't attempt to adjust it daily.

Also, do not add sanitizer or stabilizer at this point. Algae doesn't prosper on smooth surfaces (like new plaster), in alkaline conditions (like water curing new plaster), or on surfaces that are constantly being brushed (as you are about to do). Therefore, the sanitizer only has the potential to harm the plaster during the curing process, not help it.

Finally, continue to test the water for all parameters at least once per day. Better yet, test with each brushing (three times per day) and adjust the chemistry as needed in gradual increments.

As soon as the circulation begins, so must the brushing. Brush every inch of the plaster surface thoroughly and vigorously at least three times each day. Try to divide the day into equal parts, rather than brushing three times in just one part of the day.

10. **Brushing** Brushing knocks loose the plaster dust, avoiding scale buildup. It also keeps any precipitate or algae from staining the new plaster. Brush from the top of the walls down and from the shallow to the deep end, with the suction concentrated at the main drain to pull the dust and minerals into the filter. When you have finished brushing, return the suction to half main drain and half skimmer. As always, it is better to skim any dirt or leaves into the skimmer than to find them at the bottom of the pool.

11. **Filter Cleaning** Break down and clean the filter each time filtration slows and/or the filter pressure exceeds 10 psi (689 millibars) over its clean operating pressure. Plaster dust will quickly clog the filter, so be prepared to clean it as soon as the first day after starting the circulation.

12. **Vacuum** After 48 hours, vacuum the pool at least once each day before brushing. Use a brush vacuum rather than one with wheels that might leave marks on the plaster surfaces. Vacuuming will quickly fill the filter with plaster dust, so be prepared to clean the filter as needed.

13. **Final Balance** After 72 hours, begin to add sanitizer, gradually raising the residual to normal levels. As in all chemical

applications, add gradually. Because the plaster is still soft and receptive to stains, you must avoid precipitating minerals and creating stains, which can happen even with the correct amounts of chemicals. As always, the key is to add chemicals gradually and make sure the distribution is thorough by adding near a return outlet, by walking the chemical around the pool as you add it, and by brushing immediately after you add chemicals to guarantee there are no harmful concentrations of chemical in any one area. After your desired sanitizer residual has been reached, add stabilizer.

Continue these procedures for at least one week, preferably two. If you see no plaster dust coming off when you brush, the surface is curing but not yet finished. Continue the brushing routine for at least three days after you no longer observe any plaster dust.

Keep swimmers out of the pool during the break-in period. Body oil in the water can stain the plaster and add organic material, while the varying levels of chemicals during break-in can irritate skin and eyes.

Leak Repair

As previous chapters have detailed, there are many ways that a pool or spa can leak through the equipment and plumbing. There are an equal number of ways that the vessels themselves can leak.

Leak Detection Made Easy: Four Tests

The first place to look for leaks is in the exposed equipment and plumbing, but there are often visible signs of leaks that are otherwise hidden from view. Cracks in the pool or spa interior might be a sign of leaks. Tiles falling off the walls or loose coping stones often suggest structural leaks from shifting ground (which might have been caused by a water leak eroding the soil). Even cracked or lifting segments of deck might indicate the source of a pool leak. Tree roots might lift not only the deck, but the plumbing or pool wall as well. If, however, there are no such visible signs, you need to look elsewhere. Some leak detection methods will help to both verify a suspected, visible leak or locate a hidden leak.

EVAPORATION TEST

RATING: EASY

The most simple method is to fill a bucket and place it on the deck next to the pool or spa. Mark the level in the bucket with an indelible felt-tip marker and do likewise for the water level in the pool or spa. Turn off the circulation to eliminate any variables in evaporation. After several days, mark the new level of water in the bucket and pool or spa. They should evaporate an equal number of inches (or centimeters). If the pool or spa level lowers significantly more, a leak is likely. If both vessels lower a similar number of inches, then there is no leak.

DYE TESTING

RATING: EASY

As the name implies, a colored dye is disbursed in suspected areas and, as the dye disappears, the leak is found.

1. **Prep** Clean and brush the pool or spa thoroughly. Cracks can sometimes be hidden by dirt or other material settled in the crack. Pay careful attention to steps, corners, and around fittings.

   Turn the circulation off and begin the examination on a calm day. Wind rippling the surface makes it difficult to see small cracks. You might also want to squirt a little tile soap across the surface to sharpen visibility further (and you might need to repeat the process from time to time during the exam).

2. **Inspect** Examine the pool for cracks, beginning with the tile line. You might not need to go much further if you see gaps or missing tiles. Tap the tiles with the handle of a screwdriver (or gently with a hammer) to see if any fall off, are loose, or sound hollow. Note any positive locations.

   If you find probable leak locations along the tile line, don't just stop there. If tree roots, shifting ground, earthquakes, erosion, or other problems have caused leaks at the tiles, they might also exist elsewhere. It is important to find and repair all leaks.

3. **Shoot Dye** The dye test can easily be conducted using an old test kit reagent bottle or similar squeeze bottle filled with food dye (available at any grocery store). Some technicians use phenol red,

but you will need to check many locations and it is unwise to inject that much acidic chemical into the water.

Work around the pool, particularly in the locations of suspected leaks. You will need to get into the pool or spa, including diving to the main drain, to do a thorough job. At each crack or suspected area, aim the nozzle of the bottle at the crack. Squeeze a bit of dye into the area and watch it. If the dye simply swirls around the crack without being sucked in, then there is no leak in that area. If the dye is sucked into the crack, it is riding on a flow of water leaking from the pool. The speed with which the dye disappears will help you estimate the size of the leak.

4. **Observe** As with the visual inspection, continue around the entire pool looking for leaks to exhaust all possibilities. Be especially careful around skimmers, steps, rails, ladders, or other fittings. Light niches are often the source of unexplained leaks, so pay close attention to the entire light fixture area. If you are having an especially difficult time finding a leak, you might even wish to remove the light fixture and dye test directly into the niche itself, concentrating on the area where the cable passes through. Don't forget the interior of the skimmer and the main drain, as well as the return outlets.

After you have thoroughly examined the pool with the dye test, you will know what repair problems you are faced with.

DRAIN-DOWN TEST

RATING: EASY

If you have tried the evaporation test and dye testing, or if you are not a good diver and wish to skip the dye test, try the drain-down method.

1. **Prep** Turn off the equipment and mark the level of the water in the pool or spa.
2. **Mark** Mark the level again at the same time each day to establish a rate of leak. Because of normal evaporation, the level will continue to decrease indefinitely; however, the objective is to determine when the level stops lowering as a result of the leak. If, for example, you record a loss of 2 inches per day for four days, then the rate slows to

1 inch every five days, you will know the level at which the rate slowed was the level of the leak. Mark that level of rate transition.

3. **Inspect** Examine all possible leak areas along the transition level. The leak must be along this line. For example, if the water loss slowed when it reached the level of a particular return outlet, you might reasonably suspect the leak in that plumbing line. If the water slows when it lowers to the area of the light niche, the leak will likely be in there.

   The only fault with this method is that it is an indicator, not a precise tool. Since water seeks the same level in all plumbing and parts of the vessel, the water might stop at the level of a certain plumbing fixture, but actually be in an entirely different location that is coincidentally at the same level (Fig. 10-3). In any case, the new level will tell you where to look further and where you need not look.

LEAK DETECTORS AND PRESSURE TESTING

RATING: PRO

When the previous methods fail to help you locate the leak or you wish to further verify your assumptions, there are two other methods of leak location. There are electronic listening devices called geophones that can actually hear water dripping or flowing. By applying such devices around the pool and related plumbing, an operator can identify where water is moving out of the system. Because these devices are expensive and their operation requires a great deal of expe-

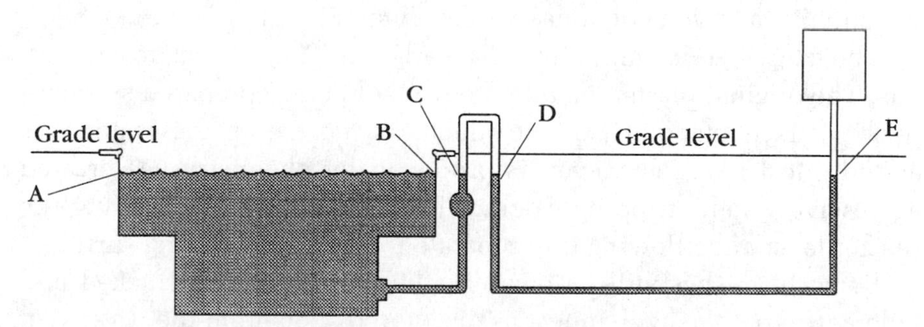

FIGURE 10-3 There are five possible leak locations for the water level depicted (marked A to E).

rience and skill, most service technicians don't buy or use them. There are numerous professionals who do, however, and they are easily found in the phone book or through referrals at your supply house.

The second method used to find leaks is pressure testing equipment. It is not difficult to pressure test a plumbing system with the knowledge already presented in this book, but the amount of time and additional equipment (plugs, adapter fittings, compressed air, and related fittings) makes this type of testing impractical for most technicians. It is wiser to contract out this work and make your money on the repairs. Many pool builders and plumbing contractors are equipped to pressure test pool or spa systems. Companies that conduct leak testing might also conduct pressure testing.

Patching and Repairing

Patching minor blemishes in plaster or replacing a few tiles is fairly easy and the products designed for these tasks are mostly foolproof. Patching techniques for vinyl liners in above-ground pools and acrylics in spas are covered in detail in the companion books in this series—*The Ultimate Guide to Above-Ground Pools* and *The Ultimate Guide to Spas and Hot Tubs* (both published by McGraw-Hill).

PLASTER

RATING: ADVANCED

When the pool is drained for any reason, you have an opportunity to look for plaster blisters or pop-offs, areas where the plaster has come away (delaminated) from the underlying surface. Such areas can be anything from the size of a quarter to dinner plates to several square feet. Beyond that and you might need to recommend replastering.

The causes of delamination, also called *calcium bleed*, are numerous. The original plaster job might have had poorly prepared surfaces, poorly mixed materials, or too much drying before the water was added. Most likely, however, improper water chemistry has created aggressive water, stripping calcium from the plaster, weakening the plaster layer and allowing it to separate from the underlying surface.

Some blemishes will be obvious, the plaster having cracked and fallen away. Others will appear as discolored spots from the water that has entered between the plaster and the subsurface. You can also look for plaster faults by tapping suspected areas with the handle of a screwdriver or with a hammer to listen for hollow sounds (assuming the pool is empty). If you are committed to making repairs, poke around the entire pool with a chisel or screwdriver to pop the blisters. They have to come off anyway, so you might as well discover them all. Once you've been around the entire pool you can then evaluate if patching is the answer or if replastering is more logical.

When you have identified blisters, chip the loose plaster away all around the area until you reach solid, dry plaster. Use a hammer, chisel, or screwdriver depending on the size of the blister. Once completely exposed, clean the area of all water and loose debris. To make sure the final patch blends in and appears even, clean up the jagged edges of the blister area by sanding the perimeter. The most effective way to smooth out these edges, especially if there are many blisters to repair, is to use a power grinder with a small [6-inch-diameter (15-centimeter)] diamond grinding wheel. The objective is to create a clean, smooth edge all around. You are now ready to patch.

The other major source of plaster problems stems from rusting rebar. If the gunite was not applied evenly over the rebar, water will rust the steel, creating red rust spots on the bottom of the pool or spa. These are very obvious against white plaster. Steel bleeds are more commonly caused by the steel wire used to tie the rebar together before the gunite is applied. If the tie wire is not closely clipped, the tails can stick up through the gunite and act as a wick to draw water and rust deeper into the rebar.

In either case, the spot of rust you see might only be the tip of the proverbial iceberg. Once the rust starts it spreads like a cancer and can affect vast areas. When you chip off the plaster, keep excavating to expose all rusted rebar. If you leave any rusting steel in the subsurface, it will leech up to the plaster and restain it.

When you have exposed all the rusted steel, cut it out with a hacksaw. Unless the area is large, you won't need to replace the rebar. If it is a large area, try to replace the same amount, size, and distribution pattern that you have removed (basically, crisscross the bar 4 inches apart). If you have excavated more than an inch or two, fill the hole with quick-set concrete and allow it to dry before going on with the surface plaster patching. The objective is to repair the hole to the level of the surrounding subsurface so that the plaster patch will be no thicker than the surrounding plaster.

The best way to learn how to work with any repair material is to practice in the workshop before attempting the repair in the field. A little wasted material is less costly (in time and money) than making mistakes on a customer's pool. All these materials are available at your supply house.

Mixing a batch of plaster for patching (or practice) and the actual application are simple.

1. **Water** Start with a clean 5-gallon (19-liter) bucket, adding 4 cups (1 liter) of water.
2. **Mixture** Slowly stir in calcium chloride (the powder used to make the mixture set up) until the liquid becomes warm to the touch. The reaction of the calcium chloride will actually make the water heat up. You will use less than a cup (250 milliliters), but keep monitoring the temperature (the chemical will not harm your fingers). Too much calcium chloride can actually melt plastic buckets, so don't be too heavy-handed. The general rule of thumb is one part calcium chloride for every ten parts cement used. You can also make patch material without it for a slower drying mixture. This might be helpful if the area to be patched is not

TRICKS OF THE TRADE: PATCH MIXING

- **An alternate method of mixing the patch material, which is especially useful if you are preparing a very dry mix for an underwater patch, is to combine the dry materials in a bucket first, then slowly add the water until the desired consistency is achieved. If you dry-mix, put the sand in the bucket first, then stir in the calcium chloride and cement (because these are lighter than the sand, they will combine more readily if they are added to the sand, rather than the other way around).**
- **For patching vertical surfaces, pool walls for example, the mixture should be thicker than if patching a horizontal surface. The texture of bread dough will make the patch material adhere and stay in place better as it dries. The looser texture for horizontal surfaces gives more flexibility and drying time when making the repair.**
- **When patching underwater (which is most common), the texture should be as dry as you can make it and still work the material. Obviously, as you fill the patch area underwater, the material will pick up additional moisture and the texture will loosen.**

underwater and it is a hot, dry day (or a windy day when the material will dry quickly anyway). Faster drying means more likelihood of cracking. Again, practice with several mixtures and compare results so you become familiar with both the art and science of plaster patching.

3. **Admix** Add 8 cups (2 liters) of white cement and 4 cups (1 liter) of sand (the supply house will have a specially mixed grain size of sand called *aggregate*). Mix the materials thoroughly with your hands, adding water if needed to achieve the texture desired for the particular patch job.

   The mixture created is called a 50/50 *admix*, even though there is twice as much cement as sand. That is because the weight of the sand is half that of the cement, meaning the final mix is actually equal weights of each.

   If the plaster in the pool or spa is colored, add the color powders slowly and keep mixing until you reach the approximate shade required.

   The most important consideration in creating patch mixture is to consider the drying time and conditions. If the plaster dries too fast, it will shrink and crack. If the material is too thick, it will not smooth out when you apply it and the result will be rough and unsightly. As noted previously, practice before going out to the job site.

   The above recipe will make enough material to fill an 18-inch-diameter (46-centimeter) patch about ½ inch (13 millimeters) deep. Remember, once you start the job, it might not be practical to mix more material and, if you do, it might not appear the same on the repaired area as the first batch. Therefore, always mix more than you need.

4. **Patch** Apply the patch with a trowel using broad strokes, trying to fill the entire area on the first application. Try to fill so that the patch material is higher than the surrounding plaster level, then remove the excess by scraping the trowel edge across the surface, making the patch clean, smooth, and level. Actually the best finishing tool for this smoothing is a straight piece of cardboard or plastic.

   The other application method is underwater, where you work by hand with a fist-size ball of material and push it into the patch area with your fingers. Again, use more than you need so you can scrape off the excess, feathering the edges into the existing plaster for a smooth finish. Don't be intimidated by working underwater, even if it means donning a face mask and fins. It saves time, money, and water over draining the pool. Unless you are a certified SCUBA diver, however, don't try to use air tanks for deeper work. In that case, drain the pool.

One last plaster patching technique is used to fill small surface cracks. Not all visible cracks allow water to penetrate the subsurface or cause delamination. Some are from too rapid drying or minor settling of the surrounding land or the pool itself. If there are no blisters, patch only the cracks.

The objective of patching small cracks is to slightly widen the crack so that it will accept patch material. You can use a small chisel to create a slight V shape along the crack. Follow the patching directions as outlined previously, either troweling on the patch material or rubbing it into the crack with your fingers and smoothing over the resulting repair with a straightedge.

Perhaps the most important aspect of any cosmetic patchwork is to have realistic expectations. Even the finest patchwork will be visible for a few weeks, simply because the new material is clean and the existing plaster has darkened with age. Colored plaster takes time to mottle and blend. If the customer understands that the reason for patching is primarily to avoid further delamination (or spreading of the rust in the case of steel bleeds) and secondarily for cosmetics, he will have more realistic expectations of the result.

COPING AND TILE

RATING: ADVANCED

When the ground around a pool shifts, which might be the result of settling or water leaking from the pool and causing erosion, the clues that this is happening might include popped tiles and coping stones. Coping stones are made of cast concrete, set edge to edge, so there is little room for expansion. A loose coping stone or one that has risen from the edge of the pool means there is some more extensive problem underneath. Rarely has a loose coping stone simply come free from the mortar in which it was originally set.

The objective then, is to remove the loose coping, excavate the underlying deck, determine the cause, and relieve the pressures that created the problem so the same coping, or more, won't come loose again. Loose tiles are also an early sign of pressures behind the location that need greater attention. The steps to diagnosing and repairing loose tile or coping stones are as follows (Fig. 10-4).

1. **Inspect** Lift off any loose coping stones and look for others by tapping each one with a hammer or the handle of a screwdriver. A hollow sound will quickly reveal other stones that might still be firmly in place, but that are not actually supported by anything.

   In many cases, you will need to remove the tile beneath the affected coping. It might have come loose already. Drain the pool to a level a few inches (or centimeters) lower than the bottom of the tile line. On larger jobs, it is easier to work inside the pool, so you will want to drain it completely.

2. **Cut** Remove the tile beneath the affected area by cutting through the grout and mortar that separates the bad section from the good. Also cut into the pool wall beneath the affected tiles. The cut should be about ½ inch (13 millimeters), made with the diamond blade on a hand-held electric grinder (described earlier). This cut

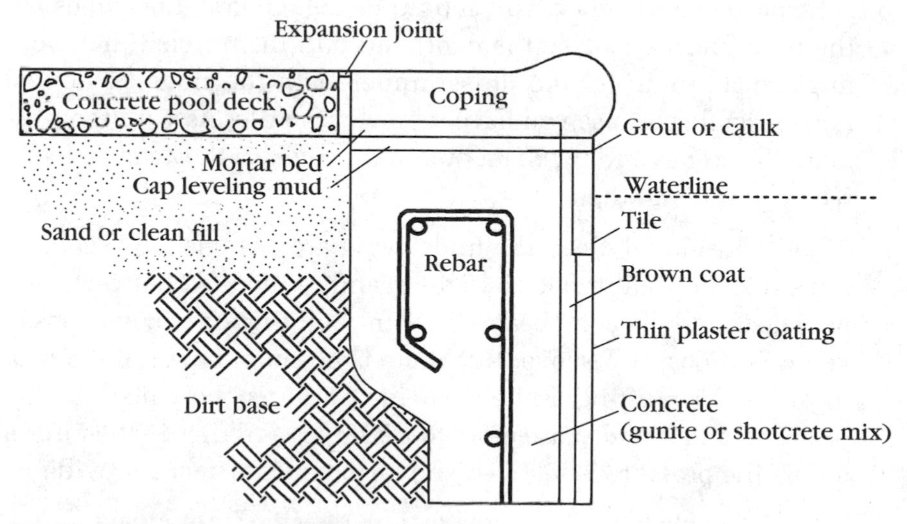

FIGURE 10-4 The coping stones and tile line.

prevents spreading cracks or chipping of the adjacent undamaged sections of plaster and tile.

3. **Chip** On small repair jobs where you have not completely drained the pool, you might want to float a piece of plywood under the work area to catch as much debris as possible. Using a broad, flat chisel, chip the tile away to expose the mortar bed and/or pool wall beneath. If you are careful and have made adequate cuts, the tile might remain intact, but don't count on it. Make sure you have replacement tiles of the same design or that you can purchase an acceptable substitute. To make reassembly easier, mark the tile or lay it on the deck behind its location so you can return it to the same spot. When you remove each tile, place scratch marks (or use an indelible felt-tip marker) to note the exact positioning of each tile to make replacement easier.

4. **Open** Remove the stones. If the coping stones were not loose already, cut the grout joint between each one to make removal easier. Use a concrete saw, available at any tool rental store, with a 12-inch (30-centimeter) diamond blade. Cut along the mortar at least 4 inches (10 centimeters) deep to free each end of the stone. The back side of the stone should not be connected with mortar, but rather with a flexible expansion joint mastic or silicone that will not hinder removal.

   Standing on the deck, you should be able to grab the stones by the nose (the side facing the pool) and pull them free. If not, you might need to drive your chisel underneath the stone, which is now possible because you have removed the tile. As with the tile, mark the stones and lay them out on the deck so they can go back where you found them.

5. **Clean** Chisel and clean the underlying area as needed. Remove old expansion joint material and examine the area between the pool deck and the bond beam. If coping and tiles have come loose, you will probably discover that there is no space between the two to allow for shifting and expansion. This is the cause of the problem, and it will cause more loose stones and tiles in the future unless the pressure is relieved and an expansion space provided.

6. **Joints** The objective is to create an expansion joint area about ½ inch (13 millimeters) wide and deep, enough to totally separate

the deck from the pool wall. Chisel or cut away any material that is pressing against the pool wall. Never cut the pool wall or bond beam to create the expansion joint.

7. **Fill** When the expansion joint is complete, fill in any dirt that might have eroded away to complete the backfill area. If severe erosion has occurred, you might need to demolish a larger portion of the deck to expose the amount of backfill lost, replacing it before continuing. In such cases, work with a general construction contractor to rebuild the deck.

8. **Patch** Prepare the plaster patch material as outlined previously. Some technicians add a latex additive to give the patching compound resiliency, but the best advice is to match the surrounding material. If the existing mortar is somewhat flexible, the new material should be as well. You can also use a premixed waterproof product like Thoroseal, applying two thin coats before resetting the stones. The advantage of using a waterproof mortar is that you prevent water from weeping or leeching into the backfill again.

9. **Reset** Clean the stones of dirt and old, loose mortar. Apply a light coat of patch material to the underside of the stone, then sufficient patch material to the mortar bed to bring the stone up to its original level. You want to raise it slightly higher than the adjacent stones so that when it is pushed into place, it will settle down to the correct level. Tap the stones in place with a rubber mallet.

   Be careful not to allow the patch material to fall into the pool. If it does, be prepared to clean it up quickly so you don't create unsightly plaster chips in the pool.

10. **Retile** Prepare a brown coat of mortar to reset the tiles. The preparation of the bed is the most important step, because there might be high spots of old grout or mortar left after the original removal of the tile. Grind these down with a hand-held electric grinder. It is better to grind too much, which can be filled with new mortar, than to leave high spots that prevent the tiles from reseating. Follow the marks you made to replace the tiles in the same locations.

    Thoroseal-type products work well and prevent water intrusion when the job is done. You might need to mix it with less water so it will hold the tiles on the vertical. Apply the tiles in the same manner as the coping stones.

11. **Grout** Regrout the stones and tiles. Grout can be premixed material purchased at the supply store or hardware store, or it can be mixed by combining one part white cement with two parts sand. Use #60 silica sand unless the joints are over ½ inch (13 millimeters) wide (then use #30). Mix the grout to a loose enough texture to ensure it will completely fill the voids between the stones. Overfill, allow the mixture to set up slightly, then wipe away the excess to a smooth, level surface. Do the same with the tiles, carefully wiping all excess off the surface of the tile. A wet sponge wiped over the finished tile and coping will remove any grout film, which will otherwise leave a discoloration looking like paste wax.

    Some pools use a colored grout on the coping stones or tile. Powdered dye is added to the grout and mixed until the color matches the existing installation. Remember, the finished job will not match until the new work has had a chance to weather and acquire the same shade as the surrounding work.

12. **Seal** Complete the expansion joint. Fill the joint with sand up to the last ½ inch (13 millimeters). Fill the last ½ inch with flexible mastic or silicone joint sealer, which can be poured as a liquid or injected like caulk. Follow the product label directions for application, especially concerning temperature and humidity ranges.

If you understand the basic underlying construction of the coping and tile area of the pool or spa, you will be able to make these basic masonry repairs. Many will not be as complicated as described, requiring only rehanging a few tiles or resetting a single loose stone. If that single tile or stone keeps coming loose, however, or if more than one are loose, follow the procedures described earlier to determine the cause and effect a long-term repair.

Remodeling Techniques

There are several other ways to remodel a pool or spa. New decks and landscaping will improve the look of an older pool (best left to a general contractor and landscaper). As for the vessel itself, the water technician can offer resurfacing of the interior, new tile, and recaulking of the expansion joints to help the look of an existing pool or spa.

Plastering and Replastering

In most areas there are numerous subcontractors who will plaster a new pool or replaster an old one. It is usually best to refer the customer to one of these professionals and earn a referral fee and the fee for the break-in service as described previously.

Because nothing is more critical to the successful maintenance of a pool or spa, it is valuable to understand the plastering process. It also allows you to evaluate a contractor and a good (or poor) plaster job.

NEW PLASTER

RATING: PRO

Let's begin with a discussion of the plastering process, then review the variations for replastering.

1. **Clean** The gunite or concrete pool shell must be clean of loose dirt, algae, and water before plaster is applied. To accomplish a contaminant-free environment, wash and scrub the shell with chlorine and water to remove organic waste. Next, wash and scrub with an acid solution of one part muriatic acid to four parts water to eliminate minerals and concrete dust (called *latiance*).
2. **Surface Prep** Plaster will not adhere to an acidic surface. Every bit of acid must be removed from the concrete shell. Neutralize the acid with trisodium phosphate (TSP) or baking soda. Use 1 pound in 5 gallons (20 grams in 1 liter) of water of either product, creating a wash. Thoroughly scrub and pump out the shell. The concrete of the shell must be moist (hydrated) but not have standing water. The shell is now ready for plastering.
3. **Hardware Prep** Remove all plumbing outlet nozzles and any removable hardware such as light fixtures or ladders. Stuff rags in the plumbing to keep plaster out of it and tape off the threads of any fittings. Finally, tape the bottom of the tile line to keep it clean.
4. **Mix** The plaster material is a mixture of one part white cement for every two parts aggregate (sand and marble dust, also called *marcite*; or sand and limestone). Calcium chloride is added to help the mixture set up more quickly, but never more than 2 percent of the weight of the cement (not the total mix; only 2

percent of the cement itself). Plaster is mixed in a cement mixing drum using a paddle mixer. Be sure that any gray cement residue from previous work is not still on the inside of the mixer because it will contaminate the color of the new plaster. Color powders are added to the mix if the final plaster is to be colored. Water is added, roughly 2 gallons for every 100 pounds (1 liter for 6 kilograms) of dry mix, until a heavy oatmeal consistency is achieved. The more water, the weaker the mix, so caution in mixing is advised. The entire mix must be completed in 10 to 12 minutes.

5. **Apply** The plaster is applied by trowel. The first coat should be scratched on thin to fill and smooth out the roughness of the concrete. Then, two more thicker coats are applied while the underlying coats are still wet. It takes several plasterers to apply the material fast enough to finish in a few hours so that the pool can be filled before the plaster dries out and cracks. Care must be taken to avoid leaving plaster on the tile or fittings. When cleaning these, avoid running water over the fresh plaster; this can leave furrows or stains.
6. **Break-in** Break-in the new plaster as described previously.

When plaster cures, the topcoat becomes almost transparent, allowing any imperfections in the shell to show through. The first coat of scratch plaster is important in leveling out such imperfections, although it cannot completely hide them. If gray lines or slightly darker spots appear at various places after the pool is full, this is usually the reason. Therefore, the finish job of the shell is as important as the scratch coat.

In locations where gunite rigs are unavailable and the concrete shell is poured by hand, such imperfections are common. In these cases, it is valuable to grind or chisel down any high spots prior to plastering.

REPLASTERING

RATING: PRO

A good plaster job over a well-made gunite shell will last up to 20 years. When plastering the shell for the first time, the plaster permeates the pores and roughness of the gunite surface, creating a firm mechanical bond. When replastering, the new plaster must adhere to the old smooth surface with either a new mechanical bond, a chemical bond, or a combination of both.

There are several ways to prepare the old surface to enhance the bonding of the new plaster. Depending on the condition of the old plaster, one or more of these preparations might be used. The surface preparation is the most important step of the replastering process, because fine workmanship will be of little value if the plaster doesn't stick to the old surface.

The first preparation method is to etch the old surface with acid. As discussed in the chemistry chapter, muriatic acid will dissolve plaster, leaving it rough and pitted and thereby promoting a better mechanical bond for the new plaster. The old plaster is washed with raw muriatic acid to deeply etch the surface.

The limit of this method, however, is that acid will not dissolve organic waste embedded in the old plaster. It does not remove loose dirt, although scrubbing and washing after application of the acid might handle that. Finally, if any acid residue is left in the porous old plaster, it can create gas when activated by the moisture in the new plaster and lift the new material from the old surface. Acid etching is usually a good step, as long as it is thoroughly cleaned away.

The next method is to chip the old surface with hammers and/or axes. Workers literally chip away at the old plaster, working around the pool creating a jagged, rough surface for the new plaster. Again, this enhances the mechanical bond for the new material.

Sandblasting literally sands away the old plaster down to the concrete. It does not, however, make the surface rough to aid mechanical bonding. Sandblasting is usually used to remove old paint from a plaster job before etching or chipping.

Finally, the removal of the old plaster can be accomplished, although expensively, by jackhammering the old material off of the pool shell. The danger of jackhammering is that it can crack the pool shell, causing structural failure when the weight of the water is added.

Having prepared the surface with one or more of these methods, a chemical bonding agent is added to the scratch coat (the first coat of plaster as described previously). The bonding agent helps the scratch coat adhere to the shell on one side, then helps the plaster adhere to the scratch coat on the other side. Called a *brown coat* because some bonding agents slightly darken the scratch coat, the chemical bond rarely lasts as long as a good mechanical bond. Any replastering job should last at least ten years, however.

Other than these preparation differences, the original plaster job and the replaster job are alike. When advising your customer on the various methods of surface preparation and their relative costs, remember the value of mechanical bonding over simple chemical bonding. Chipping or sandblasting followed by etching and cleaning can be costly, but a few hundred dollars now might double the life of a $2000 plaster job, making the extra preparation very cost-effective.

One last way to evaluate the quality of a plaster contractor is to look at a recent job site. A clean site with no plaster on the tile or white staining on the deck or walkways leading to the pool is a reflection of a good contractor. The sloppy ones are also sloppy in their plastering work, taking shortcuts that shorten the life of their work. Also, since many problems can develop with plaster that might be related to the contractor, choose one that has been in business awhile and will likely be there to back up a warranty tomorrow (or at least one that will be around to offer technical advice for any problems that you or the homeowner might have caused).

Plaster Nodules: Sometimes you will encounter small volcano-shaped formations on plaster or in the grout between tiles which are calcium hydroxide concentrations leaching out of the plaster, transmitted to the surface by pinpoint holes or other voids. They are usually easy to remove by chipping or sanding. Occasionally they are discolored and may require chipping and patching. Often, nodules will recur at the same location, often several times, until the source of the calcium hydroxide is exhausted. Nodules are not a sign of anything wrong with the plaster job or pool subsurface, just unsightly annoyances.

Fiberglass Coatings

RATING: PRO

Numerous companies now offer fiberglass coatings for your pool as an alternative to plaster (Fig 10-5). The benefits are ease of maintenance (smoother surfaces discourage algae growth and pH is easier to maintain at the desired level when no alkaline plaster is involved); long life (fiberglass surfaces are virtually indestructible and will last the life of the pool, compared to ten years, on average, for a plaster coating); and no rough spots on bare feet. Coating your pool with fiberglass is not a do-it-yourself job.

Fiberglass coatings require the same surface preparation as replastering. The best mechanical bond is achieved by roughing up the surface prior to application. Fiberglass sheeting (Fig. 10-5A) is laid on the pool interior and painted in place with chemical resins and fixatives (Fig. 10-5B).

The finished product looks great (Fig 10-5C) and there is no break-in required as there is for plaster surfaces. Refill the pool, balance the chemistry, and enjoy swimming right away. Take extra care with chemicals from this point on. If you are heavy handed with acid, for example, there is no plaster to neutralize it, so you may have corrosive water circulating through your equipment and plumbing for some time. Otherwise, fiberglass coatings are generally maintenance-free.

Inexpensive Pool Face-Lifts: Paint

RATING: PRO

The purpose of coatings of any kind over the gunite material of the shell is to prevent leaking, because concrete itself is porous. Modern pool paints are an attractive, inexpensive way to coat gunite, fiberglass,

FIGURE 10-5 Fiberglass coatings.

FIGURE 10-5 *(Continued)*

plaster, or any other interior surface with a smooth, colorful, waterproof coating.

Today, many paints will last up to five years, providing an inexpensive alternative to other coatings. Surface preparation and painting can be done by any careful technician.

As with other coatings, the success of painting depends largely on the preparation and qualities of the subsurface being covered. If the paint is to cover old plaster, it won't keep delaminating material from peeling from the shell surface. In other words, before painting, evaluate the surface being covered. If it won't last for three to five years more, then there is no value in painting over it. It is time to replaster or remove the old plaster before painting.

It is also important to have realistic expectations about paint as a surface covering. Irregularities in the subsurface will show through the paint and will still be felt underfoot in shallow areas. Colors might be vivid initially, but will fade throughout the life of the paint—especially the brighter the original color. Finally, paint will not last much longer than three to five years and might begin to peel or dissolve prior to that.

Pool paint is manufactured in three types. Chlorinated rubber is designed to be flexible when applied, therefore it is durable and appropriate for rough, previously unfinished surfaces, acting like a stretch-rubber cover. Solvent-based paints are also good on rough surfaces because they are designed for thinning with additional solvent, making them easy to spray on. Water-based epoxies are used for painting fiberglass pools or spas and are more resistant to chemical variations and temperature extremes.

Generally, cover old paint with new paint of the same type. You can use water-based epoxies to cover chlorinated rubber, but not the other way around. If you're unsure, sand off the old paint or consult the manufacturer (both good ideas anyway). Another simple test method is to rub a little of the new paint over the old painted surface. If the old material dissolves or bleeds, don't use the new product without removing the old.

Also, dark colors will cover more evenly and hold up longer than lighter colors, the worst being blue. If the customer wants a blue pool, remind him that plaster is white but the pool looks blue when the water is added and chlorinated. Paint white or a dark color.

If you are prepared to proceed, follow the paint manufacturer's label directions and these guidelines.

1. **Prep** Prepare the surface. Empty, clean, and remove any loose materials from the pool. Sand any loose plaster or high spots. If there are blisters, follow the plaster repair procedures to prepare a level surface. Remember, paint can't replaster or level out a pool surface. If the surface is fiberglass, rough sand the entire area to be painted for a good mechanical bond.

2. **Clean** Clean the finished preparation job. If you have spent many hours creating a level, paintable surface, don't waste that effort by leaving dust and dirt on it that will prevent the paint from adhering. Wash down and scrub the surface with TSP [1 cup per 1 gallon (1 milliliter per 15 milliliters) of water and 1 ounce (29

milliliters) of tile soap to cut any oils on the surface]. Make sure that you are using pure TSP, not products labeled as such but containing additives that might leave residues.

3. **Acid Wash** Acid wash the entire area to be painted, following the acid wash directions detailed later in this chapter. When acid washing for a paint job preparation, the idea is to etch the surface to create a roughness that will help the paint adhere. In regular acid washing, the idea is to clean the plaster but not to etch. In other words, for a paint job use a stronger acid mixture and leave it on the surface longer.

4. **Clean** Scrub and rinse the entire surface with TSP again to remove any trace of acid, this time without any soap. Rinse and scrub again with clean water. This is not the time to skimp on labor, because any remaining acid or dirt will prevent the paint from adhering.

5. **Dry** Allow the surface to dry thoroughly and don't plan to paint on foggy or humid days when the moisture content in the air is high. Similarly, avoid working during hot afternoons when the surface to be painted is hot. Either condition will keep the paint from bonding to the surface or cause blisters. As with all painting, follow the product directions. Mix the paint, enough to do the entire first coat. Mixing might mean simply stirring or it might mean adding catalysts according to the manufacturer's directions.

6. **Paint** Use good-quality tools to apply the paint. Thin, cheap rollers will buckle under the pressure needed to force the paint into the crevices of the roughed-up surface, especially on the first coat. A 9-inch-wide, ½-inch-nap (23-centimeter-wide, 13-millimeter-nap) lambskin roller on a sturdy frame works well. Cut in (detail paint that can't be done with a roller without overpainting) the tile, drain, light, and other detail areas with a good bristle brush first, then roll on the paint. Apply a thin but even coat. The first coat will use two to three times as much paint as subsequent coats, except when applying over fiberglass.

7. **Topcoat** After waiting the manufacturer's recommended time between coats, apply a second and, if necessary, a third coat. On all coats, but especially these finish coats, paint the entire surface at

one time. If you stop and restart, no matter how careful you are, the dividing line will still be visible with slightly different tones of color. On the last coat, sprinkle some clean sand over the wet paint on the steps and in the shallow areas to aid traction for swimmers.

8. **Cure** Let the paint cure as recommended by the manufacturer, then refill the pool or spa. Avoid using the heater for a few more days, just to be sure the paint has thoroughly cured.

TRICKS OF THE TRADE: POOL AND SPA PAINTING

Most good pool and spa paint jobs are 90 percent preparation and 10 percent execution. Here are tips to make the job easier and the results more satisfactory:

- **A smooth (patched if necessary) surface yields the best paint job, so after cleaning the surface to be painted, identify areas that need repair by running a rigid putty knife over the suspect areas such as the underside of fittings. Many paint manufacturers also make patch kits that are specially designed to facilitate painting, so use them.**
- **If the surface to be painted is very rough, use epoxy paint. You can apply several coats, building one on the other, creating a nice, smooth shell.**
- **Remember that the surface and air temperatures may differ, so don't start painting if it's approaching the warmest part of the day and the pool has been in the sun for awhile, even if the air temperature still feels cool.**
- **If painting a surface that has once been painted and you don't know what type of paint was originally used, try this test. Dab a little chlorinated rubber paint on top of the old surface and wait about a minute. Rub the area with a rag. If the old paint rubs off with the new, the original paint was also chlorinated rubber. If not, it was probably epoxy-based paint.**
- **Indoor pools should be painted only with water-based paints since they contain no noxious solvents which can create fumes. Allow 2 to 4 more days for proper drying before refilling and use fans throughout to facilitate the process.**
- **Always read the labels on the paint cans thoroughly—you'll be amazed what you can learn and the mistakes you'll avoid!**
- **Stir the paint often, especially if it's a colored paint, to ensure even appearance.**

Maintaining a Pool or Spa with a Painted Surface: Maintenance of painted surfaces is the same as other pool and spa coatings, but since there is no plaster to raise the pH, pay special attention if you need to add acid. Sometimes the painted surface appears to have a chalky coating, which is easily removed when brushed. This is not the paint "curing," as some will tell you. Chalking usually occurs when the total alkalinity has dropped too low and some minerals come out of solution and deposit on the walls of the pool, essentially scale, which is common to all pools (but usually appears only on tiles). To fix the problem, adjust water balance as previously described and use a clarifier or flocculating agent to help filter out the unwanted mineral. Brush often for several days and keep the filter running 24 hours per day to finish it off.

Acid Washing

RATING: ADVANCED

Perhaps the straightforward remodeling technique is an acid wash to clean the plaster surfaces of the pool or spa. As discussed earlier, it is necessary to drain the pool from time to time, to remove scale causing minerals and chemicals. This is a good time to acid wash the pool, to brighten the appearance and remove any deposits that might have built up.

As with other cosmetic procedures, the key to success is realistic expectations. If there is severe etching and staining, an acid wash will not smooth out rough plaster or remove every discoloration. It will make the pool look better, but it might still feel rough underfoot and show signs of metal deposits and/or scale from years of abuse. Especially with colored plaster, you must be careful not to remove so much scale and/or stain that the acid also removes substantial amounts of plaster, leaving pitting or gunite showing through.

Before acid washing, it is important to know how old the plaster is and if it has been acid washed before. If the plaster is nearing the end of its useful life (10 to 15 years), there might not be enough material to wash without stripping the surface down to the gunite. Similarly, if the pool has been acid washed two or three times already, it might be time to consider replastering, painting, or some other recoating.

1. **Shut Down** Turn off the circulation equipment and in-water lighting at the circuit breakers.

2. **Drain** Drain the pool, following the techniques described previously. The best technique to keep downtime to a minimum is to start pumping out the pool in the late afternoon. Watch your disposal technique for a long enough period to know it will evacuate all the water safely, then allow the pump to run overnight. In the morning, most residential pools will be empty and ready to acid wash. On larger pools, use multiple pumps to achieve the same results.

   Position your pump near the main drain, unless you know the lowest point of the pool to be elsewhere. I have often found that the lowest point is several feet away from the main drain, but that can be difficult to determine until the pool is virtually empty. In any case, arrive at the pool with enough time to finish pumping the last of the water. If you're down to the last 200 to 300 gallons (750 to 1000 liters), you might want to leave it in there as a dilution pond for the acid you will be using. If the pool also includes a spa, you need to drain that as well before beginning work. If you use the technique of directing the waste water to a lawn or garden for irrigation, remember to redirect the discharge hoses before you actually begin the acid wash. Direct the hoses into a deck drain or run them out to the street.

3. **Set Up** Set up your equipment on the stairs. Hose off the deck and continue to keep it wet during your work time to prevent inadvertent staining or spills. While we're on the subject, once you begin, wash your boots off each time you exit the pool. They carry acid as well.

4. **Prep** Remove the light fixture (Fig. 10-6A) and any other loose or removable hardware such as ladders, rails, or metallic return outlet nozzles. Remove the main drain cover. Clean out the main drain and light niche. If the pool is plumbed with copper, stuff the outlet pipes, skimmer, and main drain with rubber plugs or rags soaked in water and soda ash to neutralize and/or keep out as much acid as possible. Rinse (and scrub if needed) any other organic material (oils, leaves, and dirt) from the plaster, because the acid will not dissolve these and therefore will not clean the plaster beneath.

5. **Look** Inspect the pool for plaster blisters as described previously. Make any necessary repairs or patches before acid washing, so the

subsequent acid wash will clean all surfaces to approximately the same color, helping any patches to blend.

6. **Precautions** Before starting the acid wash, put on old clothes (a few small acid splashes will put holes in cloth), rubber boots, gloves, and a respirator (rated at least R25). The boots provide skin protection and insulation against shock if the pump is shorted out in the waste water around the main drain. **Always** use a respirator, gloves, and boots.
7. **Water Supply** Prepare a garden hose that can reach all parts of the pool and keep the water running. Although leaving the hose on will waste a bit of water, it will also keep the floor of the pool wet in the area you are working, neutralizing the acid as it runs toward the main drain and the pump. In this way you can control the area the acid covers and the length of acid contact.
8. **Test** Take a garden-type sprinkling can (Fig. 10-6B), add a squirt of tile soap, fill it two-thirds with water, one-third with acid (always adding the acid to the water, not the other way around). The soap will help the mixture cling to the walls of the pool and will cut the fumes somewhat. If you add the soap first, you won't need to stir or mix the solution. *The effectiveness of acid washing is a function of the strength of the mixture and the length of time it contacts the plaster.* To determine what the particular job needs, test the solution on a small area by pouring a stream from the sprinkling can. Allow it to sit on the plaster for about 30 seconds while scrubbing the area with a stiff-bristle broom or brush, then wash it off with clean water from the hose.

   If the stains haven't disappeared, you might need to leave the mixture in contact longer, strengthen the mixture, or go over the area more than once. Trial and error is the only way to become familiar with these variables, especially since you might encounter several different problems within each job. The one-third acid mix might work fine for most of the pool, but a few areas might need a 50 percent solution or even straight acid to budge tough stains. I have had to apply pure acid and leave it for up to 45 seconds on some stains to get them clean, so be prepared to spend some time experimenting on each pool to determine the needs of the job.

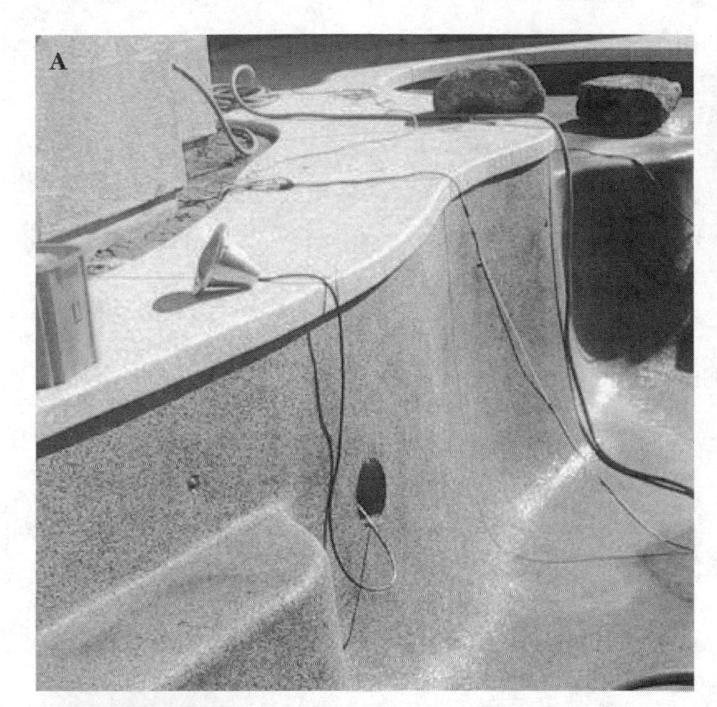

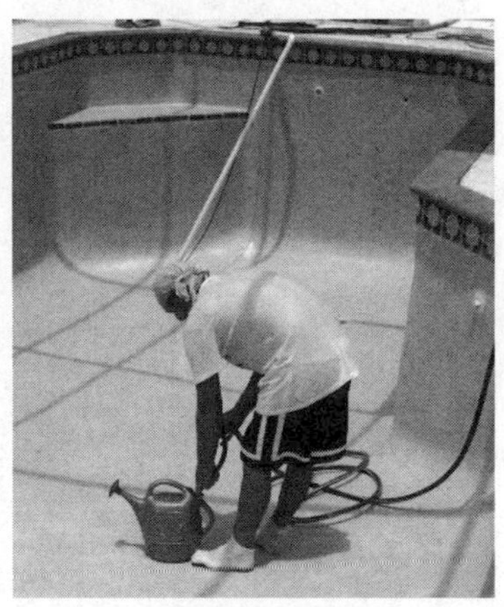

FIGURE 10-6 (A) Preparing for acid washing; (B) preparing the acid wash mixture; and (C) rinsing off excess acid.

One word of caution, however: you don't want to cure the illness, but kill the patient. It's not worth removing every last bit of discoloration or scale, just to end up with extremely rough surfaces or plaster so thin that you can see the gunite showing through. Therefore, the other reason for conducting some trial spots and solutions is to determine the effect on the plaster as well as the stains.

If you find stronger solutions and longer contact times having little effect you might have an older plaster mix composed of a silica sand base, rather than the more erosive modern mixtures of crushed marble (marcite) and calcium chloride. Don't be afraid to try stronger mixes or pure acid if that's what it takes to achieve good results.

Another reason for the test is to determine the amount of acid you will need for the job. Whatever you think you need, bring at least 4 gallons (15 liters) more (an extra case). Nothing is as aggravating as running out of acid with just a few more feet of plaster to wash. By taking note of how much you use in your test area, extrapolating that over the pool, and adding a little extra, you will generally be prepared. As a rule of thumb, a typical 20,000-gallon residential pool (76 cubic meters) usually requires about 20 gallons of acid (or about 4 liters of acid for every 4000 liters of water). You might not use it all, but it is better to be prepared.

9. **Acid Wash** Once you have determined the acid strength and contact time for the job, continue around the pool with the sprinkler can of acid solution, the scrub broom, and hose for rinsing. Keep rinse water flowing on the pool bottom when you are not actually rinsing an area that has been acid washed to neutralize acid on the pool bottom. Keep the pump operating to remove the waste from the hopper (the low point in the pool around the main drain). Even though the acid solution is somewhat neutralized by the plaster it is dissolving, it is still very potent. If you allow it to run undiluted along the bottom as you work the entire pool, when you finish you will have acid washed the bottom and the hopper several times.

   If your area establishes a minimum allowable pH for waste water, you might have to toss a cup (250 milliliters) of soda ash into the hopper and mix it with the broom from time to time to bring up the pH of the discharge you are pumping. Some jurisdictions prohibit a discharge with a pH less than 5.0 and it can take up to 8 pounds (3.6 kilograms) of soda ash to neutralize 1 gallon (3.8 liters) of acid, so make sure you have adequate soda ash on hand when you do the work (soda ash is very inexpensive).

The purpose of scrubbing with a broom is to force the solution into the pores of the plaster for thorough contact. All the scrubbing in the world won't actually remove stains, so let the acid do the work by making sure it is in complete contact with every part of the plaster. The most effective way to work is to have one person apply the acid and have another scrub. Either can rinse after they finish. If you must work alone, use a more diluted concentration, allowing you to leave it in contact with the plaster longer and giving you ample time to apply, scrub, and rinse.

10. **Safety** For safety reasons it is better to work with a helper. Acid washing not only cleans the plaster, but leaves it smoother than before as it removes scale and rough spots in the plaster. Since the smooth surfaces are also wet, it is easy to slip and fall, especially on the steeper slopes of the deep end. For this reason, I try to work around the sides of the pool, then the bottom, always standing so one foot is on the lower part of the slope and one on the higher. If you stand with both feet facing the shallow or deep end, it is easier for your feet to slip out from under you. Also if you face uphill, you are less likely to fall than if you face downhill. No matter how careful you are or how slowly you work, you will experience what I'm talking about firsthand sooner or later, so work with a helper.

11. **Tile** Take advantage of the pool being empty and clean the tiles. Under normal conditions, with water sloshing along the tile line, it is often difficult to remove all of the oils and scale on the tile. With the pool empty, scrub the tiles thoroughly, using tile soap, pumice, or the barbecue brush and, while you have it, some acid solution. If the tiles are particularly dirty, you might want to clean them before the acid wash so that any runoff washes over the plaster before it is cleaned. Use a pumice stone or block to scrub out any especially stubborn stains in the plaster.

12. **Rinse** Finally, thoroughly rinse the pool (Fig. 10-6C) and lightly scrub any remaining acid or soap residue, so that when you refill

the pool it will not have negative effects on the pH. You might want to go over the steps and shallow area, where feet contact the surface, with an electric sander and some fine to medium wet/dry sandpaper (80 to 100 grit) to smooth out any rough spots.

13. **Clean Up** Give the light fixtures and other hardware a good cleaning before reinstallation. Metal fixtures can be lightly acid washed and the plastic ones can be scrubbed with cleanser and a wire brush. Use a cup to remove any acidic water from the main drain before reinstalling the cover. Reassemble what you took apart and remove your equipment from the pool. To avoid trailing acidic mixtures across pool decks and through access areas. I rinse and then put the pump, hoses, sprinkling can, and empty acid bottles in big plastic garbage bags before removing them from the pool.
14. **Fill Up** Fill the pool. Unlike fresh plaster, there is no concern after an acid wash of bruising the plaster, so turn on fill lines and garden hoses to refill the pool as soon as possible. Before restarting the circulation, make sure the filter and strainer basket are clean so that dirt doesn't flush into the pool and onto the clean plaster.
15. **Start Up** Brush the pool while circulating the water for a few minutes, then check the chemicals and balance (as described in the section on restarting after a new plaster job). If the pH or total alkalinity are extremely low, turn off the circulation, balance the chemistry, and brush to circulate your chemicals before restarting. If you didn't do a good job removing excess acid, resulting in corrosive water, you might strip copper or other metals from the pool equipment, staining the plaster all over again.

Remodeling the Deck

One of the more dramatic improvements that can be made to an old pool is not necessarily in the pool itself, but rather a remodeling of the deck. The variety and texture of deck surfaces is limited only by your imagination.

Since deck work, including brick, wood, and the "new" surfaces discussed here, all require special training and talent to install, we will not go into detail on this topic. The point of presenting some basic information is to assist in making a choice of new deck materials. Here's a sampling of choices that are readily available today:

- Traditional cement and sand combinations, but sprayed or brushed with colors to create a more natural rock appearance.
- Cast-in-place colored concrete that is stamped by special forms to create the appearance of wood, brick, stone, or flagstone.
- Epoxy stone coverings, composed of small colored gravel held together by a clear, waterproof adhesive bonding material.
- Deck paints with nonskid texture incorporated.
- Natural cobblestone, flagstone, brick, marble, or sandstone. Real stone offers an appearance and texture that is hard to equal, but which is more costly to install.
- Wood is a low-cost alternative that provides natural beauty which changes in color as it weathers (Fig. 10-7). Wood decks must be sanded and retreated with waterproof coatings periodically to prevent cracks and splintering.

Many of these treatments are continued from the deck to the coping to make a seamless appearance around the pool. This concept is especially striking with natural materials (or those that appear natural), creating the appearance of a natural pond among the rocks.

FIGURE 10-7 **Wood deck.** *Gordon & Grant Hot Tubs.*

FAQs: REPAIRS AND OTHER SPECIAL PROCEDURES

Can I Fix Leaks Underwater?

- Yes, although the most secure patches are done on clean, dry surfaces that have been properly prepared. Small, flat areas of damaged plaster are the best candidates for underwater repair. Look for patch compounds that are designed for underwater application. Leaks around plumbing or lighting fixtures can also be repaired underwater using silicone and other patch products designed for such applications.

Can an Empty Pool Develop Cracks or Even Pop Out of the Ground?

- Yes, hydrostatic pressure (groundwater rising underneath the pool) can inflict serious structural cracks or even lift an entire pool out of the ground. Don't drain the pool for prolonged periods and never after heavy rains or if you know you live in an area with a very high water table.

How Do I Repair Leaks in My Vinyl-Lined Pool or My Acrylic Spa?

- Look for detailed repair instructions in my two other books in this series, *The Ultimate Guide to Above-Ground Pools* and *The Ultimate Guide to Spas and Hot Tubs* (both published by McGraw-Hill). Those volumes also contain hundreds of tricks of the trade and maintenance step-by-step procedures for those unique water vessels.

CHAPTER

11

Water Features

The fastest growing segment of the water industry is the water feature market. Everything presented to this point has been information that is applicable to any water vessel from the smallest portable spa to the largest swimming pool. There are a few repair and maintenance techniques that are unique to water features that are worth learning.

Fountains, koi (large goldfish) ponds, waterfalls, wading pools, and other decorative systems are grouped together under the collective term water features. Water features are essentially small swimming pools, requiring circulation, filtration, and chemical balance like any other water vessel. The differences between water features and swimming pools are based on the intended use of the water feature.

Fountains

Fountains are usually powered by small submersible pump and motor combinations and are circulated without filtration. The only way to clean a fountain is to remove as much debris as possible with a leaf rake and spa vacuum or to drain the vessel completely, scrub it out, and refill it.

Larger fountains or waterfalls are powered by larger pool-style pump and motor systems and are filtered. They include skimmers for vacuuming and often include water fill lines with automatic fillers. Since most fountains and waterfalls are shallow with extensive water surface area, evaporation is a problem. First, keeping enough water circulating is important. Second, the high rate of evaporation can leave extensive scale deposits on decorative tile or rock at the waterline.

The other hazard of fountain maintenance is wind. Not only will wind accelerate evaporation, it can blow water out of the fountain. Again, the answer is to keep the water supply adequate to prevent the pump from running dry. In extreme conditions, shut down the unit.

Building fountains and waterfalls is a specialty that has been elevated to new levels in recent years with artificial rock and unique lighting. Designers are realizing that water provides a unique living enhancement to indoor and outdoor settings and the limits are only the imagination. Shopping centers and office buildings, as well as numerous Las Vegas hotels, are among the many new users of waterfalls and fountains in architectural design (Fig. 11-1). Because of the unique engineering requirements of each design, however, there are no simple or standard outlines for designing and building waterfalls or fountains. If you want to add a water feature to a yard or existing pool, I recommend an expert to complete the job. The hydraulic requirements of lifting and moving water for fountains and waterfalls are unique and demanding.

Koi Ponds

Koi, large goldfish-like carp, have become more popular in recent years, and are kept in decorative ponds in residential and commercial settings. A colleague who sold koi to a Hollywood celebrity tells me some of them cost as much as $20,000 each.

A koi pond has special needs unlike a typical pool or spa. First, the circulation must operate 24 hours a day to sufficiently oxygenate the water. Preventive maintenance and being aware of problems before they become critical are essential. Second, because chlorine will kill the fish, it cannot be used to keep the pond clean. Lastly, koi, their food, and decomposing decorative plants generate more decaying debris than a DE or cartridge filter can handle. Therefore, koi ponds use only sand or biological filters. Sand filters were described in the chapter on filters. A biological filter is composed of several layers of gravel through which the water is circulated to create a natural filtration process.

A

B

FIGURE 11-1 (A) Water feature; (B) water feature with lily pads and koi; and (C) fountain-style water feature. *(A) California Pools & Spas, Ontario, Calif.*

C

FIGURE 11-1 (*Continued*)

Any cleaning procedure that shuts down the circulation system must be performed promptly to avoid suffocating the fish. You must also avoid disturbing settled debris which will cloud the water to the point of suffocating the fish. To clean a biological filter, the fish must be caught (use your leaf rake and telepole, but be careful not to scrape or bruise the fish) and placed in a clean barrel or drum. The drum can be filled with water from the pond before draining it so that the fish are transferred to a similar environment to lessen the stress.

Drain the pond with a submersible pump and shovel the gravel of the biological filter out of the pond. Lay the gravel out on a concrete surface or put it in barrels for cleaning. Use a hose to wash and stir the gravel until the rinse water is completely clean. Do not use any cleansing chemicals. Scrub out the pond itself, again without chemicals or detergents. Replace the gravel in the biological bed over the pond suction drains. Refill the pond.

Because municipal water supplies include sanitizers that will harm or kill the fish, you must neutralize chloramines in the water

TRICKS OF THE TRADE: KOI POND CLEANING

Rating: Easy

- **Perform full draining and cleaning annually to maintain the cleanest, clearest water.**
- **Perform cleanouts in spring. Cleaning at any time disrupts the beneficial bacteria colonies, but when water temperature exceeds 55°F (13°C), algae will grow faster than the bacteria which eats it can be reestablished. You'll be fighting green water until the bacteria catches up.**
- **Prepare. Make sure you have electricity available for the sump pump to drain the pond; check the source of replacement water; make sure chlorine neutralizers are on hand so the water is ready immediately for fish.**
- **Use a large, clean trash can to temporarily store the fish, but remember they use lots of oxygen, especially when stressed. If there are many fish (or very large ones), use multiple cans and/or aerate the water with an aquarium air pump and air stone.**
- **Be sure the cans are clean. Household trash cans may look clean but may contain residues of old detergents, solvents, or oils that will kill the fish.**
- **Don't scrub away all of the algae. The cleanout is not meant to sterilize the pond. String algae is a healthy part of the system. The cleanout is meant to remove excess debris, leaves, etc.**

before returning the fish to the pond. Neutralizers can be obtained at most garden and aquarium stores. Follow the directions on the package for the amount and procedures. Carefully return the fish to the pond, with the water from the barrel if it is clean. Using as much of the old water as possible will help the fish acclimate. Restart the circulation as soon as possible.

An aquarium or garden store will be able to suggest various water additives that will help medicate the fish to prevent infection in scrapes or injuries the fish might have sustained during the procedure. They will also provide specialty algicides that are not toxic to the fish but which will keep algae to a minimum. Of course, the fish will eat much of the algae and a properly operating biological or sand filter will keep the water clear, so don't be obsessive about algae removal from koi ponds.

The biological filter not only filters particles out of the water, but it breaks down organic materials by creating an enzyme-rich bed of bacteria that hastens decomposition of debris. It takes several days for the bacteria to reproduce and up to a month for sufficient decomposition activity to keep up with the amount of excrement, excess food, and plant material found in the typical koi pond. To aid this process, the aquarium or garden store that supplies other koi food and supplies can provide a biological filter enhancer that begins and accelerates the bacteria bloom. During the first week or two after you have cleaned the pond, it might become cloudy and debris might accumulate on the bottom, after which the biological filter will catch up with the decomposition and cleaning.

You can always help the filter by skimming and vacuuming as much debris out of the pond as possible, just as you would any other pool or spa. As noted earlier, continuous circulation is also necessary because koi absorb large amounts of oxygen from the water which must constantly be replenished. Keep a sharp eye on the circulation, quickly removing blockages or cleaning dirty sand filters before the flow slows to critical levels.

Finally, don't make the same mistake I once did. I was called to the home of actor Lou Gossett, Jr., to clean the indoor pool and as I passed through the courtyard to the house I noticed an extremely dirty fountain. On my way out, I thought I would do Lou a favor and clear up the water in the fountain for him, so I poured a gallon of liquid chlorine in the murky water as the start of a cleanup. Unfortunately, after I had poured most of the chlorine I noticed little mouths sucking air through the surface of the murky water. It turns out that deep in the brown water were some very expensive koi. I quickly tried to rescue them, clean out the fountain, and replace the water with something less toxic, but unfortunately my actions were too late. The moral of this story has value in all aspects of the water industry. Look before you pour.

Installing a Koi Pond

RATING: PRO

Ponds and pools are similar in many ways, but ponds are easy for anyone to design and install. Although some large water features are constructed with the same techniques as a pool, most backyard ponds are literally a waterproofed hole in the ground with gravel, rocks, and a circulation system.

The steps are easy and can be accomplished in a day with a little help. Aquascape Designs has made it even easier by packaging everything needed (except the shovel!) in a do-it-yourself kit. Of course, kits come with installation instructions and tips, but here are the basics that will help you determine if you want to tackle the project yourself or hire someone to do it (Fig. 11-2A):

1. **Design** Use a garden hose to lay out the perimeter of the pond (Fig. 11-2B). The beauty of backyard ponds is that the shell is created with a flexible plastic liner, so just about any horizontal or vertical configuration can be easily created. Be sure to look at your design from all angles since ponds are generally installed for aesthetic purposes and you want to choose the best location in terms of the windows, patios, or other viewing spots. If you want water lilies, be sure the area gets at least 5 hours per day of direct sun. If you're placing the pond under trees, consider the amount of work needed to remove leaves. Once you're sure, use spray paint to mark the perimeter and remove the garden hose.
2. **Lay Out the Plumbing** Consider where the pump, hoses, electrical supply, and biofilter will be located. Now is the time to determine if you want a waterfall and rockscape. Flex PVC is easier to use along the irregular lines of a pond.
3. **Excavate** Make sure the pond is deep enough (Fig. 11-2C) for fish to survive when a few inches of evaporation (or leaks) happens and, in colder climates, for winter survival. Create ledges for plants and a deeper sump in the middle for gravel to encourage bacterial growth. Keep edges sharp so the sides of the excavation don't collapse when you add rocks. Use the removed soil to create a planting berm around the pond at various levels to give a more natural appearance and to channel rain runoff away from the pond. Pipes can also be buried under this fill dirt berm.
4. **Install the Liner** To determine the size of the liner, typically 45-millimeter plastic, measure the length and width (at the widest point) of the pond. Add to *each* dimension at least twice the maximum depth of the pond plus 3 feet (1 meter). This will provide enough for contours and edges. Lay the liner into the excavation and tuck it into each ledge or contour, weighting it down with

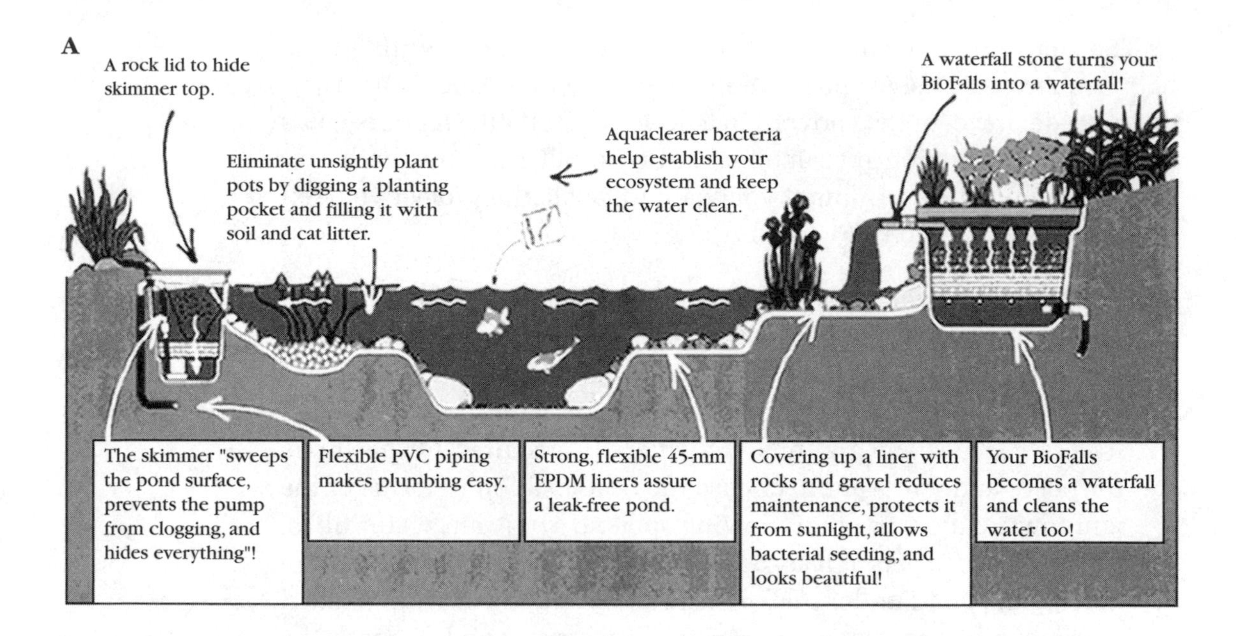

FIGURE 11-2 (A) Diagram of a backyard koi pond; (B) laying out the pond. *Aquascape Designs, Inc.*

C

D

FIGURE 11-2 (C) Excavating the pond; (D) the finished pond. *Aquascape Designs, Inc.*

gravel to keep it from shifting as you move around the pond. Some technicians lay down a heavy roofing felt or a layer of clean sand as a protective cushion for the liner, especially if the soil contains sharp stones or exposed roots.

5. **Add Rocks** Use larger rocks at key points to hold the liner securely in place and around the perimeter for the most natural look. Go back over the installation and cover the entire liner with smaller stones, gravel, or rocks. Lay several inches of gravel in the pit in the middle to encourage bacterial growth. Leave spaces for the skimmer and waterfall and add lighting as you lay in the rocks to conceal the cable and fixtures. Be sure the liner is completely covered for best appearance and protection from sunlight damage.
6. **Wash and Fill** Place a sump pump in the lowest part of the pond and begin to hose down the rocks. When the water runs clear, remove the pump and fill the pond.
7. **Build the Waterfall** If a fall is part of the design, build that atop the berm and rocks already installed.
8. **Plant** Add both water and land plants in and around the pond. Water plants remain in pots on the ledges in the pond. Pots can be concealed as the plants grow or with more gravel.
9. **Trim the Liner** Some excess plastic may extend beyond the finished product and can be trimmed or folded under the berm.
10. **Add Bacteria and Fish** It's usually best to allow the pond to circulate for a few days with bacteria in the water before adding fish (Fig. 11-2D). The fish waste load will cause algae to grow quickly and the bacteria needs to establish itself before it can keep up. This "break-in" period also ensures there are no leaks or equipment problems before adding fish. How many fish? One rule of thumb is to add no more than 1 foot (30 centimeters) of fish (total body length, either a few large fish or many small ones) per 25 square feet (2.3 square meters) of pond surface.

Rockscapes

Adding rocks around a pool or spa to create a natural look is easier than ever. Real boulders and large rocks require heavy trucks and

TRICKS OF THE TRADE: POND MAINTENANCE

Rating: Easy

Complete draining and cleanout of the pond is described above, but what about regular periodic maintenance? Well-established ponds need much less attention than pools or spas and are natural, self-sufficient ecosystems within themselves. Still, here are a few tips:

- Never feed the fish. They will become larger than the ecosystem can handle and the excess nutrients will overcome the balance.
- Remove leaves and other debris regularly to avoid ammonia buildup.
- pH should take care of itself. Never adjust it with chemicals.
- Add bacteria (available at garden centers that sell pond service items or via websites) in spring and fall to keep the ecosystem thriving. Follow package directions and buy only the amount you can use at one time (liquid products deteriorate after more than a year).
- String algae (the stuff that looks like green kitchen cleaning pads) is natural, but can grow to excess. If it does, periodically add algae "buster" products in accordance with label instructions. Bacteria eats this stuff, so you may just need to add bacteria as described above.
- Keep an eye on evaporation and add water as needed, but if you're adding more than 10 percent of the pond's total, use chlorine neutralizers. You can also fill a clean trash can with water and allow it to self-neutralize (usually within 3 or 4 days), but test the water before adding it to the pond.
- In cold climates, insulate plumbing before installation (although flex PVC won't crack if the water freezes) and consider adding a floating heater to keep the water from completely freezing. Several pond supply companies make heaters for this purpose.
- Make sure the pump turns the entire pond's water over at least every hour. Aeration is the key to a healthy pond, especially with fish, so consider adding a waterfall if the pond doesn't already have one.
- The pond circulation system must run 24 hours per day.

cranes, not to mention reinforced bond beams around the pool. Often space in a back yard is limited and prohibits use of this equipment and expense, for both the rock and machinery adds up quickly. The alternative is artificial rock (Fig. 11-3A).

Artificial rock is either hand-sculpted on site or precast. Materials used are either concrete or fiberglass, but hand-sculpted (or stamped)

A

B

FIGURE 11-3 (A) Artificial rockscape: waterfall; (B) artifical rockscape: concealed equipment area. *www.ricorock.com, by Rock Formations, Inc.*

concrete is the most common. Some manufacturers combine materials, using a plastic or fiberglass shell that creates the outline of the finished rockscape. This shell is covered with a web of rebar and then coated with concrete. Precast units are made as either boulders or flat stones, both designed to facilitate stacking. Paint finishes the look and unlike any other type of painting, with artificial rocks the idea is to avoid uniformity. A weathered, multishaded finished product looks all the more natural.

Although installing artificial rock (Fig. 11-3B) is best left to one of the many companies that specialize in this art form, the steps are basic. Hand-sculpted rocks are created with frames and wire mesh covered with concrete. Precast units are cemented together for whatever contour is desired. If rockscapes are built into the pool in the original design, the rebar of the pool structure will be extended above the pool's water surface to add a strong and secure base for adding rocks. If remodeling, stonework can be cemented or bolted onto the deck much like adding slides or railings.

Although weathering adds a realistic appearance, you don't want water penetrating the artificial surface, so annual application of a water sealer is a good idea. Otherwise, rockscapes require little maintenance. Planters can be added around the rockscape, but be sure to use plants that tolerate spray or splash from chlorinated pool water. Any time you expose more water to the air, there will be more evaporation (or outright water loss if some is sprayed out of the pool), so plan on adding water more frequently.

Artificial rockscapes can be practical as well as decorative. Many designs provide hidden storage areas (Fig. 11-3B) for pool equipment. Other designs incorporate slides, diving platforms, or grottoes for more enjoyment. Several good rockscape websites are listed in the Reference Sources and Websites section.

Waterfalls and the Vanishing Edge

Finally, a few words about waterfalls. Adding a fall to an existing pool requires a decision to add the fountain head to the circulation system, treating it essentially like any other return line, or adding a separate suction line, pump, and return to the waterfall. Many technicians are intimidated by the engineering required for installing a fountain or waterfall to a pool or water feature and miss out on valuable jobs. The truth is that the manufacturer of the display has done all this work for us!

Manufacturers of each fountain head, spray design, or waterfall weir provide data sheets explaining flow requirements, resistance (dynamic head expressed in feet or meters as previously described in this book), plumbing size requirements, and spray characteristics (such as diameter or distance of fall curve). Of course, flow and head change as the height, volume, and diameter of the discharge changes.

Actual installation of a fountain head or spray display is simple plumbing as described in other chapters, with the addition of a gate or ball valve to regulate the flow, thereby adjusting the height or diameter of the discharge. The actual fountain head can be plumbed onto standard plumbing and fittings, distributed directly off of a return line of the pool.

In the case of a dedicated system, with its own plumbing and a pump separate from the circulation system, the plumbing process requires a separate suction and return line added where it is most practical or aesthetic. The only additional concerns with a separate system are sizing the plumbing and pump.

Once again, the answer lies in the specification sheet provided by each manufacturer. For waterfalls, a chart details the required flow rate (expressed in gallons per minute) for every foot of desired projection. If, for example, you wanted the fall to curve out over the pool (curve projection) 2 feet (60 centimeters) before falling into the pool, you might require a flow rate of 35 gpm (132 liters per minute). A curve projection of 4 feet (1.2 meters) might require double that flow rate. The specifications also recommend piping size, usually suggested by the size of the adapter fitting on the fountain head or weir itself.

Pump sizing is then a matter of following the calculations of total dynamic head in the system and using a pump that can overcome that resistance and still deliver the desired flow rate. In this case, a slightly larger pump and related plumbing is better, since you can always reduce the flow with a valve, but it is impossible to increase the flow once you have completed the installation.

Not all wholesale pool-supply warehouses stock fountain heads or waterfall weirs, because the variety and sizes are endless. The warehouse staff should be able to recommend a special supply house or manufacturer that does. These supply houses and manufacturers can also supply suggestions and technical advice. You can request copies of the manufacturer's specification charts in advance to plan the work and discuss it with your customer.

One unique use of falling water is the vanishing edge (Fig. 11-4). Also called the negative edge, this waterfall employs one or more sides where the top of the wall is slightly below the water level of the rest of the pool. The water then evenly cascades over this "low wall," pouring into a catch basin. From the basin, it is pumped to the equipment and returned to the pool in a normal fashion.

Vanishing edges are a feature that must be designed into a pool in the first place or added during a major remodeling. The engineering is demanding. You must ensure that the basin can hold enough water to allow return to the pool before the pool becomes empty, without overflowing when the pump is turned off. For the technician, the vanishing edge is the ultimate skimmer, since the overflow acts as a natural removal system for leaves and dirt. The basin must be cleared of leaves or other debris regularly to avoid clogging, much like a huge skimmer.

FIGURE 11-4 Vanishing edge pool and catchment basin. *California Pools & Spas, West Covina, Calif.*

The basin must also be vacuumed, or dirt can be brushed toward the suction drain, where it will be taken to the filter. The suction drain is normally designed to accommodate a vacuum hose and the basin itself is shallow enough to reach in, remove the screen or drain cover, attach your hose and vacuum, just like the skimmer.

FAQs: WATER FEATURES

Can I Maintain a Koi Pond Myself?

- **Because living creatures are in a koi pond, you need to be extra careful in maintaining these water features. Like with any other pet, if you plan to be away for an extended period of time, hire a service or make sure a neighbor can perform the maintenance functions. Pay special attention to water balance and aeration (keeping the pump running at all times).**

Will Much Water Evaporate if I Use My Waterfall?

- **Yes, evaporation is determined in large part by the exposure of water to air. As the water tumbles over rocks (especially when warm) the surface area that is exposed to air increases, thus accelerating evaporation. Keep a close eye on the pool water level if you have a waterfall as part of the installation (or add an auto-fill unit).**

Will a Fountain, Koi Pond, or Other Water Feature Add Value to My Home?

- **Real estate markets are unpredictable, but generally speaking, a well-designed water feature does add value to the property. Size matters: make sure the water feature is not overpowering for the yard or too small to be noticed.**

CHAPTER

12

Commercial Pools

A commercial pool or spa might be defined as one that is used by the public, even if the users are members of a private club or a residential establishment. In fact, many building codes and health departments mandate rules and standards for residential pools and spas as well, but such regulation is impractical to enforce. Rules about minimum chemistry standards, signage, covers, safety equipment, and turnover rates are becoming more common in an increasing number of state and local jurisdictions.

Commercial pools and spas are generally dealt with in a fairly serious manner. Health inspectors visit the pool at least once a year in most areas and more frequently if they uncover problems. Because of liability issues, health inspectors are not shy about closing a commercial pool until violations are corrected and imposing fines if they feel compliance is not forthcoming. As a professional technician, you can lose your license (or be prevented from obtaining one) if you repeatedly fail to maintain a commercial pool or spa properly.

These factors should be considered before accepting a maintenance contract or repair at a commercial installation. You need to obtain a copy of all applicable building and health codes for your area to know what standards you must meet in your work. The apartment manager or property owner will probably not be familiar with the regulations, which are constantly being updated and changed, so it is your professional responsibility to know the rules and comply. In the event of an injury or other problem related to the pool, you will be held liable along with the property manager, so it is your job to understand the rules and advise them accordingly.

Since commercial work requires a great deal of time and attention to detail, be sure to include those factors in your job estimates. Where you might be able to provide satisfactory service to a residential customer with one or two calls per week, a commercial pool or spa might require daily visits. Especially with commercial spas, you might find that heavy bather loads create exceptional demands on the equipment, chemicals, and your time.

Types of Commercial Pools

At small apartment buildings or health clubs, the commercial installation might be a typical residential pool or spa being used by members of the public. In this case, the equipment and operational requirements will be familiar to you. The only differences in such cases will be added attention to mandated signage and some specialized equipment such as extra pressure gauges or sight glasses (detailed later).

At larger multifamily dwellings, schools, or public installations, the pool and spa might be ten times the size and volume of its residential counterpart. Competitive swimming pools, for example, are usually 50 meters (164 feet) in length by 23 meters (75 feet) in width. The depth is 4 to 6 feet (1.2 to 1.8 meters) for swimming, 6 feet (1.8 meters) for water polo, and 12 feet (3.6 meters) minimum for diving competition. Therefore, a typical Olympic-style competition pool might contain over 500,000 gallons (1,892,500 liters) of water, about 25 times the volume of a typical 18-by-30-foot (6-by-10-meter) residential pool. Commercial spas and wading pools are designed in similar multiples.

Large public pools have a skimming gutter that runs the length of the pool on each side. The surface water flows over the edge of the gutter at all times, skimming surface oils and debris, and drains into suction lines in the bottom of the gutter. The gutter is no different than the residential pool skimmer, it is simply larger.

Another type of commercial pool gaining in popularity is the freeform pool with multiple rockscapes and waterfalls. Because of the decoration and landscaping, some of these pools might seem larger than they actually are, and volume calculations become important and challenging. As with free-form or kidney-shaped residential pools, the solution to tricky volume calculations is to diagram the pool and divide it into sections, estimating the volume for each and adding the results together for a total.

Volume Calculations

Other than complex free-form volume calculations that actually become rather simple if the pool is divided into sections, the volume calculations for commercial pools is not different from residential pools. The methods described in Chap. 1 work just as well for larger pools.

There are two important differences between calculating volumes of small pools and large ones. First, the accuracy of the estimate becomes even more critical with a commercial pool, because liability and public health are involved. Because you are dealing with large numbers, even a small percentage error results in mistakes of thousands of gallons and therefore significant errors in chemical applications, turnover rates, and other maintenance calculations.

Slope Calculations

The second important difference in dealing with commercial volume calculations is that errors in estimating the slopes of the sides and floor can result in volume estimates that are off by 20 percent. Remember, 20 percent of a 500,000-gallon pool is 100,000 gallons (378,500 liters), or about five times the entire size of a typical residential pool.

As noted, the basic formulas for volume calculations remain the same. Figure 12-1 depicts the profile of a typical public pool. Calculate the volume of each section (A through D) and add them together. The additional step in commercial calculations is to adjust the estimate for the slope of the pool walls to the floor by deducting a percentage of the total.

Figure 12-2 shows a chart for a typical public pool slope. This chart will work for most installations, but it is best to determine the exact calculation by consulting the pool plans or builder. In this

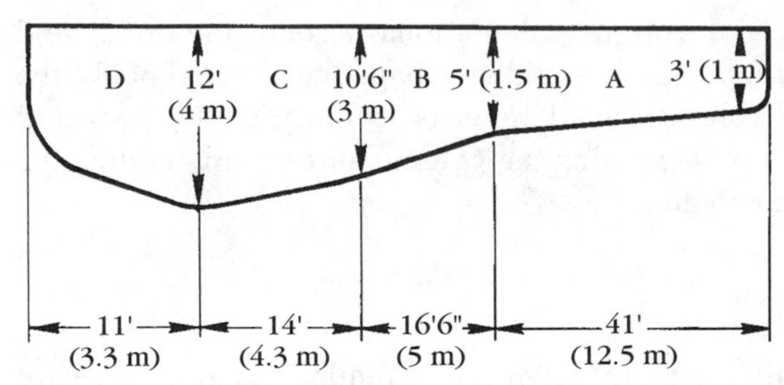

FIGURE 12-1 Volume calculations in commercial pools. *Sta-Rite Industries, Inc., Delevan, Wis.*

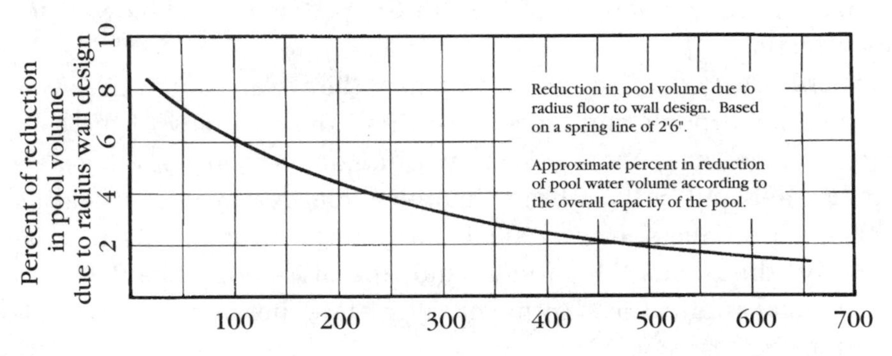

FIGURE 12-2 Slope calculations. *Sta-Rite Industries, Inc., Delevan, Wis.*

example, let's assume that the pool in Fig. 12-1 measured 450,000 gallons (1,703,250 liters) total. By reading across the bottom portion of the chart in Fig. 12-2, you will find a line for 450,000 gallons. Follow the line up to the point where the curve intersects it. Now look horizontally to the scale on the left side of the chart to determine the percentage of reduction. In this example, the number on the left side of the chart is 2, meaning 2 percent. You would therefore lower your volume calculation for this pool by 2 percent to arrive at an accurate estimate: $450{,}000 \times 0.02 = 9000$ gallons. The actual total volume therefore is about 441,000 gallons (1,669,185 liters). As you can see, a difference of 9000 gallons (34,065 liters) would make a significant difference in calculating the amount of chemicals needed to balance the water chemistry.

Bather Loads

Signage around commercial pools and spas requires maximum bather loads. Usually, the technician is the one who needs to fill in the blank on such signs. Although specifics might vary in your location, the following general rules will serve if local codes do not specify.

- One bather per 15 square feet (1.4 square meters) of surface area in portions of the pool that are 5 feet (1.5 meters) deep or less.
- One bather per 20 square feet (1.9 square meters) of surface area in portions of the pool that are more than 5 feet (1.5 meters) deep.
- Subtract 300 square feet (28 square meters) from the total surface calculation for every diving board.
- For example, the shallow end of a pool (less than 5 feet deep) is 40 feet × 20 feet = 800 square feet; 800 ÷ 15 = 53 bathers. The deep end (more than 5 feet deep) is 40 feet × 80 feet = 3200 square feet. There is one diving board, so you deduct 300 square feet, arriving at 2900 square feet; 2900 ÷ 20 = 145 bathers. The total occupancy of this pool is 53 + 145 = 198.
- Some jurisdictions require 36 square feet (3.3 square meters) per bather overall, so in a 4000-square-foot pool you would allow 111 bathers (4000 divided by 36 = 111). You may also be limited to one diver for every 100 square feet (9.3 square meters) in a diving pool.

Bather Displacement

In a public pool that might have several hundred bathers at one time, how much water is displaced by the volume of people? This question will determine flow rates of replacement water, surge chamber requirements, and other technical calculations. Even in a small pool at a public location, such as a motel or apartment building, this calculation is important. If the pool is fitted with an overflow drain to waste, the bather load will displace the water out of the pool or spa. When the bathers leave, the water level might drop below the skimmer and the pump could run dry. Obviously this would be even more likely to occur in a commercial spa.

To calculate the volume of water displaced by bathers, you must also factor in the likely amount each bather is submerged. A typical bather will displace 2 cubic feet of water (an average for adults and children). But bathers in the shallow end are usually only partially submerged, while bathers in the deep end are almost completely submerged. Therefore, shallow bathers displace 75 percent of 2 cubic feet (56 liters) each (0.75 × 2 = 1.5 cubic feet each), while in the deep end each bather displaces 90 percent of 2 cubic feet (0.9 × 2 = 1.8 cubic feet each). Obviously this is an inexact science, because on any given day you might have more or fewer bathers in each end and there might be greater or lesser percentages of the total as children who displace less water. In any event, this calculation will help determine a likely volume.

In the example pool, 53 shallow bathers × 1.5 = 79.5 cubic feet plus 145 deep bathers × 1.8 = 261 cubic feet. The total of 79.5 + 261 = 340.5 cubic feet of water displaced. At 7.5 gallons per cubic foot, you arrive at 2554 (rounded) gallons (9667 liters) displaced (340.5 × 7.5). The surge chamber (Fig. 12-3) needs to accommodate at least this volume of overflow.

Commercial Equipment

Most equipment for the commercial pool and spa will resemble the smaller versions used by residential installations. Pumps, air blowers, heaters, cleaning, and other equipment are often built of heavy-duty materials and increased capacities. The following notes detail equipment and supplies that are unique to commercial water technology.

Surge Chamber

As noted earlier, heavy bather loads displace a great deal of water. In large commercial pools, this water is not drained off to a waste drain but flows into a holding tank called a *surge chamber* (Fig. 12-3). When bathers leave the pool and the water level drops, water from the surge tank is pumped back into the pool. Since each bather leaving the pool (or splashing) takes water out of the pool on their body, replacement water is also directed into the surge chamber, where the level of the pool is monitored. Each surge chamber is designed differently, but understanding this basic concept will make the various components understandable when you examine one.

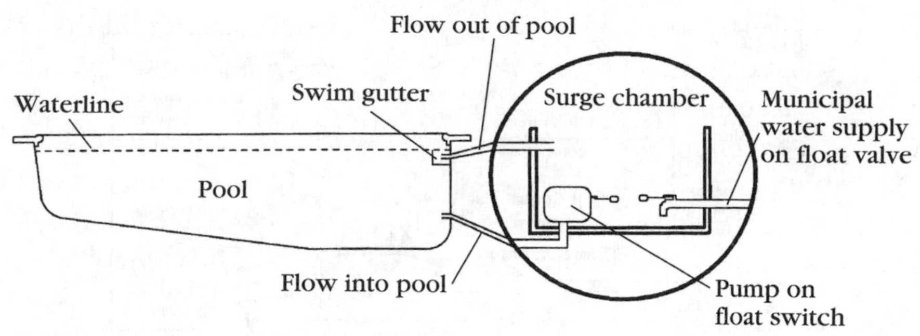

FIGURE 12-3 Surge chamber components.

Slurry Feeder

Large pools require a great deal of filter grid square footage and require a great deal of DE. The answer is a slurry tank or feeder where DE is mixed with water (it doesn't actually dissolve, but becomes suspended in the water) and is pumped at an even rate into the filter to precoat the grids.

Some slurry feeders are simply an open concrete pit in the pool equipment room where this mixing takes place. A motorized paddle mixer (called an agitator) is used to keep the DE suspended in the water, which is then pumped into the suction flow of the pump and sent to the filter grids. Other systems use a slurry tank, from which the mixture is vacuum fed to the pump, or a pressure system where the mixture is forced into the system. Modern systems use injectors and various pumps for measuring, mixing, and moving the mixture, but the objective and components are the same on each type. As with any DE precoating, the filter must first be clean (either from a teardown or backwashing), then the appropriate amount of DE is applied, followed by normal filtration.

Filter

Filters themselves will vary for commercial applications, but fall into the same categories detailed previously. DE, cartridge, and sand filters are all made for large commercial pools and spas. When one large filter is inadequate, several filters are plumbed together in series with a manifold, as in Fig. 12-4.

Cartridge filters are also manufactured for commercial use (Fig. 12-5), but as discussed in the chapter on filters, they have not been widely adopted because of the large amount of cartridge surface area required to

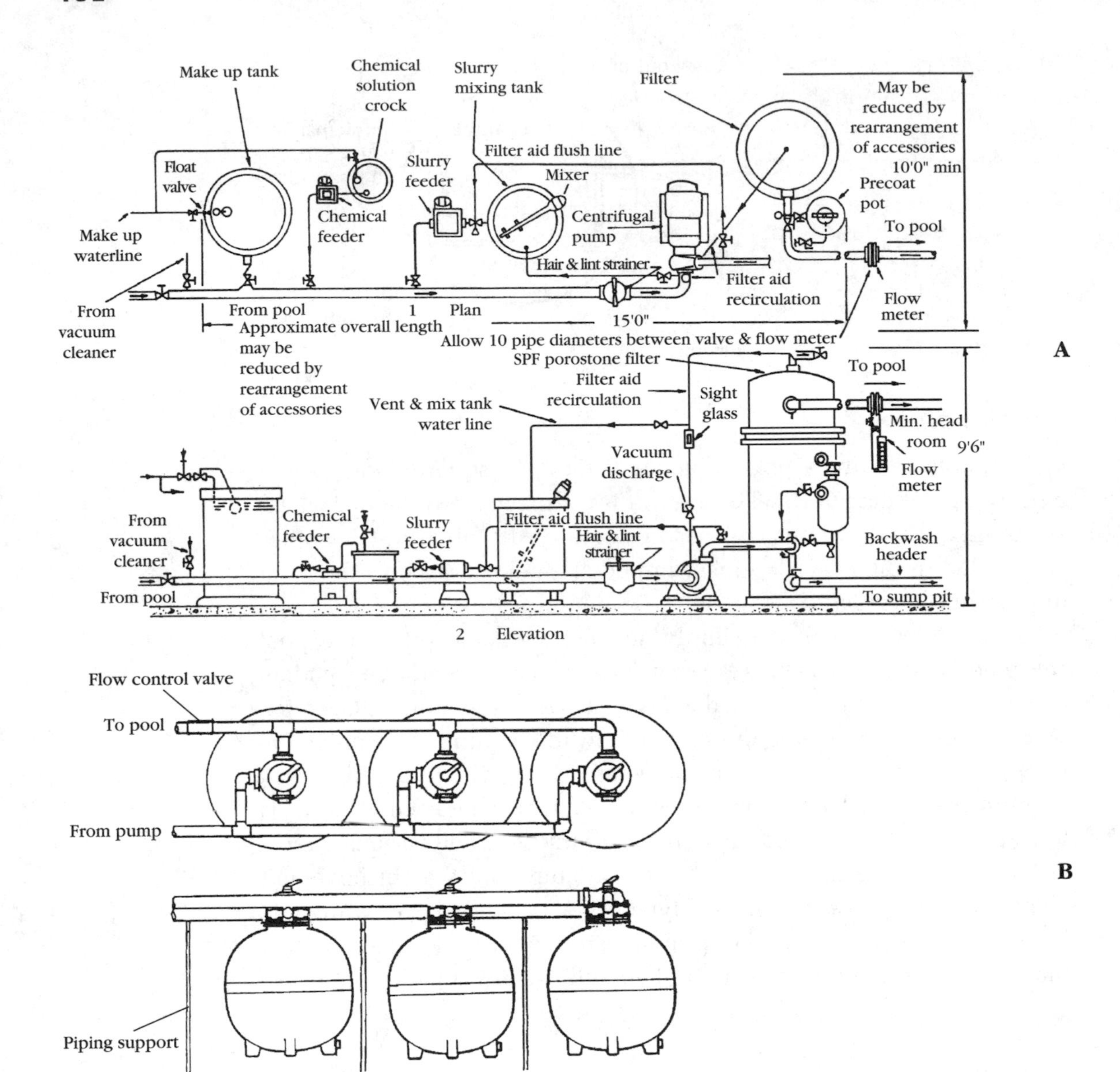

FIGURE 12-4 Commercial filters (A) in series with slurry feeder and (B) in series with manifold plumbing. *(A) U.S. Government Printing Office. (B) Jacuzzi Bros., Little Rock, Ark.*

filter large bodies of water. Cartridges can also be difficult to clean, the pleating creating crevices in which oils and dirt can become lodged.

Commercial filtration generally requires pressure gauges on the incoming and outgoing plumbing of the pump and filter. Sight glasses are often required on waste lines and the outflow plumbing of the filter.

Flow meters might be required on the heater and filter plumbing as well. Make sure these are always clean, free of leaks, readable, and reflecting accurate pressures and flows. In other words, keep the filter clean and the other equipment well-maintained and there will be no problems.

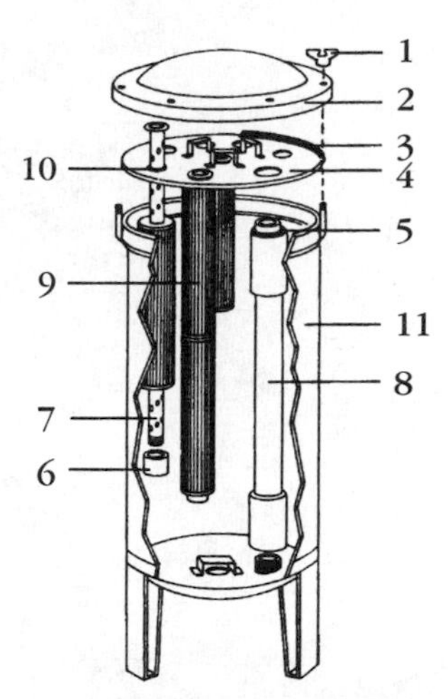

FIGURE 12-5 Commercial cartridge filter. *Harms-co, Inc.*

Gas Chlorinator

Even distribution of sanitizer is a problem in the large volumes of water found in commercial pools. Also, because bathers are likely to use the pool or spa immediately after servicing, it is important to be sure chemicals are not lingering in harmful concentrations at any point in the water.

Chlorine gas is the most efficient and least expensive sanitizer. Injecting it into the circulating water is an effective way of gradually adding chemical to a pool or spa. The cost, availability, and dangers of working with chlorine gas generally prohibit residential users from sanitizing with it, but for commercial pools it is the preferred method.

Figure 12-6 diagrams a typical chlorine gas feeding system. The gas is supplied in cylinders that discharge the chlorine gas through a pressure regulator, volume control (to control the ultimate strength of the chlorine mix), and visible volume meter to a water reservoir. When the gas is injected, it mixes with the water to form a solution that is then pumped into the circulation system for delivery to the pool or spa. By regulating the time of operation and volume of gas delivered, the sanitizing of the water can be managed. The incoming water supply might be fill water for the pool or part of the circulation water diverted for this purpose.

Modern electronic control technology is being applied to chlorine gas delivery systems, so if you plan to service a pool with a complex unit, familiarize yourself with its operation by reading the manufacturer's literature or calling them for technical support.

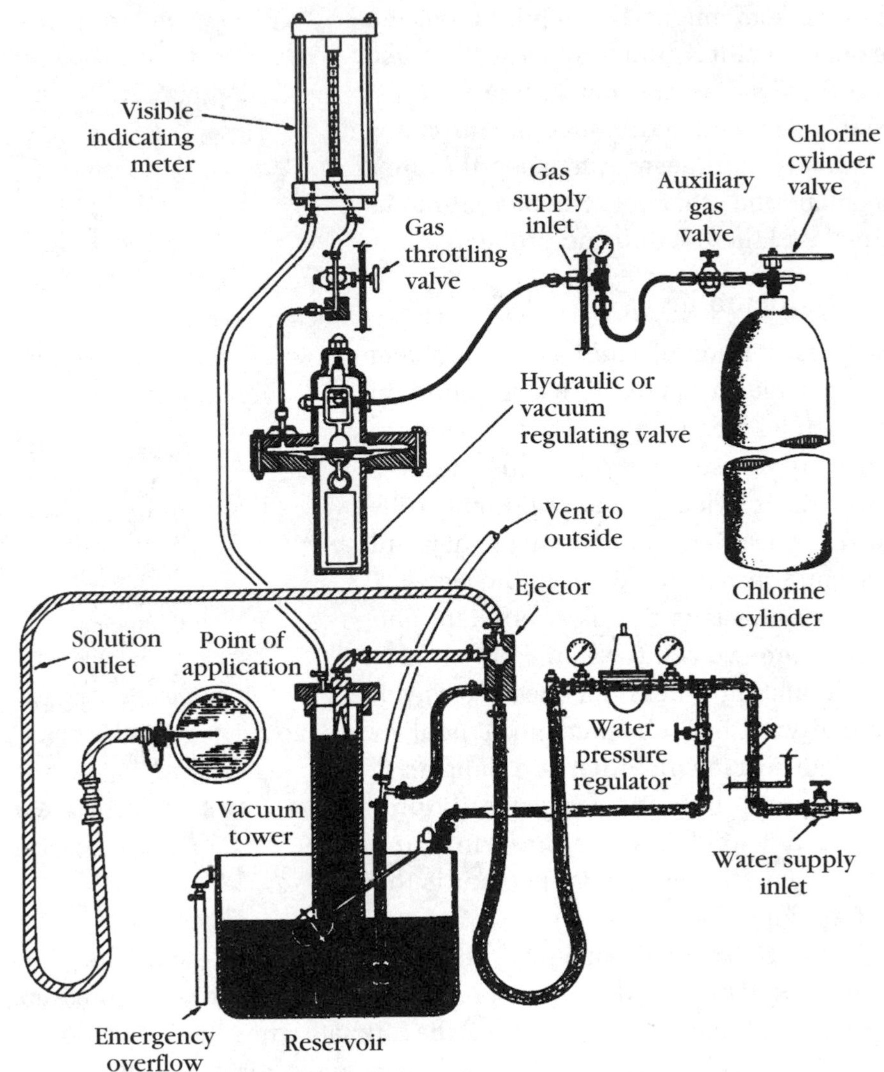

FIGURE 12-6 Chlorine gas feeder. *U.S. Government Printing Office.*

Since chlorine gas is deadly, be sure you know what you are doing before adjusting or repairing any system. Better yet, call a factory-trained technician.

Chlorine Generators

Considering how expensive it is to sanitize a large pool or spa and considering how relatively inexpensive salt, electricity, and water are, some installations make their own chlorine. Chlorine generators produce chlorine gas by electrolysis. The generator consists of a plastic tank with two compartments separated by a filter-type membrane. The larger side, which includes a positively charged electrode (called the *anode*), is filled with a salt (sodium chloride) and water solution. The smaller compartment, which includes a negatively charged electrode (called a *cathode*), is filled with distilled water or municipal water (if it is not too hard). The membrane allows electricity to pass through but restricts the chemicals produced to one side.

Chlorine gas is generated in the salt solution compartment when the electricity separates the chlorine from the sodium. The gas rises above the water and some passes through the membrane, mixing in the smaller chamber with the fresh water to form sodium hydroxide (which is drained off periodically). Hydrogen gas formed in the process is vented out of the generator. Salt is added periodically to recharge the system. The chlorine gas is then injected through plastic plumbing into the circulation plumbing. The electrical current is supplied through a transformer that converts 110- or 220-volt ac into 12-volt dc. Regulating the current regulates the volume of chlorine produced. Some variations of the chlorine generator include systems that produce the chlorine right in the circulation water flow for direct application to the pool or spa water.

High-Capacity Automatic Chlorine Feeders

Erosion feeders are becoming more commonplace in commercial pool systems (Fig. 12-7). They are valued for their low initial cost, ease of operation and maintenance, and greater degree of safety compared with any system using gas chlorine.

Essentially a residential erosion feeder system on steroids, the commercial units are made of beefy plastic with all hardware also made of noncorroding materials. Like their backyard counterparts, these feeders use chlorine (or bromine) tablets or sticks. Simple ball valves determine the rate of flow (and therefore the rate of chlorine being dissolved) through the system.

There is no maintenance required for these systems and disassembly for cleaning or adding more product is self-evident. Note the plumbing diagram (Fig. 12-4)—they are not inline with the rest of the circulation equipment because you may want to restrict the flow, thereby restricting

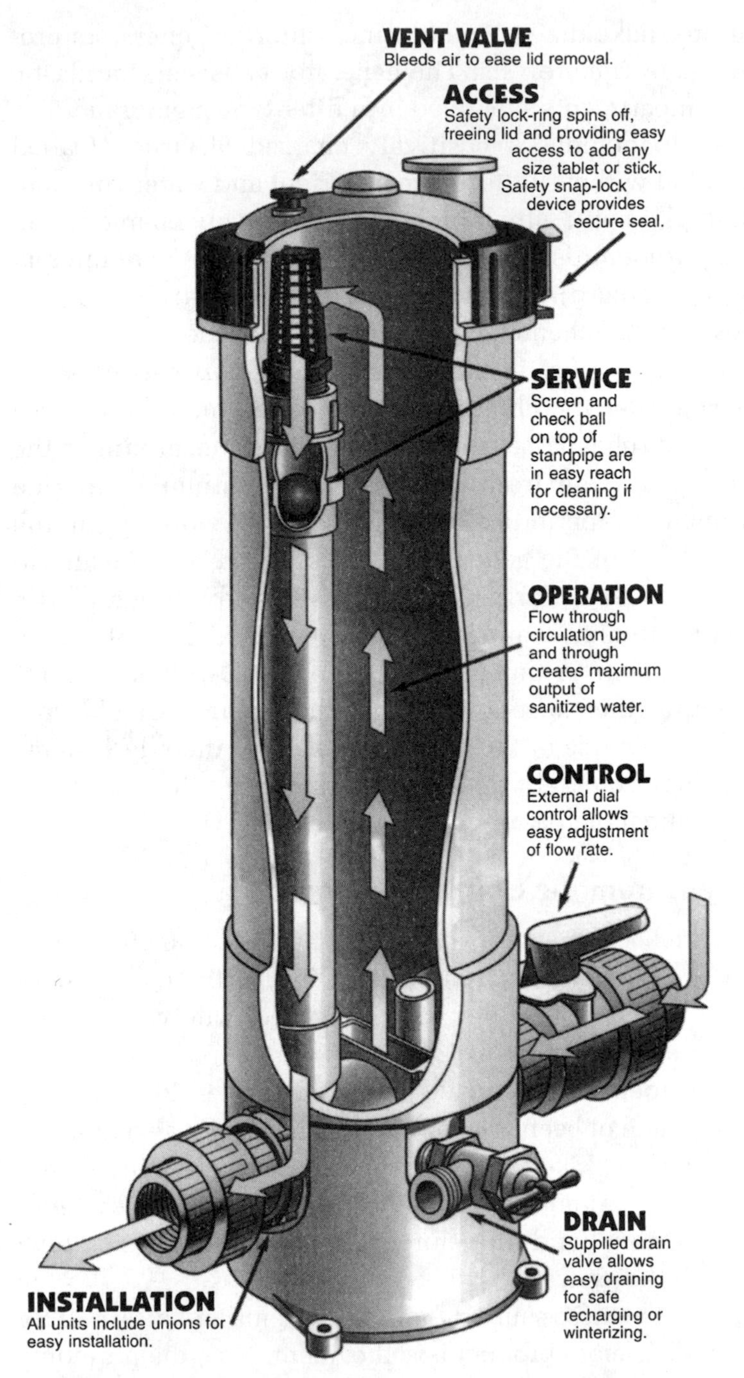

FIGURE 12-7 High-capacity chlorine feeder. *Rainbow Lifegard Products, a division of PacFab, Inc.*

the amount of chlorine being delivered. Thus, these units are plumbed with their own circulation loop (just like the smaller versions at home).

The Commercial Equipment Room

All pool or spa equipment rooms should be kept clean and neat, but this is critical in the commercial equipment room. Chemicals should never be stored near equipment, where heat or water leaks could mix with the chemicals, with toxic results.

A sump pump should be installed in a pit in the floor of the equipment room. Since leaks in commercial plumbing can be very substantial, a submersible pump with an automatic float switch should be provided that will drain water from the room to a waste line. Commercial equipment is expensive and water rising from a major leak can cost much more than a sump pump.

Safety Equipment

Commercial installations require certain safety equipment that is a good idea for residential sites as well, because pools and spas quickly become the most popular meeting place in the neighborhood.

SIGNS

Most building or health department codes require certain signage to be prominently displayed near the pool or spa. Check your local codes for specifics and required proximity to the water's edge. Many pool and spa technicians ignore these signs, but I advise you to be more careful. First of all, providing and posting these signs is a job that can generate profits—you mark up the signs like any other piece of equipment you sell. Also, you charge for the labor to install the signs.

More important, however, if you are the service technician for a commercial pool and someone is injured, the first thing their lawyer will check is if the appropriate warning signage was posted. If not, you might be held liable for negligence. At the least, if the customer gets sued, you will probably lose the account, so it pays to do this job right.

The basic signs (Figs. 12-8 and 12-9) you need (some or all, depending on your local codes) are

Maximum bather load

- This sign usually says, in simple bold, block letters, Occupant Capacity ________ or Maximum Occupants ________. You fill in

FIGURE 12-8 First-aid procedures.

the blanks, depending on the local codes. Many jurisdictions allow 15 square feet (1.4 square meters) of surface area for each bather, so a 15-by-30-foot (4.5-by-9-meter) pool would allow a maximum of 30 occupants at a time (450 square feet of surface area ÷ 15 = 30). Other rules guide spa occupancy or wading pools for kids, so check local requirements before writing the number in the blank provided. Use paint or an indelible marker so the number cannot be changed by pranksters.

Artificial respiration

- This is a sign that diagrams artificial respiration (Fig. 12-8) and first-aid techniques in case of emergency. It's a good idea to familiarize yourself with these techniques by at least studying the sign (or better yet, take a first-aid class), because you will be

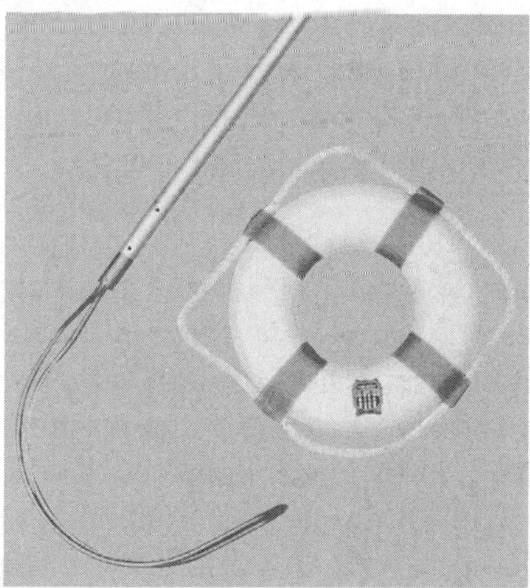

FIGURE 12-9 Signage, toss rings, and life hooks.

around pools and spas all day long and sooner or later you will run into someone who needs help.

Warning: No lifeguard/no diving

- This sign advises bathers that no lifeguard is on duty at this pool or spa and, if appropriate, no diving is allowed.

Emergency phone numbers

- This sign prominently displays 911 for general emergency calling and also provides spaces to include the phone numbers of a local emergency room or doctor, fire department, and police. Since most communities now have the 911 emergency number, some jurisdictions only require this to be posted.

Pool/spa rules

- This sign is available with residential or commercial rules and includes hours for use, age restrictions of users, commonsense rules such as No glass or No Running, etc. These rules might be generic commonsense rules or might have specific language required by the state. Arizona and Florida, for example, have their own specific requirements for this language.

Emergency shutoff

- This sign is posted next to the electrical switch that cuts all power to all equipment, in case of emergency. It simply makes users aware that the shutoff exists in case someone is injured in the pool or spa or, more often, when clothes get sucked into powerful pump suction openings in the spa.

Danger: Pool closed

- This sign is a good one to keep in your truck in case you need to close down a pool. I carry some glow-in-the-dark plastic ribbon, available at any hardware store, to tape off a pool or spa area that must be closed and also mount the sign prominently. You use this in case you have to drain a pool or spa or when chemical or repair procedures require closing a pool. The sign and tape prevents accidents and keeps you from being liable if someone jumps into an empty pool.

TRICKS OF THE TRADE: SAFE CHEMICAL STORAGE

Rating: EAsy

Because of the large volumes of potentially hazardous chemicals used in commercial pools, the facility should have a separate room (from the mechanical equipment room) for chemical storage. It must be well ventilated and locked. Appropriate warning signs should be clearly posted on the outside of the door in case any chemicals do leak and create potentially lethal, unexpected by-products such as chlorine gas. Beyond that, other precautions are warranted:

- Storage shelves (or areas on the floor) should have secondary containment equal to the volume being stored. In other words, if you're storing 50 gallons (189 liters) of liquid chlorine, place them in a plastic bin capable of holding 50 gallons of liquid in the event they leak.
- Provide adequate space, dividers, or other containment between products that do different jobs:
  - pH-lowering products
  - pH-raising products
  - Chlorine products
  - Total alkalinity adjusters
  - Flocculants or clarifiers
  - Conditioners
  - Testing reagents
  - Algicides
- Have personal protective gear readily available, such as goggles, full face shields and masks, plastic aprons, latex gloves, neoprene boots, and the first-aid kit.
- Post the CPR and other emergency procedures in a highly visible spot.
- Keep documents handy and dry such as material safety data sheets (MSDSs) for chemicals on hand, inventory (generally it's better to use older products first), and training manuals.
- Have a clear water shower and eye wash (or at least a garden hose) available in case of accidents.

Toss Rings

A toss ring (Fig. 12-9), like a life ring on a ship, is a foam plastic ring (17 to 24 inches or 40 to 60 centimeters in diameter) that has a rope attached. It should be mounted prominently near the pool for rescue purposes if someone is drowning. Most jurisdictions require the toss ring for commercial pools and spas only, but I strongly urge you to sell one to your residential customers as well.

Life Hooks

Like the toss ring, the life hook (Fig. 12-9) (a pole with a large metal hook on the end) is required to be mounted near the pool or spa to help someone out of the water if needed. The pole is 6 to 9 feet (2 to 3 meters) long with a broad, looped hook on the end (made so the hook doesn't do more harm to the victim than the drowning).

Thermometers

Some commercial pools require an in-water thermometer be available at each pool and spa and most residential customers want one too. For commercial installations, I recommend a unit that is built into the skimmer cover and takes the temperature at the skimmer. Other models usually disappear or are tampered with. You can also simply tie a thermometer to a rail or ladder, or there are tube models with a float on the top that float on the surface of the water. These usually float into the skimmer basket, which is a good placo for them to keep them out of sight. You and the health department inspector need to check the thermometer, the 200 residents of the apartment building don't.

Inline thermometers, designed like in-line flow meters, are also available for installation into the equipment plumbing or directly to the heater manifold. For you high-tech techies, digital readout, battery-operated thermometers, and pH testers are also available in floating models or with test probes that you put in the water while reading the information on a small hand-held device about the size of a calculator (see Chap. 8).

Dehumidification of Indoor Commercial Pools

RATING: PRO

Excess humidity in indoor pools causes a wide variety of problems from decayed concrete and rust to electrical shorts and respiratory ailments. It can even destroy the dehumidification equipment itself. In short, managing humidity in the indoor pool environment is a critical task.

Essentially, moist air in the pool enclosure must be drawn out and replaced with drier outside air or filtered through a dehumidifier. This is accomplished with either mechanical ventilators or specialized dehumidifying equipment.

Mechanical ventilators are the most simple to understand and operate. They are a system of fans, ducts, and dampers that are designed to draw outside air into the pool enclosure and displace the moist air inside, much the way a pool circulation system pushes and pulls water through the pool. While these systems are widely used, they do not allow much regulation of the actual humidity level. On hot summer days in very humid climates, for example, outdoor air may be more humid than the air indoors. In a very dry winter climate, the outdoor humidity may be so low that the introduced air accelerates pool evaporation and makes people indoors uncomfortable.

Dehumidifiers work by refrigerating the air, a process that generates heat which is then often diverted to warm the pool water. The net effect is a lower fuel bill than if the pool and indoor air were heated with a separate system.

While servicing dehumidification systems requires special training, some troubleshooting is common sense. For example, was the system properly sized for the pool in the first place? Building standards call for a minimum of 0.5 cubic foot (14 liters) of outdoor air per minute for every square foot (930 square centimeters) of pool and deck surface area. So if the pool and deck totaled 4000 square feet (372 square meters), the system would need to deliver at least 2000 cubic feet (57 cubic meters) per minute of fresh air. Moreover, the system must provide an additional 20 cubic feet (0.6 cubic meter) per minute of air exchange for each person in the enclosure (bathers, loungers, spectators). If moisture is a problem in your indoor pool, check the specifications of the system first.

On the other hand, a system that is too large will create drafts on swimmers, leading to complaints. Service staff may be tempted to raise the temperature of the pool's heater to compensate, which merely drives up operating costs. Humidity should generally be maintained between 50 and 60 percent for optimum operation and comfort. To avoid condensation in cold climates, however [where outdoor temperatures are lower than 40°F (4°C)], humidity should be kept closer to 40 percent. Double-pane windows will also help prevent condensation even at higher humidities.

Health Issues

RATING: PRO

The focus of attention at any commercial pool (Fig. 12-10) is health and safety. Safety is obvious, especially regarding prevention of slips and falls, chemical handling, and lifesaving in drowning incidents. Health concerns may be more subtle but these days is perhaps an even greater disaster waiting to happen if proper precautions are not taken.

LEGIONNAIRES' DISEASE

Numerous outbreaks of Legionnaires' disease (a sometimes fatal respiratory ailment) have been linked to operating whirlpool spas. Outbreaks have occurred at trade shows and on cruise ships, so no installation is immune. The Centers for Disease Control (CDC) have made recommendations to prevent these tragedies.

FIGURE 12-10 Commercial pool and spa. *California Pools & Spas, West Covina, Calif.*

TRICKS OF THE TRADE: OPERATIONAL REQUIREMENTS

Rating: Pro

As with any commercial installation, you must obtain a copy of the codes and regulations applicable in your area. The following are a few common standards to look for.

- **For health and safety reasons, most jurisdictions require that the circulation equipment be operating whenever the pool is open to bathers.**
- **Service logs must be maintained and kept available in the equipment room for inspection so that a health department inspector can verify compliance with regulations and compare servicing to water testing results. Figure 12-11 is a typical health department inspection checklist.**
- **ALWAYS close the pool or spa to users whenever you are performing cleaning or chemical service. Swimmers have been severely injured or killed when caught in vacuum hoses in the water, have tripped and fallen over service equipment on the deck, or have been harmed by chemical concentrations not yet distributed through the water. After servicing, look around the deck for wet areas, spills, or tools left behind.**
- **Many regulations require deck and pool lighting of certain specifications, whether the pool is used at night or not. It is your job to make sure that the bulbs are working and the timers are operating correctly.**
- **Make sure access to the heater temperature controls is locked. Make equally sure that time clocks and chemical supplies are locked away.**
- **Drains, gutter suction ports, deck skimmers, and other lines must be covered or fitted with a grate. Make sure these are always in place.**
- **Some jurisdictions require minimum visibility so that any injured person at the bottom will be seen. Some standards call for visibility at a certain level [such as 6 feet (2 meters)]; others call for the main drain cover to be clearly visible. Of course, if the water is cloudy it is a sign that there is some problem and you might want to close the pool and take corrective action.**

Essentially the recommendations are to clean and **replace** the filter regularly and maintain a proper chlorine residual, even if the spa is only for demonstration or decorative purposes. The CDC findings in the cruise ship incident stated, "Visual examination of the filter material showed extremely heavy organic loading. This loading remained in the filter despite reports that a routine (daily) filter backwash cycle was implemented." That facility also replaced the filter cartridge annually, but inspection and common sense should have been followed instead of a predetermined schedule.

SWIMMING POOL OFFICIAL INSPECTION REPORT

COUNTY OF LOS ANGELES DEPARTMENT OF HEALTH SERVICES
PUBLIC HEALTH PROGRAMS - ENVIRONMENTAL HEALTH
2525 Corporate Place, Rm. 150, Monterey Park CA 91754 (213) 881-4160

SITE ADDRESS		CITY			DATE	
SITE NAME	OWNER				SITE #	SUB
MAILING ADDRESS	CITY	POOL TYPE	ELE		Cl_2	pH

THE MARKED ITEMS REPRESENT HEALTH CODE VIOLATIONS AND MUST BE CORRECTED AS FOLLOWS:

☐ 1. Provide a readily accessible life ring with an attached rope of sufficient length to span the maximum width of the swimming pool.
☐ 2. Provide a readily accessible body hook permanently attached to a pole at least 12 ft. in length.
☐ 3. Post a legible sign stating "Warning - No Lifeguard On Duty" in letters at least 4" high. In addition, the sign shall state "Children under the age of 14 should not use pool without an adult in attendance".
☐ 4. Post a legible sign with a diagrammatic illustration of artificial respiration procedures.
☐ 5. Post an emergency telephone number "911".
☐ 6. Post a legible sign with the maximum occupant capacity allowed in the pool in letters at least 4" high. Swimming pools = 1 per 20 sq.ft. of pool surface area. Spa pool = 1 per 10 sq.ft.
☐ 7. Post a legible spa pool precaution sign. Consult CCR Title 24, Section 3119B.5 for verbiage.
☐ 8. Post a legible "No Diving Allowed" sign in letters at least 4" high.
☐ 9. Every spa pool with an emergency shut-off switch shall have a legible sign stating "Spa Emergency Shut-Off Switch".
☐ 10. All required signs must be clearly visible from the pool deck.
☐ 11. Maintain a free chlorine residual of at least 1.0 ppm at all times.
☐ 12. Maintain a free chlorine residual of at least 1.5 ppm when cyanuric acid is used at all times.
☐ 13. Maintain the pH between 7.2 and 8.0.
☐ 14. Maintain the level of cyanuric acid below 100 ppm.
☐ 15. Provide an approved pool water test kit which will measure a free chlorine residual.
☐ 16. Eliminate cloudiness and maintain pool water in a clean and clear condition.
☐ 17. Eliminate algae growth in the pool.
☐ 18. Vacuum pool. Eliminate dirt / leaves / debris in the pool.
☐ 19. Clean the waterline tiles.
☐ 20. Replace broken / missing pool tiles.
☐ 21. Replace broken / unreadable depth marker tiles.
☐ 22. Replace broken / missing coping.
☐ 23. Provide adequate skimming action in the pool.
☐ 24. Raise / lower water level to the mid-point of the skimmer opening.
☐ 25. Replace broken/missing skimmer strainer basket.
☐ 26. Replace broken/ missing skimmer weir assembly.
☐ 27. Replace broken / missing skimmer diverter valve assembly.
☐ 28. Limit spa pool water temperature to a maximum of 104°F.
☐ 29. Animals are prohibited in the pool and in the pool area.
☐ 30. Discontinue placing chlorine tablets in the pool skimmer(s).
☐ 31. Discontinue use of the floating chlorinator.
☐ 32. Secure / replace drain cover with an approved type which can only be removed with tools.
☐ 33. Secure / repair / replace stair handrail(s).
☐ 34. Secure / repair / replace ladder/ ladder step treads / grab rails at deep end of pool.
☐ 35. Provide / repair or replace underwater light(s).
☐ 36. Maintain underwater pool light(s) "on" during all times the pool is open for use after dark. If the pool is not separately enclosed, maintain pool light(s) on during entire nighttime hours.
☐ 37. Eliminate deck obstruction.
☐ 38. Eliminate trip and fall hazard of deteriorating / uplifting decking in pool area. Fill in expansion joint between coping and deck.
☐ 39. Repair the pool fence enclosure.
☐ 40. Provide a self-closing gate / door to pool area with self-latching hardware at least 42" above finished grade.
☐ 41. Provide a minimum turnover rate as follows: Swimming Pools = 6 hours, Swimming Pools built before October 1982 = 8 hours, Spa Pools = 0.5 hours, Wading Pools = 1 hour.
☐ 42. Operate the pool recirculation system at all times the pool is open for use and longer if necessary to maintain the water clean and clear.
☐ 43. Repair / replace the recirculation pump.
☐ 44. Repair / replace the filter.
☐ 45. Backwash the filter.
☐ 46. Provide / repair / replace the influent / effluent pressure gauge(s).
☐ 47. Provide / repair / replace the flowmeter.
☐ 48. Provide / repair / replace automatic chlorinator.
☐ 49. Maintain the automatic chlorinator filled and operational.
☐ 50. There shall be no direct connection of the pool or its recirculation system with a sanitary sewer or drainage system.
☐ 51. Clean the pool equipment room / area.
☐ 52. Correct specified items relating to the gas chlorination system.
☐ 53. Maintain rest rooms in a clean and sanitary condition.
☐ 54. Maintain shower facilities in a clean and sanitary condition.
☐ 55. Maintain dressing rooms in a clean and sanitary condition.
☐ 56. Maintain toilets, urinals, wash basins, and showers in good repair.
☐ 57. Showers and lavatories shall be provided with hot and cold water. A means to limit the hot water to a maximum of 110°F shall be provided.
☐ 58. Provide soap in soap dispensers or containers in showers.
☐ 59. Provide soap in permanently installed soap dispensers, paper towels or hot air blowers, and toilet tissue for toilets.
☐ 60. The pool(s) was not accessible or only partially accessible for inspection. Please contact the inspector to make arrangements for a complete inspection of the pool and pool area.
☐ 61. Every pool shall be under the supervision of a person who is certified as a Swimming Pool Service Technician with this Department.
☐ 62. Maintain a log of the pool operation, disinfection residual, pH and maintenance procedures.
☐ 63. Provide adequate lifeguard service.

64. OTHER

☐ 99. Pursuant to section 65545 CCR, THIS POOL IS OFFICIALLY CLOSED. This pool shall not be placed in operation until all violations have been corrected & upon specific written approval by this agency. Call to schedule a reinspection.	FIELD OFFICE	RECEIVED BY INSPECTED BY REINSPECTION ON OR AFTER

H-2057 76S962 (Rev. 1/96) 2/97

FIGURE 12-11 Health department inspection list. *County of Los Angeles.*

The CDC findings also noted that both ultraviolet and ozone treatments were effective in killing the bacteria associated with Legionnaires' disease, but only in association with a residual of chlorine or bromine. They also noted that copper and silver ion treatment and iodine were effective in reducing the bacteria.

Finally, the CDC recommended preventative measures in commercial spas such as hourly residual readings, daily superchlorination treatments, and weekly filter cartridge inspections. It also recommended reducing jet or bubble action in spas which tend to create aerosols of the water which can then be inhaled.

FECAL ACCIDENTS

No matter how careful you are as a commercial pool or spa manager, sooner or later a child or incontinent adult will cause a fecal accident in your facility. Many jurisdictions have written regulations for response, but here are some basics:

- Close the pool or spa immediately and require all bathers to leave.
- Manually remove as much fecal material as possible. If vacuuming is possible or necessary, vacuum to the sanitary sewer, not the filtration system. Disinfect the vacuum, hose, and pump before using them again.
- Disinfect the pool to a CT value of at least 9600. CT value is the concentration of chlorine (expressed in parts per million) multiplied by the time in minutes. For example, a 20 ppm chlorine residual maintained for 8 hours (480 minutes) results in a CT value of 9600 (20 ppm × 480 minutes). Any combination of chlorine concentration and time which equals at least 9600 will be satisfactory.
- Adjust the pH as needed to normal levels.
- Run the filtration system for at least four complete turnovers of the water. After this cycle, backwash to the sanitary sewer. Break down and clean the filter, disinfecting it with a dilute chlorine wash (20 parts of water to 1 part of standard pool chlorine).
- If possible, it is preferable to drain the pool to the sanitary sewer and refill. Before refilling, brush the interior with a dilute chlorine wash (20 parts of water to 1 part of standard pool chlorine).

- Restart the pool and balance the water. When chlorine residuals return to 5 ppm, the pool may be reopened.

Many local codes require reporting of fecal accidents to health authorities. Of course, prevention is the best measure, so keep an eye out for kids (or adults) who may be hiding diapers under their bathing suits and keep them out of the pool in the first place.

CHAPTER 13

50 Things Your Pool or Spa Can Do for Our Environment

When we think of damage to the environment, we tend to think about the internal combustion engine as a focal culprit. Certainly a vessel of clear, sparkling water in our backyard, with our children splashing about, doesn't immediately spring to mind. But should it?

Pools and spas are in fact a major source of environmental degradation, but they don't have to be if properly maintained. In our effort to keep them clean and at a specific temperature, we generate hazardous waste, greenhouse gases that contribute to global warming, air pollution in the form of volatile organic compounds (VOCs), and we waste the most precious natural resource on the planet—water.

But can we enjoy the many benefits of a pool or spa with a clear environmental conscience? Yes, if we follow a few basic guidelines for conservation and proper waste disposal. When it comes to a pool or spa, the same rules apply—reduce, reuse, recycle.

Chemicals

Chemicals created for the pool and spa industry can be harmful to the environment if not used and disposed of properly. Under some circumstances, chlorine converts to dioxin, now widely regarded as one of the most harmful substances on the planet. But there are ways to be sure pool and spa use of chemicals is not detrimental.

1. Chlorine and other chemicals should be purchased in amounts likely to be used within a week (or two at most) so you are more likely to use the entire amount.
2. This also means you'll use less overall, because things like chlorine, which lose potency over time, will provide more of the desired effect for less total gallons of product each year.
3. If you purchase pool supplies at a wholesaler or club discount store where large quantity packages are the only option, find a neighbor to join you and split the chemicals so each of you ends up with smaller batches.
4. Buy products that come in reusable containers (I hate to drive around with cases of empty chlorine and acid bottles, but by returning them for refilling I'm doing the planet a small favor).
5. Or buy products that come in containers that can be recycled.
6. Take out-of-date products to a hazardous waste recycling station.
7. Better yet, use them. Even products with a limited shelf life are likely to have some potency remaining after the expiration date. Since you have a test kit, you can apply the normal amount of product and test to see if you got the appropriate benefit. Your test may reveal you need to add more product for the desired result, but it's better to use up the chemicals on hand than to toss them out.

Energy Conservation

Natural gas, oil, propane, or even electricity used to heat a pool or spa generates by-products associated with global warming (yes, even the generating station that supplies your electricity probably uses fossil fuels). But just like reducing the amount of chemicals we use, reducing energy consumption not only helps the planet, it saves money too.

8. Check the pilot light on your heater. It should be a strong, blue, cone-shaped flame. If it appears "lazy" or yellow, you're wasting gas. Clean out the pilot burner or check the gas pressure.

9. Install an easy-to-read thermometer somewhere in the pool or spa so you'll always know if water temperatures are excessive. One good model is the type that is built into the skimmer lid.
10. Install solar heating.
11. Double cover your spa to trap more heat.
12. Move your spa to a warmer location, perhaps a patio that enjoys direct sunlight for part of the day or an area out of the wind.
13. Create windbreaks around the pool or spa to prevent heat loss and evaporation with fencing, shrubs, walls, or rockscapes.
14. Insulate the exterior of the spa (set it in-ground if it isn't already).
15. Don't leave the pool light on except when you're using the pool.
16. When replacing pool light bulbs, consider a lower wattage for energy conservation. The light won't be as bright, but my experience is that most people are satisfied with the results.
17. Set the heater thermostat lower (cooler).
18. Shut off the heater completely if you're not likely to use the pool or spa in the next 3 or more days.
19. Cover your pool or spa. Evaporation wastes water and heat. When only 5 gallons (19 liters) evaporate from a spa, it cools the remaining 500 gallons (1892 liters) a full degree.
20. Insulate pipes.
21. Buy energy-efficient motors, heaters, and blowers when making replacements.
22. Tune up the heater each year to maximize its performance and reduce wasted fuel.
23. Don't run pumps longer than necessary. The average pool pump uses as much energy each hour as the average window air conditioner.
24. Tune up your automatic pool cleaner so the system (including the booster pump) can be run fewer hours for the same result.
25. Operate the automatic pool cleaner every other day instead of every day.

26. If possible, locate air blowers indoors or in an insulated equipment area. Otherwise they inject cold outdoor air into the spa, forcing the heater to work even harder.

Water Conservation

Two-thirds of the planet is covered by water, but less than 5 percent of that is fresh water. Most places on earth are rapidly running out of fresh water or have polluted remaining supplies with chemicals and pesticides, rendering it unfit to drink. No matter where you live, every drop of water is precious.

27. Cover your pool or spa. OK, so I mentioned this one in the energy section, but it deserves repeating because you can lose 50 gallons (189 liters) per day from an uncovered 500-square-foot (46-square-meter) pool [that's over 18,000 gallons (68,000 liters) each year]. By reducing evaporation, you also keep total dissolved solids in line (refer to Chap. 8).
28. Although it can be lots of fun, avoid excessive splashing. Still water evaporates at half (or less) the rate of disturbed water, because wave action can double or triple the amount of surface area and contact with the air.
29. Install a separation tank to filter backwash water, so the water can be returned to the pool.
30. Backwash only when needed and then only as long as it takes for the water to run clear. As noted in Chap. 5, backwashing isn't very effective anyway and cleaning the filter properly (tear down, clean, reassemble) uses much less water.
31. Use backwash water (or any water when you drain the pool) where possible on lawns or gardens. Be sure the chlorine residual is not too high for the plants, but you can pump it to a large trash can for storage until the residual drops. If you're draining the pool, don't add chemicals for 2 or 3 days. After that, it can be used safely on lawns and plants. The typical lawn uses up to 500 gallons (1892 liters) of water for each watering, but that's not much when emptying a 30,000-gallon (113,550-liter) pool. Ask neighbors if they'd like some free lawn water rather than waste it.

32. Use the pool (or an unheated spa) instead of your home air conditioner for relief on hot days. Take a dip to cool yourself instead of cooling the entire house.

33. Check frequently for leaks. You can waste a lot of water before realizing you have a problem unless you keep a good watch on the equipment area and the water level.

34. Shut off waterfalls, fountains, and other water features when not in use. Wet slides are especially wasteful of water since the surface is usually warm and virtually all of the water that flows over them evaporates quickly.

35. Don't hose off decks; use a broom.

36. Direct rainwater to the pool for refilling.

37. Some automatic pool cleaners have a spray nozzle that tries to clean the tiles at the waterline. They don't work that well and water is sprayed out of the pool. Disable that feature if your cleaner has it.

38. Plug the overflow line when the pool is in use. Splashing and bather load will raise the level temporarily and send water to the overflow line and into the sewer. When the bathers leave, you'll be adding fresh water.

39. Add shade to the pool or spa in hot weather to reduce direct sunlight, which increases the temperature too much (leading to more evaporation) and destroys sanitizers. Try awnings or poolside umbrellas.

Recycle

Our landfills are rapidly filling up and pollutants leach from the waste into groundwater. Energy and other resources are wasted when new products are made if an existing product could be used again.

40. Save the leaves taken from your pool for mulch.

41. Buy equipment and supplies with the greatest amount of postconsumer recycled materials. Even plastic pumps and filters are now being made with recycled materials!

42. Never throw away an old piece of equipment. Take the motor to a rebuilder (who can use the parts and may even give you a few dollars for it); take the old filter or heater to a scrap metal dealer.

43. Replace the net on your skimmer instead of buying a new one. Other maintenance tools may have replacement components, so try before you buy.

44. Use discarded stockings or panty hose as catch bags on automatic cleaners or spa vacuums. The fine mesh does a great job of straining dirt and sand from the water and the material gets a second life.

Miscellaneous

Just when you thought you had done it all, there's always one or two more things you can do for your planet.

45. When painting a pool, spa, or deck, use paints with low volatile organic compounds (VOCs) to protect air quality.

46. When sanding, chipping, or grinding, control the dust and properly dispose of all residue. It may contain fine particles of old paint, metals, or chemicals which can pollute the air you are breathing in the vicinity or harm plants and pets.

47. When landscaping around the pool, avoid trees that drop lots of leaves and don't create dirt planter areas near the water where wind and rain will flush debris into the pool. Organic debris causes more chemical use, and as noted above, you're trying to cut down.

48. If you live in a cold climate and winterize, don't drain the pool entirely. Save as much water as you can for reopening the pool. It can be filtered and treated and will be as good as new.

49. In some places (such as California), storm drains flow directly to rivers and the ocean. That means that anything which drains from your pool deck into the storm drains also flows to natural water courses. So don't use detergents or harsh chemicals to clean your pool deck unless you want to be swimming with the same pollutants!

50. Conduct an environmental (and money saving!) audit of your pool monthly:

- Visually inspect the equipment area for leaks.
- Conduct the bucket evaporation test (Chap. 10) quarterly.
- Check time clocks for proper hours of operation of pumps, boosters, water features, and lights.
- Check the cover and keep it in good repair. Keep the roller or other storage device in good working order so everyone will actually use the cover.
- Check the pilot light on the heater (see above).
- Check heater thermostats to be sure no one turned them higher than necessary.
- Check proper operation of solar heating system for maximum free heating.
- Check proper operation of the automatic cleaner.

LABOR REFERENCE GUIDE

Labor rates for a qualified pool and spa technician vary in the United States from $30 to $60 per hour. Overhead, insurance, complexity of the job, and the experience of the technician all might be factors in determining the rate.

What is the *reasonable* amount of time for a job? In the automotive industry, guidebooks establish that it might take four hours for a brake job, two hours for a tune-up, six hours to rebuild a carburetor, and so on. The mechanic can then charge according to these industry standards, regardless of how long the procedure actually took him to complete. Using this method, both mechanic and customer can be assured of a fair price for the job.

In setting a standard for the amount of time a job should require, I include several factors in the calculation and leave other factors out. For example, if special parts are required, I must consider the time spent going to get those parts that aren't normally in my workshop. The time it takes to unload tools (and pack up after the job as well), set up, and prepare a job are just as much a part of the work as the repair itself. The job might require returning to make adjustments or other follow-up work, and this must be included in the initial estimate of time required.

Warranty is another factor. Heater manufacturers have factory technicians that will make repairs after the initial installation while the product is under warranty (or after the warranty expires for a fee). If I install a heater and something goes wrong, my customer doesn't want to hear that the factory technician will be along in a week or so. Therefore, even if the problem is in the equipment and not my installation, I must be prepared to make the repair or remove the equipment, exchange it, and install a replacement. This type of extra warranty must be provided for in the original labor charge. Of course, you can't charge a customer for your mistakes. I don't charge four hours for a job that takes two, just to cover the possibility that I might have goofed and will need to come back to do the work a second time.

Travel time is another consideration. In Malibu, where many of my customers live 10 miles up a winding canyon road, I have to consider the time required to stop by and make the initial estimate, the time to go back and do the work, and the time to go back and check on the results. Keep in mind also that recommended hours are total work hours for the job, including helpers (when necessary) for work such as carrying a heavy heater for a replacement procedure (and removing the old one).

Finally, consider the working conditions. If you have to remove several pieces of equipment to access the one you need to work on, or if the repair requires hanging off the edge of a steep hill working with one hand and holding on for dear life with the other, then add an appropriate number of hours to the estimate. Also, some jobs require considerable time to determine what work is needed, especially when troubleshooting heaters. That time invested must be included in the total hours for the job. In other words, when setting your rates and hours, give careful consideration to *all* aspects of the work you do or might have to do.

The following guidelines are based on many years of personal experience and supervision of employees. They are based on residential installations, not commercial. I provide them here as a reference only. They should be adjusted as needed in your circumstances. Remember, they include reasonable setup and cleanup time, as well as the actual work time I believe each job should require when performed by a technician of average skills working under average conditions. I have not listed items requiring individual estimates that might vary widely depending on size, difficulty, or condition of the pool, such as painting or breaking in plaster on a pool, leak repair, or replacing a skimmer. I have also not included estimates for repair or replacement of chlorinators or automatic pool cleaners because there are so many different makes and models, each of which has unique demands.

Procedure	Hours
Filters	
Tear down/clean filter (cartridge or DE)	1
Open sand filter, break up/clean sand	2
Tear down sand filter, change sand	4
Install new filter (remove and dispose of old unit)	4
Pumps and motors	
Replace seal, seal plate, gasket, or other nonplumbed pump components	1.5
Replace volute, strainer pot, or other plumbed component	2
Replace motor (including electrical)	2
Heaters	
Replace heater	8
Replace flange gaskets or thermostatic mixing valve	1
Troubleshoot and replace defective control circuit component(s)	1.5
Replace pilot assembly	2
Replace main gas valve or burner tray component(s)	2.5
Replace manifold headers, heat exchanger, or firebrick	3
Ream heat exchanger	4
General procedures	
Drain, acid wash, refill, and rebalance chemistry (pool)	10
Drain, acid wash, refill, and rebalance chemistry (spa)	4
Replace diving board, rail, ladder, or slide (on existing identical mount)	3
Cut floating pool cover to size, install on roller straps	2
Rebuild three-port valve	1
Replace (or replace a component in) a time clock	1
Replace an underwater light bulb	1.5
Replace an underwater light fixture	2
Replace an air blower	1.5

GLOSSARY

Not all terms in this glossary are used in the text, but because you will encounter them in other aspects of the industry or in regional variances they are included for your reference. Common words or measurement definitions not specific to the water industry are not included, but can be found in any standard dictionary. Ordinary words with colloquial industry meaning are defined only by those references.

ABS Acrylonitrile butadiene styrene. Rigid plastic pipe similar to PVC, usually manufactured black, used for drainage systems.

ac Alternating current. Electrical charge that flows from negative to positive, then reverses direction. The basic form of electricity used in most homes and businesses and, therefore, in most pool and spa equipment.

acetone A highly flammable solvent used to clean plastic surfaces and tools.

acid A liquid or dry chemical that lowers pH when added to water, such as muriatic acid.

acid demand The amount of acid required (demanded) by a body of water to lower the pH to neutral (7.0).

acidity The quality, state, or degree of being acid.

acid spotter A siphon device that attaches to a telepole for insertion into a body of water to siphon acid from a bottle (connected by means of a plastic hose). The acid can then be directed and concentrated on a particular location for stain removal.

acid wash The procedure of cleaning plaster with a solution of muriatic acid and water.

adapter bracket The part of a pump that supports the motor and connects the motor to the pump.

aerator A pipe vented to the atmosphere, sometimes with an adjustable volume control, added to a waterline to mix air and water prior to discharge.

aggregate The major component of plaster, composed of sand, marble dust, pebbles, or other solid matter.

air blower *See* blower.

air relief valve A valve on a filter that permits air to be discharged from the freeboard.

air switch A pneumatic-mechanical control device used to operate spa equipment safely. A button, located in or near the water, is depressed sending air pressure along a plastic hose to an on/off switch.

algae Airborne, microscopic plant life of many forms that grow in water and on underwater surfaces.

algicides A group of chemical substances that kill algae or inhibit their growth in water.

algistat A chemical that inhibits the growth of algae.

alkalinity The characteristic of water that registers a pH above neutral (7).

aluminum sulfate Alum. An additive for sand filters that prevents the sand from combining, hardening, and thus not filtering impurities from water.

ambient temperature The average, prevailing surrounding temperature.

American Gas Association (AGA) The national institute that sets standards for installations using natural gas as a fuel.

American National Standards Institute (ANSI) The national institute that sets standards for construction.

American Public Health Association (APHA) The national institute that sets standards for issues relating to public health and safety.

American Wire Gauge (AWG) The standard used to specify wiring of certain thickness and capacity.

ammonia Natural substance composed of nitrogen and hydrogen that readily combines with free chlorine in water forming chloramines (weak sanitizers).

amperage (amps) The term used to describe the strength of an electric current. It represents the volume of current passing through a conductor in a given time. Amps = watts ÷ volts.

anhydrous Lacking water. A characteristic of a substance that makes it readily combine with water.

anode The positively charged electrode, at which oxidation occurs, in an electrolytic process. *See also* cathode.

antisiphon valve A control device added to the domestic water supply line to prevent contaminated water from flowing backward into the pipe.

antisurge valve A check valve used in air blower plumbing to prevent water from entering the blower mechanism.

antivortex The property of a plumbing fitting that prevents a whirlpool effect when water is sucked through it. Used on main drain covers.

arc (arcing) The passage of electrical current between two points without benefit of a conductor. For example, when a wire with current is located near a metal object, the electricity might arc (pass) between the two.

automatic gas valve The valve that controls the release of natural or propane gas to a heater. Also called the combination gas valve.

available chlorine *See* chlorine, free available.

backfill Dirt, sand, or other material used to fill the gaps between a pool wall and the surrounding excavation.

backwash The process of running water through a filter opposite the normal direction of flow to flush out contaminants.

bacteria Any of a class of microscopic plants living in soil, water, organic matter, or in living beings and affecting humans as chemical reactions or viruses.

balance The term used in water chemistry to indicate that when measuring all components together (pH, total alkalinity, hardness, and temperature) the water is neither scaling nor etching. The Langlier index value of balanced water is zero.

balloon fitting A plumbing connector made of rubber or flexible plastic that readily adapts to pipe of varying sizes.

bank A chemical residual in water that is inactive but can be released by adding a catalyst.

barb fitting A plumbing fitting with exterior ribs, connected by insertion into a pipe (usually flexible pipe).

base An alkaline substance.

bather Any person using a pool or spa.

bather load The number of bathers in a pool or spa over a 24-hour period.

bather occupancy The number of bathers in a pool or spa at any one time.

bayonet Refers to a light bulb using two nipples to engage corresponding slots in the socket.

bicarbonate of soda A chemical used to raise pH and total alkalinity in water. Also called bicarb or baking soda.

biological filter A method of filtration using bacteria to expedite decomposition of organic debris, usually in conjunction with sand as a filter media.

bleach Colloquial term for liquid chlorine.

bleed To remove the air from a pipe or device, allowing water to fill the space.

blister Refers to an air pocket in a plaster surface.

blow bag Also called a drain flush or balloon bag. A device attached to a garden hose that expands under water or air pressure to seal an opening, forcing the water or air into that opening.

blower An electromechanical device that generates air pressure to provide spa jets and rings with bubbles.

bluestone *See* copper sulfate.

blue vitriol *See* copper sulfate.

boiler *See* heater.

bond beam The top of a pool or spa wall, built stronger than the wall itself to support coping stones and decking.

bonding system The wiring between electrical appliances and the ground to prevent electric shock in case of a faulty circuit. All appliances are grounded to the same wire.

booster pump A pump added to a spa system to add pressure to the jets.

break down *See* tear down.

break-in The procedure of curing and preparing new plaster for service in a pool.

break point chlorination The application of enough chlorine to water to combine with all ammonia (creating chloramines), then to destroy all chloramines, leaving a residual of free chlorine. The break point is, therefore, the point at which chlorine added to the water is no longer demanded to sanitize and is available to become residual chlorine. Some references define break point as ten parts of chlorine for every one part of ammonia present in the water when the pH is between 6.8 and 7.6.

bridging The condition existing when DE and dirt closes the intended gaps between the filter grids in a DE filter, reducing the flow rate through the filter and reducing the square footage of filter area.

broadcast Dispersal of chemicals over the widest possible surface area of a body of water.

bromine (Br_2) A water sanitizing agent. A member of the halogen family of compounds.

brown coat The undercoat of plaster applied over old plaster or as a base coat over gunite.

Btu British thermal unit. The measurement of heat generated by a fuel. The amount of heat required to raise 1 pound of water 1°F (when at or near 39.2°F).

buffer A substance that tends to resist change in the pH of a solution.

burner tray The component in the bottom of a heater that controls the burning of gas.

bushing An internal plumbing fitting that fits (threaded or slip) into a connector fitting to reduce the internal diameter so as to accommodate a smaller pipe size than the connector fitting was designed to accept. *See also* reducer.

calcium A mineral element typically found in water.

calcium bleed The condition in plaster where calcium leeches from the mixture.

calcium carbonate ($CaCO_3$) The mineral precipitated out of water, deposited on pool and spa surfaces, the major component of scale.

calcium hypochlorite ($Ca(OCl)_2$) A granular form of chlorine (widely produced under the brand name HTH), generally produced in a compound of 70 percent chlorine and 30 percent inert materials.

cam lock The device that holds or releases two halves of a telepole.

cap A plumbing fitting attached to the end of a pipe to close it completely.

capacitor A device used to store an electrical charge to be released in one short burst.

capacitor start-capacitor run motor (CSR) An ac single-phase electric motor controlled by a capacitor for both starting and running. The motor does not require an internal start switch or governor. *See also* energy efficient.

capacitor start-induction run motor (CSI) An ac split-phase electric motor with a capacitor connected in series to a separate starter winding that shuts off after the motor has reached operating speed. The capacitor winding jump starts the motor, then an induction winding takes over to run the motor. The motor requires an internal start switch and governor.

cartridge The element in a filter covered with pleats of fabric to strain impurities from water that passes through it. Generally strains out particles larger than 20 microns.

cathode The negatively charged electrode in the electrolytic process. *See also* anode.

caulking Material used in a joint to create a waterproof seal.

caustic Corrosive or etching in nature.

cavitation Failure of a pump to move water when a vacuum is created because the discharge capacity of a pump exceeds the suction ability.

centrifugal force The outward force created by an object in circular motion. The force that is used by water pumps to move water.

C frame A type of motor housing resembling a C. Adapts to a particular style of pump.

channeling Creation of a tube or channel in a filter media through which water will flow unfiltered. Channeling is caused by calcification of the media (hardening of the sand in a sand filter, for example) and is resolved by breaking up the sand and treating it with alum.

check valve A valve that permits flow of water or air in only one direction through a pipe.

chelating agent Chemical compounds that prevent minerals in solution in a body of water from precipitating out of solution and depositing on the surfaces of the container.

chine The portion of a stave (in a wooden hot tub) that extends below the croze.

chine joist The sturdy plank that supports the floor of a hot tub.

chloramine A compound of chlorine when combined with inorganic ammonia or nitrogen. Chloramines are stable and slow to release their chlorine for oxidizing (sanitizing) purposes.

chlorinator A device that delivers chlorine to a body of water.

chlorine (Cl_2) A substance made from salt that is used to sanitize water by killing bacteria. A member of the halogen family, chlorine is produced in gas, liquid, and granular form.

chlorine demand The amount of chlorine required (demanded) by a body of water to raise the chlorine residual to a predetermined level.

chlorine, free available That portion of chlorine in a body of water that is immediately capable (available) of oxidizing contaminants.

chlorine lock The term applied to chloramine formation, when ammonia is present in the water in sufficient quantity to combine with all available chlorine.

chlorine residual The amount of chlorine remaining in a body of water after all organic material (including bacteria) has been oxidized, expressed in parts per million. The total chlorine residual is the sum of all free available chlorine plus any combined chlorine (chloramine).

chlorophyta Green algae.

circuit The path through which electricity flows.

circulation system The combination of pipes, pump, and any other components through which water flows in a closed loop (from the body of water, through the components, and back to the same body of water).

clamping ring The metal band that applies pressure to and holds the lid on a filter tank or holds together two halves of a pump.

closed loop Any system that recirculates the same water, rather than constantly taking in a new supply and discharging it after only one pass through the system.

close nipple *See* nipple.

colloidal silver A compound of silver, used as an algicide.

colorimetric The name given to a chemical test procedure where reagents are added to water and change color to reflect the presence and strength of a substance. The color is compared to a color chart to evaluate the volume of that substance. Colorimetric tests are used to detect the presence of chlorine in water.

comparator The color chart or other device used to compare the color of a treated water sample with known values. Used in chemistry test kits.

compression fitting A plumbing connection that joins two lengths of pipe by sliding over each pipe and applying pressure to a gasket that seals the connection.

condensation The process by which a substance changes from a gas to a liquid (called the condensate).

conditioner A chemical that slows the decomposition of chlorine from ultraviolet light. Conditioner, usually cyanuric acid, also helps prevent spiking of pH between high and low extremes.

conductor Any substance that will carry electric current, such as a wire, metal, or the human body.

control circuit A series of safety and switching devices in a heater, all of which must be closed before electric current flows to the pilot light and automatic gas valve to ignite the heater.

coping The cap on the edge of a pool or spa mounted on the bond beam.

copper sulfate A blue granular substance added to a body of water to kill algae. Also called bluestone or blue vitriol.

corona discharge A method of producing ozone by passing electricity through oxygen and water.

coupling A plumbing fitting used to connect two lengths of pipe.

cove The curved radius between the wall of the pool and the floor.

CPVC Chlorinated polyvinyl chloride. The designation of plastic pipe that can be used with extremely hot water.

croze The milled groove in the stave of a wooden hot tub into which the floorboards are inserted.

CSI *See* capacitor start-induction run motor (CSI).

CT value Concentration of chlorine (in ppm) multiplied by time (in minutes). A formula used to determine the period a pool should be closed.

current Refers to the rate of electrical flow between two points.

cyanophyta Blue-green algae or black algae.

cyanurates Chlorine sanitizers combined with stabilizers, such as dichlor and trichlor.

cyanuric acid *See* conditioner and stabilizer.

cycle A complete turn of alternating current (ac) from negative to positive and back again (*see* hertz). Also refers to a filtration period (*see* filter run).

dedicated circuit An electrical circuit used for only one specific appliance.

delamination Separation or failure of the bond in layered materials, such as plaster to gunite or fiberglass to resin.

diameter The distance across the widest part of a circle.

diatomaceous earth (DE) A white, powdery substance composed of tiny prehistoric skeletal remains of algae (diatoms), used as a water filtration media in DE filters. DE filters can remove particles larger than 5 to 8 microns.

dichlor *See* sodium dichloro-s-triazinetrione.

dielectric An insulating material that prevents electrolysis or corrosion when applied between two surfaces of dissimilar metal.

diffuser A housing inside a pump covering an impeller that reduces speed of the water but increases pressure in the system. It helps to eliminate airlock.

discharge The flow of water out of a pipe or port.

diverter A 6-inch-long by 1½-inch-diameter plastic or bronze adapter pipe that fits into a skimmer port to facilitate connection of a vacuum hose. The diverter can divert all suction to the skimmer, closing off the main drain or vice versa.

dog *See* tripper.

DPD Diethy phenylene diamene. The chemical reagent used to detect the presence of free available chlorine in a body of water.

drain flush *See* blow bag.

dry fit To assemble PVC plumbing without glue to check that you have prepared the components correctly.

duty rating An evaluation parameter of pool and spa motors that are designed for continuous duty, meaning they could run 24 hours a day for their entire service life without stopping. The nameplate on the motor shows this by the rating, continuous duty.

dynamic head *See* head.

Dynell A type of material used to cover filter grids.

effluent The water discharging from a pipe or equipment.

elbow The term for a plumbing connector fitting with a 90-degree bend.

electrode A solid electrical conductor through which an electric current enters or leaves a cell in an electrolytic process. *See also* anode *and* cathode.

electrolysis The production of chemical changes (usually corrosion) in metal by passing an electric current through an electrolyte (usually water with mineral content).

electrometric A chemical test involving an electronic analysis meter.

element A filter grid. Also the electric heat-generating rod of an electric heater.

end bell The metal housing or cap at the end of an electric motor.

energy efficient A specific design of pool and spa motors that have heavier wire in the windings to lower the electricity wasted from heat loss. *See also* capacitor start, capacitor run.

equalizer line A pipe that balances the flow of water between two locations. Usually found in a pool skimmer, the equalizer line allows the suction line to draw water directly from the pool in the event that the water level drops below the point where suction would normally occur from the surface.

erosion system A type of chemical feeder in which granular or tablet sanitizer is slowly dissolved by constant flow of water through the device.

escutcheon plate The decorative metal ring that slips over the end of a rail or ladder improving the finished appearance of the installation.

etching Corrosion of a surface by water that is acidic or low in total alkalinity and/or hardness.

ethylenediamine tetra-acetic acid (EDTA) A reagent used for testing calcium hardness, added a drop at a time until the solution turns blue. The number of drops is compared to a chart to evaluate the hardness of the sample.

expansion joint The gap between the coping and the deck that allows for normal expansion and contraction of the materials, preventing damage that would otherwise result from pressure of the two against each other.

feathering The process of tapering the edge of a coating material, such as paint, resin, or plaster, to blend invisibly with the existing materials.

female Plumbing fittings or pipe with internal threads or connectors.

fiberoptics Underwater lighting devices that illuminate by sending light along thin plastic cable from a remote source.

fill water Water added to a pool or spa to replace water lost to evaporation or other reasons. Also called make-up water.

filter A device for straining impurities from the water that flows through it.

filter cycle *See* filter run.

filter filling The technique used to prime a pump by filling the filter with water and allowing it to flow backwards into the pump.

filter run The time between cleanings, expressed as the total running time of the system, also called the filter cycle. Care must be taken when using the terms *run* or *cycle*, because some technicians mean the number of hours the system operates each day rather than the total time between cleanings. Regionally, one term might be used for one definition and the other for the second. I use *run* for the daily operation, *cycle* for the time between cleanings.

FIP A female threaded plumbing fitting.

fireman's switch An on/off control device, mounted in a time clock, that turns off a heater 20 minutes before the time clock turns off the circulating pump and motor. This allows the heat inside the heater to dissipate before shutdown.

flange gasket The rubber sealing ring that prevents leaks between a heater and the circulation pipes.

flapper gate The part in a check valve that swings open when water is flowing in the intended direction but swings shut when water attempts to flow backward.

flare fitting A threaded plumbing fitting that requires a widening of the pipe at one end.

flash test The method of dropping chemistry test reagents directly into pool or spa water rather than into a vial containing a test amount of that water.

flex connector A coated metal pipe with threaded fittings on each end designed to bend freely for connection in tight quarters or at odd angles. Usually used to connect a gas pipe to an appliance (such as a pool heater where approved by local code) indoors, allowing the appliance to be moved without breaking the connection.

floater A chemical feeder system whereby a sanitizer tablet is placed in the device and is allowed to float around the body of water. The tablet dissolves and the sanitizer is released into the water.

float valve A plumbing device that restricts or shuts off the flow of water based on a lever that is attached to a float and that thereby rises and falls with the water level.

flocculate The process of adding a chemical to a body of water which combines with the suspended particulate matter in the water, creating larger particles that are more easily seen and removed from the water. Also called a clarifying agent or coagulant.

flow meter A device for measuring the rate of water passing through a given pipe, expressed in gallons per minute (gpm).

flow rate The volume of a liquid passing a given point in a given time, expressed in gallons per minute (gpm).

fluorine A little-used (very costly) halogen sanitizer.

flux The chemical compound applied to copper before soldering to enhance the even flow of the solder and to prevent oxidation of the metal when heating.

forty-eight (48) frame A housing design for a specific type of pool or spa pump motor.

four-pass unit A type of heat exchanger found in pool and spa heaters that directs the flow of water to be heated through the unit four times before returning it to the body of water.

freeboard The vacant vertical area between the top of the filter media and the underside of the top of the filter.

free chlorine Also called available chlorine. It is chlorine in its elemental form, not combined with other elements, available for sanitizing the water.

free-flow filter A type of sand and gravel filter that allows unpressurized passage of water through the filter media (sand).

free joint A colloquial expression in plumbing. When connecting several joints in a plumbing job, the free joints are the ones you can complete without committing to the end result; the ones that are still reversible.

freon A gas used in refrigeration and ozone production (*see* ozonator).

full-rated A characteristic of motor design meaning that the motor operates at its designated horsepower (1, 2, etc.). *See also* uprated and service factor.

fusible link (fuse link) A safety device located near the burner tray of a heater that is part of the control circuit. If the fuse link detects heat in excess of a preset limit, it melts and breaks the circuit to turn off the heater.

galvanized Iron pipe coated with zinc or another alloy to prevent corrosion.

gasket Any material (usually paper or rubber, but sometimes caulk or other pastes) inserted between two connected objects to prevent leakage of water.

gate valve A valve that restricts water flow by raising and lowering a disc across the diameter of the pipe by means of a worm drive. *See also* slide valve.

gauge The size of an electrical wire. Heavier loads can be carried on heavier-gauge wires. The numbering system of wire gauges works in reverse. A 10-gauge wire, for example, is thicker than a 14-gauge wire. Also refers to a measuring device, as in pressure gauge.

gelcoat A thin, surface, finishing coat of resin sprayed over fiberglass.

geophone An acoustic listening device used for leak detection.

glazed Surfaces that have been covered with a clear, protective coating, applied under heat, such as glazed tile.

gpm Gallons per minute.

grains per gallon (gpg) A unit of measurement used in old-style test kits rather than the more common parts per million (ppm). To convert grains per gallon to parts per million, multiply the gpg by 17.1. For example, if your old kit resulted in a reading of 10 gpg, that is equal to 171 ppm.

grid Frame covered with fabric used as a filter media; also called septa or element.

ground fault interrupter (GFI) A type of circuit breaker. A sensing device that determines when electricity in a circuit is flowing through an unintended path, usually to earth, creating a hazard of electrocution. The GFI detects current variations as low as 0.005 amp and breaks the circuit within one-fortieth of a second.

gunite A dry mixture of cement and sand that is mixed with water at the job site and sprayed onto contoured and supported surfaces to build a pool or spa, creating the shell.

halogens A family of oxidizing agents including chlorine, bromine, iodine, and fluorine.

hardness Also called calcium hardness. The amount of dissolved minerals (mostly calcium and magnesium) in a body of water.

Hartford loop Used in air blower installations. A method of plumbing the air delivery pipe above the water level of the spa, then back below the water level. The loop prevents water from siphoning into the blower.

head The measurement of pressure (expressed in feet) in a water circulation system (or parts thereof) created by friction, resistance, distance, or lift. Dynamic head + static head = total dynamic head (TDH).

heater A device that raises the temperature of water using natural gas, electricity, propane, solar, or mechanical energy for fuel. To be called a heater, the device must covert at least 70 percent of its fuel into heat (no more than 30 percent lost in venting). Over 80 percent efficiency, the device is typically called a boiler.

heat exchanger The copper tubing in a heater through which water flows. The water absorbs rising heat that is generated from the burner tray below.

heat pump A type of pool and spa heater. Like conventional heaters, the heat pump circulates water through the unit and transfers heat to the water from a fuel. Instead of using gas, electricity, or solar heat as a fuel, the heat pump takes warmth out of the air that is created by compressing a gas.

heat riser A metal pipe plumbed directly to the heater to facilitate dissipation of heat before plumbing with PVC. Also called a heat sink.

hertz Unit of measurement in ac indicating the number of cycles per second in the current as generated. In the United States, ac is generated at 60 hertz.

high-limit switch A safety device used in the control circuit of heaters. When the high-limit switch detects temperatures in excess of its preset maximum, it breaks the control circuit to shut down the heater.

high-rate sand filter A filter using sand for the filtration media designed for flows in excess of 5 gpm but less than 20 gpm (less than 15 gpm in some codes) per square foot. Strains impurities larger than 50 to 80 microns.

hopper The low bowl portion of the pool around the main drain.

horsepower (hp) The standard unit of measurement that denotes the relative strength of a mechanical device. One horsepower equals 746 watts or the power required to move 550 pounds one foot in one second.

hose bibb (also bib) The faucet to which a garden hose is attached.

hydration The process of adding moisture to a dry substance.

hydraulics The science of water movement.

hydrochloric acid (HCl) Muriatic acid.

hydrostatic pressure The force created by water that tends to push pools up out of the ground.

hydrostatic valve A check valve located in the main drain of a pool to relieve hydrostatic pressure created by rising groundwater. The valve allows groundwater into the pool, but does not let water out.

hypochlorous acid (HOCl) A form of free available chlorine resulting from the solution of an active chlorine compound added to water.

impeller Rotating part of a pump that creates centrifugal force to create suction. The impeller is said to be closed if it is shrouded (covered) on both sides of the vanes, or semiopen if it is shrouded on one side, while the interior surface of the volute creates a partial shroud on the other side.

inlet *See* return.

intermittent ignition device (IID) The electronic control and switching device used in electronic ignition heaters to operate the control circuit and automatic gas valve. Often called the brainbox.

iodine (I_2) An element used to sanitize water. A member of the halogen family like chlorine and bromine.

IPSSA Independent Pool and Spa Servicemen's Association.

J-box Short for junction box. The metal container in which wires are connected along a circuit or conduit.

jet (hydrojet) The discharge fitting through which water is returned to a spa at a high rate of speed and/or pressure.

jetted tub A bathtub fitted with hydrotherapy spa-type jets.

joint stick A paste used on threaded plumbing to prevent leaks that is available in the form of a crayon-like stick for easy application.

joist The support beam under the floor of a wooden hot tub or similar structure.

keyed shaft The shaft of a motor that has a groove for securing setscrews of the shaft extender. Used with specific designs of pumps.

kilowatt One thousand watts of electrical power. Electricity is sold by the kilowatt-hour, meaning a certain fee is charged for every 1000 watts delivered per hour.

Langlier index An indicator (like pH) that measures the scaling or etching ability of a body of water. A Langlier index of zero indicates perfect, balanced, neutral water (neither scaling nor etching).

laterals The horizontal filter grids at the bottom of a sand filter, installed in the underdrain.

latiance The mineral and cement dust formed on the surface of a pool shell when the concrete dries. It must be removed prior to plastering.

lazy flame A natural or propane gas flame in the burner tray of a pool or spa heater that burns in slow, wavering licks or stuttering, rather than the normal strong, clear blue flame burning straight upward.

leafmaster A brand name; leafmaster is a term applied to any device that vacuums large debris from a pool by means of water pressure created with a garden hose.

leaf rake A large open net secured to a frame that attaches to a tele-pole that is used to skim debris from the surface of the water.

lignin The white, pulpy cellulose material that binds the organic material of wood together. When a wooden hot tub is overchlorinated, stripping out the lignin, the tub will appear to be growing a white fur that will brush off and clog the filter.

line A wire conducting electricity. Also refers to a run of pipe carrying water.

lithium hypochlorite (LiOCl) Organic, granular, highly soluble form of chlorine, produced at around 35 percent available chlorine.

load An appliance that uses electricity.

macintosh bag filter An old style of filter (no longer in use) that uses a canvas bag as the filter media. Also called a coffin filter. Not manufactured in the past 30 years and never was legal for commercial installation.

main drain The suction fitting located in the lowest portion of a body of water. The principal intake for the circulation system.

main valve The flow control device in the combination gas valve of a pool or spa heater that regulates the flow of gas to the burner tray.

male Any plumbing fitting or pipe with external threads or connectors.

manifold An assembly or component that combines several other components together. A pipe fitting with several lateral outlets for connecting one pipe with others.

manometer A device used for measuring the pressure of gas, expressed in inches of water column.

marcite A term used regionally (in Florida, for example) for plaster, generally to denote that there is marble dust content in the plaster mix.

media Any material used to strain impurities from water that passes through it. DE and the fabric covering a cartridge are both examples of filter media.

micron A unit of measurement equal to 0.000001 meter or 0.0000394 inch. For example, a grain of table salt is approximately 100 microns in diameter.

MIP A male threaded plumbing fitting.

mission clamp A trade name. A mission clamp is a rubber connector hub-type fitting, secured with a stainless steel clamping band, used in plumbing to join two lengths of pipe of similar or dissimilar diameter.

mogul base The threaded end of a light bulb that is approximately twice the diameter of the typical household bulb end. Used in old-style pool light fixtures.

mottling A difference in shades within a given color. The term is used mostly in plaster to denote cloudy patches, blotches, or streaks of uneven color.

muriatic acid Also called hydrochloric acid. This chemical is the most commonly used substance for reducing pH and total alkalinity in water.

nameplate The label on a motor that details the operating and manufacturer's specifications.

NEC National Electrical Code. The generally accepted standard for electrical installations and procedures.

neutral The pH reading at which the substance being measured is neither acidic nor alkaline. Neutral pH is 7.0.

niche The housing built into the wall of a pool or spa to accommodate a light fixture.

nipple A short length (less than 12 inches) of pipe threaded at each end. If the nipple is so short that the entire length is threaded, it is called a close nipple.

no-hub connector A rubber (or neoprene) plumbing fitting used to connect two lengths of pipe, attached with hose clamps or other pressure devices.

NPT National Pipe Thread. The generally accepted standard specifications for threaded plumbing pipe and fittings.

nut driver A tool like a screwdriver but fitted with a female socket that surrounds a given size of nut for loosening or tightening.

open loop A type of solar heating system that circulates the water being heated from the pool or spa, through the solar panels, and back to the pool or spa.

organic Any material that is naturally occurring (not manufactured), such as leaves, sweat, oil, or urine.

O-ring A thin rubber gasket used to create a waterproof seal in certain plumbing joints or between two parts of a device, such as between the lid and the strainer pot on a pump.

ORP Oxidation reduction potential. A unit of measure of sanitizer ability in water, measured with an electronic ORP meter.

OTO Orthotolidine. The test reagent used in detecting the presence of chlorine and, by the resulting color of the OTO, the amount of chlorine, expressed in parts per million.

ozonator A device using electricity and oxygen to create ozone and deliver it to a body of water for sanitizing purposes. Ozone is produced in two ways for pool and spa use. One method, called the corona discharge method, is to shoot an electrical charge through oxygen and water, creating ozone. The other method is to electrify freon (the gas used in air conditioners and refrigerators) to produce ultraviolet light. In both cases, the production happens in a chamber in an ozonator built into the circulation system.

ozone (O_3) Three atoms of oxygen, creating a colorless, odorless gas used for water sanitation.

peristaltic chlorinator An automatic chlorine dispenser that applies pressure on a tube to create a vacuum, thereby creating suction of liquid chlorine from a tank.

pH The relative acidity or alkalinity of soil or water, expressed on a scale of 0 to 14, where 7 is neutral, 0 is extremely acidic, and 14 extremely alkaline.

phaeophyta Brown or yellow algae.

phenol red (phenolsulfonephthalein) The most widely used chemical reagent to measure the pH in a sample of water.

photosynthesis The natural process by which green plants, such as algae, produce food from sunlight. Using the green chemical chlorophyll, photosynthesis converts carbon dioxide and water into carbohydrates (sugars) for nourishment of the plant, that then discharges oxygen as a by-product of the process.

pi A mathematical constant equal to 3.14.

pilot The small gas flame that ignites the burner tray of a heater.

pilot generator The device that converts heat from the pilot light into electricity to power a control circuit on a heater. Also called a power pile or thermocouple.

pipe dope A paste used to prevent leaks in threaded plumbing.

plaster A hand-applied combination of white cement, aggregates, and additives that covers the shell of a gunite pool or spa to waterproof and add beauty. Plaster can also be colored.

plug A plumbing fitting used to close a pipe completely by inserting it (slip or threaded) into a female fitting or pipe end.

port An opening, as in a discharge port being the opening through which water flows out of a pipe or system.

positive seal The type of multiport valve that directs all of the water flow to one direction, allowing no flow in the other lines. A non-positive seal directs most of the water to one line, allowing some water to bypass into the other lines.

potassium monopersulfate An oxidizing chemical used to sanitize water. Used to supersanitize without using chlorine. Also catalyzes bromine.

potentiometer Also called a pot. A device that senses variance in temperature. Part of the thermostat assembly in a pool or spa heater.

power pile *See* pilot generator.

ppm Parts per million. The measurement of a substance within another substance; for example, 2 ounces of chlorine in 1 million ounces of water would equal 2 ppm.

precipitate An insoluble compound formed by chemical action between two or more normally soluble compounds. When water can no longer dissolve and hold in solution a compound, it is said to precipitate out of solution.

precoat The process of applying DE to grids in a DE filter after cleaning but before restarting normal circulation and filtration.

pressure gauge A device that registers the pressure in a water or air system, expressed in pounds per square inch (psi).

pressure sand filter A type of sand and gravel filter in which the water is strained through the filter media under pressure (as opposed to a free-flow filter). *See also* high-rate and rapid sand filters.

pressure switch A safety device in a heater control circuit that senses when there is inadequate water pressure (usually less than 2 psi) flowing through a heater (which might damage the heater) and breaks the control circuit, thereby shutting down the heater.

prime The process of initiating water flow in a pump to begin circulation by displacing air in the suction side of the circulation system.

psi Pounds per square inch.

pumice A natural soft (yet abrasive) stone substance (similar to lava rock) used to clean pool tiles.

pump A mechanical device driven by an electric motor that moves water.

pump curve The performance efficiency evaluation of a pump plotted on a graph comparing size, flow rate, and resistance (head).

PVC Polyvinyl chloride. The type of plastic pipe and fittings most commonly used in pool and spa plumbing.

quats Quaternary ammonium compounds. An organic ammonia compound used as an algicide in water.

radius One-half of the diameter of a circle.

rapid sand filter A filter using sand as a filter media that is designed for water flows not to exceed 3 gpm (5 gpm in commercial installations) per square foot.

reagent A liquid or dry chemical that has been formulated for water testing. A substance (agent) that reacts to another known substance, producing a predictable color.

rebar An abbreviation for reinforcement bar. Steel rods laid in concrete (such as a pool shell) to add strength.

rebound Gunite material that does not adhere when sprayed to form a pool or spa shell, falling off as pebbles and removed from the finished job.

reducer An external plumbing fitting that connects two pipes of different diameter. *See also* bushing.

residual The amount of a substance remaining in a body of water after the demand for that substance has been satisfied.

resin A liquid plastic substance (with the consistency of honey) applied to fiberglass fabric to strengthen and stiffen it.

retaining rod The metal rod in the center of certain filters on which is attached a retainer ring to hold grids in place.

retainer The plastic disc that fits over the top of a set of filter grids to hold them in place with the aid of a retainer rod.

return The line and/or fitting through which water is discharged into a body of water. Also called an inlet.

riser *See* heat riser. Also refers to any vertical run of pipe.

robotic pool cleaner A self-contained, electric-powered device that vacuums a pool.

rotor The rotating part of an electric motor that turns the drive shaft that itself is driven by magnetism produced between it and the stator. Also the diverter in the backwash valve of certain types of filters.

run Any horizontal length of pipe. *See also* filter run.

safety barrier A fence, wall, or other obstruction around a pool or spa to prevent entry by children or pets.

salt chlorine generator A device that converts salt in pool water into chlorine by means of an electrolytic process.

sand filter A filtration device using sand as the filter (straining) media. *See also* rapid sand filter and high-rate sand filter.

sanitizer Any chemical compound that oxidizes organic material and bacteria to provide a clean water environment.

saturation The point at which a body of water can no longer dissolve a mineral and hold it in solution.

scale Calcium carbonate deposits that form on surfaces in contact with extremely hard water. Water in this condition is said to be scaling or precipitating.

scratch coat The first layer of plaster applied to smooth out uneven surfaces prior to application of a fine or finish coat of plaster. To ensure a good mechanical bond between this base coat and the final coat, the plaster is scratched while still wet to create furrows.

screed A large leveling tool, often made from a straight piece of wood, used to level out a wet concrete slab by drawing it over the surface.

seal A device in a pump that prevents water from leaking around the motor shaft.

seal plate The component in a pump in which the seal is situated.

sediment trap A short length of vertical pipe attached below a run of gas supply pipe to catch any contaminants prior to the gas entering the heater.

separation tank A container used in conjunction with a DE filter to trap DE and dirt when backwashing.

septum *See* grid.

sequestering agent Used to remove metals by the filter. *See also* chelating agent.

service factor A measure of the reserve ability of an electric motor beyond its stated ability. For example, a 1-hp motor might actually function at 1.25 hp. *See also* up-rated and full-rated.

shaft extender A bronze fitting added to the shaft of a motor to lengthen the shaft to accommodate the design of the pump being used. The extender is held in place with three setscrews fastened tightly against the motor shaft.

shell *See* gunite.

shock treatment *See* superchlorination.

shotcrete A mixture of sand, concrete, and water that is delivered to the job site premixed to create a pool or spa shell (*see also* gunite). Shotcrete is applied through an air-powered nozzle.

shutoff head Zero gallons per minute. The point at which there is so much resistance to the free movement of water in a circulation system that the force of the pump cannot overcome the resistance and there is no movement of water.

sight glass A clear glass or plastic section of pipe that allows viewing of the water in the line. Used when backwashing filters to know when the discharge water is clean.

silica The type of sand used in filters and concrete, actually quartz. The size of grain is measured and numbered, as in #20 silica sand.

single-phase A single electrical circuit of alternating current.

single pole, single throw (SPST) switch The most basic on/off electrical switch, handling one line circuit (single pole) each time the switch is thrown. Single pole, double throw (SPDT) also operates one line circuit, but it alternately electrifies two different load circuits.

sink *See* heat riser.

skid pack A metal frame onto which is mounted the equipment needed to operate a portable spa, usually a pump and motor, filter, heater, blower, and control devices. Also called a spa pack.

skimmer A part of the circulation system that removes debris from the surface of the water by drawing surface water through it.

slide valve A guillotine-like plumbing device that restricts or shuts off the flow of water in a line. In essence, a gate valve that slides a disc up or down across the flow in the pipe. *See also* gate valve.

slip fitting A plumbing fitting that joins to a pipe without threads, but that slides into a prefitted space.

slurry A thin, watery mixture.

soda ash (Na_2CO_3) Sodium carbonate. A white powdery substance used to raise the pH of water.

sodium bicarbonate ($NaHCO_3$) *See* bicarbonate of soda.

sodium bisulfate ($NaHSO_4$) A chemical compound used to lower the pH and total alkalinity of water (dry acid).

sodium dichloro-s-triazinetrione ($C_3N_3O_3Cl_2Na$) Dichlor. A granular, stabilized form of chlorine sanitizer, generally about 60 percent available.

sodium hypochlorite (NaOCl) A liquid solution containing approximately 15 percent chlorine.

sodium thiosulfate A chemical used to neutralize chlorine in a test sample prior to testing for pH, without which a false reading might result.

soft water So-called soft water is very low in calcium and magnesium (less than 100 ppm) and is therefore considered aggressive or hungry and likely to dissolve those minerals when it comes in contact with plaster.

solar panel A metal, glass, or plastic enclosure, usually 4-by-8 feet by a few inches thick, through which water flows absorbing heat from the sun. The basic component of a solar heating system.

solder A soft metal wire (usually lead or alloys) used as a bonding agent when plumbing with copper pipe and fittings (an action called sweating).

spalling The breaking away of plaster in thin layers.

spiking In water chemistry, when readings of a parameter alternately register extremely high then extremely low.

spin filter A style of DE filter in which the grids rotate to facilitate washing the dirt and DE off the grid surfaces.

squared A number multiplied by itself. For example, $4 \times 4 = 4^2 = 16$.

square flange A casing style of certain motors, designed to adapt to certain pumps.

stabilizer Any compound that tends to increase water's resistance to chemical change. *See also* conditioner.

stack The vent pipe of a heater. A heater installed indoors requires such a stack, while one installed outdoors uses a flat-top vent and is called stackless.

stanchion The vertical support pipe of a railing.

standing pilot A heater ignition device in which the gas flame (to ignite the main burner) is always burning.

static head The amount of resistance (pressure) created from nonmoving restrictions in a water circulation system.

stator The stationary part in an electric motor that contains the wound wires (winding) that generates magnetism to drive the rotor.

strainer basket A plastic mesh container that strains debris from water flowing through it inside the strainer pot.

strainer pot The housing on the intake side of a pump that contains a strainer basket and serves as a water reservoir to assist in priming.

street fitting Any plumbing fitting that has one male end and one female end.

submersible pump A pump and motor that can be submerged to pump out or recirculate a body of water. Also called a sump pump.

superchlorinate Periodic application of extremely high levels of chlorine (in excess of 10 ppm) to completely oxidize any organic material in a body of water (including bacteria) and leave a substantial chlorine residual. Procedure performed to sanitize elements in water that might resist normal chlorination. Also called shocking or shock treatment.

surge chamber A tank designed to temporarily hold the water displaced by bathers in a pool, returning the water when the bathers exit the pool. Designed into commercial pools where there are heavy bather loads.

sweating The soldering together of copper pipe and fittings. Also refers to the formation of condensation on the exterior of pipes and equipment containing water, caused by temperature differences between the water in the pipe and the surrounding air.

tear down To disassemble a piece of equipment for service or repair. Specifically used in reference to filter cleaning (as opposed to backwashing or other temporary cleaning), a filter teardown means complete disassembly and cleaning. Also called break down.

Teflon tape A thin fabric provided on a roll used to coat threaded plumbing fittings to prevent leaks.

telepole A metal or fiberglass pole that extends to twice its original length, the two sections locking together. The telepole is used with most pool and spa cleaning tools.

T fitting A plumbing fitting shaped like the letter T that connects pipes from three different sources.

therm The unit of measurement you read on a gas bill; is 100,000 Btu/hour of heat.

thermal overload protector A temperature-sensitive switch on a motor that cuts the electric current in the motor when a preset temperature is exceeded.

thermistor The sensor that sends temperature information to the thermostat.

thermocouple *See* pilot generator. Also called a thermopile.

thermostat A part of the heater control circuit. An adjustable device that senses temperature and can be set to break the circuit when a certain temperature is reached. It then closes the circuit when the temperature falls below that level.

Thoroseal A brand name of waterproof concrete commonly used in pool and spa repairs.

three-port valve A plumbing fitting used to divert flow from one direction into two other directions.

time clock An electromechanical device that automatically turns an appliance on or off at preset intervals.

titration A chemical test method to determine the amount of a substance in a sample of water. A sample is colored, then drops of the titrant are added to the sample. When the sample turns clear or changes to another predicted color, the result of the test is determined by counting the number of drops of titrant that were required to create that change. For example, an acid demand test is performed with titration. The number of drops of titrant required in that test determines the amount of acid to add to the water.

torque Application of an amount of force against an object. Usually refers to the amount of force required to tighten a fastener to a specified requirement.

total alkalinity The measurement of all alkaline substances (carbonates, bicarbonates, and hydroxides) in a body of water.

total dissolved solids (TDS) The sum of all solid substances dissolved in a body of water, including minerals, chemicals, and organics.

total dynamic head (TDH) The total amount of resistance (backpressure) created by restriction to free flow of water in a circulation system. TDH is composed of vacuum (restrictions to suction) and pressure (restrictions to returning water to the pool). TDH is expressed in feet.

transformer A device that changes electric current from one voltage to another. Commonly used in electronic ignition heaters where the supply of 110 or 220 volts is converted (transformed) to 24 volts for use in the heater control circuit.

trichloro-s-triazenetrione ($C_3N_3O_3Cl_3$) Trichlor. A dry form of stabilized chlorine, produced in granular or tablet form at around 90 percent available chlorine.

tripper The small metal clamp that fits in the clock face of a time clock to activate the clock (on/off) at preset times. Also called a dog.

trisodium phosphate A detergent used to clean filter grids and cartridges, it breaks down oils that acid washing alone cannot. Commercially sold as TSP.

turbidity Cloudiness.

turnover rate The amount of time required for a circulation system to filter 100 percent of the water in a particular body of water.

UL Underwriters Laboratory. The agency that sets standards for and certifies the safety of electrical and mechanical devices.

underdrain The assembly in the bottom of a sand filter that connects laterals and plumbing for water filtration.

union A plumbing fitting connecting two pipes by means of threaded male and female counterparts on the end of each pipe.

uniseal A type of motor case designed to fit a specific style of pump.

up-rated The full-rated motor operates to its listed capacity. The up-rated motor has a similar horsepower rating but will function to even higher standards if called upon. *See also* service factor.

vacuum A device used to clean the underwater surfaces of a pool or spa by creating suction in a hose line.

valve A device in plumbing that controls the flow of water.

vanishing edge pool A pool or water feature with one or more sides that are below the overall surface water level. Water spills over the vanishing edge into a catch basin below for recirculation to the pool. Also known as negative edge or rim flow design.

Visoflame tube A trade name of the Teledyne Laars heater pilot lighting system.

volt The basic unit of electric current measurement expressing the potential or pressure of the current. Volts = watts ÷ amps.

volute The volute is the housing that surrounds the impeller and diffuser in a pump, channeling the water to a discharge pipe.

water column pressure The unit of measurement used in evaluating gas pressure with a manometer.

water feature A fall, rockscape, or other decorative design using moving water.

water hammer The phenomenon of variations in water pressure in a closed pipe that results from too rapid acceleration or deceleration of flow. Momentary pressures in excess of normal static pressure producing a hammering sound.

watt The way appliances are measured for power consumption. One watt is equal to the volume of one amp delivered at the pressure of one volt. Watts = amps × volts.

weir The barrier in a skimmer over which water flows. A floating weir raises and lowers its level to match the water level in a pool or spa. Another type is shaped like a barrel and floats up and down inside the skimmer basket.

winterizing The process of preparing a pool or spa to prevent damage from freezing temperatures and other harsh weather conditions.

REFERENCE SOURCES AND WEBSITES

Virtually every equipment and supply manufacturer produces valuable literature with repair and maintenance information, technical guidelines, and exploded diagrams of their products. Collect literature from websites, at trade shows, from your wholesale supply house, and anywhere else you can.

Many manufacturers conduct factory training seminars on the repair and maintenance of their products. Trade periodicals publish a list of these educational opportunities. Take advantage of them. Doug Steimele of California Pools and Spas says

> *The service technician should feel free to call a reputable pool builder in the area for advice. Most builders would rather dispense some free advice than have uninformed technicians giving pool owners inaccurate information. Ask, ask, ask!*

Excellent technical articles, tips, and news regarding new products can be found in the following periodicals:

Aqua (ask for the excellent annual "Buyers' Guide" when you subscribe)
aquamagazine.com

Service Industry News
P.O. Box 5829
San Clemente, CA 92674
949-366-9981

Pool and Spa News (ask for the excellent annual "Directory Issue" when you subscribe)
poolspanews.com

The following publications were used for reference in preparing this book:

Water Analysis Handbook. Loveland, Colo.: Hach Company, 1992.

Manas, Vincent T. *National Plumbing Code Handbook.* New York: McGraw-Hill, 1957.

Uniform Swimming Pool, Spa and Hot Tub Code. International Association of Plumbing and Mechanical Officials, 1988.

Taylor, Charlie. *Everything You Always Wanted to Know about Pool Care but Didn't Know Where to Ask.* Chino, Calif.: Service Industry Publications, 1989.

United States Navy. *Basic Electricity.* Washington, D.C.: U.S. Government Printing Office, 1969.

Wood, Robert W. *Home Electrical Wiring Made Easy.* 2d ed. Blue Ridge Summit, Pa.: TAB Books, 1993.

Websites

Almost every manufacturer has a website these days, many with excellent how-to tips and diagrams. I have relied on many websites in preparing this book and here are a few of my favorites:

aqua-flo.com	User-friendly pool and spa equipment specs, photos, pdf format downloads
aquamagazine.com	Good links to industry resources
baquacil.com	Dozens of products for pool chemistry; good FAQs with details on biguanicides
brett-aqualine.com	Great downloadable pdf files for spa troubleshooting, wiring
delairgroup.com	Great above-ground pool photos and "buyers guide"
eren.doe.gov/rspec	Very creative, useful site about saving water and energy in pools and spas
fafco.com	Step-by-step layout of solar pool heating, lots of tech data
gpspool.com/dealers	Good list of manufacturers

heatsiphon.com	Pool heating analyzer, lots of tech info and explanations of heat pumps
hurlcon.com.au	Australian pool/spa equipment manufacturer of excellent products and tech info
jandy.com	Good selection of pool and spa equipment, especially Teledyne Laars heaters; good tech info and training seminars
kd.com	Great "university" of water maintenance topics; good description of portable pools, including assembly
magnetek.com	Great technical library once you get through to the consumer pool and spa section; cool "Energy Savings Predictor" software
muskin.com	Portable pools; great movie intro and user-friendly site
polarispoolsystems.com	Good site for automatic pool cleaners, with FAST diagrams and photo downloads
pondlady.com	Great site on pond installation, care, critters. Very easy to navigate, with practical photos
poolspa.com	Limited links to other manufacturers
poolspanews.com	Great searchable site with lots of tech help and links
raypak.com	Easy to download spec sheets on pool and spa heaters; training seminars
ricorock.com	Good photos of artificial rockscapes
rock-n-water.com	Great photos of artificial rockscapes; very cool site with training seminars
solar-tec.com	Good site for intro to solar; pricing, all user-friendly

spatop.com	Everything you could ever need in spa covers and clever accessories
splashsuperpools.com	Cool site; good photos of cool above-ground pools
starite.com	Good site with easy to get owner's manuals and great chart that tells you how long each will take to download at your modem speed
swimmingpools.com	Lots of good pool "how-to's" including vinyl liner installation

The *Pool and Spa News* annual "Directory Issue" includes a complete list of pool-related websites.

Professional Associations

There are many great professional resources to find support, information, and training or to learn more about industry standards. Their names describe what they do and many of them contributed to this book.

Associated Swimming Pool Industries of Florida
3701 N. Country Club Drive, #2107
Aventura, FL 33180-1721
305-937-0960

California Spa and Pool Industry Education Council
980 Ninth Street
Sacramento, CA 95814
916-447-4113

The Chlorine Institute, Inc.
1300 Wilson Blvd
Arlington, VA 22209
703-741-5760
chlorineinstitute.org

Independent Pool and Spa Service Association
17715 Chatsworth Street, #203
Granada Hills, CA 91344
818-360-9505
ipssa.com

Master Pools Guild
9601 Gayton Road, #207
Richmond, VA 23233-4963
800-392-3044
masterpoolsguild.com

National Plasterers Council
2811 Tamiami Trail, Suite P
Port Charlotte, FL 33952
941-766-0634
npconline.org

National Recreation and Park Association Aquatic Section
650 W. Higgins Road
Hoffman Estates, IL 60195
847-843-7529
aquaticsnrpa.org

National Spa and Pool Institute
2111 Eisenhower Avenue
Alexandria, VA 22314
703-838-0083
nspi.org

National Spa and Pool Institute of Canada
7370 Bramalea Road, #4
Mississauga, ON L5S 1N6, Canada
905-676-1591
nspi.ca

National Swimming Pool Foundation
10803 Gulfdale, #300
San Antonio, TX 78216
210-525-1227
NSPF.com

Professional Pool Operators of America
P.O. Box 164
Newcastle, CA 95658
916-663-1265
ppoa.org

Swimming Pool Trades and Contractors Association
1804 W. Burbank Boulevard
Burbank, CA 91506-1315
818-845-7565

United Pool and Spa Association
P.O. Box 13223
Jacksonville, FL 32206
904-353-4403

World Waterpark Association
P.O. Box 14826
Lenexa, KS 66285-4826
913-599-0300
waterparks.com

INDEX

48-frame type, 106, 522
90-degree check valve, 75

above-ground pools, 12–19, 326
ABS (acrylonitrile butadiene styrene), 511
ac (alternating current), 152, 379, 511
ac circuits, testing, 162–163
acetone, 511
acid
  brown-red water from, 361
  cyanuric, 330, 336, 337, 351, 370, 380
  demand/balance and, 330, 426, 511
  described, 511
  etching with, 443, 520
  ethylenediamine tetra-acetic acid, 369, 520
  hydrochloric, 524
  hypochlorous, 332, 525
  muriatic, 336, 348, 373, 380, 443, 527
  pH balance and, 426
  relative acidity, 347, 349, 528
  safety guidelines for, 348, 452, 455
  spilling of, 381, 387
  testing for, 362, 367, 368, 370
  tips for using, 348
  trichlor cyanurate and, 338
acid spotter, 511
acid washing, 448, 450–456, 511
acidity, 330, 347, 349, 511, 528
acrylic tubs/spas, 15–16, 429, 432, 458
acrylonitrile butadiene styrene (ABS), 511
adapter bracket, pumps, 100–101, 511
admixtures (admix), 435
aeration, 469, 474
aerator, 512
AGA (American Gas Association), 512
aggregates of concrete/plaster, 435, 441, 512
agitator, 481
air, in filters, 426
air-activated (pneumatic) controls, 278, 512
air blowers, 72, 501, 502, 514
air leaks, 145, 280–281. *See also* leak detection/repair
air pollution, 499
air relief valves, 184–185, 203, 512
air switches, 278–281, 512
alarms, pool, 299
algae, 353–361
  algicides for, 358–360, 512
  black algae (cyanophyta), 353, 354–355, 518
  chlorine booster treatment, 360
  colloidal (suspended) silver algicide, 359, 517
  copper sulfate (liquid copper) algicide, 358–359, 518
  described, 353, 512
  effect on water chemistry, 353

algae *(continued)*
 elimination techniques for, 355–360
 favorable conditions for, 353
 fiberglass coatings and, 444
 filters and, 355, 356, 359–360
 forms of, 354–355
 green algae (chlorophyta), 354
 koi ponds and, 463
 new plaster and, 428
 pH and, 353, 360, 361
 pink/red algae, 355
 polymer algicides, 359
 quaternary ammonium compounds (quats), 359–360
 recurring, 381
 removing with brush, 353–357
 sanitizers and, 353, 355–356
 in spas/hot tubs, 360
 spot removal of, 357
 sunlight and, 353
 superchlorination (shock treatment) for, 340
 trichlor treatment, 337, 356, 357
 ultraviolet (UV) light and, 342–343
 water circulation and, 353, 356
 winterized pools and, 409–410, 411
 yellow algae (phaeophyta), 354, 528
algicides, 358–360, 512
algistat, 512
alkalinity
 demand/balance and, 330
 described, 512
 effect of algae on, 353
 relative alkalinity, 347, 349, 528
 soda ash for, 348, 454, 455, 532
 spiking and, 349
 testing for, 349–350, 368
 tips for adding, 348, 349
 total alkalinity, 349–350 , 368, 426–427, 535
 water hardness and, 350
alternating current (ac), 152, 379, 511
alum (aluminum sulfate), 177, 198, 512
aluminum sulfate (alum), 177, 198, 512
aluminum wall panels, 15
ambient temperature, 512
American Gas Association (AGA), 512
American National Standards Institute (ANSI), 512
American Public Health Association (APHA), 512
American Wire Gauge (AWG), 156, 512
ammonia
 anhydrous, 340, 360
 breakpoint chlorination, 339, 340, 514–515
 buildup of, 469
 chloramines, 338–340, 361, 365–366, 516
 described, 512
 dirt/debris and, 469
 eliminating from pool, 339–340
 odor, 343
 quaternary ammonium compounds (quats), 359–360
 vs. chlorine, 339, 340, 343
 amperage (amps), 104, 106, 149, 512
amps (amperage), 104, 106, 149, 512
anhydrous, 513
anhydrous ammonia, 340, 360
animals
 chewing air hoses, 281
 falling in pool, 417
 gnawing on wires, 260
 in motor housing, 141
 nesting in equipment, 213, 214, 260, 267, 400

anode, 345–346, 485, 513
ANSI (American National Standards Institute), 512
antifreeze, 413–416
antisipihon valves, 85–86, 513
antisurge valves, 513
antivortex drain covers, 43, 513
antivortex effect, 513
APHA (American Public Health Association), 512
arc (arcing), 149, 513
arcing (arc), 149, 513
artificial respiration signage, 488–490
ash, 404
automated controls, 283–289, 290
automatic bypass valve, 77, 78
automatic combination gas valves, 256–257
automatic gas valve, 217–219, 513
automatic pool cleaners, 299–309, 327, 501, 503. *See also* pool cleaners
automatic valve systems, 66–68
AWG (American Wire Gauge), 156, 512
awnings, 503

backfill, 513
backwash hoses, 182–183
backwash process, 513
backwash valves, 167, 179–182, 199–203
backwash water, 502
backwashing
  DE filters, 179, 188–190
  sand filters, 196
  schedule for, 205
  uses for, 167, 179, 399, 513
  water conservation and, 502
bacteria
  beneficial (in ponds), 463–469
  biological filters and, 514
  described, 513
  killing/oxidizing, 331–347, 361, 531, 534
  Legionnaire's disease, 497
  odors and, 361
  sanitizers for. *See* sanitizers
  ultraviolet (UV) light and, 342
baking soda, 380, 426, 441, 514
balanced water makeup, 330, 331, 372, 513
ball valves, 71–72, 344, 485
balloon fittings, 57, 513
bank, 513
barb fittings, 513
barriers, 298–299, 326, 351
base, 513
bather loads, 353, 479, 503, 514
bather occupancy, 514
bathers
  body oil from, 190, 361, 392, 398, 418, 528
  defined, 513
  fecal accidents, 497–498
  harmful chemicals and, 483, 495
  lotion/soap from, 190, 350, 361, 392, 418
  precautions for children, 164, 298–299, 326, 531
  precautions when servicing pool, 495
  pump/skimmer caution, 164
  signage for, 419–420, 487–491
  splashing and, 502, 503
  surge chamber and, 480–481, 534
  vacuum hose caution, 495
  water displacement, 479–480
bathtubs, jetted, 97, 149, 525. *See also* spas
batteries, 151–152, 290
bayonet, 514
bib (hose bib), 524
bicarbonate of soda, 380, 426, 441, 514

biguanicides (PHMB), 346–347, 380, 381, 540
biological filters, 173, 460, 462–464, 514
black algae (cyanophyta), 353, 354–355, 518
bleach, 514
bleed, 514
blister, 514
blow bag, 143–145, 514
blow bag pump priming, 143–144
blowers, 72, 501, 502, 514
blue-green water, 360
blue vitriol. *See* copper sulfate
bluestone. *See* copper sulfate
boiler. *See* heaters
bond beam, 13, 27–31, 514
bonding system, 514
bonding wires, 135, 151, 160
booster pump clock, 272
booster pump systems, 300–309
booster pumps/motors for spas, 148–149, 514
break down. *See* tear down
breaker box, 135
breaker switch, 135
breakers. *See* circuit breakers
breakpoint chlorination, 339, 340, 514–515
brick, 30–31
bridging, in DE filters, 192, 515
British thermal unit (Btu), 515
broadcast, 515
bromine, 342, 346, 366, 380, 381, 515
brown coat, plaster, 443, 444, 515
brown-red water, 361
brushing. *See also* cleaning/servicing guidelines
  chemical distribution and, 408, 428
  importance of, 408
  plaster dust, 427, 428
  procedures for, 408–409
  removing algae, 353–357
  tile cleaning, 398–399
  wall brushes, 387, 388
Btu (British thermal unit), 515
Btu/kilowatt ratings, 228
bubble solar covers, 320–323
buffer, 515
building permits, 21
bulbs, for pool lighting, 311, 314–319, 343, 501
bumping, 167–168
burner tray, 208, 209–211, 252, 515
bushing, 408–409, 515

C frame motors/pumps, 95, 105–106, 516
calcium, 515
calcium bleed, 432, 515
calcium carbonate, 423, 427, 515
calcium chloride, 426, 427, 434, 441
calcium deposits. *See* scale deposits
calcium hardness, 350, 426, 523
calcium hypochlorite, 337, 341, 515
cam lock, 515
cap, 515
capacitor, 103–105, 140, 515
capacitor start, capacitor run (CSR) pump motors, 104–105, 515
capacitor start, induction run (CSI) pump motors, 104, 516, 518
carbon buildup, 257–258
carbon monoxide hazards from heaters, 235, 236
cartridge, 516
cartridge filters. *See also* filters
  cleaning, 198–199, 200
  commercial pools, 481–483
  microns, 177

overview, 173–174
size of, 173, 174–179
cathode, 345–346, 485, 516
caulking, 516
caustic, 516
cavitation, 112, 141, 516
CDC (Centers for Disease Control), 494–497
Centers for Disease Control (CDC), 494–497
centrifugal force, 93–94, 516
centrifugal-type pumps, 93–94
chalking, 450
channel lock pliers, 45–46
channeling, 197, 516
check valves, 42, 72–75, 186, 243, 516
chelating agent, 360, 425, 516
chemical listing/application, 408
chemical safety
acid, 348, 452, 455
chemical storage, 491
chlorine, 336, 348, 381
chlorine gas, 334–335, 341, 484–485, 491
danger of mixing chlorine with muriatic acid, 336, 348, 381
DPD reagents, 366–367
guidelines, 336
muriatic acid, 336, 348, 381
pool chemical mixing guidelines, 380
propane gas, 223
tips for, 380
water testing and, 366–367, 370–371
chemistry of water, 329–381. *See also* treatment delivery systems
acid in. *See* acid
algae/algeacides and. *See* algae
alkalinity in, 349–350, 426–427, 512, 535
alternatives to chlorine, 342–347
ammonia, 338–340, 343, 514–515
bacteria in. *See* bacteria
balanced water makeup, 330, 331, 372, 513
bather load and, 353
biguanicides (PHMB) as sanitizers in, 346–347, 380, 381, 540
blue-green water, 360
breakpoint chlorination, 339, 340, 514–515
bromine, 342, 346, 366, 380, 381, 515
brown-red water, 361
calcium hardness, 350, 426, 523
calculating water volume, 3–8
chloramines, 338–340, 361, 365–366, 516
chlorine in. *See* chlorine
chlorine lock, 338, 339
chlorine residual in, 330, 333, 340, 353, 517
cloudy water, 360, 364–365
components of, 331–353
conditioner, 407, 491, 518
corrosion of metal parts and, 361, 379
cyanurates, 336, 337–338, 350, 518
cyanuric acid, 330, 336, 337, 351, 370, 380
dangers of. *See* chemical safety
delivery systems for. *See* treatment delivery systems
demand/balance in, 330, 426, 511
dichlor (sodium dichloro-s-triazinetrione), 337, 338, 341, 533
dirt/debris and, 352
discoloration of hair, skin, nails, 361
equilibrium in, 372
erosion systems, 376

chemistry of water *(continued)*
evaporation and, 352
eye irritation, 361
FAQs, 381
floaters for, 337–338, 346, 375, 521
hardness in, 330, 350, 361, 369, 523
heater problems and, 251
heavy metals, 369, 424
how often to add chemicals, 381
ionizers as sanitizers in, 345–346
Langlier index, 371–373
metal contaminants, 369, 424
odors, 361
ozone/ozonators as sanitizers in, 343–345, 381
parameters of, 330–331
pH in. *See* pH levels
plaster, breaking in, 426–428
pool chemical mixing guidelines, 380
potassium monopersulfate sanitizers in, 346, 380, 529
precipitate in, 330, 529
rain and, 352–353
safety guidelines. *See* chemical safety
sanitizers in. *See* sanitizers
saturation, 330, 371–372, 531
saturation point, 330, 371–372
scale and, 361
skin irritation, 361
stabilizer, 351, 352, 427, 428, 533
superchlorination, 337–340, 346, 354, 374, 411, 534
temperature effect on, 331, 352
testing. *See* testing water
total alkalinity, 349–350, 368, 426–427, 535
total dissolved solids (TDS), 350–351, 369, 372, 535
trichlor cyanurate, 337–338, 341, 356, 357, 536
turbidity, 364–365
ultraviolet (UV) light and, 342–343
weather effects on, 352–353, 410–411
wind effect on, 352
children, precautions for, 164, 298–299, 326, 531
chine, 516
chine joist, 516
chloramines, 338–340, 361, 365–366, 516
chlorinated polyvinyl chloride (CPVC), 518
chlorinators, 376–378, 516
chlorine, 332–342
alternatives to, 342–347, 381
availability of, 333
breakpoint chlorination, 339, 340, 514–515
by-products of, 333
calcium hypochlorite, 337, 341, 515
calculating parts per million, 8–9
chemical processes of, 332
chloramines and, 338–340, 361, 365–366, 516
chlorine lock and, 338, 339
convenience of, 333
cost of, 333, 341
cyanurates, 336, 337–338, 350, 518
cyanuric acid, 330, 336, 337, 351, 370, 380
danger of mixing with muriatic acid, 336, 348, 381
demand for, 332–333, 517
described, 517
dry form of, 332, 336–338
environmental concerns, 499–500

evaluation criteria for, 333–334
free available, 517
gas form of, 334–335, 341, 483–487, 491
hypochlorous acid (HOCl), 332, 525
liquid form of, 331, 335–336, 373–374
lithium hypochlorite (LiOCl), 338, 526
mechanical delivery devices, 375–377
odors, 361
OTO chlorine-level test kit, 365–366
pH levels and, 332–334
PHMB sanitizers and, 346–347
pool covers and, 320
product comparison, 340–341
product overview, 340–341
pump delivery systems for, 377, 378
residual levels of, 330, 333, 340, 353, 517
salt chlorine generators, 377–379
shock treatment, 339–340, 346, 354, 374, 411, 534
sodium hypochlorite, 335, 341, 533
spilling of, 381
stability of, 333
superchlorination, 337–340, 346, 354, 374, 411, 534
tabs/tablets, 337–338, 375
testing for, 365–367
trichlor cyanurate, 337–338, 341, 356, 357, 536
ultraviolet light and, 333
vs. ammonia, 339, 340, 343
vs. ultraviolet light, 342–343
weather effect on, 352–353
chlorine boosters, 360
chlorine feeders, 485–487
chlorine gas, 334–335, 341, 483–487, 491
chlorine generators, 484–485
chlorine-level testing, 365–367
chlorine lock, 338, 339, 517
chlorine residual, 330, 333, 340, 353, 517
chlorophyta (green algae), 354, 517
circuit breakers
connections, 152–153
fuses, 155
ground fault interrupter (GFI), 157–158, 422, 523
illustrated, 153
overview, 153–155
pool draining and, 419
replacing, 155
resetting, 140, 153, 154–155, 158
safety guidelines, 135, 140, 154, 155, 157–158
tripping of, 140–141, 153–154
troubleshooting, 154–155
circuit continuity, 318
circuits
ac, 162–163
closed, 150
control, 212–221, 258–266, 518
dc, 162–163
dc battery, 151–152
dedicated, 518
described, 149, 517
open, 150
overloaded, 154–155
relay, 159–160, 283–288
short, 149, 151
circular pools, 5–7
circulation equipment, 425
circulation system, 517
clamping ring, 517
cleaners, pool. *See* pool cleaners
cleaning filters
cartridge filters, 198–199, 200

cleaning filters *(continued)*
cleaning/teardown schedule, 205
DE filters, 188–196
general guidelines for, 176, 205, 427
sand filters, 189, 196–198
time between cleanings, 176
cleaning/servicing guidelines, 383–418. *See also* pool cleaners; repairs/maintenance
"bathtub ring," 350, 361, 399, 418
bicarbonate of soda for cleaning, 380, 426, 441, 514
brushing. *See* brushing
deck/cover cleaning, 394–395, 503
equipment area, 400
equipment check for, 399–400
FAQs, 418
filters. *See* cleaning filters
heavy-duty cleaning, 404
leaf rake, 387, 397, 526
metal fixtures, 456
plastic fixtures, 456
pool chemistry servicing. *See* water chemistry
pool winterizing, 409–418
pumice stones for, 394, 398, 455, 530
quick start guide, 395
service logs, 495
service/maintenance schedule, 407
spas, 388, 393, 403, 409
surface skimming, 396–397
telepoles for, 383–387
test kits, 363, 365–368, 371, 393
testing water chemistry. *See* testing water
thermometers, 393
tile cleaning, 392–393, 398–399, 455
tools for, 383–394
with TSP, 441, 448
vacuuming. *See* vacuuming
water features, 409
water level checks, 396
waterline area, 399, 407, 418, 503
cleanup, after pool construction, 32
clogged pumps/filters, 99, 119, 120
close nipple. *See* nipple
closed circuits, 150
closed-face impeller, 98–100, 133
closed loop systems, 78–79, 517
cloudy water, 360, 364–365
colloidal silver, 517
colloidal (suspended) silver algicide, 359, 517
colorimetric testing, 362, 370, 517
commercial equipment, 480–498. *See also* equipment
chlorine feeder, 485–487
chlorine gas delivery systems, 483–484
chlorine generators, 484–485
cutting power to, 490
filters, 481–483
safety equipment, 487–492
slurry feeders, 481, 482
surge chamber, 480–481, 534
commercial equipment rooms, 487
commercial pools/spas, 475–495
bather displacement, 479–480
bather loads, 479
chemical storage, 491
considerations, 475–476
dehumidification of, 492–494
depth of, 476
fecal accidents, 497–498
filters, 481–483
health issues, 494–498
in-water thermometer, 492
Legionnaires' disease, 494–497

operational requirements, 495
regulations, 475–476, 495
safety equipment, 487–492
signage, 487–491
size, 476
slope calculations, 477–478
types of, 475–476
violations, 475
volume calculations, 477–480
water displaced by bathers, 479–480

comparator, 517
competitive swimming pools, 476
compression fittings, 55–57, 241–242, 518
compression nut, 55, 57
compression ring, 55–57
concrete dust, 441
concrete pools, 11–12
concrete sealer, 30
condensation, 518
conditioners, 407, 491, 518. *See also* cyanuric acid
conductors, 150, 151, 518
connectors
crimp, 157
female, 520
flex, 521
male, 526
no-hub, 57, 527
straight, 113

construction, pool, 20–35
containers, 1–3
continuity, 318
"Continuous Duty" rating, 107
control circuit, 212–221, 258–266, 518
control circuit switches, 258–266
coping, 31–32, 518
coping stones, 30, 440
copper, 369
copper plumbing, 53–57
copper soldering (sweating), 53, 534
copper sulfate (liquid copper) algicide, 358–359, 518
corona discharge, 518
corrosion, 361, 379
cost considerations, 10–11
couplings, 48, 518
cove, 518
covers
antivortex, 513
cleaning, 394–395
drain, 43, 513
electric, 324–326
evaporation and, 268, 501, 502
foam, 323
heaters and, 227, 268
motor, 146
pool/spa, 268, 320–326, 417, 501, 502
sheet vinyl, 323–324
solar covers, bubble type, 320–323

CPVC (chlorinated polyvinyl chloride), 518
crimp connector, 157
croze, 518
CSI (capacitor start, induction run) pump motors, 104, 516, 518
CSR (capacitor start, capacitor run) pump motors, 104–105, 515
CT value, 497, 518
current, 150, 518
cyanophyta (black algae), 353, 354–355, 518
cyanurates, 336, 337–338, 350, 518
cyanuric acid, 330, 336, 337, 351, 370, 380. *See also* conditioners; stabilizers
cycle, 150, 518

dc battery circuits, 151–152
dc circuits, testing, 162–163
dc (direct current), 151–152, 379

DE (diatomaceous earth), 165–167
  clogged impellers and, 99
  described, 165, 519
  filter cleaning and, 194–196
DE filters, 165–169. *See also* filters
  backwashing, 179, 188–190
  bridging in, 192, 515
  cleaning, 188–196
  microns, 177
  size of, 169, 175–178
  spin, 169
  vertical grid, 167–169
debris. *See* dirt/debris
decks
  building/remodeling, 456–457
  cleaning, 394–395, 503
  construction of, 31–32
  drains for, 422
decorative ponds. *See* ponds
dedicated circuit, 518
dehumidifiers, 492–494
delamination, 432, 519
demand/balance, water chemistry, 330
design, pool, 20–22
diameter, 519
diatomaceous earth. *See* DE (diatomaceous earth)
dichlor (sodium dichloro-s-triazinetrione), 337, 338, 341, 533
dielectric, 519
diethyl phenylene diamene (DPD), 366–367, 519
diffuser, 519
dioxin, 500
direct current (dc), 151–152, 379
dirt/debris
  ammonia buildup and, 469
  backwashing and, 190
  filters and, 176, 178, 190, 204–205
  in heaters, 400
  pool cleaners for removing. *See* pool cleaners
  skimmers for removal of, 37–41
  solar panels, 85
  strainer pots for, 119–122, 403, 534
  water chemistry and, 352
  winterized pools and, 409–410, 411
discharge, 519
discharge head, 112
discoloration of hair, skin, nails, 361
diverter, 63–65, 519
diverter units, 40, 41
diving, signage for, 490
diving boards, 291–295, 326
dogs. *See* trippers
DPD (diethyl phenylene diamene), 366–367, 519
drain-down test for leaks, 430–431
drain flush. *See* blow bag
drain grate, 43
drain manifold, 171
drain pipe, 171
draining pools/spas
  high-volume pump-out units, 146–147
  hydrostatic pressure and, 42
  low-volume pumps/motors, 147–148
  procedure for, 419–422
drains
  covers for, 43, 114, 513
  deck, 422
  gravity, 421
  main, 27, 41–43, 526
  spa, 43
  storm, 504
  underdrain, 536
dry fit, 519
duty rating, pump motors, 107–108, 519

dye testing for leaks, 429
dynamic head, 112. *See also* head
Dynell, 519

EDTA (ethylenediamine tetra-acetic acid), 369, 520
effluent, 519
elbow, 519
electric cover systems, 324–326
electric heaters, 222–223, 224
electric robot pool cleaners, 299–300, 530
electrical equipment, 326
electrical panel, 152–155
electrical specifications, 106–107
electrical switch, 490
electrical theory, 150–152
electrical volume, 153
electrical wiring/connections
  ac (alternating current), 152, 379, 511
  amperage (amps), 104, 106, 149, 512
  arc/arcing in, 149
  circuit breakers in. *See* circuit breakers
  circuits in. *See* circuits
  conductors in, 150, 151
  crimp connector, 157
  current in, 150
  cycles in, 150
  dc (direct current), 151–152, 379
  electrical panel in, 152–155
  gauge of wire in, 150, 156–157
  ground fault interrupters (GFIs), 157–158, 422, 523
  grounding of appliances, 151
  heaters, 245–246
  hot line, 150
  insulators, 151
  lines in, 150
  loads in, 150, 153, 526
  multimeter, 161–163
  neutral line, 150
  open circuits, 150
  overloaded circuits, 154–155
  pool/spa construction, 28
  pump motor specifications, 106–107
  relay circuit, 159–160, 283–288
  safety guidelines for, 135, 140, 155–161, 315, 380, 422
  shock hazards, 159–161, 315, 380, 422
  short circuits, 151
  switches. *See* switches
  testing techniques for, 161–163
  volts in, 150
  watts in, 150
  wiring in, 156–157
electricity
  basics, 149–152
  closed circuits, 150
  cost of operation (pumps/motors), 148
  energy conservation, 105, 500–502
  open circuits, 150
  safety guidelines, 135, 140, 155–161, 315, 380, 422
  shock hazards, 159–161, 315, 380, 422
  terminology, 149–150
  testing techniques, 161–163
  water and, 161
electrode, 519
electrolysis, 69, 519
electromechanical timers, 269–273, 275–277
electrometric testing, 364, 520
electronic control circuit. *See* control circuit
electronic ignition heaters, 219–221, 222, 245–246, 256
electronic thermostats, 263–265
electronic timers, 274–275
element, 520

emergency phone numbers, 490
emergency shutoff signage, 490
end bell type motors, 103, 104, 106, 137
end bells, 520
energy conservation, 105, 500–502
environmental checklist, 504–505
environmental issues, 499–505
  chemicals, 499–500
  energy conservation, 105, 500–502
  energy efficient motors, 104, 105, 520
  energy efficient pumps, 163
  general tips, 504–505
  recycled materials, 500, 503–504
  solar heating systems, 76–85, 91, 223–225, 501
  water conservation, 421–422, 502–503
equalizer line, 41, 520
equilibrium, 372
equipment, 269–327. *See also* commercial equipment
  covers, 320–326
  cutting power to, 490
  dehumidifiers, 492–494
  diving boards, 291–295
  electrical, 326
  FAQs, 326
  flow meters, 289–291
  ladders, 297–298
  lighting, 309–319
  maintaining. *See* repairs/maintenance
  pool cleaners. *See* pool cleaners
  rails, 297–298
  recycling, 504
  remote controls, 277–289
  repairing. *See* repairs/maintenance
  replacement components, 504
  safety, 487–492
  safety barriers, 298–299, 326, 531
  slides, 295–297
  time clocks, 269–277
  winterizing, 411–412, 415–416
equipment area, 400
equipment checks, 399–400
equipment rooms, 487
equipment set, 32
erosion feeders, 485–487
erosion systems, 376, 520
escutcheon plate, 520
etching, 520
ethylenediamine tetra-acetic acid (EDTA), 369, 520
evaporation
  fountains/waterfalls, 460, 474
  pool/spa covers and, 268, 501, 502
  splashing and, 502
  testing for leaks, 429–430
  total dissolved solids (TDS) and, 350–351
  water chemistry and, 352
  wind and, 352, 501
excavation, 21–26
expansion joints, 31–32, 520
eye/skin irritation, 361

fan/heat control, motors, 103
FAQs
  advanced plumbing systems, 91
  basic pool plumbing, 59
  cleaning/servicing, 418
  filters, 205
  general (pools/spas), 36
  heaters, 268
  motors, 164
  pool equipment, 326
  pumps, 164
  repairs, 458
  special procedures, 458
  water chemistry, 381
  water features, 474

feathering, 520
fecal accidents, 497–498
female fittings, 44, 520
female openings, 47
fences. *See* barriers
fiberglass, 409
fiberglass coatings, 444–445
fiberglass pools/spas, 15–17
fiberoptics, 318–319, 520
fill water, 520
filter cycle, 521
filter filling, 521
filter filling pump priming, 144
filter grid manifold, 168
filter media
  DE, 165–169
  sand, 169–173
filter run, 176, 521
filter size
  cartridge filters, 173, 174–179
  DE filters, 169, 175–178
  determining, 174–179, 205
  sand filters, 172, 175–178
filters, 165–205
  air in, 426
  algae and, 355, 356, 359–360
  biological, 173, 460, 462–464, 514
  calculations for, 113–114
  cartridge. *See* cartridge filters
  choosing, 174–185, 205
  cleaning. *See* cleaning filters
  clogged, 99, 119, 120
  commercial pools, 481–483
  cracks in, 203
  DE. *See* DE filters
  described, 165, 520
  dirt/debris and, 176, 178, 190, 204–205
  FAQs, 205
  head in, 113–114
  installing, 186–188
  leaks, 199–205
  lids on, 203
  loose grids, 141
  makes/models, 174–185
  normal operating pressure, 183
  price considerations, 176
  pumps and, 114
  purging air from, 426
  replacing, 186–188
  sand. *See* sand filters
  size of. *See* filter size
  types of, 165–174
  vacuuming to, 401–404
  winterizing, 416
FIP fitting, 521
fireman's switch, 245–246, 265, 521
fish, 106, 461, 463, 464, 474
fish ponds, 173, 177, 460–468, 469, 474
fittings
  balloon, 57, 513
  barb, 513
  compression, 55–57, 241–242, 518
  female, 44, 520
  FIP, 521
  flare, 521
  male, 44, 47, 526
  MIP, 527
  pipe, 44, 48–49
  PVC electrical, 281
  slip, 48, 532
  street, 534
  T, 113, 534
  threaded, 44, 48, 243
flange gasket, 521
flapper gate, 521
flapper gate valve, 74–75, 521
flare fitting, 521
flash test, 521
flex connector, 521
float, 89
float valve, 521
floaters, 337–338, 346, 375, 521
flocculate, 521

flooded motors, 135, 146
flow control valve, heater, 208–209
flow loss, 109
flow meters, 289–291, 483, 522
flow rate
  described, 109, 522
  heaters, 114
  overview, 109–111
  pumps/motors and, 105, 109–111, 114–118
fluorine, 522
fluorocarbons, 226
flux, 54, 55, 522
foam covers, 323
foamy water, 361
fountains, 459–460, 462, 474, 503
four-pass unit, 522
free chlorine, 522
free-flow sand/gravel filters, 173, 522
free joint, 522
freeboard, 184, 522
Freon, 226, 522, 528
frozen water, 409, 410, 412–417
full-rated, 522
fungus (pink/red), 355
fuse links (fusible links), 213–214, 263, 522
fuses, 155, 263
fusible links, 213–214, 263, 522

gallons per minute (gpm), 523
galvanized, 522
garden hoses, 388–392
gas
  chlorine, 334–335, 341, 483–487, 491
  hydrogen, 485
  natural, 219, 221–222, 256, 268
  propane, 221–222, 223, 268
  safety guidelines, 223, 246, 334–335, 341, 484–485, 491
  shutoff valves, 244–245
gas chlorinator, 334–335, 341, 483–484
gas connections for heaters, 244–245
gas-fueled heaters, 207–222
gas valves, 209, 211
gaskets
  described, 522
  lubrication for, 64–65
  plumbing, 63, 64, 65
  pumps, 120–121, 130
gate valves, 71–72, 73, 523
gauge, wire, 150, 156–157, 523
gauge assemblies, 203–204
gelcoat, 523
geophones, 431–432, 523
GFIs (ground fault interrupters), 157–158, 422, 523
glazed, 523
global warming, 499, 500
glossary, 511–537
glue, PVC, 50–52
Gordon, Len, 278
gpg (grains per gallon), 369, 523
gpm (gallons per minute), 523
grains per gallon (gpg), 369, 523
granular sanitizer products, 374–375
gravel/sand filters, 169–173
gravity drain plumbing, 421
green algae (chlorophyta), 354, 517
grid, 523
ground fault, 151
ground fault interrupters (GFIs), 157–158, 422, 523
grounding, 151
grout, 440
gunite, 28, 523

hacksaw, 46
hair/lint trap, 96, 119, 120
halogens, 342, 523. *See also* bromine

hardness of water, 330, 350, 361, 369, 523
Hartford loop, 523
HCL (hydrocholoric acid), 524
head
  calculating, 112–114
  described, 523
  discharge, 112
  dynamic, 112
  filters and, 113–114
  heaters and, 114
  overview, 108–109
  pumps/motors and, 108–109
  shutoff, 112, 532
  skimmers and, 114
  static, 112
  suction, 110–112
  total dynamic head (TDH), 112, 115–118, 535
head loss, 109
health department inspection list, 495, 496
health/safety guidelines
  carbon monoxide hazards from heaters, 235, 236
  chemical. *See* chemical safety
  chlorine gas, 334–335, 341, 484–485, 491
  circuit breakers, 135, 140, 154, 155, 157–158
  codes/regulations, 495
  commercial pools/spas, 492–498
  copper soldering, 59
  draining pools, 419–420
  electricity, 135, 140, 155–161, 315, 380, 422
  fecal accidents, 497–498
  fumes, 59
  gas supply, 246
  heaters, 209, 235, 236, 246
  importance of, 494
  Legionnaires' disease, 496–497
  life hook, 492
  lighting, 317
  motors, 135, 147–148
  pool vacuum hazards, 495
  precautions for children, 164, 298–299, 326, 531
  precautions when servicing pool, 495
  precautions when using skimmers, 39
  propane gas, 223
  pumps, 135, 147–148
  PVC glue, 59
  signage, 487–491
  submersible pump/motor hazard, 147–148
  toss rings, 492
  vacuum hose caution, 495
heat dissipation (heat sink), 102–103
heat exchanger
  carbon buildup on, 257–258
  described, 524
  reversible water connections, 238–239
  water flow issues, 246–251
heat exchanger tubes, 208, 209
heat/fan control, motors, 103
heat loss, 70
heat pumps, 225–227, 524
heat riser, 524
heat sink (heat dissipation), 102–103
heaters, 207–268. *See also* heating pool
  automated controls, 290
  automatic combination gas valves, 256–257
  automatic gas valve, 217–219
  basic principles, 207, 208
  Btu/kilowatt ratings for electric heaters, 228
  burner tray, 208, 209–211, 252, 515
  calculations for, 114, 230–233

heaters *(continued)*
carbon monoxide hazards from, 235, 236
checking for debris, 400
checking operation of, 400
clearances around, 235–237
compression fittings, 241–242
conserving energy, 500–502
control circuit in, 212–221, 258–266
cost to run, 232–233, 268
covers and, 227, 268
described, 524
electric-fueled, 222–223, 224
electrical connections, 245–246
electronic ignition, 219–221, 256
FAQs, 268
fireman's switch in, 245–246, 265, 521
flow control valve, 208–209
flow rate, 114
fuses, 263
fusible links in, 213–214, 263, 522
gas connections for, 244–245
gas-fueled, 207–222
head in, 114
heat exchanger tubes, 208, 209
heat pumps, 225–227
high-limit switches, 216–217, 263
IID, 218–221, 256, 265–266, 525
inadequate ventilation, 257–258
input vs. output ratings of, 229
installing, 233–246
lighting/relighting pilot in, 253–256
locations for, 234, 235
makes and models of, 228
millivolt, 210, 211–212, 261
natural vs. propane gas, 221–222
noise from, 240
oil-fueled, 227–228
on/off switch in, 215, 216, 263–265
overview, 70
pilot light, 211–212, 253–256, 528
plumbing for, 69–70, 72, 234, 240–243
PP (powerpile) gas valves, 219
pressure switches, 217, 261–262
preventive maintenance, 267–268
pumps and, 114
quick startup guide for, 247
repairs to, 236, 246–266
reset button on, 266
reverse flow and, 69–70
reversible water connection to, 238–239
safety guidelines, 209, 235, 236, 246
selection of, 228–233
shutoff valves, 244–245
size of, 228–232, 233
solar-fueled, 223–225
solid-state controls in, 266
spas, 223, 231–233
stack vs. stackless, 235
standing pilot system, 210, 211–212, 255–256
temperature control in, 209
TH gas valves, 219
therm, as unit of cost in, 232–233, 534
thermocouple gas valves, 211–212
thermostat in, 215, 216, 263–265, 290, 501
threaded fittings, 243
ton ratings for heat pumps, 227
tools for, 237
TP gas valves, 219

TR gas valves, 219
troubleshooting, 246–266
venting for, 235, 239–240
water chemistry and, 251
water flow, 243, 246–251
winterizing, 416
heating pool. *See also* heaters
solar heating systems, 76–85, 91
time required for, 268
hertz, 107, 524
high-limit switches, 216–217, 263, 524
high-rate sand filters, 169–173, 524
high-volume pump-out units, 146–147
HOCl (hypochlorous acid), 332, 525
hopper, 524
horsepower (hp), 524
horsepower ratings, 100, 103, 105–119
hose bibb (bib), 524
hoses
animals chewing, 281
backwash, 182–183
garden, 388–392
pool vacuum, 183, 495
hot line, 150
hot tubs. *See also* spas
algae in, 360
fiberglass, 15
wood, 19–20, 526
housing design, pump motors, 105–106
hp (horsepower), 524
HTH-type products, 337, 515
humidity, excess, 492–494
hydration
breaking in plaster, 423
described, 524
total dynamic head (TDH) in, 112, 115–118, 535
hydraulics, 108–118. *See also* water flow
described, 108, 524
dynamic head in, 112
filters and, 113–114
flow rate in, 109–111
head in, 109–112
heaters and, 114
horsepower and, 108–119
importance of, 109
motors, 108–119
pipe fitting calculations and, 112–113
poolside hardware and, 114
pumps, 108–119
static head in, 112
suction head in, 110–112
turnover rate in, 114–115
hydrochloric acid (HCL), 524
hydrogen gas, 485
hydrogen peroxide, 346
hydrojet (jet), 149, 525
hydrostatic pressure, 42, 458, 524
hydrostatic valve, 42, 525
hypochlorous acid (HOCl), 332, 525

IID (intermittent ignition device), 218–221, 256, 265–266, 525
impellers
clogged, 99
closed-face, 98–100, 133
diatomaceous earth and, 99
horsepower rating, 100
jammed, 140
open-face, 98–100
overview, 98–100
pumps, 98–100, 130–133, 139–142, 525
removing, 122, 131–133
semiopen-face, 98–100
speed, 116
vanes, 98–100
volute, 97–98, 99
water's-eye view of, 97

Independent Pool and Spa Serviceman's Association (IPSSA), 525
indoor pools, 449, 492–494
inflatable pools, 18–19
inlet. *See* return
insect nests, 256, 267, 276
insulators, 151
Intelliflo pump, 163
intermittent ignition device (IID), 218–221, 256, 265–266, 525
internal flow control valve, 251, 252
iodine, 525
ionizers, 345–346
IPSSA (Independent Pool and Spa Serviceman's Association), 525
iron, 361, 369

J-box, 525
jackhammering, 443
jet (hydrojet), 149, 525
jetted tubs, 97, 149, 525. *See also* spas
joint stick, 44–45, 525
joist, 525
junction box, 156

keyed shafts, pump, 101–102, 525
kilowatt, 525
koi, 106, 461, 463, 464, 474
koi ponds, 173, 177, 460–468, 469, 474

labor reference guide, 507–509
ladders, 297–298
lamps. *See* lighting
landscaping, 504
Langlier index, 371–373, 525
laterals, 171, 525
latiance, 441, 525
lazy flame, 263, 526
lead solder, 55
leaf canister, 403
leaf rake, 387, 397, 526
leaf vacuum, 390–392
leafmaster, 388, 390–392, 404–406, 526
leak detection/repair, 428–440
  air leaks, 145, 280–281
  coping stones, 436–440
  copper plumbing and, 53
  drain-down test, 430–431
  dye testing, 429–430
  evaporation, 429–430
  filters, 199–205
  fixing underwater, 458
  geophones, 431–432
  O-rings and, 72
  packing glands and, 72
  patching plaster cracks, 432–436
  piston valves, 180
  pressure testing, 431–432
  rotary backwash valves, 180, 201
  solar heating systems, 85
  testing for leaks, 428–432
  three-way valves, 65–66
  tiles, 436–440
  water conservation and, 503
leaves, saving for mulch, 503
Legionnaires' disease, 494–497
life hooks, 489, 492
lifeguard, 490
lighting, 309–319
  120/240 volt, 310–312
  bulb replacement, 314–316
  bulbs for, 311, 314–319, 343, 501
  cords for, 311, 313, 315
  fiberoptics, 318–319
  fixture assembly, 316
  installation guidelines for, 312–314
  lamps, 311
  light fixture corrosion, 361
  low-voltage, 318

mogul bases for, 312, 527
niche receptacle for, 310, 527
safety guidelines, 317
troubleshooting, 316–318
watertight seals for, 317
lignin, 526
lime deposits, 250, 330
line, 150, 526
line voltage, 140
lint trap, 96, 119, 120
LiOCl (lithium hypochlorite), 526
liquid copper (copper sulfate) algicide, 358–359
lithium hypochlorite (LiOCl), 338, 526
load, 150, 153, 526
lock, chlorine, 338, 339
low-volume pumps/motors, 147–148
lubrication, gaskets, 64–65
lubrication, silicone, for pumps, 133

macintosh bag filter, 526
main drains, 27, 41–43, 526
main valve, 526
maintenance/repairs. *See also* cleaning/servicing guidelines
FAQs, 458
heaters, 236, 267–268
leak repair. *See* leak detection/repair
painted surfaces, 450
plaster repair, 432–436
pool cleaner repair, 306–309
pumps/motors, 119–146
schedule for, 407
service/maintenance schedule, 407
valves, 65–68, 67–68
male fittings, 44, 47, 526
manifold, 526
manometer, 526
marcite, 441, 526
marcite (plaster) coatings, 33–34, 529
admixtures (admix), 435
alkaline lime in, 330
application of, 441–442
breaking-in, 34, 423–428, 514
brown coat, 443, 444, 515
delamination, 432
patching/repairing, 432–436
plaster nodules, 444
plaster pools, 11–12
plastering process, 441–442
problems with, 432–433
replastering, 441, 442–444
sandblasting, 443, 444
scratch coat, 443
water chemistry, 426–427
material safety data sheet (MSDS), 380
maximum bather load signage, 487–488
mechanical delivery devices, 375–377
mechanical thermostats, 263–265
media, 526
metal chelation agent, 360
metal corrosion, 361, 379
micron, 177, 527
mildew, 361
mildew odor, 361
millivolt heaters, 210, 211–212, 261
MIP fitting, 527
mission clamps, 57, 527
mogul base, 312, 527
motor/pump health checklist, 119
motorized valve systems, 66–68
motors, 103–106. *See also* pumps
booster, 148–149
breaker trips, 140–141
C frame type, 95, 105–106, 516
capacitor in, 103–105, 140
capacitor start, capacitor run (CSR), 104–105, 515

motors *(continued)*
capacitor start, induction run (CSI), 104, 516, 518
conserving energy, 501
cost of operation, 148
covers for, 146
duty rating for, 107–108, 519
dynamic head in, 112
electrical safety, 135, 147–148
electrical specifications, 106–107
electricity required for, 103–104
end bell type, 103, 104, 106, 137
energy efficient, 104, 105, 520
exploded view of, 95
fan/heat control in, 103
FAQs, 164
flooded, 135, 146
flow rate of, 105, 109–111, 114–118
head rate of, 108–109
high-volume pump-out units, 146–147
horsepower rating of, 103, 105–119
housing design, 105–106
hums but won't run, 139–140
hydraulics, 108–119
installation/wiring diagram for, 107
low-volume, 147–148
maintenance/repairs, 119–146
manufacturer/date for, 107
nameplate on, 106, 527
new installation, 135–137
noise from, 141–142, 164
overview, 103–104
phasing of, 107
programmable, 163
ratings of, 106
reinstalling, 133–135
removing, 133–135
replacing, 137–139
rotational speed, 100, 104
serial number, 107
service factor, 106, 107, 531
shaft/shaft extender in, 101–102, 131, 132, 532
for spas, 148–149
speed, 100, 104, 148
split phase, 104
square flange, 96, 105–106, 533
startup amperage vs. running amperage, 104
static head in, 112
submersible, 146–148, 421–422, 534
suction head in, 110–112
switchless, 104–105
tools for, 120
troubleshooting, 139–142
types of, 104–105
vibration in, 141–142
voltage rating, 105
water damage to, 139, 140
waterfalls, 459
windings of, 103, 104, 139
winterizing, 415–416
won't start, 139–140
motting, 527
MSDS (material safety data sheet), 380
multimeter, 161–163
multiport backwash valves, 181–182, 200–203
muriatic acid, 336, 348, 373, 380, 443, 527
mustard algae, 354
musty odor, 361

nameplate, pump motors, 106, 527
NaOCl (sodium hypochlorite), 335, 341, 533
National Electric Code (NEC), 527
National Pipe Thread (NPT), 48, 527

natural gas, 219, 221–222, 256, 268
NEC (National Electric Code), 527
neutral line, 150
neutral pH, 527
niche receptacle for lighting, 310, 527
nipple, 527
nitrogen, 340, 512, 516
no-hub connectors, 57, 527
nodules, plaster, 444
noise
  heaters, 240
  pumps/motors, 141–142, 164
  valves, 75
nonpositive valves, 63
NPT (National Pipe Thread), 48, 527
nut driver, 527

O-rings
  described, 528
  filters, 203
  leak detection/repair and, 72
  pumps, 120, 121–122
odors, 335, 361
oil-fueled heaters, 227–228
Olympic-style competition pool, 476
on/off switch, heaters, 215, 216, 263–265
one-touch units, 288–289
open circuits, 150
open-face impeller, 98–100
open loop systems, 76–77, 527
operational requirements, 495
organic, 528
ORP (oxidation reduction potential), 528
orthotolidine (OTO), 365–366, 528
OTO chlorine-level test kit, 365–366
OTO (orthotolidine), 365–366, 528
overflow line, 503
overloaded circuits, 154–155
oxidation reduction potential (ORP), 528
ozonators, 343–345, 528
ozone, 343–345, 381, 528
ozone systems, 343–345

packing gland, 72
painting pool surfaces, 445–450, 504
parts per million (ppm), 8–9, 529
patching plaster cracks, 432–436
peristaltic chlorinator, 377, 378, 528
permits, pool construction, 21
pH levels, 347–349
  adjusting, 426
  algae and, 353, 360, 361
  chlorine and, 332–334
  described, 347, 528
  neutral, 527
  sanitizers and, 333
  scale deposits and, 330, 347
  testing for, 367, 426
  water chemistry, 347–349
phaeophyta (yellow algae), 354, 528
phasing, pump motors, 107
phenol red pH testing, 349, 367, 528
PHMB (polyhexamethylene biguanicide), 346–347, 380, 381, 540
photosynthesis, 528
pi, 528
pilot generator, 211, 218–219, 255, 259–261, 528
pilot light, 211–212, 253–256, 528
pilots for heaters, 253–256
pink algae, 355
pipe connection fittings, 44, 48–49
pipe cutters, 46, 47
pipe dope, 44–45, 529
pipe run, 44, 531

pipe wrenches, 45–46
pipes. *See also* plumbing
  diameter, 48, 58
  drain, 171
  flow through, 58
  horizontal runs, 49
  measuring, 44, 48
  size of, 58
  stanchion, 171
  underground, 49
  vertical runs, 49
  winterizing, 409
piston backwash valves, 179–180, 200
plans/permits, pool construction, 21, 22
plaster (marcite) coatings, 33–34, 529
  admixtures (admix), 435
  alkaline lime in, 330
  application of, 441–442
  breaking-in, 34, 423–428, 514
  brown coat, 443, 444, 515
  delamination, 432
  patching/repairing, 432–436
  problems with, 432–433
  replastering, 441, 442–444
  sandblasting, 443, 444
  scratch coat, 443
  water chemistry, 426–427
plaster nodules, 444
plaster pools, 11–12
plastering process, 441–442
plug, 529
plumbing. *See also* pipes
  advanced systems, 61–91
  basic systems, 37–59
  copper, 53–57
  FAQs, 59
  gas connections, 244–245
  general guidelines, 43–47
  gravity drain, 421
  heaters, 69–70, 72, 234, 240–243
  leak detection/repair, 428–440
  miscellaneous fittings, 57
  normal flow, 69
  planning for, 27
  PVC in, 47–52, 59, 281
  reverse flow, 69–70, 100
  shutoff valves, 71–73, 84, 91, 112
  sizing of, 58
  skimmer, 27, 40–41
  solar heating systems, 76–85, 91
  three-port valves, 61–68
  tools for, 59
  water level controls, 85–90
  winterizing, 413–415
plumbing layout, 37, 38
plumbing systems, 37–59
plumbing unions, 70–71
pneumatic (air-activated) controls, 278, 512
polyhexamethylene biguanicide (PHMB), 346–347, 380, 381, 540
polymer algicides, 359
polymer wall panels, 15
polyvinyl chloride (PVC), 47, 530
ponds
  bacteria, beneficial, 463–469
  filters, 173
  koi/fish, 173, 177, 460–468, 469, 474
pool chemicals. *See* water chemistry
pool cleaners, 299–309. *See also* cleaning/servicing guidelines
  automatic, 299–309, 327, 501, 503
  booster pump systems, 300–309
  installing, 301–305
  operation of, 305–306
  repairing, 306–309
  robotic, 299–300, 530
  suction-side systems, 309

vacuum head, 300, 301–309
pool closed signage, 490
pool equipment. *See* equipment
pool lights. *See* lighting
pool paint, 445–450, 504
pool service, 418. *See also* cleaning/servicing guidelines
pool/spa rules, 490
pools. *See also* spas
above-ground, 12–19, 326
adding shade to, 503
alarms for, 299
algae in. *See* algae
bacteria in. *See* bacteria
"bathtub ring," 350, 361, 399, 418
blocking access to, 417. *See also* safety barriers
calculating water volume, 3–8
chemicals for. *See* water chemistry
commercial. *See* commercial pools/spas
concrete, 11–12
considerations for, 10–11
construction of, 20–35
cost considerations, 10–11
covers for, 268, 320–326, 417, 501, 502
depth of, 3–8, 292, 476
design of, 20–22
draining. *See* draining pools/spas
empty, 419
environmental considerations. *See* environmental issues
excavation, 21–26
extras, 11
FAQs, 36
fiberglass, 15–17
filling with water, 424–425
indoor, 449, 492–494
inflatable, 18–19
key features, 10–11
lighting for. *See* lighting
plaster, 11–12
portable, 18–19
reopening after winter, 418
safety barriers around, 298–299, 326, 531
safety equipment for, 487–492
shapes of, 3–8, 477–478
size, 10
types of, 9–20
typical, illustrated, 1, 2
vanishing edge pools, 471–474, 536
vinyl-lined, 12–15
wading, 174, 403, 459, 476, 488
wood, 19–20, 526
poolside hardware, 114
port, 529
portable pools, 18–19
positive seal, 529
positive seal valve, 63
potassium monopersulfate, 346, 380, 529
potentiometer, 529
pounds per square inch (ppi), 530
power pile. *See* pilot generator
power supply, 152
PP (powerpile) gas valves, 219
ppm (parts per million), 8–9, 529
precipitates in water, 330, 529
precoat, 529
pressure, hydrostatic, 42
pressure gauges, 183, 184–185, 203–204, 529
pressure safety valve, 243
pressure sand/gravel filters, 169–173, 529
pressure switches, 217, 261–262, 529
pressure testing for leaks, 431–432
price considerations, 10–11
prime, 529
priming pumps, 142–145

problems. *See also* repairs/maintenance
  air switch units, 280–281
  automated controls, 290
  circuit breakers, 154–155
  diving boards, 292–295
  gaskets, 120–121
  heater repairs, 236, 246–266
  leaks. *See* leak detection/repair
  lighting problems, 316–318
  motors, 139–142
  O-ring problems, 121–122
  plaster (marcite) coatings, 432–433
  remote controls, 290
  reverse flow problems, 100
  solar heating problems, 86
  time clocks, 276–277
  twist timers, 273–274
professional associations, 542–543
programmable pumps/motors, 163
propane gas, 221–222, 223, 268
psi (pounds per square inch), 530
pumice stones, 394, 398, 455, 530
pump curve, 530
pump delivery systems, 377, 378
pump/motor health checklist, 119
pump performance curve, 110, 111
pump strainer basket, 141, 399, 407
pumps, 93–164. *See also* motors
  48 type, 106
  adapter bracket in, 100–101, 511
  air leaks in, 145
  blow bag, 143–145, 514
  booster, 148–149
  breaker trips, 140–141
  C frame type, 95, 105–106, 516
  calculations for, 112–114
  caution for swimmers, 164
  cavitation in, 112, 141, 516
  centrifugal-type pumps, 93–94
  clogged, 99, 119, 120
  conserving energy, 501
  cost of operation, 148, 164
  described, 530
  discharge head in, 112
  draining pools with, 146–148
  dynamic head in, 112
  electrical safety, 135, 147–148
  energy efficient, 163
  exploded view of, 95, 96
  FAQs, 164
  filters and, 113–114
  flow rate of, 105, 109–111, 114–118
  gaskets in, 120–121, 130
  hair/lint trap for, 96, 119, 120
  head rate of, 108–109
  heat dissipation (heat sink), 102–103
  heaters and, 114
  high-volume pump-out units, 146–147
  horsepower rating of, 105–119
  hydraulics, 108–119. *See also* hydraulics
  impeller in, 98–100, 130–133, 139–142, 525
  Intelliflo pump, 163
  keyed shafts in, 101–102
  low-volume, 147–148
  maintenance/repairs, 119–146
  motors for. *See* motors
  new installation, 135–137
  noise from, 141–142, 164
  O-rings in, 120, 121–122
  pipe fitting calculations for, 112–113
  poolside hardware and, 114
  priming of, 142–145
  programmable, 163
  reinstalling, 133–135
  removing, 133–135

replacing, 137–139
reverse flow problems, 100
running dry, 102, 164
seal plate in, 100–102, 531
seals in, 102–103, 122–133, 531
self-priming, 95, 98, 110
shaft/shaft extender in, 101–102, 131, 132, 532
shroud of, 98–99
shutoff head in, 112, 532
silicon lubricant for, 133
size of, 118–119
for spas, 148–149
speed, 100, 104, 115, 148
square flange in, 96, 105–106
Sta-Rite, 100, 131, 139
static head in, 112
strainer pot/basket for, 96–97, 119–120
submersible type, 146–148, 419–422, 534
suction head in, 110–112
T-handles on, 145–146
tools for, 120
total dynamic head (TDH) in, 112, 115–118, 535
troubleshooting, 119–146
turnover rate in, 114–115
uniseal flange type, 106
vanes of impeller in, 98–100
vibration in, 141–142
volute chamber in, 97–98, 99
waterfalls, 472–474
winterizing, 409, 411–416
PVC/plastic pipe plumbing, 47–52
advantages of, 281
attaching to metal pipes, 59
electrical conduit/fittings, 281
gas connections, 244–245
gluing joints in, 50–52, 59
overview, 47–49
plumbing methods, 50–52
PVC (polyvinyl chloride), 47, 530
PVC primer, 50
quaternary ammonium compounds (quats), 359–360, 530

radius, 530
rails, 297–298, 361
rainwater, 352–353, 396, 503
rapid sand filters, 169, 530
ratings, pump motors, 106
Raypak Unitherm Governor, 251
reagents, 362–370, 380, 530
rebar (steel), 11–12, 27, 433, 530
rebound, 28, 530
rectangular pools, 3–5
recycled materials, 500, 503–504
red fungus, 355
reducer, 530
reel system, 322–323
references, 539–543
reinforcing bar (rebar) in concrete, 11–12, 27, 433, 530
relay circuit, 159
relays, electrical, 159–160, 283–288
remodeling techniques, 440–457
acid washing pool surfaces, 450–456
deck work, 456–457
fiberglass coatings, 444–445
painting pool surfaces, 445–450, 504
plaster/replastering, 441–444
remote controls, 277–290
air switches, 278–281
hard wired, 283–289
troubleshooting, 290
wireless, 281–283
repairs/maintenance. *See also* cleaning/servicing guidelines; troubleshooting
FAQs, 458
heaters, 236, 267–268
leak repair. *See* leak detection/repair

repairs/maintenance *(continued)*
  painted surfaces, 450
  plaster repair, 432–436
  pool cleaner repair, 306–309
  pumps/motors, 119–146
  schedule for, 407
  service/maintenance schedule, 407
  valves, 65–68, 67–68
replastering process, 442–444
reset button
  heaters, 266
  remote control units, 290
residual, 368, 530
resin, 530
retainer, 530
retaining rod, 530
return, 530
return outlets, 114, 348, 370
reverse flow, 69–70, 100
riser, 530
robotic pool cleaners, 299–300, 530
rock, 30–31
rockscapes, 468–471
rollers, 322–323
rotary backwash valves, 180–181, 200–203
rotor, 531
rules, pool/spa, 490
rust, 65, 253–256, 433

safety barriers, 298–299, 326, 531
safety controls, 209
safety equipment, 487–492
safety/health guidelines
  carbon monoxide hazards from heaters, 235, 236
  chemical. *See* chemical safety
  chlorine gas, 334–335, 341, 484–485, 491
  circuit breakers, 135, 140, 154, 155, 157–158
  codes/regulations, 495
  commercial pools/spas, 492–498
  copper soldering, 59
  draining pools, 419–420
  electricity, 135, 140, 155–161, 315, 380, 422
  fecal accidents, 497–498
  fumes, 59
  gas supply, 246
  heaters, 209, 235, 236, 246
  importance of, 494
  Legionnaires' disease, 496–497
  life hook, 492
  lighting, 317
  motors, 135, 147–148
  pool vacuum hazards, 495
  precautions for children, 164, 298–299, 326, 531
  precautions when servicing pool, 495
  precautions when using skimmers, 39
  propane gas, 223
  pumps, 135, 147–148
  PVC glue, 59
  signage, 487–491
  submersible pump/motor hazard, 147–148
  toss rings, 492
  vacuum hose caution, 495
salt, 335, 350, 377, 379
salt chlorine generators, 377–379, 531
sampling procedures for testing, 370–371
sand, 196–198
sand filters, 169–173. *See also* filters
  cleaning, 189, 196–198
  described, 531
  free-flow sand/gravel filters, 173, 522
  microns, 177
  pressure sand/gravel filters, 169–173, 529

size of, 172, 175–178
sandblasting, 443, 444
sanitizers, 331–347
algae and, 353, 355–356
biguanicides (PHMB), 346–347, 380, 381, 540
bromine, 342, 346, 366, 380, 381, 515
chlorine. *See* chlorine
described, 531
granular products, 374–375
ionizers as, 345–346
ozone/ozonators, 343–345, 381
potassium monopersulfate, 346, 380, 529
saturation, 330, 371–372, 531
saturation point, 330, 371–372
scale deposits
brick, 30–31
demand/balance and, 330
described, 31, 531
eliminating, 361
lime deposits, 250, 330
pH and, 330, 347
rock/stone, 30–31, 350
solar panels, 85
tile, 31, 350
water hardness and, 350
at waterline, 350, 361, 460
schedule number, 47
scratch coat, 443, 531
screed, 531
seal plate, pumps, 100–102, 531
seals, pumps, 102–103, 122–133, 531
sediment trap, 531
self-priming pumps, 95, 98, 110
semiopen-face impeller, 98–100
separation tank, 168, 531
septum. *See* grid
sequestering agent, 531. *See also* chelating agent
service and maintenance schedule, 407
service factor, pump motors, 106, 107, 531
service logs, 495
servicing/cleaning guidelines, 383–418. *See also* pool cleaners; repairs/maintenance
"bathtub ring," 350, 361, 399, 418
bicarbonate of soda for cleaning, 380, 426, 441, 514
brushing. *See* brushing
deck/cover cleaning, 394–395, 503
equipment area, 400
equipment check for, 399–400
FAQs, 418
filters. *See* cleaning filters
heavy-duty cleaning, 404
leaf rake, 387, 397, 526
metal fixtures, 456
plastic fixtures, 456
pool chemistry servicing. *See* water chemistry
pool winterizing, 409–418
pumice stones for, 394, 398, 455, 530
quick start guide, 395
service logs, 495
service/maintenance schedule, 407
spas, 388, 393, 403, 409
surface skimming, 396–397
telepoles for, 383–387
test kits, 363, 365–368, 371, 393
testing water chemistry. *See* testing water
thermometers, 393
tile cleaning, 392–393, 398–399, 455
tools for, 383–394
with TSP, 441, 448
vacuuming. *See* vacuuming
wall brushes, 387, 388
water features, 409

servicing/cleaning guidelines *(continued)*
  water level checks, 396
  waterline area, 399, 407, 418, 503
shaft, pumps, 101–102
shaft extender, pumps, 101–102, 131, 132, 532
shafts, keyed, 101–102, 525
sheet vinyl covers, 323–324
shell. *See* gunite
shock hazards, 159–161, 315, 380, 422
shock treatment. *See* superchlorination
short circuits, 149, 151
shotcrete, 532
shoulders, 71
shroud, pumps, 98–99
shutoff head, 112, 532
shutoff valves
  gas, 244–245
  plumbing, 71–73, 84, 91, 112
sight glasses, 185, 532
signage, 419–420, 487–491
silica, 532
silicon lubricant, 133
single-phase, 532
single pole, double throw (SPDT) switch, 159
single pole, single throw (SPST) switch, 159, 532
sink. *See* heat riser
skid pack, 532
skimmer basket, 120, 399, 404, 407, 492
skimmers, 37–41
  access to, 38
  described, 37, 532
  diatomaceous earth and, 194–195
  head and, 114
  illustrated, 38, 39
  plumbing for, 27, 40–41
  precautions, 39
  purpose of, 37
  spa, 38
  suction in, 39, 40–41
  types of, 37, 39, 41
  variances, 40–41
  weirs, 38–39
  winterizing, 412–413, 416
skimming, 396–397
skin discoloration/irritation, 361
slide valves, 72, 532
slides, 295–297, 326, 503
slip fittings, 48, 532
slope calculations, 477–478
slurry, 195, 532
slurry feeders, 481, 482
slurry tanks, 481
soap, tile, 392–393, 397–399
soda ash, 348, 454, 455, 532
sodium bicarbonate, 380, 426, 441, 514
sodium bisulfate, 533
sodium chloride, 485
sodium dichloro-s-triazinetrione (dichlor), 337, 338, 341, 533
sodium hypochlorite (NaOCl), 335, 341, 533
sodium thiosulfate, 533
soft water, 533
solar covers, bubble type, 320–323
solar heating systems, 76–85, 91, 223–225, 501
solar panels, 66, 76, 78, 91, 223–225, 416, 533
solder, 533
solid-state controls, 266
soot, black, 317, 404
sooting, 257–258, 268
spa jets, 149
spa skimmers, 38
spalling, 533

spas. *See also* hot tubs; pools
acrylic tubs/spas, 15–16, 429, 432, 458
air blowers, 72, 501, 502, 514
algae in. *See* algae
blocking access to, 417
booster pumps/motors for, 148–149
calculating water volume, 3–8
chemicals for, 499–500. *See also* water chemistry
cleaning, 388, 393, 403, 409
commercial. *See* commercial pools/spas
construction of, 20–35
covers, 268, 320–326, 417, 501, 502
design of, 20–22
draining. *See* draining pools/spas
drains in, 43
environmental issues, 499–505
FAQs, 36
fiberglass, 15, 16, 17
filling with water, 424–425
foamy water, 361
heaters, 223, 231–233
inflatable, 18–19
insulating, 501
lighting, 309–319
location of, 501
reopening after winter, 418
safety barriers around, 298–299, 326, 531
solid-state controls, 266
suction, 43
types of, 9–20
typical, illustrated, 1, 2
vacuuming, 393, 403
vinyl-lined, 12–15
SPDT (single pole, double throw) switch, 159
spiking, 349, 533
spin filter, 533
splashing, 502, 503
split phase pump motors, 104
SPST (single pole, single throw) switch, 159, 532
square flange pump motors, 96, 105–106, 533
square pools, 3–5
squared, 533
Sta-Rite pumps, 100, 131, 139
stabilizers, 351, 352, 427, 428, 533. *See also* cyanuric acid
stack, 533
stack heater, 235
stackless heater, 235
stanchion, 533
stanchion pipe, 171
standing pilot device, 533
standing pilot heater, 210, 211–212, 255–256
static head, 112, 533
stator, 534
steel bars (rebar), 11–12, 27, 433, 530
steel bleeds, 433
steel wall panels, 13
stone, 30–31, 350, 440
storm drains, 504
strainer basket, 119–120, 403, 534
strainer pot/basket, pumps, 96–97, 119–120
strainer pots, 119–122, 534
street fitting, 534
submersible pumps/motors, 146–148, 419–422, 534
suction head, 110–112
suction-side pool cleaners, 309
sump pumps, 487
sunlight, 333, 352, 353. *See also* ultraviolet (UV) light
superchlorination (shock treatment), 337–340, 346, 354, 374, 411, 534
surface skimming, 396–397
surge chamber, 480–481, 534

sweating (copper soldering), 53, 534
sweep head pool cleaners, 300–301
swimmers
  body oil from, 190, 361, 392, 398, 418, 528
  defined, 513
  fecal accidents, 497–498
  harmful chemicals and, 483, 495
  lotion/soap from, 190, 350, 361, 392, 418
  precautions for children, 164, 298–299, 326, 531
  precautions when servicing pool, 495
  pump/skimmer caution, 164
  signage for, 419–420, 487–491
  splashing and, 502, 503
  surge chamber and, 480–481, 534
  vacuum hose caution, 495
  water displacement, 479–480
swing gate valve, 74–75
switch diagrams, 159
switches
  air, 278–281, 512
  breaker, 135
  control circuit, 258–266
  electrical, 159, 490
  fireman's, 245–246, 265, 521
  heaters. *See* heaters
  high-limit, 216–217, 263, 524
  pressure, 217, 261–262, 529
  SPDT (single pole, double throw), 159
  SPST (single pole, single throw), 159, 532
  type/location, 159
switchless motors, 104–105

T fittings, 113, 534
T-handles, pumps, 145–146
TDH (total dynamic head), 112, 115–118, 535
TDS (total dissolved solids), 350–351, 369, 372, 535
tear down, 534
Teflon tape, 44, 45, 534
telescoping poles (telepoles), 383–387, 534
temperature. *See also* heaters
  ambient, 512
  controlling, 209–211, 216, 495
  effect of weather on, 352
  effect on water chemistry, 331, 352
  solar heating, 80–81
test kits (water testing), 363, 365–368, 371, 393
testing water, 362–373. *See also* water chemistry
  acid in, 362, 367, 368, 370
  alkalinity in, 349–350, 368
  chlorine-level testing, 365–367
  color charts for, 368
  colorimetric testing in, 362, 370, 517
  cyanuric acid, 370
  diethyl phenylene diamene (DPD) testing, 366–367, 519
  electrometric testing, 364, 520
  ethylenediamine tetra-acetic acid (EDTA), 369
  grains per gallon (gpg), 369, 523
  hardness in, 350, 366, 369, 372–373
  heavy metals in, 369
  Langlier index, 371–373
  metal contaminants, 369
  OTO chlorine-level test kit, 365–366
  pH in, 367
  phenol red pH testing, 349, 367, 528
  reagents, 362–370, 380, 530
  sampling procedures for, 370–371
  schedule for, 408

test kits for, 363, 365–368, 371, 393
test strips for, 268, 363–364
tips for, 368
titration testing in, 362
total dissolved solids (TDS), 350–351, 369, 372
turbidity in, 364–365
TH gas valves, 219
therm, as unit of cost in heaters, 232–233, 534
thermal overload protector, 103, 535
thermistor failure, 290
thermistors, 264, 288, 289, 290, 535
thermocouple gas valves, 211–212
thermometers, 393, 492, 501
thermostats
electronic, 263–265
heaters, 215, 216, 263–265, 290, 501
mechanical, 263–265
Thoroseal concrete sealer, 30, 535
threaded fittings, 44, 48, 243
three-port valves, 61–68, 535
tile brushes, 392–393
tile soap, 392–393, 397–399
tiles
all-tile pool, 29
cleaning, 392–393, 398–399, 455
grout, 440
leak detection/repair, 436–440
scale deposits on, 31
scratches in, 399
at waterline, 28–29, 503
time clocks, 269–277
checking operation of, 400
cleaning, 276
described, 269, 535
electromechanical, 269–273, 275–277
electronic timers, 274–275
pool cleaner booster pump, 272
replacing, 275–276
settings, 277
trippers, 277
troubleshooting, 276–277
twist timers, 271, 273–274
titrant, 362
titration, 362, 535
titration testing, 362
ton rating for heat pumps, 227
tongue, 157
torque, 103, 535
toss rings, 489, 492
total alkalinity, 349–350, 368, 426–427, 535
total dissolved solids (TDS), 350–351, 369, 372, 535
total dynamic head (TDH), 112, 115–118, 535
TP gas valves, 219
TR gas valves, 219
transformers, 535–536
travel time, 508
tread kits, for diving boards, 295
treatment delivery systems, 373–379
chlorine liquid, 373–374
erosion systems, 376
floaters, 337–338, 346, 375, 521
granular sanitizer products, 374–375
mechanical delivery devices, 375–377
muriatic acid, 373
pump delivery systems, 377, 378
salt chlorine generators, 377–379
tabs/tablets, 375
tree roots, 428, 429, 468
trees, 234, 237, 465, 504
trichlor cyanurate, 337–338, 341, 356, 357, 536

trichlor (trichloro-s-triazenetrione), 337–338, 536
trichloro-s-triazenetrione (trichlor), 337–338, 536
trippers, 270, 272, 277, 536
trisodium phosphate (TSP), 347, 441, 447–448, 536
troubleshooting. *See also* repairs/maintenance
  air switch units, 280–281
  automated controls, 290
  circuit breakers, 154–155
  diving boards, 292–295
  gaskets, 120–121
  heater repairs, 236, 246–266
  leaks. *See* leak detection/repair
  lighting problems, 316–318
  motors, 139–142
  O-ring problems, 121–122
  plaster (marcite) coatings, 432–433
  remote controls, 290
  reverse flow problems, 100
  solar heating problems, 86
  time clocks, 276–277
  twist timers, 273–274
TSP (trisodium phosphate), 347, 441, 447–448
turbidity, 364–365, 536
turnover rate, 114–115, 536
twist timers, 271, 273–274

UL (Underwriters Laboratory), 536
ultraviolet (UV) light. *See also* sunlight
  algae and, 342
  bacteria and, 342
  chlorine and, 333, 352
  ozonators, 343
  PVC and, 48
  sanitizers, 343
  water chemistry, 342–343, 352
umbrellas, 503
underdrain, 536
Underwriters Laboratory (UL), 536
unions, 70–71, 536
uniseal, 536
uniseal flange frame type, 106
up-rated, 536
UV light. *See* ultraviolet (UV) light

vacuum head (vac sweep) cleaners, 300, 301–309, 388–390
vacuuming, 400–407
  to filter, 388, 401–404
  to leafmaster, 388, 390–392, 404–406
  plaster dust, 427
  schedule for, 418
  spas, 393, 403
  wading pools, 403
vacuums, pool
  described, 536
  hazards, 495
  hoses, 183, 495
  leaf vacuums, 390–392
  spa, 393
valve actuator, 67
valves
  air relief, 184–185, 203, 512
  antisipihon, 85–86, 513
  antisurge, 513
  automatic bypass, 77, 78
  automatic gas valves, 66–68, 217–219, 513
  backwash, 167, 179–182, 199–203
  ball, 71–72, 344, 485
  check, 42, 72–75, 186, 243, 516
  described, 536
  flapper gate, 74–75, 521
  float, 521
  flow control, 208–209
  gas, 209, 211

gate, 71–72, 73, 523
hydrostatic, 42, 525
internal flow control, 251, 252
lubrication, 64–65
main, 526
maintenance, 67–68
motorized/automated, 66–68
multiport, 181–182, 200–203
noisy, 75
nonpositive, 63
one-way, 42
overheating, 66
piston, 179–180, 200
positive seal, 63
pressure safety, 243
repairing, 65–68
rotary, 180–181
shutoff, 71–73, 91, 244–245
slide, 72
sticky, 64–65
swing gate, 74–75
thermocouple gas, 219
three-port, 61–68
warping, 66
Y, 61–68
vanes, impeller, 98–100
vanishing edge pools, 471–474, 536
ventilation, 240, 257–258
venting, 235, 239–240
vibration, pumps/motors, 141–142
vinyl-lined pools, 12–15
Visoflame tube, 536
VOCs (volatile organic compounds), 499
volatile organic compounds (VOCs), 499
voltage converter, 152
voltage rating, 105, 151
volts, 150, 536
volume calculations, 8, 477–480
volute, 97–98, 99, 536
volute chamber, pumps, 97–98, 99
wading pools, 174, 403, 459, 476, 488
wall brush, 387, 388
warranty, 507–508
washing pool surfaces, 450–456
water
algae in. *See* algae
backwash, 502
bacteria in. *See* bacteria
balanced, 330, 331, 372, 513
blue-green color, 360
brown-red color, 361
chemistry of. *See* water chemistry
chocolate-brown color, 346–347
circulation, 425
cloudy, 360, 364–365
collecting under pool, 42
copper in, 369
depth, 3–8, 292, 476
electricity and, 161
environmental issues, 499–500
fecal matter in, 497–498
foamy, 361
frozen, 409, 410, 412–417
hardness, 330, 350, 361, 369, 523
heavy metals in, 361, 369, 407
iron in, 361, 369
parts per million (ppm) calculation, 8–9
path of, 1–3
rate of flow, 58
saturation, 330, 371–372, 531
saturation point, 330, 371–372
soft, 533
testing. *See* testing water
volume, calculating, 3–8
water chemistry, 329–381. *See also* treatment delivery systems
acid in. *See* acid
algae/algeacides and. *See* algae
alkalinity in, 349–350, 426–427, 512, 535

water chemistry *(continued)*
alternatives to chlorine, 342–347
ammonia, 338–340, 343, 514–515
bacteria in. *See* bacteria
balanced water makeup, 330, 331, 372, 513
bather load and, 353
biguanicides (PHMB) as sanitizers in, 346–347, 380, 381, 540
blue-green water, 360
breakpoint chlorination, 339, 340, 514–515
bromine, 342, 346, 366, 380, 381, 515
brown-red water, 361
calcium hardness, 350, 426, 523
calculating water volume, 3–8
chloramines, 338–340, 361, 365–366, 516
chlorine in. *See* chlorine
chlorine lock, 338, 339
chlorine residual in, 330, 333, 340, 353, 517
cloudy water, 360, 364–365
components of, 331–353
conditioner, 407, 491, 518
corrosion of metal parts and, 361, 379
cyanurates, 336, 337–338, 350, 518
cyanuric acid, 330, 336, 337, 351, 370, 380
dangers of. *See* chemical safety
delivery systems for. *See* treatment delivery systems
demand/balance in, 330, 426, 511
dichlor (sodium dichloro-s-triazinetrione), 337, 338, 341, 533
dirt/debris and, 352
discoloration of hair, skin, nails, 361
environmental issues, 499–500
equilibrium in, 372
erosion systems, 376
evaporation and, 352
eye irritation, 361
FAQs, 381
floaters for, 337–338, 346, 375, 521
hardness in, 330, 350, 361, 369, 523
heater problems and, 251
heavy metals, 369, 424
how often to add chemicals, 381
ionizers as sanitizers in, 345–346
Langlier index, 371–373
metal contaminants, 369, 424
odors, 361
ozone/ozonators as sanitizers in, 343–345, 381
parameters of, 330–331
pH in. *See* pH levels
plaster, breaking in, 426–428
pool chemical mixing guidelines, 380
potassium monopersulfate sanitizers in, 346, 380, 529
precipitate in, 330, 529
rain and, 352–353
safety guidelines. *See* chemical safety
sanitizers in. *See* sanitizers
saturation, 330, 371–372, 531
saturation point, 330, 371–372
scale and, 361
skin irritation, 361
stabilizer, 351, 352, 427, 428, 533
superchlorination, 337–340, 346, 354, 374, 411, 534
temperature effect on, 331, 352

testing. *See* testing water
total alkalinity, 349–350, 368, 426–427, 535
total dissolved solids (TDS), 350–351, 369, 372, 535
trichlor cyanurate, 337–338, 341, 356, 357, 536
turbidity, 364–365
ultraviolet (UV) light and, 342–343
weather effects on, 352–353, 410–411
wind effect on, 352
water column pressure, 536
water conservation, 421–422, 502–503
water features, 459–474
cleaning, 409
conserving water, 503
decorative ponds. *See* ponds
described, 537
FAQs, 474
fountains, 459–460, 462, 474, 503
rockscapes, 468–471
vanishing edge pools, 471–474, 536
wading pools, 174, 403, 459, 476, 488
waterfalls. *See* waterfalls
water fill systems, 85–90
water flow. *See also* hydraulics
blockage, 251
check valves, 72–75
heaters, 243, 246–251
water hammer, 537
water level checks, 396
water level controls, 85–90
water quality index, 371–373, 525
water system, overview, 1–3
water treatment. *See* treatment delivery systems
waterfall weirs, 472
waterfalls, 471–474
cleaning, 459–464, 469, 473–474
conserving water, 503
evaporation and, 460, 474
koi ponds, 460–468, 469, 474
motors, 459
pumps, 472–474
rockscapes, 468–471
vanishing edge pools, 471–473, 536
waterfalls, 471–474
waterline
"bathtub ring," 418
cleaning, 399, 407, 418, 503
discoloration around, 418
scale deposits on, 350, 361, 460
spas, 4, 6
tile at, 28–29, 503
watt, 150, 537
weather
effect on water chemistry, 352–353
rain, 352–353, 396, 503
sun, 333, 352, 353
temperature, 352
wind. *See* wind
websites, 540–542
weirs, 38–39, 537
wet slides, 503
wind
effect on water chemistry, 352
fountains and, 460
water evaporation and, 352, 460, 501
windbreaks, 501
windings, pump motors, 103, 104, 139
winterizing, 409–418, 537
wire sizing chart, 156
wire wrapping, 157
wireless remote controls, 281–283
wires
bonding, 135, 151, 160
gauge, 150, 156–157, 523

wiring in electricity, 156–157. *See also* electrical wiring/connections
wood pools, 19–20, 526
Y valves, 61–68
yellow algae (phaeophyta), 354, 528